Fish Feed Formulation with Agroindustrial Wastes and Byproducts

Fish Feed Formulation with Agroindustrial Wastes and Byproducts

Edited by

Hrudayanath Thatoi
Siksha 'O' Anusandhan Deemed to be University, Centre for Industrial Biotechnology Research, School of Pharmaceutical Sciences, Bhubaneswar, India

Pradeep K. Das Mohapatra
Raiganj University Department of Microbiology, Raiganj, India

Partha Bandyopadhyay
FINRAY BIOTECH PRIVATE LIMITED, Ahmedabad, India

Sourav Chattaraj
Siksha 'O' Anusandhan Deemed to be University, Centre for Industrial Biotechnology Research, School of Pharmaceutical Sciences, Bhubaneswar, India

ELSEVIER

ACADEMIC PRESS
An imprint of Elsevier

Academic Press is an imprint of Elsevier
125 London Wall, London EC2Y 5AS, United Kingdom
50 Hampshire Street, 5th Floor, Cambridge, MA 02139, United States

Notices

Knowledge and best practice in this field are constantly changing. As new research and experience broaden our understanding, changes in research methods, professional practices, or medical treatment may become necessary.

Practitioners and researchers must always rely on their own experience and knowledge in evaluating and using any information, methods, compounds, or experiments described herein. In using such information or methods they should be mindful of their own safety and the safety of others, including parties for whom they have a professional responsibility.

To the fullest extent of the law, neither the Publisher nor the authors, contributors, or editors, assume any liability for any injury and/or damage to persons or property as a matter of products liability, negligence or otherwise, or from any use or operation of any methods, products, instructions, or ideas contained in the material herein.

ISBN: 978-0-443-41728-3

For information on all Academic Press publications visit our website at https://www.elsevier.com/books-and-journals

Publisher: Mara Conner
Acquisitions Editor: Vaishali Singh
Editorial Project Manager: Ritika Dagar
Production Project Manager: Jenela Priscy
Cover Designer: Arumugam Kothandan

Typeset by TNQ Tech

Contents

Contributors

Meddich Abdelilah
Centre d'Agrobiotechnologie et de Bioingénierie, Unité de Recherche Labellisée CNRST (Centre AgroBiotech-URL-CNRST-05), Université Cadi Ayyad, Marrakesh, Morocco

Dibyendu Acharya
Aquaculture Laboratory, The University of Burdwan, Department of Zoology, Burdwan, West Bengal, India

Apaarna
Department of Microbiology, Graphic Era (Deemed to be University), Dehradun, Uttarakhand, India

Shabana Arjmand
Government Degree College, Srinagar, Kashmir, India

Addisu Assefa
Applied Microbiology, Madda Walabu University, Bale Robe, Ethiopia

S. Bandeppa
Division of Soil Science, ICAR-Indian Institute of Rice Research, Hyderabad, Telangana, India

Partha Bandyopadhyay
Finray Biotech Private Ltd, Ahmedabad, Gujarat, India

Urna Banerjee
Kolkata, West Bengal, India

Mukul Machhindr Barwant
Department of Botany, Sanjivani Arts Commerce and Science College Kopargoan, Ahmednagar, Maharashtra, India

Ayusman Behera
P.G. Department of Zoology, Maharaja Sriram Chandra Bhanja Deo University, Baripada, Odisha, India

Hanane Boutaj
Laboratory of Life and Health Sciences, Faculty of Medecine and Pharmacy of Tangier, Abdelmalek Essaadi University, Tetouan, Morocco; Centre d'Agrobiotechnologie et de Bioingénierie, Unité de Recherche Labellisée CNRST (Centre AgroBiotech-URL-CNRST-05), Université Cadi Ayyad, Marrakesh, Morocco

Gökhan Boyno
Department of Plant Protection, Faculty of Agriculture, Van Yuzuncu Yil University, Van, Türkiye

Sourav Chattaraj
Centre for Industrial Biotechnology Research, School of Pharmaceutical Sciences, Siksha 'O' Anusandhan Deemed to be University, Bhubaneswar, Odisha, India

Anyesha Roy Choudhary
Department of Biotechnology, Odisha University of Technology and Research, Bhubaneswar, Odisha, India

Younes Rezaee Danesh
Department of Plant Protection, Faculty of Agriculture, Van Yuzuncu Yil University, Van, Turkey; Department of Plant Protection, Faculty of Agriculture, Urmia University, Urmia, Iran

Sukadev Das
P.G. Department of Zoology, Maharaja Sriram Chandra Bhanja Deo University, Baripada, Odisha, India

Suman Das
Department of Microbiology, Bankura Sammilani College, Bankura, West Bengal, India

Abhisek Das
Environmental Science Laboratory, School of Applied Sciences, Kalinga Institute of Industrial Technology (Deemed to Be University), Bhubaneswar, Odisha, India

Swagat Kumar Das
Department of Biotechnology, Odisha University of Technology and Research, Bhubaneswar, Odisha, India

Manisha Dash
Environmental Science Laboratory, School of Applied Sciences, Kalinga Institute of Industrial Technology (Deemed to Be University), Bhubaneswar, Odisha, India

Supriya Dash
Department of Biotechnology, Odisha University of Technology and Research, Bhubaneswar, Odisha, India

Pradeep K. Das Mohapatra
Department of Microbiology, Raiganj University, Raiganj, West Bengal, India

Mona F.A. Dawood
Botany and Microbiology Department, Faculty of Science, Assiut University, Assiut, Egypt

Harisankar Dey
Applied Phycology Laboratory, Post Graduate Department of Botany, Ramakrishna Mission Vivekananda Centenary College, Kolkata, West Bengal, India

Noureddine El Aouad
Laboratory of Life and Health Sciences, Faculty of Medecine and Pharmacy of Tangier, Abdelmalek Essaadi University, Tetouan, Morocco

Cherkaoui El Modafar
Centre d'Agrobiotechnologie et de Bioingénierie, Unité de Recherche Labellisée CNRST (Centre AgroBiotech-URL-CNRST-05), Université Cadi Ayyad, Marrakesh, Morocco

Athanasios Exadactylos
Department of Ichthyology and Aquatic, Environment, School of Agricultural Sciences, University of Thessaly, Volos, Greece

Arindam Ganguly
Department of Microbiology, Bankura Sammilani College, Bankura, West Bengal, India

Mousumi Ganguly
Department of Nutrition, Bankura Sammilani College, Bankura, West Bengal, India

Koushik Ghosh
Aquaculture Laboratory, The University of Burdwan, Department of Zoology, Burdwan, West Bengal, India

Koushik Guchhait
Department of Biochemistry, Medical College of Wisconsin, Milwaukee, WI, United States

B.E. Guerra Sierra
Universidad de Santander-Facultad de Ciencias Exactas Naturales Y Agropecuarias, Bucaramanga, Colombia

Divya Gunsola
Department of Microbiology, Graphic Era (Deemed to be University), Dehradun, Uttarakhand, India

Seyed Hossein Hoseinifar
Department of Fisheries, Gorgan University of Agricultural Sciences and Natural Resources, Gorgan, Golestan, Iran

Navneet Joshi
Department of Biosciences, School of Liberal Arts and Sciences, Mody University of Science and Technology, Lakshmangarh, Rajasthan, India

Riyan Kaibarta
Department of Microbiology, Bankura Sammilani College, Bankura, West Bengal, India

Mehak Kanwar
Centre for Industrial Biotechnology Research, School of Pharmaceutical Sciences, Siksha 'O' Anusandhan Deemed to be University, Bhubaneswar, Odisha, India

Rachan Karmakar
Department of Environmental Science, Graphic Era (Deemed to be University), Dehradun, Uttarakhand, India

Bahman Khoshru
Soil and Water Research Institute (SWRI), Agricultural Research, Education and Extension Organization (AREEO), Karaj, Iran

Konstantinos V. Kotsanopoulos
Department of Ichthyology and Aquatic, Environment, School of Agricultural Sciences, University of Thessaly, Volos, Greece

Rahul Kumar
Department of Microbiology, Graphic Era (Deemed to be University), Dehradun, Uttarakhand, India

K.P. Kumaraguru Vasagam
Department Nutrition, Genetics and Biotechnology Division, Central Institute of Brackishwater Aquaculture (CIBA), Chennai, Tamil Nadu, India

Salma Latique
Laboratory of Drugs Sciences, Faculty of Medicine, Pharmacy and Dentistry, Sidi Mohamed Ben Abdellah University, Fes, Morocco

Debasree Lodh
Department of Botany, Holy Cross College, WEST, Agartala, Tripura, India

Sakho Mahamadou
Laboratory of Life and Health Sciences, Faculty of Medecine and Pharmacy of Tangier, Abdelmalek Essaadi University, Tetouan, Morocco

Bijay Kali Mahapatra
ICAR-Central Institute of Fisheries Education, Kolkata, West Bengal, India

Joydev Maity
Department of Fishery Sciences, Vidyasagar University, Midnapore, West Bengal, India

Sanjeeb Kumar Mandal
Department of Biotechnology, Chaitanya Bharathi Institute of Technology, Hyderabad, Telangana, India

Petros V. Martsikalis
Department of Ichthyology and Aquatic, Environment, School of Agricultural Sciences, University of Thessaly, Volos, Greece

Snehasish Mishra
School of Biotechnology, Campus 11, KiiT Deemed University, Bhubaneswar, Odisha, India

Debasis Mitra
Department of Microbiology, Graphic Era (Deemed to be University), Dehradun, Uttarakhand, India

Pratikhya Mohanty
School of Biotechnology, Campus 11, KiiT Deemed University, Bhubaneswar, Odisha, India

Gargee Mohanty
P.G. Department of Zoology, Maharaja Sriram Chandra Bhanja Deo University, Baripada, Odisha, India

Saswati Mohanty
School of Biotechnology, Campus 11, KiiT Deemed University, Bhubaneswar, Odisha, India

Shibani Mohapatra
Center for Biotechnology, Siksha 'O' Anusandhan (Deemed to Be University), Bhubaneswar, Odisha, India

Arpan Mondal
Applied Phycology Laboratory, Post Graduate Department of Botany, Ramakrishna Mission Vivekananda Centenary College, Kolkata, West Bengal, India

Pratap Mukhopadhyay
ICAR-CIFA, Bhubaneswar, Odisha, India

Abhisek Mukhopadhyay
Aquaculture Laboratory, The University of Burdwan, Department of Zoology, Burdwan, West Bengal, India

Srikanth Muthyala
Department of Biotechnology, Chaitanya Bharathi Institute of Technology, Hyderabad, Telangana, India

Adeyemi Nurudeen Olatunbosun
Department of Plant Physiology and Crop Production, Federal University of Agriculture, Alabata, Ogun State, Nigeria

Alok Kumar Panda
Environmental Science Laboratory, School of Applied Sciences, Kalinga Institute of Industrial Technology (Deemed to Be University), Bhubaneswar, Odisha, India

Fannie Isela Parra Cota
Campo Experimental Norman E. Borlaug-CIRNO. Instituto Nacional de Investigaciones Forestales, Agrícolas y Pecuarias (INIFAP), Ciudad Obregón, Sonora, Mexico

Marika Pellegrini
Department of Life, Health and Environmental Sciences, University of L'Aquila, L'Aquila, Italy

Disha Penchala
Department of Biotechnology, Chaitanya Bharathi Institute of Technology, Hyderabad, Telangana, India

Sunita Pradhan
Department of Fishery Sciences, Vidyasagar University, Midnapore, West Bengal, India

Gadde Padma Priyasha
Department of Biotechnology, Chaitanya Bharathi Institute of Technology, Hyderabad, Telangana, India

Dilfuza Jabborova Pushkinovna
Medicinal Plants Genetics and Biotechnology, Institute of Genetics and Plant Experimental Biology, Uzbekistan Academy of Sciences, Tashkent, Uzbekistan

Youssef Ait Rahou
Centre d'Agrobiotechnologie et de Bioingénierie, Unité de Recherche Labellisée CNRST (Centre AgroBiotech-URL-CNRST-05), Université Cadi Ayyad, Marrakesh, Morocco

Karthikeyan Ramalingam
School of Life Sciences, B.S Abdur Rahman Crescent Institute of Science and Technology, Chennai, Tamil Nadu, India

Prasanta Rath
Environmental Science Laboratory, School of Applied Sciences, Kalinga Institute of Industrial Technology (Deemed to Be University), Bhubaneswar, Odisha, India

Kübra Sağlam
Department of Food Engineering, Faculty of Agriculture, Tekirdağ Namik Kemal University, Tekirdağ, Turkey

Sukanya Sahu
Department of Biotechnology, Odisha University of Technology and Research, Bhubaneswar, Odisha, India

Sergio de los Santos-Villalobos
Instituto Tecnológico de Sonora, Ciudad Obregón, Sonora, Mexico

Gustavo Santoyo
Instituto de Investigaciones Químico-Biológicas, Universidad Michoacana de San Nicolás de Hidalgo, Morelia, Mexico

Tanmoy Sarkar
School of Agriculture, Swami Vivekananda University, Kolkata, West Bengal, India

Shraddha Bhaskar Sawant
Bihar Agricultural University, Bhagalpur, Bihar, India

Sovan Sengupta
Applied Phycology Laboratory, Post Graduate Department of Botany, Ramakrishna Mission Vivekananda Centenary College, Kolkata, West Bengal, India

Dipanwita Chaudhuri Sil
Department of Botany, Holy Cross College, WEST, Agartala, Tripura, India

Balwant Singh
Department of Botany K. S. Saket PG College, Dr. Ram Manohar Lohia Avadh University, Ayodhya, Uttar Pradesh, India

Puneet K. Singh
School of Biotechnology, Campus 11, KiiT Deemed University, Bhubaneswar, Odisha, India

G. Sri Thatchayani
HCL Technologies, ELCOT IT Park, Madurai, Tamil Nadu, India

Lopamudra Subudhi
Center for Biotechnology, Siksha 'O' Anusandhan (Deemed to Be University), Bhubaneswar, Odisha, India

Tanaya Sukul
Aquaculture Laboratory, The University of Burdwan, Department of Zoology, Burdwan, West Bengal, India

B. Sumithra
Department of Biotechnology, Chaitanya Bharathi Institute of Technology, Hyderabad, Telangana, India

Saroj K. Swain
Central Institute of Freshwater Aquaculture (ICAR-CIFA), Bhubaneswar, Odisha, India

Nooruddin Thajuddin
School of Life Sciences, B.S Abdur Rahman Crescent Institute of Science and Technology, Chennai, Tamil Nadu, India

Hrudayanath Thatoi
Centre for Industrial Biotechnology Research, School of Pharmaceutical Sciences, Siksha 'O' Anusandhan Deemed to be University, Bhubaneswar, Odisha, India

Sumaiya Begum Thuraputheen
School of Life Sciences, B.S Abdur Rahman Crescent Institute of Science and Technology, Chennai, Tamil Nadu, India

Kiran Yellappa Vajanthri
Department of Biotechnology, Chaitanya Bharathi Institute of Technology, Hyderabad, Telangana, India

Devvret Verma
Department of Biotechnology, Graphic Era (Deemed to be University), Dehradun, Uttarakhand, India

CHAPTER

1 Comprehensive perspective and challenges of non-conventional feed ingredients, the ease and profitability of obtaining and processing them, and their nutritional content

Rahul Kumar[1], Rachan Karmakar[2], Kübra Sağlam[3], S. Bandeppa[4], Tanmoy Sarkar[5], Addisu Assefa[6], Arindam Ganguly[7], Dilfuza Jabborova Pushkinovna[8], Adeyemi Nurudeen Olatunbosun[9], Gustavo Santoyo[10] and Debasis Mitra[1]

[1]*Department of Microbiology, Graphic Era (Deemed to be University), Dehradun, Uttarakhand, India;* [2]*Department of Environmental Science, Graphic Era (Deemed to be University), Dehradun, Uttarakhand, India;* [3]*Department of Food Engineering, Faculty of Agriculture, Tekirdağ Namik Kemal University, Tekirdağ, Turkey;* [4]*Division of Soil Science, ICAR-Indian Institute of Rice Research, Hyderabad, Telangana, India;* [5]*School of Agriculture, Swami Vivekananda University, Kolkata, West Bengal, India;* [6]*Applied Microbiology, Madda Walabu University, Bale Robe, Ethiopia;* [7]*Department of Microbiology, Bankura Sammilani College, Bankura, West Bengal, India;* [8]*Medicinal Plants Genetics and Biotechnology, Institute of Genetics and Plant Experimental Biology, Uzbekistan Academy of Sciences, Tashkent, Uzbekistan;* [9]*Department of Plant Physiology and Crop Production, Federal University of Agriculture, Alabata, Ogun State, Nigeria;* [10]*Instituto de Investigaciones Químico-Biológicas, Universidad Michoacana de San Nicolás de Hidalgo, Morelia, Mexico*

1. Introduction

The global livestock and aquaculture industries are increasingly challenged to respond to the rising demand for animal protein, while also dealing with fears of feed resource scarcity, increasing production costs, and environmental stewardship (Makkar, 2018; Chattaraj et al., 2026). Traditional feed ingredients, such as maize, soybean meal, and FM, have been the foundation of animal nutrition for centuries; however, their persistence has serious implications for competition with human food production, land use change, loss of biodiversity, and greenhouse gas emissions

Fish Feed Formulation with Agroindustrial Wastes and Byproducts. https://doi.org/10.1016/B978-0-443-41728-3.00010-1

(Dalin and Rodríguez-Iturbe, 2016). Moreover, market volatility, climate change-related agricultural productivity disruptions, and geopolitical tensions create additional risks for feed supply chain exposure (Sheeraz et al., 2024; Kumar et al., 2024). Here, the quest for unconventional feed ingredients that are underutilized, sustainable, and nutritionally suitable has taken significant ground as a tactic toward decreasing dependence on traditional feed resources and building principles of circular economy in animal husbandry (Chattaraj et al., 2024b,c). This chapter discusses new horizons of non-traditional feed ingredients, particularly plant proteins, insect meals, single-cell proteins (SCPs), and agro-industrial byproducts, and critically assesses their possible contributions to sustainable systems of animal nutrition. Plant proteins isolated from plants such as soybean meal, sunflower meal, cottonseed meal, and other legumes have been studied as a source of replacement or supplementation of conventional protein feeds. Although soybean meal is still a leading plant protein because of its high protein content and well-balanced amino acid profile, it is linked to issues of deforestation and habitat destruction during production (Chattaraj et al., 2024a; Choręziak et al., 2025). Other alternative plant-based options, such as sunflower meal and cottonseed meal, have good nutritional value but have antinutritional factors such as gossypol or need processing for better digestibility.

Legumes such as peas, lupins, and faba beans add more diversity to sources of protein, with advantages including nitrogen fixation and improvement in soil quality; however, they also have challenges in the form of antinutritional factors and nutrient composition variability (Martineau-Côté et al., 2022; Kumar et al., 2025a,b,c,d). Insect meals are a new and highly sustainable alternative protein source that has been growing in interest among researchers over the last few years. Insects such as black soldier fly larvae (*Hermetia illucens*), mealworms (*Tenebrio molitor*), and crickets (*Acheta domesticus*) are very promising because of their high lipid and protein contents, good amino acid composition, and potential to make organic waste streams useful biomass (Van Peer et al., 2021). Black soldier fly larvae, in particular, are very effective agro-industrial waste bioconverters with the potential for circularity and valorization of waste gains. In addition to their nutritive value, insect diets include functional ingredients, such as antimicrobial lipids (e.g., lauric acid), which could impart health-enhancing properties to animals (Salgado-Ramos et al., 2024). Regulatory systems, safety evaluations, and consumer acceptance remain key areas of concern for the generalization of insect-based feeds. SCPs, including yeasts, microalgae, and bacteria, represent another avenue for the search for alternative feedstocks (Zhuang et al., 2024). SCPs can be produced by biotechnological means from various substrates, ranging from industrial wastes to lignocellulosic biomass or gaseous carbon sources such as methane and carbon dioxide (Kumar et al., 2025a; Gupta et al., 2025). *Saccharomyces cerevisiae* and *Candida utilis* are the most utilized yeasts in the form of feed supplements, providing not only good-quality proteins, but also probiotic and immunomodulatory functions. Microalgae, such as *Spirulina* and *Chlorella* contain high levels of protein, essential fatty acids, vitamins, and bioactive compounds, in addition to the advantages of carbon sequestration and biomass production on non-agricultural land (Chattaraj and Das Mohapatra, 2024; Chaudhary et al., 2024). Bacteria such as *Methylococcus capsulatus* and *Bacillus subtilis* are another new SCP source that can produce high-yielding proteins through gas fermentation technologies (Li et al., 2024). Despite their promise, the mass commercialization of SCPs is hampered by the cost of production, scalability, and regulatory permits. Agro-industrial waste, such as rice bran, wheat bran, fruit and vegetable processing waste, and oilseed cakes, provides a closed-loop solution by recycling waste streams into valuable feedstock (Ramachandran et al., 2024; Kumar et al., 2025b).

These wastes are frequently in surplus, under-exploited, and help reduce feed costs while avoiding waste disposal problems. For instance, wheat and rice bran are high in dietary fiber and moderate in protein, whereas residues in fruits and vegetables contain fermentable carbohydrates, antioxidants, and functional compounds (Tlais et al., 2020). However, their uneven quality profiles, perishable nature, and the presence of possible contaminants such as mycotoxins, pesticide residues, or pathogens require stringent quality control, processing, and safety assurance for successful feed utilization. The use of unconventional feed ingredients, such as plant proteins, insect meals, single-cell proteins, and agro-industrial by-products, has tremendous potential to improve the sustainability, resilience, and circularity of animal production systems worldwide (Muniz et al., 2024). However, there are issues with ensuring quality consistency, optimizing nutritional formulations, safety, and regulatory environments. This chapter presents an integrated overview of the state of knowledge, technological trends, and outlook for future developments in non-conventional feed ingredients from the perspective of guiding research priorities, policies, and field applications in the feed sector. By critically analyzing the nutritional value, environmental advantages, and constraints of such alternative feed materials, this article adds its voice to the general debate on developing climate-resilient, resource-responsive, and cost-effective animal nutrition policies for the future.

2. Nutritional profiles and processing considerations

2.1 Nutritional content

The nutrient composition and processing characteristics of unconventional feed ingredients are key determinants of their acceptability, safety, and performance in animal nutrition (Babatunde et al., 2021). In contrast to conventional feedstuffs, such as maize, soybean meal, and fishmeal, unconventional ingredients have significant variations in nutrient levels, digestibility, anti-nutritional factors, and functional bioactive compounds. Knowledge of these dimensions is crucial for incorporating such feeds into well-formulated rations that satisfy the nutritional needs of livestock and aquaculture animals without impairing their health, performance, or product quality (EFSA, 2008). This chapter presents a critical examination of the nutritional value characteristics and processing aspects of prominent groups of non-conventional feed components, such as plant proteins, insect meals, SCPs, and agro-industrial byproducts (Table 1.1).

Table 1.1 Nutritional composition and anti-nutritional factors of selected non-conventional feed ingredients.

Ingredient type	Protein (%)	Lipid (%)	Fiber (%)	Major anti-nutritional factors (ANFs)	References
Soybean meal	44–48	1.5–2.5	5–7	Trypsin inhibitors, phytates, tannins	Saah et al. (2020)
Sunflower meal	28–42	1.5–3.5	20–30	Phenolic compounds, chlorogenic acid, fiber	Wildermuth et al. (2016)
Cottonseed meal	36–41	1–2	15–25	Gossypol, cyclopropenoid fatty acids	Obert et al. (2007)
Legumes (general)	20–35	1–5	10–20	Tannins, lectins, saponins	Sultana et al. (2017)
Black soldier fly larvae meal	40–45	30–35	7–9	Chitin, potential microbial contaminants	Bessa et al. (2021)
Mealworm meal	48–52	25–30	6–8	Chitin, possible allergenic proteins	Broekman et al. (2017)
Cricket meal	60–70	15–25	5–7	Chitin, allergenic proteins	Yang et al. (2024)
Yeasts (SCP)	45–55	1–5	3–8	High nucleic acid content	Schulz and Oslage (1976)
Microalgae (SCP)	40–60	10–20	5–15	Polysaccharides, pigments, some toxins in specific species	Moreira et al. (2022)
Bacteria (SCP)	50–70	2–8	5–10	High nucleic acids, endotoxins (strain-dependent)	Manzo et al. (2020)
Rice bran	12–15	15–20	10–14	Phytates, tannins	Wisetkomolmat et al. (2022)
Wheat bran	14–17	4–6	40–50	Phytates, non-starch polysaccharides	Sztupecki et al. (2023)
Fruit and vegetable wastes	5–15	1–5	10–30	Variable, includes oxalates, phenolics	Sagar et al. (2018)

2.2 Insect meals

Insect diets, especially those from black soldier fly larvae (BSFL), mealworms, and crickets, offer a unique nutritional profile well suited to the protein and lipid needs of livestock, poultry, and aquaculture animals. Black soldier fly larvae meal, for example, has higher than 40% crude protein and as much as 35% fats, with an excellent amino acid profile high in lysine, threonine, and methionine (Mancini et al., 2021). Furthermore, BSFL contains bioactive molecules such as lauric acid, which has antimicrobial activity that can contribute to gut health. Mealworm meal from Tenebrio Molitor contains more than 40% protein and some lipid component rich in polyunsaturated fatty acids (Mancini et al., 2021). Cricket meals also have a high protein content (>60%) and high amounts of micronutrients, such as iron and zinc (Udomsil et al., 2019). However, insect foods can show variability in nutrient content depending on the rearing substrate, stage of development, and processing. Processing operations such as drying (oven drying and freeze drying), grinding, and defatting are widely used to improve shelf stability, minimize microbial contamination, and homogenize nutrient content. The inclusion of insect feeds in animal nutrition has shown favorable outcomes with regard to feed conversion ratio, growth rate, and immune function modulation; however, production scale-up, safety (e.g., microbial contamination, allergens, digestibility of chitin), and standardization of nutrient composition are areas that require further study.

2.3 Plant-based proteins

Plant proteins such as soybean meal, sunflower meal, cottonseed meal, and legumes are high among non-conventional sources of protein because of the availability, cost-effectiveness, and sustainability of fishmeal substitutes and conventional protein feed alternatives. Despite its widespread adoption, soybean meal is constrained in some amino acids such as methionine and harbors anti-nutritional compounds such as trypsin inhibitors, lectins, and oligosaccharides that can compromise digestibility and intestinal health (Di et al., 2024). Processing technologies, such as heat treatment, extrusion, or enzymatic supplementation, are typically used to enhance the nutritional quality of soybean meal by inactivating these anti-nutritional factors. Sunflower meal provides an average protein quality of more than 28% with a rather high fiber content, which can restrict its utilization in monogastric diets, but renders it suitable for ruminants (Nedelkov, 2023). Cottonseed meal, with protein content normally 36%–40%, is accompanied by gossypol, a toxic polyphenolic compound that damages reproductive performance and organ function in non-ruminants (Mageshwaran, 2021). Processing methods, such as solvent extraction, heat treatment, and iron supplementation, have been employed to eliminate gossypol and increase safety. Legumes such as peas, lupins, and faba beans have high protein content (20%–30%) and digestible carbohydrates, but can be rich in tannins, trypsin inhibitors, and vicine/convicine, and require treatments such as dehulling, soaking, germination, or fermentation to improve nutritional quality (Saha et al., 2022). In general, plant proteins provide good amino acid profiles, fiber, and energy but need to be well formulated to support balanced diets and avoid anti-nutritional effects (Fig. 1.1).

2.4 Single-cell proteins

Single-cell proteins, including yeasts, microalgae, and bacteria, are microbial biomass that possess an unusually high protein content, between 40% and more

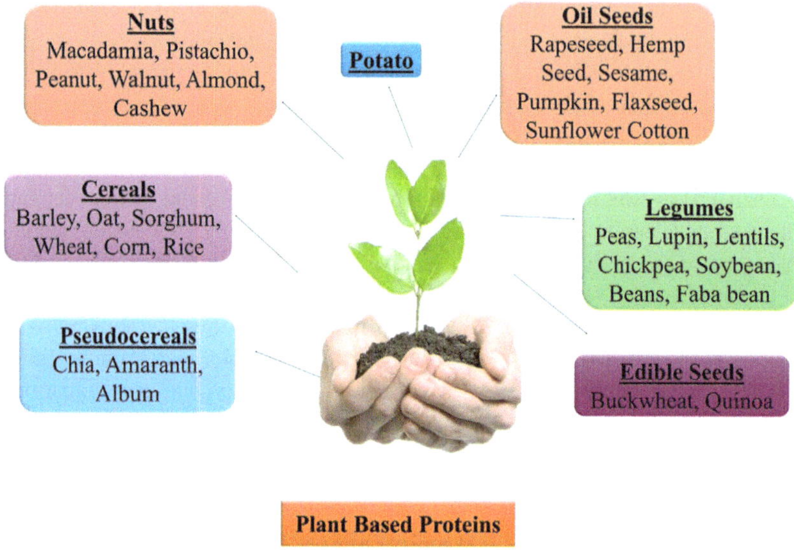

FIGURE 1.1 Sources of plant-based proteins.

Exploring the diverse sources of plant-based proteins, from legumes to nuts and seeds.

than 70% on a dry matter basis, depending on their source and processing (Onyeaka et al., 2022). Yeasts, such as *Saccharomyces cerevisiae* and *Candida utilis* are commonly used as protein supplements in animal feed, with 40%–55% protein, B-complex vitamins, and immunomodulating β-glucans (Grigore et al., 2025). The mannans and glucans constituting their cell walls could also play beneficial roles in gut health, but can restrict digestibility in some species, necessitating enzymatic hydrolysis or autolysis to enhance nutrient access. Microalgae, such as *Spirulina* and *Chlorella* contain high levels of protein (up to 60%–70%), polyunsaturated fatty acids (notably omega-3s such as EPA and DHA), carotenoids, and minerals (Remize et al., 2021). Nonetheless, inflexible cell walls in microalgae require mechanical or enzymatic cell disruption to increase nutrient release. Bacterial proteins, such as those from *Methylococcus capsulatus* or *Bacillus subtilis* have been able to reach protein contents of over 70%, possessing balanced amino acid profiles with the possibility of large-scale production through gas fermentation processes (Sun et al., 2022). Treatment of SCPs includes cell harvesting (centrifugation and filtration), drying (spray-drying and freeze-drying), and, in a few instances, disruption of cell walls for enhanced digestibility. Problems such as nucleic acid levels (which cause hyperuricemia in non-ruminants), cost of production, and safety issues must be resolved to achieve wider commercial use of SCPs in animal nutrition (Richmond and Hu, 2013).

Agro-industrial by-products are a heterogeneous category of feed materials with widely fluctuating nutritional composition. Rice bran, which is a by-product of rice milling, is composed of 12%–16% protein, 15%–20% fat, and 10%–15% fiber, but due to its high oil content, it may promote rancidity and require stabilization processes such as heat treatment or antioxidant supplementation to extend shelf life (Spaggiari et al., 2021). Wheat bran, commonly used in monogastric and ruminant diets, contains moderate protein (12%–16%) and high fiber (approximately 10%–12%), creating a good energy source as well as a prebiotic substrate (Cheng et al., 2022). Fruit and vegetable by-products, including citrus pulp, tomato pomace, and carrot peels, vary widely in content, but typically contain carbohydrates, fiber, vitamins, and antioxidants. However, their moisture levels (frequently above 70%) are problematic for storage and transport, and they need to be dried, ensiled, or fermented to preserve and maintain microbial safety. Mycotoxins, pesticide contaminants, and microbial spoilage are hazards associated with agro-industrial byproducts that require rigorous quality control procedures, correct storage, and feed safety analysis prior to their incorporation into animal diets (Shah et al., 2025). Processing technologies, such as ensiling, extrusion, pelleting, and chemical treatment, can enhance the stability, palatability, and nutritional uniformity of these byproducts. The nutritional content and processing aspects of non-conventional feed ingredients are pivotal for realizing their potential as environmentally friendly alternatives in animal nutrition (Biswas and Purohit, 2024). Although plant-based proteins, insect meals, SCPs, and agro-industrial byproducts provide various nutrients, their effective incorporation into feed systems involves solving anti-nutritional factors, making them more digestible, ensuring safety, and providing constant quality through proper processing technologies. Improvements in feed formulation, bioprocessing, and quality standards are crucial to maximize the utilization of these alternative resources for the establishment of climate-smart, resource-efficient, and economically productive animal production systems (Wang et al., 2022).

3. Processing techniques

The use of non-traditional feed materials in animal diets requires different processing methods to improve their nutritional content and digestibility, remove anti-nutritional compounds, and maintain feed safety and shelf life. Such methods are important for overcoming the limitations of many alternative feeding resources, such as inconsistencies in nutrient content, contamination with toxins or other substances, and vulnerability to spoilage. Among the most widely used processing techniques, heat treatment, fermentation, drying and grinding, and extrusion are notable for their effectiveness and versatility in different feed matrices (Table 1.2).

3.1 Heat treatment

Heat processing techniques, such as roasting, steaming, boiling, and autoclaving, are major contributors to detoxifying plant feed ingredients through denaturation of heat-labile anti-nutritional compounds, such as trypsin inhibitors, lectins, and gossypol (Abu Hafsa et al., 2022). For instance, soybean meal and cottonseed meal tend to be heat-treated to minimize trypsin inhibitor and gossypol levels, respectively, to

Table 1.2 Economic and processing considerations for non-conventional feed ingredients.

Ingredient type	Ease of procurement	Processing requirements	Cost implications	References
Insect meals	Moderate; dependent on developing insect farming infrastructure and substrate availability	Requires controlled rearing, drying, grinding, and sometimes defatting	Higher initial investment; costs decreasing with scale	Jucker and Belluco (2020)
Plant-based proteins	High; widely available through established agricultural supply chains	Usually requires heat treatment, grinding, and solvent extraction	Generally low cost due to large-scale production	Gençdağ et al. (2021)
Single-cell proteins (SCPs)	Moderate to low; dependent on bioreactor capacity and substrate supply	Requires fermentation, drying, and strict quality control	High production costs; potential to decrease with technology	Sekoai et al. (2024)
Agro-industrial byproducts	High; abundant as waste from processing industries	Usually it requires drying, grinding, and sometimes detoxification	Low cost or free raw material, but processing costs vary	Singh et al. (2021)

improve protein digestibility and safety for monogastric animals. Heat treatment destroys pathogenic microorganisms and enhances feed hygiene. However, high temperatures can also degrade heat-labile nutrients such as vitamins and some amino acids (particularly lysine), necessitating the optimization of process conditions to prevent detoxification while retaining nutrients (Katu et al., 2025). Insect meal manufacturing also uses controlled heat to dry larvae and minimizes microbial contamination without affecting the protein quality. Heat treatment is a basic process used to enhance the nutritional content and safety of most nontraditional feedstuffs.

3.2 Fermentation

Fermentation, a biological treatment method using desirable microorganisms, such as lactic acid bacteria, yeasts, and fungi, is gaining recognition for its capacity to increase the nutritional and functional features of alternative feeds (Kårlund et al., 2020). This can enhance protein digestibility, break down anti-nutritional factors, such as tannins and phytates, and enrich the substrate with enzymes, vitamins, and bioactive metabolites (Chattaraj et al., 2023) (Chattaraj et al., 2025). For example, fermentation of legume seeds or agro-industrial residues, such as fruit and vegetable wastes, can improve the availability of nutrients and minimize fiber content, improving nutrient uptake in animals. Fermentation is also a natural preservation process that decreases pH and suppresses spoilage microorganisms, thereby increasing the shelf life (Teshome et al., 2022). In single-cell protein production, fermentation plays a vital role in the production of yeast and bacterial biomass by processing variable substrates into high-quality feed ingredients. Fermentation is beneficial, but needs to be strictly controlled for microbial strains, substrate formulation, temperature, and time to obtain consistent outcomes and prevent contamination (Skowron et al., 2022).

3.3 Drying and grinding

Drying is a critical post-harvest processing operation for minimizing the moisture level of feed ingredients to below levels where microbial growth and enzymatic spoilage can be prevented, thus improving storage stability. Methods including sun drying, oven drying, freeze drying, and spray drying are utilized depending on the type of feed material and the operational scale (Parniakov et al., 2022). For insect foods, drying processes should maintain protein integrity while providing safety; for instance, freeze-drying maintains nutritional value but is expensive, while oven-drying provides convenience but can jeopardize the loss of nutrients if not closely regulated. After drying, grinding serves to minimize the particle size and enhance feed homogeneity, palatability, and digestibility. Fine grinding maximizes surface area, allowing enzymatic activity and microbial fermentation in the gut (Niu et al., 2023). However, excessive grinding can result in dust formation and handling issues. The combination of drying and grinding is highly applicable to agro-industrial by-products and plant meals, which improves their compatibility with formulated feeds (de Oliveira et al., 2025).

3.4 Extrusion

Extrusion processing is the process of pushing the feed material through a hot pressurized barrel by a screw mechanism, exposing it to high temperatures and shear forces for brief periods. Extrusion processing is extensively used to create pelleted or expanded feed products with increased digestibility, lower anti-nutritional factors, and better palatability (Ali et al., 2024). Extrusion is especially efficient for denaturing proteins, gelatinizing starches, and sterilizing feed, which makes it ideal for plant proteins and agro-industrial residues. Extrusion also enables incorporation of multiple ingredients into a homogeneous final product with attractive physical properties (Lazou, 2024). In the processing of insect meal, extrusion can be employed to create insect meal-based feed pellets with improved storage stability and ease of handling. Extrusion temperatures are high enough to inactivate trypsin inhibitors, lectins, and other unwanted compounds and to eliminate microbial contaminants (Nikmaram et al., 2015). However, extrusion involves a large capital outlay and substantial energy inputs, and attention needs to be paid to ensure that the process parameters are optimized to minimize nutrient degradation, particularly for heat-sensitive vitamins and fatty acids. Processing methods, such as heat treatment, fermentation, drying and grinding, and extrusion, are invaluable tools for unleashing the full potential of non-conventional feed ingredients (Siddiqui et al., 2025). Every process has special strengths in terms of enhancing nutritional quality, safety, and shelf life, and the selection of processes usually relies on the feed material, target animal species, and economic considerations. The ongoing innovation and improvement of these processing technologies, coupled with innovations in feed formulation and quality control, will play a critical role in incorporating sustainable alternative feed resources into animal nutrition systems (Akintan et al., 2024).

4. Challenges in utilizing non-conventional feed ingredients

Although they hold promising possibilities for sustainability and protein diversity in animal feeds, non-traditional feedstuffs have a few challenges that need to be overcome before their effective and extensive use (Chisoro et al., 2023). These challenges include nutritional heterogeneity, occurrence of anti-nutritional factors (ANFs), digestibility problems, palatability limitations, and regulatory issues. Recognition and alleviation of these restraints are critical for maximizing feed formulation and promoting animal health and productivity (Akintan et al., 2025).

4.1 Nutritional variability

One of the most significant challenges in the use of non-conventional feed ingredients is the inherent variability in their nutritional contents. Species differences, rearing or cultivation practices, substrate or soil quality, harvesting time, and postharvest handling can lead to considerable variations in protein levels, amino acid patterns, lipid levels, vitamins, and minerals (Dawadi et al., 2022). For instance, insect feed can differ in nutrient content based on the substrate used to feed larvae, whereas vegetable feeds, such as sunflower or cottonseed, exhibit seasonal variations (Kröncke and Benning, 2022). Similarly, SCPs produced from various microbial strains or fermentation substrates have varying nutrient profiles. Such diversity makes it difficult to formulate feed and requires constant nutritional analysis and standardization to maintain a uniform diet (Dwivedi et al., 2017). This also creates difficulties in forecasting animal performance and formulating diets to address nutritional needs.

4.2 Presence of anti-nutritional factors (ANFs)

Most non-conventional feedstuffs contain natural components that can interfere with nutrient utilization, metabolism, or animal health (Abdoun et al., 2022). Soybean and cottonseed meals often contain ANFs in the form of trypsin inhibitors, tannins, phytates, saponins, and gossypol, which decrease protein digestibility and mineral availability, or lead to toxicity at elevated concentrations (Vujetić et al., 2025). Similarly, insect meals can include chitin, a structural polysaccharide that can hinder digestibility in monogastric animals. Certain SCPs can generate excess nucleic acids or endotoxins based on the microbial strain employed. ANFs require the application of suitable processing methods, such as heat treatment, fermentation, or the addition of enzymes, to eliminate or decrease these substances and enhance nutritional utilization (Suhag et al., 2021). Without removing ANFs, suboptimal animal growth, immune suppression, or metabolic disorders can occur (Fig. 1.2).

4.3 Digestibility issues

The digestibility of nutrients from unconventional feed ingredients tends to be lower than that of conventional feed because of the physical and biochemical barriers (Te Pas et al., 2021). An example is the high fiber content in certain plant meals and chitin in insect exoskeletons, which decrease protein and energy digestibility in non-ruminants. SCPs contain cell wall materials resistant to digestion, which restricts the release of nutrients (Colombo et al., 2023). These digestibility limitations can lower feed efficiency and increase feed cost. Enhancing digestibility through processing, such as grinding, fermentation, or enzyme supplementation, is vital for improving nutrient bioavailability of nutrients (Adebo et al., 2022). In addition, species-specific digestive physiology should be considered when developing diets containing these ingredients because ruminants can utilize fibrous materials more efficiently than monogastric animals.

4.4 Palatability concerns

Animal acceptability of non-conventional feed ingredients is governed by taste, smell, texture, and visual appeal. Novel flavors or odors, bitterness due to residual ANFs, and harsh texture can result in decreased voluntary feed consumption and subsequent growth and productivity (Rodrigues et al., 2024). For instance, insect meals can possess pungent or unpleasant odors if they are not processed or stored properly. Vegetable-based diets can cause bitterness or astringency, whereas SCPs may possess unusual flavors based on fermentation products. Increasing palatability by the formulation of feed tactics, flavor masking, or mixing with traditional components is usually required to promote sufficient feed consumption (Gaikwad et al., 2024). Additionally, the process of dietary adaptation over time may be necessary to enhance acceptance.

4.5 Regulatory hurdles

The application of new feed ingredients, such as insect meals and SCPs, is under regulatory review to protect the safety of animals, humans, and the environment (Sogari et al., 2023). Approval steps include stringent analysis of ingredient composition, possible contaminants (heavy metals, mycotoxins, and pathogens), allergenic potential, and environmental concerns. Regulatory systems differ extensively between regions and are in constant flux, hindering commercialization and increasing compliance expenses (Leuz, 2010). For instance, insect feeds are subject to limitations in specific jurisdictions on the substrates used for insect production and feed use. SCPs derived from genetically engineered microbes can also be subjected to further regulatory hurdles. Alignment of standards and definitive guidelines is

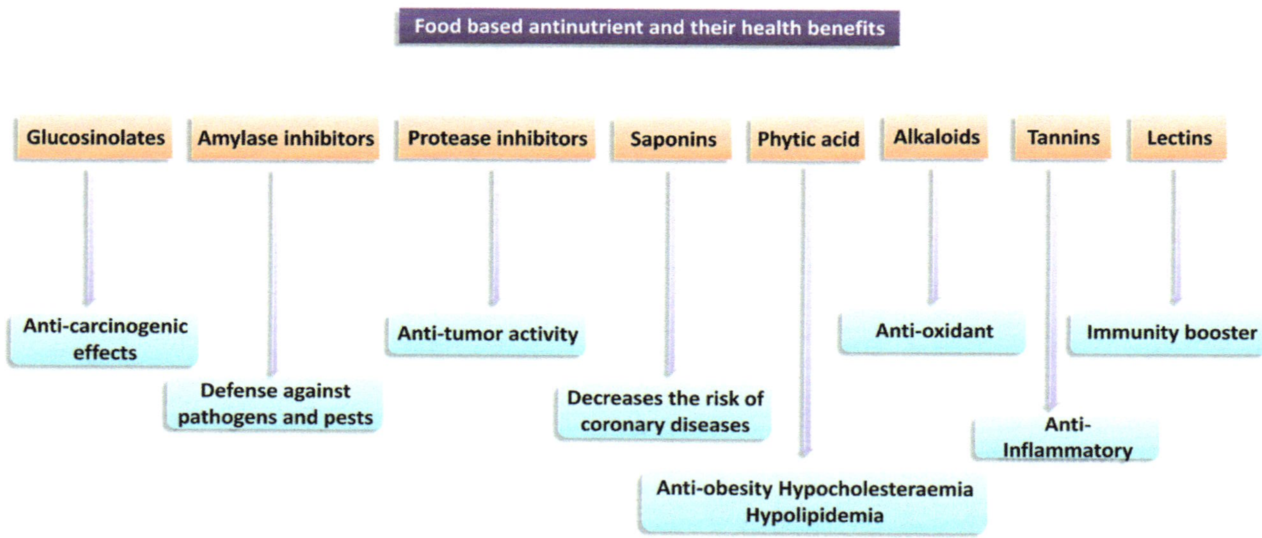

FIGURE 1.2 Food-based antinutrient factors and their applications.

Exploring the role of food-based antinutrient factors and their potential applications in nutrition and health.

critical for ease of market access and stakeholder trust-building (Gualandris et al., 2015). The use of unconventional feed materials poses a set of interrelated challenges regarding nutritional homogeneity, anti-nutritional components, digestibility, palatability, and compliance with regulations. To overcome these challenges, improvements in processing technologies, strict quality control, focused research, and policy formulation are instrumental in realizing their full potential as sustainable and economically competitive feed substitutes within contemporary animal production systems (Qaim, 2020).

5. Strategies to overcome challenges

To fully realize the potential of non-conventional feed ingredients and successfully incorporate them into animal feed, it is necessary to address the multifarious problems described above (Dhandayuthapani et al., 2020). In this regard, several strategies were developed and continue to be improved to minimize variability in nutritional content, anti-nutritional factors, digestive issues, palatability, and regulatory hurdles (Samtiya et al., 2020). Supplementation, processing improvements, blend formulation, and research and development are discussed below as key approaches.

5.1 Supplementation

Supplementation is one of the most basic but efficient measures to neutralize nutritional imbalances or deficiencies of non-conventional feed ingredients. This is achieved by adding essential amino acids, vitamins, minerals, enzymes, or other bioactive substances to harmonize the nutrient spectrum and improve the overall quality of the diet. For instance, supplementation with lysine or methionine is required when employing insect meal or plant proteins that lack specific essential amino acids (Leinonen et al., 2019). Correspondingly, enzyme supplements, such as phytase or protease, may contribute to the degradation of anti-nutritional factors, such as phytates, and enhance nutrient bioavailability. Probiotic or prebiotic supplementation can also be added to improve gut welfare and feed utilization (Amenyogbe et al., 2020). Supplementing the nutritional deficiency of these new feeds enhances animal growth performance and feed efficiency without undermining their cost-effectiveness.

5.2 Processing enhancements

Advances in processing technology are crucial for enhancing the safety and nutritional value of nonconventional feedstuffs. Heat treatment, fermentation, drying, grinding, and extrusion are commonly used to minimize anti-nutritional compounds, increase digestibility, and extend shelf life (Kuswaha et al., 2021). Heat processing can inactivate enzyme inhibitors and toxic materials present in most plant feeds, whereas fermentation with healthy microbes can hydrolyze complex fibers and ANFs, incorporating extra nutrients and bioactive metabolites into the feed. Drying and grinding enhance stability during storage and ease of mixing in feed rations, whereas extrusion processing can increase the availability of nutrients by breaking cell walls and denaturing unwanted substances (John et al., 2024). Texture and flavor profiling are also improved by processing enhancements, which alter texture and flavor profiles (Wang et al., 2023). Ongoing innovations in such methods, including the employment of enzyme-assisted or nano-processing methods, will continue to add value to non-conventional feed ingredients.

5.3 Blending ingredients

Mixing unconventional feed ingredients with conventional feedstuffs is a successful method for maximizing feed formulation, overcoming nutritional limitations, and enhancing palatability. The mixing of plant proteins, insect meals, and SCPs with the traditional proteins of soybean meal or fishmeal is used by nutritionists to create balanced diets that satisfy the needs of various animal species and production phases (Safdari et al., 2025). Blending serves to dilute the anti-nutritional factor concentration, balance amino acid profiles, and cover up unwanted flavors or odors naturally present in some novel ingredients. Blending also eases the process of gradual change for animals, adjusting to new feed ingredients while avoiding intake (Kumar et al., 2022). Additionally, blending aids in cost savings in that it allows for the addition of a lower-cost alternative protein without affecting diet quality. Strategic combinations of ingredients based on local availability and production targets can significantly increase the feasibility and acceptability of non-conventional feeds.

5.4 Research and development

Commitment to ongoing research and development (RandD) is essential for countering the challenges associated with the use of non-conventional feed ingredients (Charles Rajesh Kumar and Majid, 2020). Successive studies have aimed to streamline cultivation and rearing procedures to minimize variability in nutrition and maximize consistency in yield. Science has also investigated new processing technologies, including enzymatic treatments, microbial fermentation advancements, and nanotechnology-based applications to improve the bioavailability of nutrients and lower anti-nutritional factors (Itrat et al., 2025). Studies on animal gastrointestinal physiology and microbiome interactions have led to enhanced digestibility and feed utilization. RandD also aids in formulating regulatory guidelines and safety tests to enable commercialization and consumer acceptance (Rawal et al., 2024). Multidisciplinary research collaboration involving animal nutrition, microbiology, food science, and environmental sustainability is driving the development and mainstream adoption of alternative proteins. RandD investment not only overcomes existing constraints but also releases new functional attributes and uses, which are stimuli for innovation in producing sustainable animal feed (Khan et al., 2021). Overcoming the challenges of non-traditional feed ingredients requires a multistranded approach. Supplementation provides nutritional completeness; processing technologies add quality and safety to feed; blending enhances formulation efficiency and acceptability; and ongoing research problems innovation and regulation (Kang and Park, 2025; Kumar et al., 2025c). Collectively, these approaches offer a strong platform to facilitate sustainable and economically justifiable utilization of new feed sources to secure the resilience and sustainability of global animal production systems.

6. Conclusion

Unconventional feed ingredients, such as insect meals, plant proteins, and single-cell proteins, present exciting alternatives to traditional feedstuffs, responding to key issues of sustainability, resource usage efficiency, and escalating worldwide demand for animal protein. Their varied nutritional profiles, combined with environmentally friendly production, make them central elements in animal nutrition in the future. However, their large-scale use is hampered by issues such as nutritional heterogeneity, the occurrence of anti-nutritional compounds, digestibility constraints, palatability issues, and changing regulatory environments. Overcoming these hurdles through focused supplementation, sophisticated processing methods, strategic mixing with traditional ingredients, and ongoing research and development are crucial for delivering their full potential. As technological innovations continue and regulations evolve, non-traditional feed ingredients will become increasingly feasible, making a key contribution to resilient and sustainable livestock production systems. Incorporating these alternate proteins can be an important way to improve global food security with less environmental harm, in line with the general sustainable agriculture goals.

References

Abdoun, K., Alsagan, A., Altahir, O., Suliman, G., Al-Haidary, A., Alsaiady, M., 2022. Cultivation and uses of *Moringa oleifera* as non-conventional feed stuff in livestock production: a review. Life 13 (1), 63.

Abu Hafsa, S.H., Hassan, A.A., Elghandour, M.M.M.Y., Barbabosa-Pliego, A., Mellado, M., Salem, A.Z.M., 2022. Dietary anti-nutritional factors and their roles in livestock nutrition. Springer Science and Business Media LLC, pp. 131–174. https://doi.org/10.1007/978-3-031-07496-7_4.

Adebo, J.A., Njobeh, P.B., Gbashi, S., Oyedeji, A.B., Ogundele, O.M., Oyeyinka, S.A., Adebo, O.A., 2022. Fermentation of cereals and legumes: impact on nutritional constituents and nutrient bioavailability. Fermentation 8 (2). https://doi.org/10.3390/fermentation8020063.

Akintan, O., Gebremedhin, K.G., Uyeh, D.D., 2024. Animal feed formulation—connecting technologies to build a resilient and sustainable system. Animals 14 (10), 1497.

Akintan, O.A., Gebremedhin, K.G., Uyeh, D.D., 2025. Linking animal feed formulation to milk quantity, quality, and animal health through data-driven decision-making. Animals 15 (2), 162.

Ali, I.M., Forsido, S.F., Kuyu, C.G., Ahmed, E.H., Andersa, K.N., Chane, K.T., Regasa, T.K., 2024. Effects of extrusion process conditions on nutritional, anti-nutritional, physical, functional, and sensory properties of extruded snack: a review. Food Sci. Nutr. 12 (11), 8755–8761. https://doi.org/10.1002/fsn3.4472.

Amenyogbe, E., Chen, G., Wang, Z., Huang, J.S., Huang, B., Li, H., 2020. The exploitation of probiotics, prebiotics and synbiotics in aquaculture: present study, limitations and future directions. Aquac. Int. 28 (3), 1017–1041. https://doi.org/10.1007/s10499-020-00509-0.

Babatunde, O.O., Park, C.S., Adeola, O., 2021. Nutritional potentials of atypical feed ingredients for broiler chickens and pigs. Animals 11 (5), 1196.

Bessa, L.W., Pieterse, E., Marais, J., Dhanani, K., Hoffman, L.C., 2021. Food safety of consuming black soldier fly (*Hermetia illucens*) larvae: microbial, heavy metal and cross-reactive allergen risks. Foods 10 (8). https://doi.org/10.3390/foods10081934.

Biswas, D., Purohit, S.R., 2022. Nutritional profile, antinutritional factors, and digestibility of proteins in non-conventional leaf protein sources. ACS Food Sci. Technol. 4 (5), 1017–1029. https://doi.org/10.1021/acsfoodscitech.3c00676.

Broekman, H.C.H.P., Knulst, A.C., de Jong, G., Gaspari, M., den Hartog Jager, C.F., Houben, G.F., Verhoeckx, K.C.M., 2017. Is mealworm or shrimp allergy indicative for food allergy to insects? Mol. Nutr. Food Res. 61 (9). https://doi.org/10.1002/mnfr.201601061.

Charles Rajesh Kumar, J., Majid, M.A., 2020. Renewable energy for sustainable development in India: current status, future prospects, challenges, employment, and investment opportunities. Energy Sustain. Soc. 10 (1). https://doi.org/10.1186/s13705-019-0232-1.

Chattaraj, S., Behera, B.K., Das Mohapatra, P.K., 2024a. Effective valorization of Chlorella biomass and Brewers spent grain substituting fish meal and soybean meal in the diet of herbivorous fish Cirrhinus reba for higher growth, digestibility and sustainable cultivation. J. Appl. Phycol. 36 (1), 217–232.

Chattaraj, S., Chattaraj, M., Mitra, D., Ganguly, A., Thatoi, H., Das Mohapatra, P.K., 2025. 16S amplicon sequencing of the gastrointestinal microbiota of Cirrhinus reba and isolation of an autochthonous probiotic using culture based approaches. Syst. Microbiol. Biomanuf. 5 (1), 156–170.

Chattaraj, S., Das Mohapatra, P.K., 2024. Immunostimulatory and antagonistic potential of the methanolic extract of Oedogonium intermedium SCB in Cirrhinus reba challenged with Aeromonas hydrophila. Aquac. Int. 32 (3), 2523–2548.

Chattaraj, S., Ganguly, A., Mitra, D., Mitra, D., Das Mohapatra, P.K., 2023. Study of intestinal bacteria of Cirrhinus reba and characterization of a new probiotic bacteria: an initiative to save the threatened species of Cirrhinus. Appl. Biochem. Microbiol. 59 (6), 833–849.

Chattaraj, S., Mitra, D., Chattaraj, M., Ganguly, A., Thatoi, H., Mohapatra, P.K.D., 2024b. Brewers' spent grain as fish feed ingredient: evaluation of bio-safety and analysis of its impact on gut bacteria of Cirrhinus reba by 16S Metagenomic sequencing. Curr. Res. Microb. Sci. 7, 100286.

Chattaraj, S., Mitra, D., Ganguly, A., Thatoi, H., Mohapatra, P.K.D., 2024c. A critical review on the biotechnological potential of Brewers' waste: challenges and future alternatives. Curr. Res. Microb. Sci. 6, 100228.

Chattaraj, S., Sahoo, S.S., Behera, S., Thatoi, H., 2026. Sustainable aquafeed development using Spirulina: effects on growth performance of Catla catla. Algal Res 93, 104447. https://doi.org/10.1016/j.algal.2025.104447.

Chaudhary, S., Chaturvedi, P., Chaudhary, D., Kumar, R., 2024. Extraction of biofuels and valuable products (essential fatty acids) from microalgae: the greenhouse gas emissions. In: Sustainable production innovations: bioremediation and other biotechnologies, pp. 345–366.

Cheng, W., Sun, Y., Fan, M., Li, Y., Wang, L., Qian, H., 2022. Wheat bran, as the resource of dietary fiber: a review. Crit. Rev. Food Sci. Nutr. 62 (26), 7269–7281. https://doi.org/10.1080/10408398.2021.1913399.

Chisoro, P., Jaja, I.F., Assan, N., 2023. Incorporation of local novel feed resources in livestock feed for sustainable food security and circular economy in Africa. Front. Sustain. 4, 1251179.

Choręziak, A., Rosiejka, D., Michałowska, J., Bogdański, P., 2025. Nutritional quality, safety and environmental benefits of alternative protein sources—an overview. Nutrients 17 (7), 1148. https://doi.org/10.3390/nu17071148.

Colombo, S.M., Roy, K., Mraz, J., Wan, A.H.L., Davies, S.J., Tibbetts, S.M., Øverland, M., Francis, D.S., Rocker, M.M., Gasco, L., Spencer, E., Metian, M., Trushenski, J.T., Turchini, G.M., 2023. Towards achieving circularity and sustainability in feeds for farmed blue foods. Rev. Aquac. 15 (3), 1115–1141. https://doi.org/10.1111/raq.12766.

Dalin, C., Rodríguez-Iturbe, I., 2016. Environmental impacts of food trade via resource use and greenhouse gas emissions. Environ. Res. Lett. 11 (3), 035012. https://doi.org/10.1088/1748-9326/11/3/035012.

Dawadi, E., Magar, P.B., Bhandari, S., Subedi, S., Shrestha, S., Shrestha, J., 2022. Nutritional and post-harvest quality preservation of mushrooms: a review. Heliyon 8 (12). https://doi.org/10.1016/j.heliyon.2022.e12093.

de Oliveira, R.M.S.F., dos Santos, H.M.L., Fernandes, S.S., Egea, M.B., 2025. Exploring plant agro—industrial by—products as a source of fibrous food ingredients: a review of extraction methods and technological properties. J. Food Sci. 90 (7). https://doi.org/10.1111/1750-3841.70408.

Dhandayuthapani, K., Malathy, S., Mulla, S.I., Gupta, S.K., 2020. An insight into the potential application of microalgae in pharmaceutical and nutraceutical production. In: Algae: multifarious applications for a sustainable world. Springer Singapore, India, pp. 135–179. https://doi.org/10.1007/978-981-15-7518-1_7.

Di, D., He, S., Zhang, R., Gao, K., Qiu, M., Li, X., Sun, H., Xue, S., Shi, J., 2024. Exploring the dual role of anti-nutritional factors in soybeans: a comprehensive analysis of health risks and benefits. Crit. Rev. Food Sci. Nutr. 1–18. https://doi.org/10.1080/10408398.2024.2430757.

Dwivedi, S.L., Lammerts van Bueren, E.T., Ceccarelli, S., Grando, S., Upadhyaya, H.D., Ortiz, R., 2017. Diversifying food systems in the pursuit of sustainable food production and healthy diets. Trends Plant Sci. 22 (10), 842–856. https://doi.org/10.1016/j.tplants.2017.06.011.

European Food Safety Authority (EFSA), 2008. Animal welfare aspects of husbandry systems for farmed Atlantic salmon—scientific opinion of the panel on animal health and welfare. EFSA J. 6 (7), 736.

Gaikwad, V., Rautela, A., Kedare, G., Sarkar, S., Pal, R.P., Sharma, A., 2024. Role of flavoured additives in ruminant production feed additives and supplements for ruminants. Springer Singapore, India, pp. 265–280. https://doi.org/10.1007/978-981-97-0794-2_12.

Gençdağ, E., Görgüç, A., Yılmaz, F.M., 2021. Recent advances in the recovery techniques of plant-based proteins from agro-industrial by-products. Food Rev. Int. 37 (4), 447–468.

Grigore, D.M., Mircea, M.L., Pogurschi, E.N., 2025. Toward sustainable broiler production: evaluating microbial protein as supplementation for conventional feed proteins. Agriculture 15 (14), 1486.

Gualandris, J., Klassen, R.D., Vachon, S., Kalchschmidt, M., 2015. Sustainable evaluation and verification in supply chains: aligning and leveraging accountability to stakeholders. J. Oper. Manag. 38, 1–13. https://doi.org/10.1016/j.jom.2015.06.002.

Gupta, N., Wijenayake, W.P.T., Roy, D., Kumar, R., Rangot, M., Chugh, P., Santoyo, G., Chattaraj, S., Bhaskar, S.S., Andjelković, S., Guerra-Sierra, B.E., Mitra, D., 2025. Smart recycling and sustainable lignocellulosic waste management. In: Value addition and utilization of lignocellulosic biomass: through novel technological interventions. Springer Nature, India, pp. 221–250. https://doi.org/10.1007/978-981-96-2786-8_9.

Itrat, N., Israr, B., Arif, S., Narjis, M., Asghar, S., Ali, A., 2025. Nutrient absorption dynamics and food contamination. Springer Science and Business Media LLC, pp. 101–131. https://doi.org/10.1007/978-3-031-94582-3_3.

John, R., Bhardwaj, R., Janamatti, A.T., Singh, P., 2024. Milling, processing, and storage of oats strategies and effect on nutritional properties. In: Oat (Avena sativa): production to plate. CRC Press, pp. 148–188. https://doi.org/10.1201/9781003263302-5.

Jucker, C., Belluco, S., 2022. Developing of low-impact diets for the mass rearing of Acheta domesticus. J. Insects Food Feed 6, S61.

Kang, H.J., Park, Y.K., 2025. Food for special medical purposes: a global review of usage trends, regulatory frameworks, and key growth drivers. Clin. Nutr. Res. 14 (2), 146.

Kårlund, A., Gómez-Gallego, C., Korhonen, J., Palo-Oja, O.M., El-Nezami, H., Kolehmainen, M., 2022. Harnessing microbes for sustainable development: food fermentation as a tool for improving the nutritional quality of alternative protein sources. Nutrients 12 (4). https://doi.org/10.3390/nu12041020.

Katu, J.K., Tóth, T., Varga, L., 2025. Enhancing the nutritional quality of low-grade poultry feed ingredients through fermentation: a review. Agriculture 15 (5), 476.

Khan, N., Ray, R.L., Kassem, H.S., Hussain, S., Zhang, S., Khayyam, M., Asongu, S.A., 2021. Potential role of technology innovation in transformation of sustainable food systems: a review. Agriculture 11 (10), 984.

Kröncke, N., Benning, R., 2022. Self-selection of feeding substrates by Tenebrio molitor larvae of different ages to determine optimal macronutrient intake and the influence on larval growth and protein content. Insects 13 (7), 657. https://doi.org/10.3390/insects13070657.

Kumar, K., Dey, A., Rose, M.K., Dahiya, S.S., 2022. Impact of dietary phytogenic composite feed additives on immune response, antioxidant status, methane production. Antioxidants 11 (2). https://doi.org/10.3390/antiox11020325.

Kumar, R., Kamboj, N., Mitra, D., 2024. Microbial innovations in agriculture: interdisciplinary approaches to leveraging microbes for food sustainability and security. Indian J. Microbiol. Res. 11 (3), 129–139.

Kumar, R., Farda, B., Mignini, A., Djebaili, R., Koolman, L., Paul, A., Mondal, S., Joel, J.M., Pandit, A., Panneerselvam, P., Pellegrini, M., Mitra, D., 2025a. Microbial solutions in agriculture: enhancing soil health and resilience through bio-inoculants and bioremediation. Bacteria 4 (3), 28.

Kumar, R., Gupta, G., Hussain, A., Rani, A., Thapliyal, A., Gunsola, D., Mitra, D., 2025b. Pioneering zero—waste technologies utilization and its framework on sustainable management: international, national and state level. Discov. Appl. Sci. 7 (3), 224.

Kumar, R., Kamboj, N., Mitra, D., Ray, A., 2025c. Strengthening local communities through advanced recycling technology for sustainable growth. In: Mitra, D., Choudhury, T., Madan, A., Chattaraj, S., Pellegrini, M. (Eds.), AI technologies for enhancing recycling processes. IGI Global Scientific Publishing, pp. 393–420.

Kumar, R., Kamboj, N., Mitra, D., Verma, D., Choudhury, T., 2025d. Bacterial strategies: novel pathways for sustainable development through wastewater. In: Mitra, D.,

Choudhury, T., Madan, A., Chattaraj, S., Pellegrini, M. (Eds.), AI technologies for enhancing recycling processes. IGI Global Scientific Publishing., pp. 421–440

Kuswaha, R., Sigh, V., Sigh, M., Puranik, V., Kaur, D., 2021. Effect of food processing on nutritional and anti-nutritional components. Chall. Oppor. Nutr. Envirom. Agric. 1, 97–111.

Lazou, A.E., 2024. Food extrusion: an advanced process for innovation and novel product development. Crit. Rev. Food Sci. Nutr. 64 (14), 4532–4560. https://doi.org/10.1080/10408398.2022.2143474.

Leinonen, I., Iannetta, P.P.M., Rees, R.M., Russell, W., Watson, C., Barnes, A.P., 2019. Lysine supply is a critical factor in achieving sustainable global protein economy. Front. Sustain. Food Syst. 3. https://doi.org/10.3389/fsufs.2019.00027.

Leuz, C., 2010. Different approaches to corporate reporting regulation: how jurisdictions differ and why. Account. Bus. Res. 40 (3), 229–256. https://doi.org/10.1080/00014788.2010.9663398.

Li, Y.P., Ahmadi, F., Kariman, K., Lackner, M., 2024. Recent advances and challenges in single cell protein (SCP) technologies for food and feed production. Npj Sci. Food 8 (1). https://doi.org/10.1038/s41538-024-00299-2.

Mageshwaran, V., 2021. An overview of gossypol and methods of its detoxification in cottonseed meal for non-ruminant feed applications. Indian J. Nat. Prod. Resour. 12 (3), 348–358.

Makkar, H.P.S., 2018. Review: feed demand landscape and implications of food-not feed strategy for food security and climate change. Animal 12 (8), 1744–1754. https://doi.org/10.1017/s175173111700324x.

Mancini, S., Mattioli, S., Paolucci, S., Fratini, F., Dal Bosco, A., Tuccinardi, T., Paci, G., 2021. Effect of cooking techniques on the in vitro protein digestibility, fatty acid profile, and oxidative status of mealworms (*Tenebrio molitor*). Front. Vet. Sci. 8. https://doi.org/10.3389/fvets.2021.675572.

Manzo, G., Hind, C.K., Ferguson, P.M., Amison, R.T., Hodgson-Casson, A.C., Ciazynska, K. A., Weller, B.J., Clarke, M., Lam, C., Man, R.C.H., Shaughnessy, B.G.O., Clifford, M., Bui, T.T., Drake, A.F., Atkinson, R.A., Lam, J.K.W., Pitchford, S.C., Page, C.P., Phoenix, D.A., Lorenz, C.D., Sutton, J.M., Mason, A.J., 2020. A pleurocidin analogue with greater conformational flexibility, enhanced antimicrobial potency and in vivo therapeutic efficacy. Commun. Biol. 3 (1). https://doi.org/10.1038/s42003-020-01420-3.

Martineau-Côté, D., Achouri, A., Karboune, S., L'Hocine, L., 2022. Faba bean: an untapped source of quality plant proteins and bioactives. Nutrients 14 (8), 1541.

Moreira, J.B., Vaz, B.d.S., Cardias, B.B., Cruz, C.G., Almeida, A.C.A.d., Costa, J.A.V., Morais, M.G.d., 2022. Microalgae polysaccharides: an alternative source for food production and sustainable agriculture. Polysaccharides 3 (2), 441–457. https://doi.org/10.3390/polysaccharides3020027.

Muniz, E.d.N., Montenegro, R.T.d.Q., da Silva, D.N., D'Almeida, A.P., Gonçalves, L.R.B., de Albuquerque, T.L., 2024. Advances in biotechnological strategies for sustainable production of non-animal proteins: challenges, innovations, and applications. Fermentation 10 (12). https://doi.org/10.3390/fermentation10120638.

Nedelkov, K.V., 2023. A new approach for processing and use of sunflower meal. Bulgarian J. Agric. Sci. 29 (2), 384–389.

Nikmaram, N., Kamani, M.H., Ghalavand, R., 2015. The effects of extrusion cooking on antinutritional factors, chemical propertiesand contaminating microorganisms of food. Int. J. Farm. Allied Sci. 4 (4), 352–354.

Niu, G., Zhang, T., Cao, S., Zhang, X., Tao, L., 2023. Effect of corn grinding methods and particle size on the nutrient digestibility of Chahua chickens. Animals 13 (14), 2364. https://doi.org/10.3390/ani13142364.

Obert, J.C., Hughes, D., Sorenson, W.R., McCann, M., Ridley, W.P., 2007. A quantitative method for the determination of cyclopropenoid fatty acids in cottonseed, cottonseed meal, and cottonseed oil (Gossypium hirsutum) by high-performance liquid chromatography. J. Agric. Food Chem. 55 (6), 2062–2067. https://doi.org/10.1021/jf0617871.

Onyeaka, H., Anumudu, C.K., Okpe, C., Okafor, A., Ihenetu, F., Miri, T., Odeyemi, O.A., Anyogu, A., 2022. Single cell protein for foods and feeds: a review of trends. Open Microbiol. J. 16 (1). https://doi.org/10.2174/18742858-v16-e2206160.

Parniakov, O., Mikhrovska, M., Wiktor, A., Alles, M., Ristic, D., Bogusz, R., Nowacka, M., Devahastin, S., Mujumdar, A., Heinz, V., Smetana, S., 2022. Insect processing for food and feed: a review of drying methods. Dry. Technol. 40 (8), 1500–1513. https://doi.org/10.1080/07373937.2021.1962905.

Qaim, M., 2020. Role of new plant breeding technologies for food security and sustainable agricultural development. Appl. Econ. Perspect. Pol. 42 (2), 129–150. https://doi.org/10.1002/aepp.13044.

Ramachandran, P., Rai, S., Sangeeta, S., Khati, A., 2024. Waste management techniques for various food processing industries. In: Wealth out of food processing waste: ingredient recovery and valorization. CRC Press, pp. 1–40. https://doi.org/10.1201/9781003269199-1.

Rawal, S., Kaur, H., Bhathan, S., Mittal, D., Kaur, G., Ali, S.A., 2024. Ruminant gut microbiota: interplay, implications, and innovations for sustainable livestock production. Springer Science and Business Media LLC, pp. 205–228. https://doi.org/10.1007/978-3-031-54372-2_7.

Remize, M., Brunel, Y., Silva, J.L., Berthon, J.Y., Filaire, E., 2021. Microalgae n-3 PUFAs production and use in food and feed industries. Mar. Drugs 19 (2). https://doi.org/10.3390/MD19020113.

Richmond, A., Hu, Q., 2013. Microalgae for human and animal nutrition handbook of microalgal culture: applied phycology and biotechnology. Wiley, pp. 461–503. https://doi.org/10.1002/9781118567166.ch25.

Rodrigues, S.S.Q., Leite, A., Vasconcelos, L., Teixeira, A., 2024. Exploring the nexus of feeding and processing: implications for meat quality and sensory perception. Foods 13 (22), 3642. https://doi.org/10.3390/foods13223642.

Saah, B.F., Tene, S.T., Boungo, G.T., Klang, J.M., 2020. Optimization of the reduction of phytates and trypsin inhibitors of soybeans (Glycine max L.): effect of soaking and cooking. Int. J. Food Nut Sci. 7, 70–76.

Safdari, Z., Ariffin, M.K.A., Supeni, E.E., Sultan, M.T.H., 2025. Fulfilling feed demands: industrial production of Saccharomyces cerevisiae as key protein source for aquaculture, poultry and livestock. Pertanika J. Sci. Technol. 33 (2), 939–968. https://doi.org/10.47836/pjst.33.2.18.

Sagar, N.A., Pareek, S., Sharma, S., Yahia, E.M., Lobo, M.G., 2018. Fruit and vegetable waste: bioactive compounds, their extraction, and possible utilization. Compr. Rev. Food Sci. Food Saf. 17 (3), 512–531. https://doi.org/10.1111/1541-4337.12330.

Saha, D., Patra, A., Prasath, V.A., Pandiselvam, R., 2022. Anti-nutritional attributes of faba-bean faba bean: chemistry, properties and functionality. Springer International Publishing, pp. 97–122. https://doi.org/10.1007/978-3-031-14587-2_5.

Salgado-Ramos, M., Castagnini, J.M., Dar, B.N., Gharibzahedi, S.M.T., Phimolsiripol, Y., Rimac-Brnčić, S., Martínez-Culebras, P.V., Altintas, Z., Barba, F.J., 2024. Edible insects as sustainable and nutrient-rich food sources: exploring innovations and advancements for future food practices–A comprehensive review. Food Rev. Int. 40 (9), 3103–3128. https://doi.org/10.1080/87559129.2024.2341725.

Samtiya, M., Aluko, R.E., Dhewa, T., 2020. Plant food anti-nutritional factors and their reduction strategies: an overview. Food Prod. Process. Nutr. 2 (1), 6.

Schulz, E., Oslage, H.J., 1976. Composition and nutritive value of single-cell protein (SCP). Anim. Feed Sci. Technol. 1 (1), 9–24. https://doi.org/10.1016/0377-8401(76)90003-1.

Sekoai, P.T., Roets-Dlamini, Y., O'Brien, F., Ramchuran, S., Chunilall, V., 2021. Valorization of food waste into single-cell protein: an innovative technological strategy for sustainable protein production. Microorganisms 12 (1). https://doi.org/10.3390/microorganisms12010166.

Shah, A.M., Zhang, H., Shahid, M., Ghazal, H., Shah, A.R., Niaz, M., Naz, T., Ghimire, K., Goswami, N., Shi, W., Xia, D., Zhao, H., 2025. The vital roles of agricultural crop residues and agro-industrial by-products to support sustainable livestock productivity in subtropical regions. Animals 15 (8). https://doi.org/10.3390/ani15081184.

Sheeraz, M., Iqbal, N., Sajid, M., 2024. A nexus between climate change risk and financial performance of agricultural businesses in developing economy. In: Climate change and finance: navigating the challenges and opportunities in capital markets. Springer Nature Switzerland, pp. 255–278.

Siddiqui, S.A., Kamran, F., Stankov, S., Rathod, N.B., Teixeira-Costa, B.E., Fidan, H., Bhat, M.I., Sofi, S.A., 2025. Unveiling the diversity of non conventional proteins - from sources, extraction, technofunctionality, nutraceutical potential to advancement in food applications—a systematic review. Waste Biomass Valoriz. 16 (1), 29–51. https://doi.org/10.1007/s12649-024-02614-0.

Singh, R., Das, R., Sangwan, S., Rohatgi, B., Khanam, R., Pedda Ghouse Peera, S.K., Das, S., Lyngdoh, Y.A., Langyan, S., Shukla, A., Shrivastava, M., Misra, S., 2021. Utilisation of agro-industrial waste for sustainable green production: a review. Environ. Sustain. 4 (4), 619–636. https://doi.org/10.1007/s42398-021-00200-x.

Skowron, K., Budzyńska, A., Grudlewska-Buda, K., Wiktorczyk-Kapischke, N., Andrzejewska, M., Wałecka-Zacharska, E., Gospodarek-Komkowska, E., 2022. Two faces of fermented foods—the benefits and threats of its consumption. Front. Microbiol. 13. https://doi.org/10.3389/fmicb.2022.845166.

Sogari, G., Amato, M., Palmieri, R., Hadj Saadoun, J., Formici, G., Verneau, F., Mancini, S., 2023. The future is crawling: evaluating the potential of insects for food and feed security. Curr. Res. Food Sci. 6. https://doi.org/10.1016/j.crfs.2023.100504.

Spaggiari, M., Dall'Asta, C., Galaverna, G., Bilbao, M.D.del C., 2021. Bran By-Product: From valorization strategies to nutritional perspectives. Foods 10 (1), 85. https://doi.org/10.3390/foods10010085.

Suhag, R., Dhiman, A., Deswal, G., Thakur, D., Sharanagat, V.S., Kumar, K., Kumar, V., 2021. Microwave processing: a way to reduce the anti-nutritional factors (ANFs) in food grains. LWT 150. https://doi.org/10.1016/j.lwt.2021.111960.

Sultana, R., Singh, R.S., Ratnakumar, P., Verma, N., Chaturvedi, S.K., Chaudhary, A.K., Sameer, C.V., Siddiqui, M.W., 2017. Pulse secondary metabolites: a perspective on human and animal health. In: Plant secondary metabolites: volume I: biological and therapeutic significance. Apple Academic Press, pp. 215–242. https://doi.org/10.1201/9781315366326.

Sun, W., Shahrajabian, M.H., Lin, M., 2022. Research progress of fermented functional foods and protein factory-microbial fermentation technology. Fermentation 8 (12), 688.

Sztupecki, W., Rhazi, L., Depeint, F., Aussenac, T., 2023. Functional and nutritional characteristics of natural or modified wheat bran non-starch polysaccharides: a literature review. Foods 12 (14), 2693. https://doi.org/10.3390/foods12142693.

Te Pas, M.F.W., Veldkamp, T., de Haas, Y., Bannink, A., Ellen, E.D., 2021. Adaptation of livestock to new diets using feed components without competition with human edible protein sources—a review of the possibilities and recommendations. Animals 11 (8). https://doi.org/10.3390/ani11082293.

Teshome, E., Forsido, S.F., Rupasinghe, H.P.V., Olika Keyata, E., 2022. Potentials of natural preservatives to enhance food safety and shelf life: a review. Sci. World J 2022. https://doi.org/10.1155/2022/9901018.

Tlais, A.Z.A., Fiorino, G.M., Polo, A., Filannino, P., Cagno, R.D., 2020. High-value compounds in fruit, vegetable and cereal byproducts: an overview of potential sustainable reuse and exploitation. Molecules 25 (13). https://doi.org/10.3390/molecules25132987.

Udomsil, N., Imsoonthornruksa, S., Gosalawit, C., Ketudat-Cairns, M., 2019. Nutritional values and functional properties of house cricket (*Acheta domesticus*) and field cricket (Gryllus bimaculatus). Food Sci. Technol. Res 25 (4), 597–605. https://doi.org/10.3136/fstr.25.597.

Van Peer, M., Frooninckx, L., Coudron, C., Berrens, S., Álvarez, C., Deruytter, D., Verheyen, G., Van Miert, S., 2021. Valorisation potential of using organic side streams as feed for *Tenebrio molitor*, *Acheta domesticu*s and *Locusta migratori*a. Insects 12 (9), 796. https://doi.org/10.3390/insects12090796.

Vujetić, J., Spasevski, N., Dragojlović, D., 2025. Processing techniques of removing antinutrients from oilseed cakes as by-products intended for animal feeding. Food Feed Res. 52 (1), 37–51.

Wang, T., Yu, Z., Ahmad, R., Riaz, S., Khan, K.U., Siyal, S., Chaudhry, M.A., Zhang, T., 2022. Transition of bioeconomy as a key concept for the agriculture and agribusiness development: an extensive review on ASEAN countries. Front. Sustain. Food Syst. 6. https://doi.org/10.3389/fsufs.2022.998594.

Wang, X., McClements, D.J., Xu, Z., Meng, M., Qiu, C., Long, J., Jin, Z., Chen, L., 2023. Recent advances in the optimization of the sensory attributes of fried foods: appearance, flavor, and texture. Trends Food Sci. Technol. 138, 297–309. https://doi.org/10.1016/j.tifs.2023.06.012.

Wildermuth, S.R., Young, E.E., Were, L.M., 2016. Chlorogenic acid oxidation and its reaction with sunflower proteins to form green—colored complexes. Compr. Rev. Food Sci. Food Saf. 15 (5), 829–843.

Wisetkomolmat, J., Arjin, C., Satsook, A., Seel-audom, M., Ruksiriwanich, W., Prom-u-Thai, C., Sringarm, K., 2022. Comparative analysis of nutritional components and phytochemical attributes of selected Thai rice bran. Front. Nutr. 9. https://doi.org/10.3389/fnut.2022.833730.

Yang, J., Zhou, S., Kuang, H., Tang, C., Song, J., 2024. Edible insects as ingredients in food products: nutrition, functional properties, allergenicity of insect proteins, and processing modifications. Crit. Rev. Food Sci. Nutr. 64 (28), 10361–10383. https://doi.org/10.1080/10408398.2023.2223644.

Zhuang, Z., Wan, G., Lu, X., Xie, L., Yu, T., Tang, H., 2024. Metabolic engineering for single-cell protein production from renewable feedstocks and its applications. Adv. Biotechnol. 2 (4). https://doi.org/10.1007/s44307-024-00042-8.

CHAPTER

Utilization of fruit and vegetable wastes in fish feed formulation

2

Sukanya Sahu, Anyesha Roy Choudhary, Swagat Kumar Das and Supriya Dash
Department of Biotechnology, Odisha University of Technology and Research, Bhubaneswar, Odisha, India

1. Introduction

Fish feed is the most important part of the aquaculture cycle, around 60%–70% of the total cost spent in one production cycle is spent on feed procurement (Sari and Yulisma, 2017). The rapid expansion of aquaculture has increased the demand for high-quality feed, traditionally dominated by fish meal and soybean meal (Chattaraj et al., 2026a,b). Aquaculture production is expected to continue growing to meet the increasing global demand for fish as a protein source. However, the sustainability of conventional feed ingredients remains a major concern due to overfishing, land degradation, and high feed production costs. The dependence on fish meal and soybean meal has also contributed to ecological imbalances, leading to the exploration of alternative, cost-effective, and environmentally friendly feed options. Several conditions must be met in the selection of alternative feed ingredients, including these ingredients that have sufficient nutritional content for fish growth, are relatively inexpensive, have large amounts of availability, are guaranteed their sustainability, and do not compete with humans in their use (Chattaraj et al., 2024; Suprayudi et al., 2011).

2. Fish nutritional requirements

A few freshwater and marine species, primarily carp, the Nile tilapia, catla, Atlantic salmon, milkfish, gilthead seabream, giant yellow croaker, and European seabass, dominate the world's fish aquaculture production (The State of World Fisheries and Aquaculture 2024, 2024). A significant amount of dietary protein—nearly 50%, depending on life stage—is needed by carnivorous species such as Atlantic salmon, croaker, seabream, and seabasses (Kapoor and Feng, 2022). Nearly 40% of protein is needed by herbivorous fish, such as cyprinids (common carp *Cyprinus carpio*, grass carp *Ctenopharyngodon idellus*, and *Catla catla*) (Kaushik, 1995). For the best growth outcomes, a diet's essential amino acid (EAA) profile is just as important as its crude protein content. The crude protein requirements can be

Fish Feed Formulation with Agroindustrial Wastes and Byproducts. https://doi.org/10.1016/B978-0-443-41728-3.00019-8

reduced if a high-quality (balanced EAA:NEAA ratio) protein source is provided. Both freshwater and marine fish share the same 10 EAA: Arginine (Arg), Histidine (His), Isoleucine (Ile), Leucine (Leu), Lysine (Lys), Methionine (Met), Phenylalanine (Phe), Threonine (Thr), Tryptophane (Trp), Valine (Val). However, specific amino acid requirements change with the carnivory level of species. Fishes of higher trophic level have significantly lower Lysine (Lys), Threonine (Thr), and Phenylalanine (Phe)+Tryptophane (Trp) requirements when compared with those of lower trophic level (Xing et al., 2024).

Fish rely mostly on lipids for energy, yet freshwater and marine species differ significantly in this regard. While marine fish have higher requirements, freshwater fish normally need a lipid content of 3%–12% dry matter. While some diadromous fish, such as Atlantic salmon, and herbivorous fish, such as common carp, can produce EPA and DHA through α-linolenic acid (ALA) supplied by the diet, marine species generally cannot manufacture EPA and DHA. All of these fatty acids are regarded as necessary, and the need for EPA is usually greater than that for DHA (Tocher et al., 2019).

Energy is necessary for the growth and metabolic functions of both freshwater and saltwater fish. The metabolic oxidation of the proteins, lipids, and carbohydrates inside produces the energy (Nutrient Requirements of Fish and Shrimp, 2011). A growing fish develops new tissues, and some of the energy it receives from its food is used to create store molecules (mostly triglycerides and glycogen) after it has synthesized structural biomolecules (proteins, nucleic acids, and membrane lipids).

Fish embryogenesis, skeletal integrity and growth, reproduction, immune system response, vision, appetite, and flesh quality are all significantly impacted by vitamins. Vitamin deficiencies can cause clubbed gills and other organ damage, anemia, cataracts, exophthalmia, irregular swimming, hyperirritability, convulsions, and other mental dysfunctions, even if vitamin requirements are minimal (Mai et al., 2022). The major nutritional components of fish feed and their role has been summarized in Fig. 2.1.

3. Nutritional content of fruit and vegetable waste

Fruit and vegetable waste emerges as a promising alternative due to its abundance, nutritional value, and potential to reduce feed costs. These waste materials, which include peels, pulps, and seeds, are often discarded despite their rich nutrient composition. These materials contain carbohydrates, proteins, lipids, vitamins, and minerals beneficial for fish growth. The major types of fruit and vegetable waste along with their nutritional components has been summarized in Table 2.1.

The use of fruit and vegetable waste in fish feed can contribute to waste reduction, sustainable aquaculture, and improved fish health. Additionally, incorporating fruit and vegetable waste into fish diets aligns with circular economy principles by repurposing agricultural and food industry by-products. The selection of fish feed

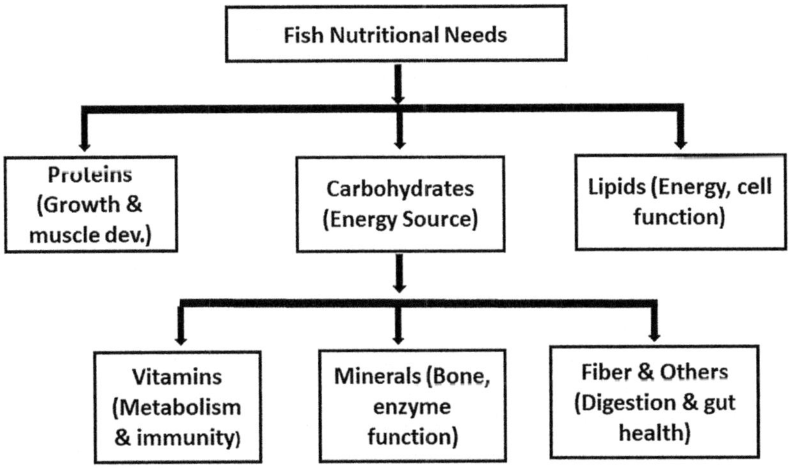

FIGURE 2.1

Nutritional requirement of fish.

Table 2.1 Types of fruits and vegetable waste and their nutritional components.

Fruit and vegetable waste	Nutritional components
Banana peels	Rich in fiber, potassium, and antioxidants
Citrus peels	Contain flavonoids, vitamin C, and essential oils with antimicrobial properties
Carrot waste	High in beta-carotene and fiber
Tomato pomace	Source of lycopene, protein, and fiber
Apple pomace	Contains pectin, fiber, and phenolic compounds

ingredients has several criteria that must be met, including that the material must have sufficient nutritional content to meet the nutritional needs of fish, feed must have low fiber content so that it is easily digested to increase feed efficiency, ingredients must not contain poison, this material is easily available and has a cheap economic value (Devani and Basriati, 2015).

4. Tomato processing by-products

Peels, seeds, some pulp, and fibrous portions—often referred to as tomato bagasse or pomace—are by-products of tomato processing. This trash, which is primarily made up of fiber, nonetheless exhibits intriguing levels of proteins after drying. The main lipidic components are unsaturated fatty acids, specifically oleic acid (C18:1n-9), α-linolenic acid (C18:3n-3), and linolenic acid (C18:2n-6). Carotenoids (such as lycopene and β-carotene) and phenolic compounds, which are known to have enhancer activities, are among the important substances found in tomato by-products (Coelho et al., 2023).

5. Pumpkin processing by-products

Pumpkin peels and seeds, which are typically discarded during pumpkin processing, contain essential nutrients. While the seeds are mostly composed of proteins and lipids, the peels have a higher amount of fiber, followed by protein and, at a lower level, lipids. The main components of pumpkin seed oil are unsaturated fatty acids, specifically LA and C18:1n-9, along with C16:0 and C18:0. However, comparatively significant levels of Val, Leu, and Lys are found in pumpkin skins and seeds. Vitamin C, beta-carotene, minerals (namely P and Fe), and phenolic compounds are all abundant in pumpkin skins. Together with Fe, Ca, and Na, the most prevalent minerals in seeds are K and Zn. Pumpkin seeds also contain considerable amounts of tocopherol and phenolic compounds, particularly flavonoids (Hussain et al., 2021).

6. Potato processing by-products

Potato peels are a byproduct of the potato processing, starch, flour, and canning industries. This residue's nutritional makeup varies depending on the type of potato and where it comes from, but it mostly consists of proteins, carbs, and a small quantity of fat. The primary constituents of the potato peel lipidic fraction are C18:1n-9, LA, ALA, C16:0, and C18:0. In addition to minerals such as K, P, Ca, Mg, and Fe, potato peels are said to be a good source of phenolic compounds with antibacterial and antioxidant qualities (Vaitkevičienė, 2019).

7. Banana processing by-products

Nutrients such as carbs, protein, dietary fiber, and vitamins are abundant in banana by-products, including overripe or damaged fruits, peels, pseudo-stems, leaves, and blooms. While fruits are high in carbs, peels and leaves have a significant amount of fats and proteins. The EAAs Arg, Leu, Phe, Ile, Trp, Lys, His, Thr, Met, and Val are found in banana peels. In addition to other unsaturated FA, including LA and C18:1n-9, banana peels are an intriguing source of n-3 PUFA, specifically ALA and DHA. Numerous phytochemicals, including as phenolic compounds, carotenoids, vitamin C, vitamin E, and vitamin B complex, are found in banana wastes (Kumari et al., 2023).

8. Fruits and vegetables by-products processing and transformation

8.1 Milling and grinding

Milling and grinding are mechanical operations that convert fruit and vegetable byproducts into smaller particles, making them easier to handle and incorporate into aquafeed formulas. Different matrices can be reduced to powder using a variety of techniques, including attrition mills, hammers, balls, and rollers. The equipment used is determined by criteria such as the hardness, moisture content, and desirable particle size of the fruits and vegetable byproducts. It is important to keep in mind that grinding can change the physicochemical characteristics of fruits and vegetable by-products, especially dietary fiber, which is crucial to their functional qualities, and reduce their particle size (Gali et al., 2020).

8.2 Drying

Drying is a fundamental method employed to reduce the moisture content of fruits and vegetables by-products, making them suitable for long-term storage and incorporation into aquafeed formulations. Drying significantly prevents the activity of microorganisms and enzymes, thereby extending the shelf-life of fruit and vegetable by-products (Mphahlele et al., 2019).

Drying is the process of removing moisture from fruits and vegetable by-products via evaporation, which is commonly accomplished by exposing the material to heat and airflow. Depending on the kind of by-product, the required moisture content, and the viability of the business, different drying methods are used. Air drying, solar drying, freeze drying, and mechanical drying (such as rotary dryers and fluidized bed dryers) are common drying techniques.

8.3 Fermentation

The indigenous bacteria found naturally in raw food ingredients are the main force behind the biotechnological fermentation process. By improving the microbiological stability of food matrices, the initial objective was to increase the shelf life of food items, especially those based on fruits and vegetables, allowing for long-term storage at room temperature. Recent discoveries have revealed that food fermentation not only has nutritional and health benefits, such as enriching foods with vitamins, promoting prebiotic/probiotic effects, and improving digestibility, but it also contributes to increased food safety through allelopathic activity against harmful bacteria and fungi contaminants (Licandro et al., 2020), with obvious advantages when applied to fruits and vegetable by-products.

8.4 Enzymatic treatment and chemical processing

Enzymatic and chemical treatments are commonly utilized in the processing of fruits and vegetable byproducts to improve their usefulness as aquafeed ingredients.

Despite differences in their mechanisms, benefits, and drawbacks, both approaches are essential for increasing nutrient availability, lowering anti-nutritional factors, and promoting feed quality. Biological catalysts such as cellulases, proteases, and phytases are used in the enzymatic treatment. Cellulases, for instance, are efficient in hydrolyzing cellulose into fibrous byproducts, increasing aquatic species' access to energy-rich polysaccharides (El-Fakharany et al., 2024). Conversely, proteases break down proteins into amino acids and peptides, increasing the nutritional content and digestibility of by-products high in protein, such as pea pods and tomato pomace (Kesbiç et al., 2022).

Chemical processing, on the other hand, offers an alternative approach to improving the quality of fruits and vegetables by-products for aquafeed. This method often involves using acids, alkalis, or other chemical agents to modify the structure of byproduct components. Acid hydrolysis, for instance, enhances the solubility of proteins and carbohydrates, making them more digestible and suitable for aquatic species (Cecchi et al., 2023).

8.5 Emerging technologies

Emerging technologies have transformed the processing and transformation of fruit and vegetable by-products, resulting in novel methods to improve their use as aquafeed ingredients. These technologies concentrate on enhancing functional qualities, maximizing nutrition retention, and tackling sustainability issues.

Emerging technologies are transforming fruit and vegetable by-product processing by improving efficiency, safety, reducing waste, and making by-product utilization more sustainable and effective. The emerging technologies include nonthermal processing, advanced thermal technologies, bio- and chemical technologies (Fig. 2.2). The mode of action and advantages of different methods are summarized in Table 2.2.

The nonthermal processing method includes techniques such as High-Pressure Processing (HPP), ultrasound, pulsed electric field (PEF) technology and cold plasma techniques. Out of these methods, PEF technology is gaining popularity because of its ability to promote nutrition recovery from fruit and vegetable waste (Nowosad et al., 2021). PEF breaks down cell membranes and makes it easier to absorb intracellular nutrients by delivering brief bursts of high-voltage electricity to cells. According to recent studies, citrus fruits and peel treated with PEF boosted polyphenol extraction to 50% and increased orange, pomelos, and lemon juice yields by 25%, 37%, and 59%, respectively (El Kantar et al., 2018). Because PEFs may function at low temperatures, they minimize nutritional loss while using less energy. This makes them especially beneficial.

However, each processing techniques has its unique advantages and disadvantages with regards to cost, energy use, product quality, and environmental impact etc. The various challenges and advantages of different food processing techniques have been summarized in Fig. 2.3.

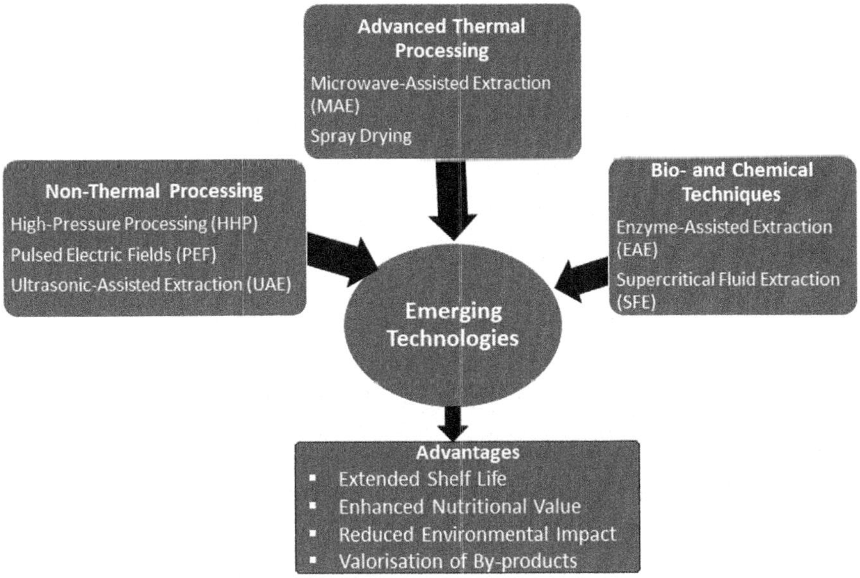

FIGURE 2.2

Emerging food processing technologies.

9. Effects on fish growth

It indicates that partial replacement of conventional feed ingredients with fruit and vegetable waste can enhance fish growth and immunity.

- **Growth performance**: Citrus peel inclusion improved feed conversion ratios in tilapia (Mandal et al., 2018).
- **Immunity enhancement**: Tomato pomace exhibited immunostimulatory effects in carp (Goddard et al., 2015).
- **Gut health improvement**: Fiber rich apple pomace positively influenced gut microbiota in trout (García-Pérez et al., 2019) fruit and vegetable waste.

10. Conclusion

The integration of fruit and vegetable waste in fish feed formulation represents a sustainable and economically viable approach to aquaculture nutrition. Aquaculture fish is an excellent source of high-quality protein, fat-soluble vitamins, and essential n-3 fatty acids such as EPA and DHA, as well as vital minerals such as calcium and phosphorus, which help prevent osteoporosis, cardiovascular diseases, degenerative conditions, and skin disorders. As the world undergoes a "blue transformation," aquaculture is emerging as the most environmentally sustainable food production

Table 2.2 Emerging technologies used for processing of fruit and vegetable waste.

Name of the method	Technology involved	Working principle
High-pressure processing (HPP)	Nonthermal processing	High pressure increases cell permeability to release bioactive compounds
Ultrasound	Nonthermal processing	Uses cavitation for extracting active ingredients.
Pulsed electric field (PEF) technology	Nonthermal processing	Uses high-voltage electricity to break cells
Cold plasma	Nonthermal processing	Utilizes both physical disruption and biochemical damage caused by reactive particles such as ions, radicals, and electrons
Microwave and ohmic heating	Advanced thermal technologies	Generates heat rapidly within the microbial cell via electromagnetic waves leading to cell membrane damage and increased permeability
Freeze-drying	Advanced thermal technologies	Damage the cell wall by forming ice crystals that expand and contract, causing lysis
Enzyme-assisted extraction (EAE)	Bio- and chemical technologies	Uses enzymes to break down cell walls and release valuable compounds
Supercritical fluid extraction (SFE)	Bio- and chemical technologies	It ruptures the cells due to the rapid changes in pressure and the fluid's unique solvent properties

sector. However, the intensification of fed aquaculture presents challenges, with the development of nutritionally efficient, equitable, and economically and environmentally sustainable aquafeeds being a key priority. This study highlights the potential of fruit and vegetable by-products as cost-effective and sustainable ingredients in aquafeed formulations. By-products such as peels, pomace, and seeds are rich in essential nutrients and bioactive compounds that can improve fish health and growth performance while simultaneously reducing feed costs and minimizing environmental impact. The incorporation of fruit and vegetable by-products into aquafeed requires efficient processing and transformation techniques. Innovative methods such as HPP and ultrasound-assisted extraction have been shown to effectively preserve the nutritional quality of these by-products, making them suitable for use in aquafeeds. Additionally, traditional techniques such as milling and grinding offer viable processing options, each with its own advantages and limitations depending on the specific material and intended application. These methods can also be combined within a biorefinery approach to optimize utilization. Overall, these processing strategies present valuable opportunities to enhance the fruit and vegetable by-products value chain.

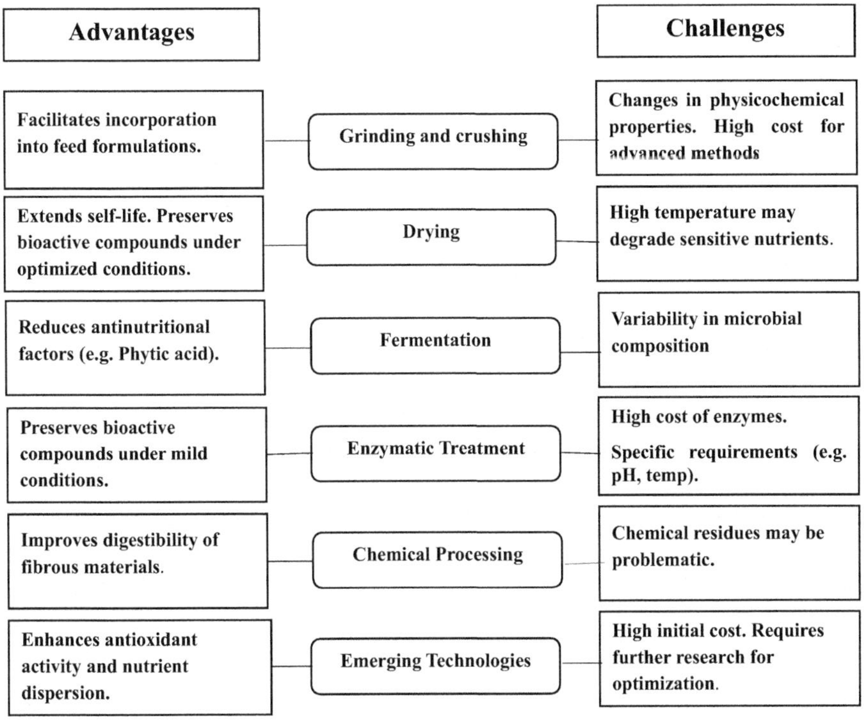

FIGURE 2.3

Advantages and challenges of different food processing techniques.

This review highlights the need for continued research and development to optimize the processing, formulation, and application of F&V by-products in aquaculture. Moving forward, integrating these by-products into aquafeed formulations not only supports the sustainable growth of the aquaculture sector but also contributes to reducing food waste and advancing global food security.

References

Cecchi, L., Khatib, M., Bellumori, M., Civa, V., Domizio, P., Innocenti, M., Balli, D., Mulinacci, N., 2023. Industrial drying for agrifood by-products re-use: cases studies on pomegranate peel (*Punica granatum* L.) and stoned olive pomace (pâtè, *Olea europaea* L.). Food Chem. 403, 134338. https://doi.org/10.1016/j.foodchem.2022.134338.

Chattaraj, S., Mitra, D., Chattaraj, M., Ganguly, A., Thatoi, H., Mohapatra, P.K.D., 2024. Brewers' spent grain as fish feed ingredient: evaluation of bio-safety and analysis of its impact on gut bacteria of Cirrhinus reba by 16S Metagenomic sequencing. Curr. Res. Microb. Sci. 7, 100286.

Chattaraj, S., Ganguly, A., Mitra, D., et al., 2026a. Sustainable valorization of algal biomass in fish nutrition and dietetics: a critical review and scientometric study. Waste Biomass Valor. https://doi.org/10.1007/s12649-026-03619-7.

Chattaraj, S., Sahoo, S.S., Behera, S., Thatoi, H., 2026b. Sustainable aquafeed development using Spirulina: Effects on growth performance of Catla catla. Algal Res. 104447.

Coelho, M.C., Rodrigues, A.S., Teixeira, J.A., Pintado, M.E., 2023. Integral valorisation of tomato by-products towards bioactive compounds recovery: human health benefits. Food Chem. 410, 135319. https://doi.org/10.1016/j.foodchem.2022.135319.

Devani, V., Basriati, S., 2015. Optimization of artificial fish feed nutrients using a multi-objective (goal) programming model. J. Sci. Technol. Ind. 12, 255–261.

El Kantar, S., Boussetta, N., Lebovka, N., Foucart, F., Rajha, H.N., Maroun, R.G., Louka, N., Vorobiev, E., 2018. Pulsed electric field treatment of citrus fruits: improvement of juice and polyphenols extraction. Innov. Food Sci. Emerg. Technol. 46, 153–161. https://doi.org/10.1016/j.ifset.2017.09.024.

El-Fakharany, E.M., Saleh, A.K., El-Maradny, Y.A., El-Sayed, M.H., Alali, I., Alsirhani, A. M., Alalawy, A.I., Alhawiti, A.S., Alatawi, I.S., Mazi, W., El-Gendi, H., 2024. Comprehensive insight into recent algal enzymes production and purification advances: toward effective commercial applications: a review. Int. J. Biolog. Macromol. 283. https://doi.org/10.1016/j.ijbiomac.2024.137783. Available from: https://www.sciencedirect.com/science/journal/01418130.

Gali, L., Bedjou, F., Velikov, K.P., Ferrari, G., Donsì, F., 2020. High-pressure homogenization-assisted extraction of bioactive compounds from Ruta chalepensis. J. Food Measure. Character. 14 (5), 2800–2809. https://doi.org/10.1007/s11694-020-00525-x.

García-Pérez, M., et al., 2019. Apple pomace in rainbow trout diets: effects on digestibility and gut microbiota. Aquac. Nutr. 25 (1), 89–98.

Goddard, J.S., et al., 2015. Tomato pomace in aquaculture feeds: effects on growth and immunity. Aquac. Rep. 7, 101–109.

Hussain, A., Kausar, T., Din, A., Murtaza, M.A., Jamil, M.A., Noreen, S., Rehman, H.u., Shabbir, H., Ramzan, M.A., 2021. Determination of total phenolic, flavonoid, carotenoid, and mineral contents in peel, flesh, and seeds of pumpkin (*Cucurbita maxima*). J. Food Proc. Preserv. 45 (6). https://doi.org/10.1111/jfpp.15542. Available from: http://onlinelibrary.wiley.com/journal/10.1111/(ISSN)1745-4549.

Kapoor, R., Feng, H., 2022. Characterization of physicochemical, packing and microstructural properties of beet, blueberry, carrot and cranberry powders: the effect of drying methods. Powder Technol. 395, 290–300. https://doi.org/10.1016/j.powtec.2021.09.058.

Kaushik, S.J., 1995. Nutrient requirements, supply and utilization in the context of carp culture. Aquaculture 129 (1–4), 225–241. https://doi.org/10.1016/0044-8486(94)00274-r.

Kesbiç, O.S., Acar, Ü., Hassaan, M.S., Yılmaz, S., Guerrera, M.C., Fazio, F., 2022. Effects of tomato paste by-product extract on growth performance and blood parameters in common carp (*Cyprinus carpio*). Animals 12 (23). https://doi.org/10.3390/ani12233387. Available from: http://www.mdpi.com/journal/animals.

Kumari, P., Gaur, S.S., Tiwari, R.K., 2023. Banana and its by-products: a comprehensive review on its nutritional composition and pharmacological benefits. eFood 4 (5). https://doi.org/10.1002/efd2.110. Available from: onlinelibrary.wiley.com/journal/26663066.

Licandro, H., Ho, P.H., Nguyen, T.K.C., Petchkongkaew, A., Nguyen, H.Van, Chu-Ky, S., Nguyen, T.V.A., Lorn, D., Waché, Y., 2020. How fermentation by lactic acid bacteria can address safety issues in legumes food products? Food Contr. 110, 106957. https://doi.org/10.1016/j.foodcont.2019.106957.

Mai, K., Ai, Q.H., Waagbø, R., Zhou, X.Q., Feng, L., 2022. Vitamins fish nutrition. Elsevier, China, pp. 57–179. https://doi.org/10.1016/B978-0-12-819587-1.00014-8. Available from: https://www.sciencedirect.com/book/9780128195871.

Mandal, R., et al., 2018. Citrus peel as a potential ingredient in aquafeeds: a review. Fish. Sci. 74 (3), 432–440.

Mphahlele, R.R., Pathare, P.B., Linus Opara, U., 2019. Drying kinetics of pomegranate fruit peel (cv. Wonderful). Sci. Afr. 5, e00145. https://doi.org/10.1016/j.sciaf.2019.e00145.

Nowosad, K., Sujka, M., Pankiewicz, U., Kowalski, R., 2021. The application of PEF technology in food processing and human nutrition. J. Food Sci. Technol. 58 (2), 397–411. https://doi.org/10.1007/s13197-020-04512-4.

Nutrient Requirements of Fish and Shrimp, 2011. Available from: https://doi.org/10.17226/13039.

Sari, I.P., Yulisma, M., 2017. Growth rate and efficiency of nila feed (*Oreochomis niloticus*) maintained in periodically fed tarp ponds. Indo. Swamp Aquacul. J. 5, 45–55.

Suprayudi, M.A., Dimahesa, W., Jusadi, D., Setiawati, M., Ekasari, J., 2011. Supplementation of sheep rumen fluid crude enzyme in vegetable ProteinBased to touch up the growth of Tilapia (*Oreochromis niloticus*). Indo. J. Ichthyol. 11, 177–183.

The State of World Fisheries and Aquaculture 2024, 2024.

Tocher, D.R., Betancor, M.B., Sprague, M., Olsen, R.E., Napier, J.A., 2019. Omega-3 long-chain polyunsaturated fatty acids, EPA and DHA: bridging the gap between supply and demand. Nutrients 11 (1). https://doi.org/10.3390/nu11010089. Available from: https://www.mdpi.com/2072-6643/11/1/89/pdf.

Vaitkevičienė, N., 2019. A comparative study on proximate and mineral composition of coloured potato peel and flesh. J. Sci. Food Agric. 99 (14), 6227–6233. https://doi.org/10.1002/jsfa.9895.

Xing, S., Liang, X., Zhang, X., Oliva-Teles, A., Peres, H., Li, M., Wang, H., Mai, K., Kaushik, S.J., Xue, M., 2024. Essential amino acid requirements of fish and crustaceans, a meta-analysis. Rev. Aquacul. 16 (3), 1069–1086. https://doi.org/10.1111/raq.12886. Available from: http://onlinelibrary.wiley.com/journal/10.1111/(ISSN)1753-5131.

CHAPTER

Incorporating oilseed cakes and meals in aquafeeds

3

Mukul Machhindr Barwant[1], Balwant Singh[2], Dipanwita Chaudhuri Sil[3], Debasree Lodh[3], Fannie Isela Parra Cota[4], Apaarna[5], Kübra Sağlam[6], Gökhan Boyno[7], Younes Rezaee Danesh[7, 8], Arindam Ganguly[9], Sourav Chattaraj[10], Sergio de los Santos-Villalobos[11] and Debasis Mitra[5]

[1]*Department of Botany, Sanjivani Arts Commerce and Science College Kopargoan, Ahmednagar, Maharashtra, India;* [2]*Department of Botany K. S. Saket PG College, Dr. Ram Manohar Lohia Avadh University, Ayodhya, Uttar Pradesh, India;* [3]*Department of Botany, Holy Cross College, WEST, Agartala, Tripura, India;* [4]*Campo Experimental Norman E. Borlaug-CIRNO. Instituto Nacional de Investigaciones Forestales, Agrícolas y Pecuarias (INIFAP), Ciudad Obregón, Sonora, Mexico;* [5]*Department of Microbiology, Graphic Era (Deemed to be University), Dehradun, Uttarakhand, India;* [6]*Department of Food Engineering, Faculty of Agriculture, Tekirdağ Namik Kemal University, Tekirdağ, Turkey;* [7]*Department of Plant Protection, Faculty of Agriculture, Van Yuzuncu Yil University, Van, Turkey;* [8]*Department of Plant Protection, Faculty of Agriculture, Urmia University, Urmia, Iran;* [9]*Department of Microbiology, Bankura Sammilani College, Bankura, West Bengal, India;* [10]*Centre for Industrial Biotechnology Research, School of Pharmaceutical Sciences, Siksha 'O' Anusandhan Deemed to be University, Bhubaneswar, Odisha, India;* [11]*Instituto Tecnológico de Sonora, Ciudad Obregón, Sonora, Mexico*

1. Introduction

It is desirable to employ appropriate fishmeal alternatives as aquaculture feed to guarantee sustainable and profitable fish farming. As possible alternatives to fishmeal, cakes made from soybean, linseed, rapeseed, and sunflower seeds have been reported in previous literatures (Chattaraj et al., 2024a,b, 2026a). Oilseed cake was used to replace 25% (w/w) of the reference fishmeal formulation. This resulted in pellets with nutritional profiles comparable to those of the reference fish feed, but with less noticeable expansion, higher sedimentation velocity, decreased water stability, and abrasion resistance. The screw arrangement of the extruder was changed by increasing the number of reverse components and kneading blocks to improve the radial expansion of pellets containing oilseed cakes, thereby reducing the sedimentation velocity. The use of finer ground oilseed cake particles can also result in a decrease in sedimentation velocity. The world population has doubled during this time, and the amount of fish consumed per person has consistently increased from approximately 10 kg in the 1960s to approximately 20 kg annually in 2012. Global wild fish fishing cannot be increased in response to the

Fish Feed Formulation with Agroindustrial Wastes and Byproducts. https://doi.org/10.1016/B978-0-443-41728-3.00006-X

growing demand for fish, because overfishing jeopardizes the sustainability of the natural fish supply. Consequently, aquaculture fish production has increased significantly over the past few decades, and the need for fishmeal, the main component of fish feed used in aquaculture, has also increased. Numerous studies have examined SM as a partial substitute for fishmeal (Tyapkova et al., 2016).

Aquaculture has emerged as one of the fastest-growing food production sectors globally, contributing significantly to the global supply of animal proteins (Chattaraj et al., 2025, 2026b). As capture fisheries stagnate owing to overexploitation and environmental changes, aquaculture provides a promising alternative to meet the rising demand for seafood. However, the sustainability of this industry depends largely on the quality and sources of aquafeeds, particularly protein inputs. Traditionally, fishmeal and fish oil, derived from wild-caught forage fish, have been the cornerstones of aquafeeds because of their high protein content, digestibility, and balanced amino acid profiles (Tacon and Metian, 2015). Despite these advantages, the finite availability and high cost of fishmeal pose major constraints to the continued expansion of aquaculture, prompting the search for more sustainable and economically viable protein alternatives.Among the diverse plant-based ingredients investigated, oilseed cakes and meals—by-products of oil extraction from crops such as soybean (*Glycine max*), groundnut (*Arachis hypogea*), sunflower (*Helianthus annuus*), sesame (*Sesamum indicum*), mustard (*Brassica* spp.), and linseed (*Linum usitatissimum*)—have gained significant attention. These products are not only abundant and cost-effective, but are also rich in protein, essential amino acids, lipids, and micronutrients that can support the growth and health of various aquaculture species (Hardy, 2010; Francis et al., 2001).

The growing interest in oilseed meal is driven by multiple factors. First, they align with the goals of sustainable aquaculture by reducing reliance on marine resources. Second, they offer economic benefits by lowering feed costs, particularly in regions where oilseed cultivation and processing are prevalent. Third, many oilseed-derived feeds exhibit functional properties such as antimicrobial and antioxidant effects, which may enhance fish immunity and feed efficiency (Hong et al., 2004). Despite their promise, the use of oilseed meal in aquafeeds is not a challenge. A major constraint is the presence of anti-nutritional factors (ANFs), such as phytic acid, tannins, saponins, glucosinolates, and protease inhibitors. These compounds can impair nutrient digestibility, reduce feed palatability, and compromise fish health (Francis et al., 2001). For instance, phytic acid chelates important minerals, such as calcium, zinc, and iron, making them unavailable for absorption. Trypsin inhibitors interfere with protein digestion, whereas glucosinolates in mustard cakes can have goitrogenic effects in fish (Krogdahl et al., 2010). Various processing techniques have been developed to mitigate the adverse effects of ANFs and improve the nutritional quality of oilseed meal. These methods include thermal treatment, fermentation, enzymatic hydrolysis, and solvent extraction. Heat treatment inactivates thermolabile inhibitors, such as trypsin inhibitors and lectins, whereas fermentation using microbes such as *Aspergillus oryzae* can enhance digestibility and reduce bitter taste compounds (Hong et al., 2004). Enzymatic supplementation, particularly with phytase, can break down phytic acid and enhance mineral bioavailability (Schafer et al., 2009). Another consideration for the use of oilseed meal is species-specific tolerance and digestibility. Omnivorous fish, such as tilapia (*Oreochromis* spp.), common carp (*Cyprinus carpio*), and catfish (*Clarias* spp.), have demonstrated a greater ability to utilize plant-based proteins, including oilseed meals, compared to carnivorous species, such as salmonids and marine finfish (El-Sayed, 1999; Turchini et al., 2009). This variation is attributed to differences in the digestive physiology, enzyme activity, and amino acid requirements. For carnivorous species, strategic supplementation with limiting amino acids such as methionine and lysine is often necessary when using oilseed-based feeds (Barrows et al., 2008). In economic terms, the adoption of oilseed meal in aquafeeds offers substantial benefits. The relatively low cost and local availability of oilseed byproducts make them accessible options for small- and medium-scale fish farmers, particularly in developing countries (Hasan et al., 2007). Environmentally, these meals reduce the pressure on wild fish stocks and contribute to waste valorization in agro-industrial systems, promoting a circular economy approach (Mungkung et al., 2013). Life cycle assessments (LCA) have shown that plant-based feeds have a lower environmental footprint in terms of greenhouse gas emissions, eutrophication, and water use than fishmeal-based diets (Naylor et al., 2009). Innovations in plant breeding and biotechnology hold promise for enhancing the nutritional value of oilseed crops and minimizing their antinutritional profiles. Genetic engineering may allow the development of low-ANF oilseed varieties or varieties with enhanced levels of specific amino acids. Additionally, novel feed processing technologies combined with functional feed additives such as prebiotics, probiotics, and enzymes can further optimize the utilization of oilseed meals in aquaculture diets (Chattaraj et al., 2023, 2024c; Glencross, 2009).

2. Nutritional composition of oilseed cakes and meals

Oilseed cakes and meals are high-protein byproducts that result from the extraction of oil from various oilseeds, including soybean (*Glycine max*), groundnut (*Arachis hypogaea*), sunflower (*Helianthus annuus*), cottonseed (*Gossypium* spp.), rapeseed mustard (*Brassica* spp.), and linseed (*Linum usitatissimum*). These by-products are widely used as feed ingredients because of their significant nutritional value and relatively low cost. However, their composition can vary depending on factors such as the type of seed, method of oil extraction (mechanical pressing vs. solvent extraction), and degree of dehulling. A thorough assessment of their nutritional components is essential to maximize their value in animal nutrition and explore their potential use in sustainable agriculture (Fig. 3.1).

2.1 Protein content and amino acid profile

Protein is the most significant component of oilseed meal, contributing to both nutritional and functional properties.

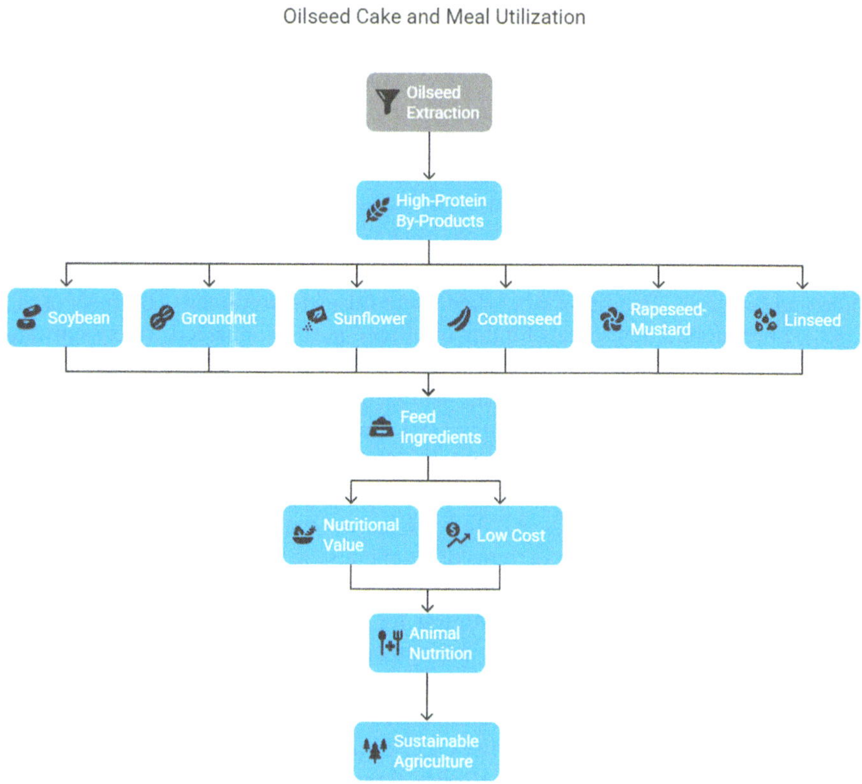

FIGURE 3.1

Oilseed Cake and Meal Utilization.

2.1.1 Crude protein content

The crude protein (CP) content of oilseed meals varies significantly across species and is influenced. The crude protein (CP) levels found in oilseed meals differ markedly depending on the species and the methods used for processing. Focusing on primary oilseeds, soybean meal is often viewed as the benchmark, with solvent-extracted meals typically having approximately 44%–48% CP, while mechanically expelled meals can reach up to 52% CP (NRC, 2012). Its extensive application in livestock and poultry feeds can be attributed to its high protein quality and digestibility. using various processing techniques. Among the major oilseeds, groundnut cake generally has a crude protein (CP) content of approximately 45%–50%, which varies based on the remaining oil level and quality of the kernels (Devendra, 1988). It is especially high in arginine, although it may be at risk for aflatoxin contamination. Cottonseed meal typically contains 35%–40% CP, but this can vary according to the level of gossypol present, which is toxic in its raw form (Hutagalung, 1981) (Table 3.1). Sunflower meal shows a broad CP range of 28%–41%, which is influenced by the removal of hulls before processing. Dehulled sunflower meals

Table 3.1 Nutritional composition of common oilseed cakes and meals.

Oilseed cake/meal	Protein content (% dry weight)	Major amino acids	Lipid content (%)	Carbohydrates and fiber (% dry weight)	Vitamins and minerals	Bioactive compounds	Bioactive compounds
Soybean meal	44%–50% (Karr-Lilienthal et al., 2005)	Lysine, arginine	1%–3% (Rakshit et al., 2019)	Carbohydrates: 30%–35%; fiber: 5%–7% (Rakshit et al., 2019)	B-vitamins, Ca, P (Liu, 2011)	Isoflavones: genistein, daidzein (Liu, 2011)	Isoflavones: genistein, daidzein
Rapeseed (canola) meal	35%–42% (Newkirk, 2009)	Methionine, lysine	1%–2% (Tripathi and Mishra, 2007)	Carbohydrates: 25%–30%; fiber: 10%–12% (Tripathi and Mishra, 2007)	Vitamin E, P, Se (Tripathi and Mishra, 2007)	Glucosinolates (Tripathi and Mishra, 2007)	Glucosinolates
Sunflower cake	28%–40% (Natesan and Muthukrishnan, 2014)	Arginine, phenylalanine	1%–2% (Baldwin et al., 2010)	Carbohydrates: 20%–25%; fiber: 15%–20% (Baldwin et al., 2010)	Mg, K, vitamin E (Baldwin et al., 2010)	Chlorogenic acid (Baldwin et al., 2010)	Chlorogenic acid
Cottonseed cake	30%–35% (Nagalakshmi et al., 2016)	Arginine, leucine	6%–8% (Nagalakshmi et al., 2016)	Carbohydrates: 20%–25%; fiber: 12%–15% (Nagalakshmi et al., 2016)	Iron, magnesium, B-vitamins (Nagalakshmi et al., 2016)	Gossypol (Nagalakshmi et al., 2016)	Gossypol
Sesame cake	35%–38% (Elleuch et al., 2011)	Methionine, cysteine	5%–7% (Elleuch et al., 2011)	Carbohydrates: 25%–30%; fiber: 10%–12% (Elleuch et al., 2011)	Calcium, phosphorus, vitamin B1 (Elleuch et al., 2011)	Sesamin, sesamolin (Elleuch et al., 2011)	Sesamin, sesamolin

Continued

Table 3.1 Nutritional composition of common oilseed cakes and meals.—*cont'd*

Oilseed cake/meal	Protein content (% dry weight)	Major amino acids	Lipid content (%)	Carbohydrates and fiber (% dry weight)	Vitamins and minerals	Bioactive compounds	Bioactive compounds
Groundnut (peanut) cake	45%–50% (Singh and Oswalt, 1995)	Lysine, arginine	6%–8% (Singh and Oswalt, 1995)	Carbohydrates: 20%–25%; fiber: 10%–12% (Singh and Oswalt, 1995)	Niacin, vitamin E, iron (Singh and Oswalt, 1995)	Resveratrol (Singh and Oswalt, 1995)	Resveratrol
Linseed (flaxseed) cake	32%–35% (Russo and Reggiani, 2015)	Histidine, threonine	8%–10% (Russo and Reggiani, 2015)	Carbohydrates: 20%–22%; fiber: 15%–18% (Russo and Reggiani, 2015)	Omega-3, Zn, Mg (Russo and Reggiani, 2015)	Lignans (Russo and Reggiani, 2015)	Lignans
Mustard cake	30%–38% (Tripathi and Mishra, 2007)	Methionine, tryptophan	3%–5% (Tripathi and Mishra, 2007)	Carbohydrates: 20%–25%; fiber: 10%–15% (Tripathi and Mishra, 2007)	Selenium, vitamin E (Tripathi and Mishra, 2007)	Allyl isothiocyanate (Tripathi and Mishra, 2007)	Allyl isothiocyanate
Safflower cake	25%–30% (Jayasena and Nasar-Abbas, 2011)	Threonine, valine	1%–2% (Jayasena and Nasar-Abbas, 2011)	Carbohydrates: 30%–35%; fiber: 18%–20% (Jayasena and Nasar-Abbas, 2011)	Potassium, iron, vitamin B3 (Jayasena and Nasar-Abbas, 2011)	Polyphenols (Jayasena and Nasar-Abbas, 2011)	Polyphenols
Palm kernel cake	16%–18% (Devendra, 2009)	Low in lysine, rich in arginine	8%–10% (Devendra, 2009)	Carbohydrates: 35%–40%; fiber: 16%–18% (Devendra, 2009)	Magnesium, phosphorus (Devendra, 2009)	Phenolic acids (Devendra, 2009)	Phenolic acids

generally offer higher protein and lower fiber contents (Rama Rao et al., 2006). Rapeseed-mustard meal usually has a CP content of 32%–38%, but its usage is often restricted due to the presence of glucosinolates (Tripathi and Mishra, 2007).

2.1.2 Amino acid profile

The biological value of oilseed meals for monogastric animals is primarily influenced by their amino acid profiles, with significant variations observed across the different types. Soybean meal is distinguished by its **balanced amino acid composition**, particularly **its high lysine content** (6.0%–6.5% of crude protein), which is critical for poultry and swine nutrition. While it provides substantial threonine, **methionine levels remain a limiting factor,** often requiring supplementation in formulated diets (NRC, 2012).

Groundnut meal is characterized by elevated arginine concentrations (~11% crude protein), but deficiencies in lysine and methionine restrict its independent use in poultry feed systems. This imbalance necessitates blending with complementary protein sources to meet dietary requirements (Kaur et al., 2012). Sunflower meal serves as a valuable source of sulfur-containing amino acids, such as methionine and cysteine, making it an effective partner for lysine-rich ingredients, such as soybean meal. However, its relatively low lysine content limits its standalone efficacy, underscoring the need for strategic formulation to optimize the amino acid balance in animal feed (Rama Rao et al., 2006). These amino acid profiles emphasize the need to combine different oilseed meals or to supplement with synthetic amino acids in order to fill specific nutritional gaps and improve feed efficiency. Cottonseed meal is a good source of methionine; however, lysine remains a limiting amino acid. Additionally, gossypol in cottonseed meal binds to lysine, reducing its bioavailability (Hutagalung, 1981) (Table 3.1).

Rapeseed-mustard meal contains high levels of sulfur-containing amino acids, but is deficient in lysine and isoleucine, which limits its nutritional value when used alone (Tripathi and Mishra, 2007). These variations necessitate strategic formulation when oilseed meal is used in animal feed to balance essential amino acids and ensure optimal growth and productivity.

2.2 Lipid, carbohydrate, and fiber content

2.2.1 Lipid content

The residual lipid content in oilseed meals serves a dual purpose: it boosts the energy value of the feed and enhances palatability. However, when lipid content is too high, it can negatively affect storage stability and increase the risk of rancidity, particularly in humid environments. Soybean meal generally contains 1.5%–2.5% residual lipids in solvent-extracted forms, while mechanically expelled soybean meal may have as much as 6%–7% residual oil (Feedipedia, 2021). Groundnut cake, especially when cold-pressed, can retain 8%–10% oil, which significantly increases its energy contribution (Devendra, 1988). Sunflower and cottonseed meals typically contain 1%–3% residual oil, and linseed meal contains approximately 4%–6%, which also provides a source of omega-3 fatty acids (Göhl, 1982). The fatty acid composition of these meals is also noteworthy; linseed meal is particularly high in α-linolenic acid (an omega-3 fatty acid), whereas groundnut and sunflower meals are rich in oleic and linoleic acids (Table 3.1).

2.2.2 Carbohydrate content

Carbohydrates in oilseed meals are mainly present as non-starch polysaccharides and oligosaccharides, which contribute to the overall energy value, but can also influence digestibility. Soybean meal contains approximately 30% carbohydrates, with 9%–12% oligosaccharides such as raffinose and stachyose, which are known to cause flatulence and digestive issues in non-ruminant animals (Choct et al., 2010). Groundnut meal typically contains 25%–30% carbohydrates, primarily digestible sugars and cellulose (Devendra, 1988). Sunflower and cottonseed meals are notable for their high content of structural carbohydrates, such as hemicellulose and cellulose, especially when the meals are not dehulled, and some oilseed meals contain anti-nutritional factors, such as phytic acid, particularly in soybean and rapeseed meals, which can chelate minerals and reduce their bioavailability (Ravindran et al., 1995) (Table 3.1).

2.2.3 Fiber content

Fiber content is a key factor in determining the suitability of oilseed meals for monogastric versus ruminant diets. When unhulled, sunflower meal can contain up to 27% crude fiber, limiting its digestibility and energy value for poultry (Rama Rao et al., 2006). Cottonseed meal contains 12%–15% crude fiber, which includes lignin, further reducing its digestibility (Göhl, 1982). Soybean meal, especially dehulled, is relatively low in fiber (around 4%–7% crude fiber), making it ideal for poultry and swine (NRC, 2012). Rapeseed meal fiber content ranges between 11% and 14%, and newer low-fiber varieties have been developed for broader inclusion in feeds (Tripathi and Mishra, 2007). Neutral detergent fiber (NDF) and acid detergent fiber (ADF) values offer more precise measurements of the indigestible components. These fiber fractions influence the rate of digestion, gut health, and nutrient absorption. High-fiber meals are better suited for ruminants that possess microbial fermentation capacity in their rumens, and rephrasing the write-up without changing references. Oilseed cakes and meals are crucial ingredients in animal feed because they offer high levels of protein, a favorable amino acid profile, and energy from residual lipids. However, their use must be carefully balanced, considering factors such as fiber content, the presence of anti-nutritional compounds, and potential amino acid deficiencies. Techniques such as dehulling, heat treatment, and enzyme supplementation are commonly used to enhance the nutritional quality and address these challenges. Ongoing research is essential to further refine how these by-products are included in both monogastric and ruminant diets, to support more sustainable and efficient animal agriculture (Table 3.1).

2.3 Vitamins, minerals, and bioactive compounds

Oilseed cakes and meals, produced as by-products of oil extraction, are widely utilized as feed ingredients for livestock nutrition. Beyond serving as protein-rich feed sources, they are increasingly being recognized for their valuable nutritional components, such as vitamins, minerals, and bioactive compounds. This chapter examines the nutritional characteristics of oilseed cakes and meals, with particular emphasis on their vitamin and mineral contents, as well as the presence of bioactive compounds that may offer health benefits. The discussion also covers the importance of these nutrients for both animal and human health and highlights the variability in their composition among different types of oilseed sources.

Oilseed cakes and meals, such as soybean meal, canola meal, sunflower cake, cottonseed meal, linseed cake, and groundnut meal are important agricultural byproducts. Although they have traditionally been valued for their high protein content, recent research has emphasized their broader nutritional significance. Vitamins and minerals play crucial roles in metabolic processes, and the bioactive compounds found in these meals contribute to antioxidant activity, immune system support, and disease prevention (Saini and Keum, 2018). The vitamin content in oilseed meals varies depending on the type of seed and processing methods used. Fat-soluble vitamins, particularly vitamin E (including tocopherols and tocotrienols), are commonly found in high amounts in sunflower, soybean, and canola meals.

2.3.1 Vitamin E

Sunflower and soybean meal are abundant sources of α-tocopherol and γ-tocopherol, both of which are powerful antioxidants that protect cellular structures from oxidative damage (Blekas et al., 1995).

2.3.2 B-complex vitamins

B-complex vitamins, including thiamine (B1), riboflavin (B2), niacin (B3), and folate (B9), play vital roles in energy metabolism and red blood cell production. Soybean and groundnut meals are particularly rich in B-complex vitamins (Nzikou, 2009). Oilseed meals are valuable sources of essential macro- and micro-minerals, including calcium (Ca), phosphorus (P), magnesium (Mg), iron (Fe), zinc (Zn), and selenium (Se), which play important structural, catalytic, and regulatory roles in biological systems (Vist, 1996).

2.4 Minerals in oilseed cakes and meals

2.4.1 Macro-minerals

Linseed and cottonseed meals are particularly notable for their calcium and magnesium content. Canola meal stands out because of its relatively high phosphorus content (Bell, 1993).

2.4.2 Micro-minerals

Groundnut and soybean meal are rich in iron and zinc, respectively. Se, which is crucial for the function of antioxidant enzymes, is present in trace amounts in linseed and rapeseed meal (Tripathi and Mishra, 2007).

2.4.3 Bioactive compounds

In addition to their vitamin and mineral content, oilseed cakes and meals contain a diverse array of phytochemicals and secondary metabolites that enhance their nutraceutical value.

2.4.4 Phenolic compounds

Meals from sunflower, rapeseed, and sesame seeds are particularly rich in phenolic acids such as chlorogenic acid, caffeic acid, and ferulic acid. These compounds are known for their antioxidant, anti-inflammatory, and antimicrobial activities (Carré and Rozan, 2004).

2.4.5 Flavonoids

Soybean and flaxseed meals contain flavonoids, such as quercetin, kaempferol, and isorhamnetin, which are known for their strong antioxidant properties and potential anticancer effects (Rochfort and Panozzo, 2007).

2.4.6 Phytosterols

Canola and sunflower meals are especially high in phytosterols, including campesterol and beta-sitosterol, which are associated with cholesterol-lowering effects and cardiovascular health benefits (Piironen et al., 2000).

2.4.7 Lignans

Linseed (flaxseed) meal is an excellent source of lignans such as secoisolariciresinol diglucoside (SDG), which is associated with hormonal regulation and a reduced risk of hormone-related cancers (Adolphe et al., 2010).

3. Factors affecting nutrient composition

The levels of vitamins, minerals, and bioactive compounds in oilseed meals are influenced by several factors, including the genetic makeup of the seed, agronomic practices, harvesting time, storage conditions, and the method of oil extraction, such as cold pressing and solvent extraction. These variables can lead to significant differences in the nutrient content and quality of oilseed meals.

4. Implications and applications

Owing to their rich and varied nutrient profiles, oilseed meals are increasingly being recognized as functional ingredients in both animal feed and human nutrition. The presence of bioactive compounds enhances their value, making them attractive candidates for use in functional foods, dietary supplements, and therapeutic products. Their incorporation can improve not only the nutritional adequacy of diets, but also offer added health benefits, such as antioxidant and disease-preventive effects. Oilseed cakes and meals are under-utilized sources of vitamins, minerals, and bioactive compounds. Their inclusion in food and feed formulations supports nutritional sufficiency, and promotes health and disease prevention. Future research should focus on refining processing techniques to better preserve these nutrients, and further investigate their bioavailability and specific functional roles in health.

5. Common oilseed cakes and meals used in aquafeeds

With the increasing need for sustainable and cost-effective alternatives to fishmeal in aquafeeds, plant-derived protein sources, such as oilseed cakes and meals, have attracted considerable interest. These by-products from the vegetable oil industry supply the essential nutrients required for the optimal growth and development of aquatic species. This chapter provides an overview of the nutritional properties, anti-nutritional factors, processing strategies, and practical uses of major oilseed cakes and meals in aquaculture, with a focus on soybean meal, mustard and rapeseed meal, groundnut cake, linseed meal, and cottonseed meal.

5.1 Soybean meal (SBM)

5.1.1 Origin and processing

Soybean meal is the most widely utilized plant protein source in aquaculture. It is produced by extracting oil from soybean seeds (*Glycine max*), typically through solvent extraction, which increases protein content by removing most of the oil (NRC, 2011).

5.1.2 Nutritional composition

SBM contains approximately 44%–48% crude protein and is rich in essential amino acids, such as lysine and arginine, although it is relatively low in methionine (Gatlin et al., 2007). It also contains a significant amount of carbohydrates (approximately 30%) and minerals (approximately 6%–7%).

5.1.3 Anti-nutritional factors

Raw SBM contains several antinutritional factors, including trypsin inhibitors, lectins, phytic acid, and saponins, which can hinder protein digestion and mineral uptake. Most of these compounds are inactivated during heat processing (Francis et al., 2001).

5.1.4 Applicationin aquafeeds

SBM is extensively used as a fishmeal substitute in the diets of species such as carp, tilapia, and catfish. However, high inclusion rates in feeds for carnivorous fish, such as salmonids, can lead to enteritis, so its use is often balanced with amino acid supplements and functional additives to optimize performance (Refstie et al., 2000).

6. Mustard and rapeseed meal

6.1 Origin and processing

Meals from rapeseed (*Brassica napus*) and mustard (*Brassica juncea*) are produced after oil extraction primarily in South Asia and parts of Europe. Both mechanical and solvent extraction methods are commonly used.

6.2 Nutritional composition

These meals generally provide 32%–38% protein, moderate lipid levels (1%–3%), and a relatively high fiber content (up to 14%). Canola, a low-glucosinolate variety of rapeseed, is considered to be safer and more suitable for feed applications (Newkirk et al., 2003).

6.3 Anti-nutritional factors

The main anti-nutritional concern of mustard and rapeseed meals is their high glucosinolate content, which can negatively affect thyroid function and reduce feed palatability. Tannins and phytic acid are also present at moderate levels (Tripathi and Mishra, 2007).

6.4 Applicationin aquafeeds

Through detoxification processes, such as heat treatment and fermentation, mustard and rapeseed meals have been effectively used in fish diets, especially for omnivorous species such as carp (Mukhopadhyay and Ray, 1999).

7. Groundnut cake (GNC)

7.1 Origin and processing

Groundnut cake is produced as a by-product of oil extraction from groundnuts (*Arachis hypogea*), usually through mechanical expelling or solvent extraction.

7.2 Nutritional composition

Groundnut cake generally contains 40%–50% crude protein and is particularly rich in arginine; however, it is deficient in lysine and sulfur-containing amino acids, as described by Devendra. Its lipid content (6%–12%) is higher than that of most other oilseed meals, providing additional energy in aquafeed formulations (Devendra, 1988).

7.3 Anti-nutritional factors and safety concerns

Aflatoxin contamination is a major concern for groundnut cake, as *Aspergillus flavus* can proliferate under poor storage conditions, leading to significant safety risks

(Moses, 1984). Other antinutritional factors include phytic acid and low levels of tannins, which can reduce nutrient availability.

7.4 Application in aquafeeds

Groundnut cake is widely used in aquafeeds for tilapia and carp because of its cost-effectiveness and high protein content. However, strict quality control and detoxification protocols are necessary.

8. Linseed meal (LSM)

8.1 Origin and processing

Linseed meal is a by-product of oil extraction from flaxseed (*Linum usitatissimum*) and is typically produced through mechanical pressing or solvent extraction.

8.2 Nutritional composition

Linseed meal contains 30%–38% protein and is particularly notable for its high omega-3 fatty acid content, particularly alpha-linolenic acid (ALA). It also contains mucilage, which may affect digestibility and has a high fiber content of approximately 10% (Kaushik et al., 1995).

8.3 Anti-nutritional factors

Linseed meal contains cyanogenic glycosides, mainly linamarin, which can release hydrogen cyanide when hydrolyzed. Appropriate heat treatment is effective in significantly reducing these compounds, as noted by Tripathi and Mishra (2007).

8.4 Application in aquafeeds

Although less commonly used than other oilseed meals, linseed meal has potential in aquafeed formulations owing to its beneficial lipid profile. It is suitable for herbivorous and omnivorous fish species and can enhance n-3 PUFA content in fish flesh (Olsen and Hasan, 2012). Recent studies have shown that fermentation and the use of probiotics can further improve the nutritional quality of linseed meal, making it a viable partial replacement for fishmeal in carp diets, with up to 30% inclusion of fermented linseed oil cake supporting optimal growth and nutrient utilization (Banerjee et al., 2023; Ghosh et al., 2019).

9. Cottonseed meal (CSM)

9.1 Origin and processing

Cottonseed meal was obtained after oil extraction from the seeds of *Gossypium* spp. It is produced by mechanical expulsion or solvent extraction and is widely available in regions with extensive cotton cultivation.

9.2 Nutritional composition

Cottonseed meal contains 30%–45% crude protein but is limited to lysine. It also has a high fiber content and contains significant amounts of free gossypol, a toxic polyphenolic compound (Robinson and Li, 2008).

9.3 Anti-nutritional factors

Gossypol is the primary limiting factor for the use of cottonseed meal in aquafeeds, as it can adversely affect the liver and reproductive health. Detoxification strategies, such as heating, iron supplementation, or the use of glandless cotton varieties, can help reduce toxicity (Jones, 1991).

9.4 Application in aquafeeds

When properly processed to reduce gossypol content, cottonseed meal has been used to replace fishmeal in the diets of omnivorous species, such as tilapia and carp. However, inclusion levels must be carefully managed to prevent gossypol-related toxicity (Li and Robinson, 2006). Oilseed cakes and meals, including groundnut cake, linseed meal, and cottonseed meal, offer promising and sustainable alternatives to fishmeal in aquaculture nutrition. Each ingredient has unique nutritional strengths and limitations; however, its effective use depends on appropriate processing and formulation strategies to address anti-nutritional factors. Advancements in feed technology, improved processing (e.g., fermentation of linseed meal), and crop genetic improvements will further enhance the role of these plant-based protein sources in future aquafeeds (Devendra, 1988; Moses, 1984; Kaushik et al., 1995; Tripathi and Mishra, 2007; Olsen and Hasan, 2012; Robinson and Li, 2008; Jones, 1991; Li and Robinson, 2006; Banerjee et al., 2023).

10. Processing techniques and their impact on nutritional value

Processing methods have a major impact on the nutritional content and safety of oilseed meal and cake when incorporated into aquafeeds. Several methods, including heat treatment, fermentation, enzymatic processing, and oil extraction, are important for enhancing nutrient availability, decreasing antinutritional factors (ANFs), and increasing digestibility among aquatic animals. Processing methods are important for improving the nutritional quality and safety of oilseed cakes and meals used in aquafeed formulations. The method of oil extraction has significant effects on the residual nutrient content and digestibility of the by-products. The two principal methods used are mechanical pressing (expeller pressing) and solvent extraction. While mechanical pressing allows for more residual oil, contributing to a higher energy content, it can also lead to greater fiber and anti-nutritional factor (ANF) content. Solvent extraction, commonly with hexane, is effective in removing more of the oil and has less fat content, but risks causing loss of heat-sensitive nutrients and subsequent detoxification (Francis et al., 2001; Gatlin et al., 2007). Thermal processing is usually used to inactivate thermo-labile ANFs such as trypsin inhibitors, lectins, and urease. Efficient thermal processing enhances protein digestibility and improves feed efficiency through the denaturation of anti-nutritional proteins and ease of nutrient availability (Krogdahl et al., 2010). However, excessive heat can break down critical amino acids such as lysine and decrease the quality of proteins, making optimal thermal control mandatory (Burel et al., 2000). For instance, heat-treated soybean meal has been reported to have better palatability and digestibility in carp and tilapias, without sacrificing amino acid integrity under controlled processing conditions (El-Sayed, 1999). Fermentation and enzymatic treatment have proven to be successful biotechnological methods for improving the nutritional quality of oilseed meal. Solid-state and submerged fermentation by desirable microorganisms (e.g., Aspergillus niger, Bacillus subtilis, Lactobacillus spp.) hydrolyzes complex carbohydrates, breaks down ANFs, and enhances amino acid availability (Mukhopadhyay and Ray, 1999; Hong et al., 2004). Similarly, the use of exogenous enzymes such as phytase, cellulase, and protease in feed ingredients or pre-treatment contributes significantly toward improving the digestibility of phosphorus and protein by degrading phytic acid, non-starch polysaccharides, and protein-ANF complexes (Cao et al., 2007). These processing treatments not only enhance the nutritional quality of oilseed meals but also facilitate their sustainable use in aquafeeds, enhance fish growth performance and health, and decrease dependence on marine-sourced ingredients.

10.1 Oil extraction methods

Oil extraction is the first step in the manufacture of oilseed meal and cake. The process adopted, solvent extraction or mechanical pressing, influences the residual oil content and nutrient value of the by-product. Mechanical pressing, and more importantly cold pressing, has higher residual oil and bioactive content retention, such as tocopherols and phytosterols, but lower protein concentrations. Solvent extraction (most commonly with hexane) produces meals with lower fat and higher protein concentrations that can be used for high-protein aquafeed production. Solvent residues and processing-induced structural protein damage may lower palatability and digestibility (NRC, 2011). In addition, extrusion-based pretreatment before oil extraction may enhance protein digestibility through the weakening of cell walls and inactivation of ANFs, such as trypsin inhibitors and phytic acid (Makkar et al., 2016). Thus, the selection of the oil extraction process should be a compromise between the efficiency of oil recovery and the nutritional and functional properties of the residual meal.

10.2 Heat treatment and detoxification

Most oilseed meals harbor ANFs and toxins, such as gossypol (cottonseed), phorbol esters (Jatropha), glucosinolates (mustard), and protease inhibitors (soybean). Heat treatment is commonly used to detoxify substances and improve their nutritional value. Toasting, autoclaving, and extrusion cooking are efficient in heat-inactivated heat-labile ANFs, such as trypsin inhibitors, lectins, and goitrogens. For example, well-toasted soybean meal drastically lowers trypsin inhibitor activity while maintaining amino acid integrity (Kakade et al., 1974; Francis et al., 2001). Similarly, moist heat (steam treatment) can decrease the free gossypol content of cottonseed meal, making it safer for addition to aquafeeds (Zhang et al., 2007). Nevertheless, overprocessing through heat can cause Maillard reactions, which decrease the availability of lysine and lower protein digestibility (Hoffman et al., 2019). Thus, it is essential to optimize heat parameters to maximize detoxification without compromising nutritional value.

10.3 Fermentation and enzymatic processing

Biotechnological treatment by microbial fermentation and enzymatic hydrolysis is becoming increasingly popular as an eco-friendly method for upgrading the nutritional value of oilseed meals and cakes. Fermentation by desirable microbes, such as Bacillus subtilis, Aspergillus niger, or lactic acid bacteria, lowers ANFs, enhances protein solubility, and enhances the levels of bioactive peptides and B-vitamins (Adeyemo et al., 2020; Hong et al., 2004). Fermented soybean meal, for instance, exhibits better digestibility and lower allergenicity in fish, such as tilapia and carp. Enzyme processing through proteases, cellulases, or phytases increases the hydrolysis of complex carbohydrates and proteins, and enhances nutrient availability. The use of phytase specifically reduces phytate-bound phosphorus, enhances mineral bioavailability, and reduces phosphorus pollution from aquaculture (Sugiura et al., 2001). Carbohydrase supplementation also benefits the degradation of non-starch polysaccharides, which hinders the utilization of energy in non-ruminant species (Refstie et al., 2005). Overall, the incorporation of fermentation and enzymatic treatment in feed-manufacturing systems can greatly enhance the value and safety of oilseed meals in aquafeeds.

10.4 Anti-nutritional factors and strategies for their mitigation

The use of oilseed cakes and meals in aquafeeds is a potential means to curb reliance on traditional fishmeal. Nevertheless, their application tends to be restricted by anti-nutritional factors (ANFs), which can compromise the utilization of nutrients, lower palatability, and even unfavorably impact fish growth and health. Insight into the nature, action, and control of such compounds is necessary for rational feed formulation. Oilseed meals and cakes, which are good sources of protein and energy in aqua feed, tend to have a variety of anti-nutritional factors (ANFs) that may restrict their use and have deleterious effects on fish performance and health. The ANFs involved were protease inhibitors (for example, trypsin and chymotrypsin inhibitors), phytic acid, tannins, saponins, glucosinolates, lectins, and non-starch polysaccharides (NSPs). Their concentrations vary with the type of oilseed, processing method, and genetic background of the plant (Krogdahl et al., 2010; Francis et al., 2001). Protease inhibitors block digestive enzyme functions, resulting in reduced protein utilization. Phytic acid chelates important minerals, such as calcium, iron, and zinc, rendering them unavailable for absorption, whereas tannins complex with proteins and carbohydrates, lowering their digestibility. Saponins and glucosinolates may damage intestinal function and trigger physiological stress reactions, whereas lectins are disruptive to the gut epithelium, blunting nutrient absorption and immunity (Francis et al., 2001; Lim et al., 2001). They can cause undesirable feed conversion ratios, growth depression, weakened immune systems, and enhanced susceptibility to diseases in fish species, particularly in carnivorous fish such as salmonids and sea bass, which are sensitive to plant diets (Torstensen et al., 2008; Krogdahl et al., 2010).

Various detoxification methods and feeding strategies have been developed to counteract the adverse effects of ANF. Heat treatment can successfully inactivate thermolabile substances such as trypsin inhibitors and lectins. Prevention of overprocessing is important to prevent the destruction of amino acids and the loss of nutritional value (Burel et al., 2000). Fermentation with useful microbes, such as Bacillus subtilis, Lactobacillus spp., and Aspergillus niger, has been reported to break phytic acid, tannins, and saponins, increase protein digestibility, and incorporate bioactive peptides (Mukhopadhyay and Ray, 1999; Hong et al., 2004). Enzymatic processing, especially with phytase, carbohydrases, and proteases, can enhance nutrient availability and minimize the effects of phytic acid and NSPs (Cao et al., 2007). Feed formulation techniques, including the utilization of mixed protein sources, supplementation with amino acids, and addition of functional feed ingredients such as prebiotics and probiotics, can also counteract the nutritional losses resulting from ANFs. Utilization of low-ANF oilseed crop cultivars and maximizing inclusion rates in the diet are other strategies that enable safer and more effective utilization of oilseed meal in aquaculture (Gatlin et al., 2007; Glencross et al., 2007).

10.5 Types of anti-nutritional compounds

Oilseed-derived feed ingredients contain various ANFs, which interfere with nutrient digestion, absorption, and metabolism. These include-

Protease inhibitors (e.g., trypsin and chymotrypsin inhibitors in soybean and groundnut meals) hinder protein digestion (Francis et al., 2001). **Phytic acid**, found in almost all oilseeds, binds to minerals such as zinc, iron, and calcium, thereby reducing their bioavailability (Kumar et al., 2010). **Lectins**, primarily found in legumes, disrupt the gut epithelium and interfere with nutrient absorption. **Saponins** present in soybean and alfalfa can cause damage to the intestinal mucosa and reduce feed intake (Francis et al., 2002). **Glucosinolates** in mustard and rapeseed meal degrade into isothiocyanates and nitriles, which may be toxic to aquatic animals (Schöne et al., 1997). **Gossypol**, a polyphenolic compound found in cottonseed, is toxic in its free form and can affect reproductive and immune functions in fish (Zhang et al., 2007). **Phorbol esters** found in Jatropha curcas seed cake are highly toxic and must be removed prior to use in aquafeeds (Devappa et al., 2010).

10.6 Effects on fish health and growth

The presence of ANFs in aquafeeds negatively affects growth performance, feed conversion, and general health of aquatic animals. Protease inhibitors inhibit protein digestibility and amino acid availability, resulting in stunted growth and high feed conversion ratios (Krogdahl et al., 2010). Phytic acid not only decreases mineral bioavailability, but also causes phosphorus waste, which is a critical environmental issue in aquaculture (Sugiura et al., 2001). Lectins and saponins can interfere with gut integrity, causing inflammation, decreased absorptive function, and immunosuppression. Gossypol and glucosinolates exhibit systemic toxicity and influence liver function and endocrine control (Naylor et al., 2009). In addition, extended feeding of ANF-supplemented diets has been shown to cause oxidative stress and interfere with antioxidant enzyme mechanisms in fish, making them more susceptible to disease (Lim and Lee, 2009).

10.7 Detoxification techniques and feed formulation strategies

Several strategies have been developed to reduce the levels or effects of ANFs in oilseed meal used in aquafeeds.

10.7.1 Physical and chemical processing

Heat treatment (e.g., autoclaving and roasting) is efficient in inactivating heat-labile ANFs, such as protease inhibitors and lectins (Makkar et al., 2016). Aqueous extraction and soaking can be used to remove water-soluble ANFs such as saponins and phytic acid. Chemical treatments with alkaline or acidic solutions can decrease gossypol and glucosinolate levels; however, cost and safety are limiting factors (Adeparusi and Jimoh, 2011).

10.7.2 Biological processing

FERMENTATION: With the aid of microbes such as Bacillus, Aspergillus, and Lactobacillus, ANFs are lowered and nutrient bioavailability, palatability, and protein quality are enhanced (Adeyemo et al., 2020; Hong et al., 2004).

ENZYME SUPPLEMENTATION: Phytases, among others, increase the availability of phosphorus and lower the anti-nutritional effect of phytic acid is lowered (Cao et al., 2007). Proteases and carbohydrases can improve the digestibility of protein- and fiber-rich oilseed meal.

10.7.3 Feed formulation approaches

Mixing of ingredients: Mixing oilseed meals with fishmeal or animal proteins may reduce the effect of ANFs and equalize amino acid profiles. The addition of feed additives, such as organic acids, chelated minerals, and probiotics, neutralizes the adverse effects of ANFs and promotes gut well-being (Kumar et al., 2010). Optimized inclusion rates: Safe inclusion rates specific to the species must be established for every oilseed product. For instance, soybean meal may be included at rates of up to 30%–40% in tilapia and carp diets, with little or no negative consequences if adequately processed (Gatlin et al., 2007). A multi-faceted strategy incorporating physical, biochemical, and nutritional strategies is needed to counteract the effects of ANFs and facilitate the safe application of oilseed cakes and meals in sustainable aquafeed production.

10.8 Effects of oilseed-based diets on fish growth and health

The use of oilseed cakes and meals in aquafeeds provides a sustainable option for fishmeal by taking advantage of the high protein content and abundant availability of oilseed by-products. Nonetheless, the physiological effects of vegetable ingredients on growth, immunity, and overall fish health are complex and species-specific. This section deals with the role of oilseed-based diets in altering digestibility, feed efficiency, immune functions, gut microbiota, and metabolic well-being. The inclusion of oilseed cakes and meals in aquafeeds has gained increasing interest as a sustainable substitute for fishmeal. However, their effects on fish growth and health are influenced by variables such as digestibility, feed conversion efficiency (FCE), immune modulation, and gut microbiota interactions. Digestibility and FCE are important markers of feed quality and utilization. Plant proteins such as those from soybean, rapeseed, and sunflower meals tend to be less digestible than fishmeal because they contain fiber, anti-nutritional factors (ANFs), and unbalanced amino acid compositions (Francis et al., 2001; Glencross et al., 2007). Nevertheless, processing methods such as heat treatment, enzymatic supplementation, or microbial fermentation can considerably enhance oilseed meal digestibility and enhance protein and energy utilization in fish such as *Oreochromis niloticus* and *Cyprinus carpio* (Refstie et al., 2004; El-Sayed, 1999). The immune reactions of fish to oilseed-based diets are affected by both the nutritional content and bioactive compounds of plant ingredients. Moderate oilseed meal inclusion, such as soybean meal, can maintain normal immune function. Excessive inclusion levels, especially in carnivorous fish, have been reported to stimulate gut inflammation, suppress leukocyte activity, and impair disease resistance as a result of ANFs such as saponins, lectins, and phytic acid (Krogdahl et al., 2010; Torstensen et al., 2008). Nonetheless, the addition of functional additives, such as probiotics, prebiotics, and immunostimulants, or bioprocessing techniques (e.g., fermentation), can alleviate such adverse effects and improve mucosal immunity and pathogen resistance (Gatlin et al., 2007; Hong et al., 2004). Oilseed diets also affect the gut microbiota composition, which in turn affects nutrient metabolism, immunity, and health. Plant-based diets have been found to change the diversity and abundance of beneficial gut microbes, including Lactobacillus and Bacillus species, which are linked to improved digestion and immune function (Ringø et al., 2016). However, the sudden or massive substitution of fishmeal with plant proteins can stimulate the growth of harmful or opportunistic bacteria, resulting in gut dysbiosis and compromised health (Urayama et al., 2021). The addition of fermented oilseed meals and dietary enzymes stabilizes the intestinal microbiome and improves metabolic homeostasis (Bakke-McKellep et al., 2007). Although oilseed-based diets provide a sustainable alternative for aquaculture as a source of protein, successful implementation calls for an appropriate consideration of formulation strategies, processing methods, and health-promoting interventions to achieve maximum growth performance, immune robustness, and gut integrity.

10.9 Digestibility and feed conversion efficiency

Nutrient digestibility and the feed conversion ratio (FCR) are key measures of aquaculture feed performance. Oilseed cakes, such as soybean meal, groundnut meal, and rapeseed meal, contain high protein levels, but also potentially indigestible carbohydrates, fiber, and anti-nutritional factors (ANFs) that impede digestibility and nutrient assimilation (Francis et al., 2001). The addition of raw oilseed meal tends to decrease the apparent digestibility coefficients (ADC) of dry matter and crude protein due to the presence of non-starch polysaccharides (NSPs), phytic acid, and protease inhibitors (Krogdahl et al., 2010). However, when suitably processed via fermentation or enzymatic hydrolysis, oilseed meals exhibit enhanced digestibility and lower FCR in fish species, such as tilapia, carp, and catfish (El-Sayed, 1999; Hernández et al., 2010). For example, soybean meal is acceptable at feeding levels up to 30%–40% in the diets of Nile tilapia (*Oreochromis niloticus*) without impacting growth or FCR, provided that the meal is not only supplemented with methionine and lysine (Gatlin et al., 2007), but also set in a species-specific formula that can safely tolerate plant-based feeds.

10.10 Immune responses and disease resistance

Oilseed-based diets stimulate and suppress the immune system in fish. While the inclusion of ANFs such as saponins, tannins, and lectins can dampen immunity and enhance the susceptibility of fish to disease, bioactive ingredients in oilseed meals can also regulate immunity if wisely employed (Kumar et al., 2010). It has been observed that the incorporation of enzyme-treated or fermented oilseed cakes can upgrade innate as well as adaptive immune parameters like lysozyme activity, respiratory burst, and antibody production in fish (Zhou et al., 2006). For example, dietary supplementation of heat-treated cottonseed meal with vitamins and antioxidants improved resistance to bacterial infection in hybrid catfish (*Clarias gariepinus* × Heterobranchus bidorsalis). Moreover, oilseed meals, such as flaxseed and sesame, have essential fatty acids and lignans that possess immunostimulatory and anti-inflammatory activities (Brennan, 2009). Such bioactives can increase cytokine expression, enhance mucosal immunity, and suppress oxidative stress markers, thereby leading to improved disease resistance.

10.11 Gut microbiota and metabolic effects

Gut microbiota is critical for nutrient metabolism, immune modulation, and general health in fish. Diets containing oilseed meal have a notable effect on the gut microbial profile, tending to favor the growth of beneficial microbes if processed and formulated appropriately (Ringø et al., 2016). The inclusion of soybean meal in aquafeeds, especially at high levels, has been linked to gut enteritis in Atlantic salmon (*Salmo salar*), characterized by microbial dysbiosis and inflammation (Urán et al., 2008). However, fermented oilseed products tend to enrich microbial diversity and promote populations of lactic acid bacteria and other health-promoting microbes that contribute to improved gut health and metabolic activity (Ng et al., 2009). In addition, oilseed-derived oligosaccharides and dietary fiber fractions can function as prebiotics, stimulating the generation of short-chain fatty acids (SCFAs) and enhancing intestinal barrier function. Increased SCFA production is associated with enhanced nutrient absorption and lipid metabolism and decreased intestinal inflammation (Liu et al., 2014). Metabolically, oilseed-based diets affect protein turnover and lipid deposition. For instance, canola or sunflower meals have been reported to lower the accumulation of visceral fat and regulate hepatic enzyme activity in carp and tilapia (Siddhuraju and Becker, 2001). Such effects are dose-dependent and require proper supplementation to prevent nutrient imbalances.

10.12 Economic and environmental benefits of oilseed based aquafeeds

10.12.1 Cost effectiveness compared to fishmeal

In the context of aquaculture and growing feed prices, the cost-effectiveness of oilseed-based aquafeeds versus fishmeal-based feeds will be an important consideration in today's aquaculture environment; when fishmeal stocks are low, it is difficult to believe that fishmeal can continue to be valued at levels. Oilseed-based aquafeeds are being exploited for their financial advantages in aquaculture, given their abundance, availability, and low price, especially soybean, canola (rapeseed), and sunflower (Hossain et al., 2023). With high protein content, fishmeal-based aquafeeds have historically been the benchmark in aquaculture, but due to dwindling wild fish stocks, rising energy prices, pressure on fishermen to not fish due to environmental concerns, and the world's current inflation, fishmeal prices are on the rise and thus are increasingly more difficult to purchase for the aquaculture feed industry. Given that feed costs account for 50%–70% of overall aquaculture costs, production systems that utilize oilseeds and oilseed meals as protein sources are poised to increase, with many species of fish, when practical, now being fed oilseed-based protein cakes (oilseed meals) instead of fishmeal. At such times that fish meal is needed individually, researchers showed that replacing fishmeal with oilseeds, especially soybean meal, significantly reduced feed costs by 30–40% in some cases, and had no impact on fish growth and productivity when diets were appropriately re-done to consider balanced diets limiting amino acids like methionine and lysine (Serra et al., 2024). Advancements in feed formulation technology and resultant techniques, such as heat treatment, fermentation, and supplementation with enzymes, have also dealt with some of the anti-nutritional issues surrounding protein feed application of oilseed meals to feed, making oilseeds more accessible and with higher consumer digestibility and utility (Daniel, 2018). Thus, the integration of oilseed-based food in aquaculture not only provides a promising solution for increasing costs and environmental concerns but also contributes to industry stability and economic viability.

10.12.2 Sustainability and reduced environmental impact

The shift from fishmeal to oilseed-based aquafeeds represents a major step toward increasing stability and reducing the environmental footprint of global aquaculture. Traditional fishmeal, obtained from wild-caught forage fish, puts immense pressure on marine ecosystems and contributes to the loss of biodiversity, and fuel-intensive harvesting and processing methods increase greenhouse gas emissions due to processing methods. Conversely, oilseeds, especially soybeans, canola, and sunflower food, are land-based, renewable, and widely available alternatives that significantly reduce ecological effects. Recent life cycle assessment studies have shown that oilseed-based feeds produce much less greenhouse gas emissions, require less water, and exhibit less eutrophication and acidification capacity than fishmeal-based diets (Parker et al., 2018; Tignani et al., 2024). Soybean meal, in particular, has been shown to reduce carbon footprints in feed formulation when sourced from responsible supply chains with low deforestation risk. In addition, oilseed cultivation can be integrated into sustainable agricultural systems to promote crop rotation, soil health, and carbon sequestration, especially when deforestation practices and regenerative methods are employed. Recognition of oilseeds-product-like soybean food, a residual of soybean oil extraction, supports circular economy principles using agro-industrial waste, which reduces waste and improves feed stability metrics (Naylor et al., 2021). In addition, adopting oilseed products, which are residues from vegetable oil extraction, supporting waste valorization, and circular economy models, reduces the environmental burden by increasing resource efficiency. Nutritional challenges in oilseeds have decreased due to the presence of anti-nutrient factors in food, fermentation, enzyme supplementation, and genetic crop reforms, which increase the availability of nutrients and reduce environmental loads from uneaten feed and waste. Additionally, replacing fishmeal with plant-based ingredients contributes to the preservation of wild fish populations and marine food web integrity, aligning with global conservation goals. As the aquaculture sector continues to grow to meet the demands of increasing global protein, adopting oilseed-based aquafeeds is a scalable and environmentally responsible solution with the United Nations Sustainable Development Goals, especially SDG 12 (responsible consumption and production), SDG13 (climate action), and SDG 14 (Life Below Water) (Malcorps et al., 2019; Colombo and Giovanni, 2021). Since the Aquaculture sector is attributed to more than 60% of global fish consumption by 2030, comprehensive use of oilseed-based feed presents a practical and ecologically sound guide to meet the demands of protein, reducing the aquatic biodiversity and reducing environmental degradation.

10.12.3 Role in reducing overfishing of marine resources

There is an increase in the demand for sustainable aquafeeds that do not overly exploit marine resources because of increasing aquaculture production. The integration of oilseed-based ingredients is recognized as an important strategy for rapidly promoting stability in accelerated aquafeeds. Traditionally, aquafeeds are highly dependent on fishmeal and fish oil, which are obtained from wild-caught forage fish, such as anchovies, sardines, and menhaden species, which contribute significantly to marine ecosystems and thus play important ecological roles in marine food webs. This heavy dependence has contributed to the overexploitation of marine resources, which intensifies concerns about the biodiversity of the ocean and long-term fishery sustainability (Colombo and Giovanni, 2021). In response to concerns about its impact, the aquaculture industry focused on oilseed crops such as soybean, canola, and sunflower, which are nutritionally good and at the same time renewable. These oilseeds are real plants containing essential amino acids, proteins, and fatty acids that can be used for the development of fish species, such as salmon, tilapia, and carp, which, in turn, will support the growth of immune and overall health in aquatic animals. The aquaculture sector can now minimize its demand for wild fish by the application of oilseed products; thus, the ecosystem is relieved as less pressure is exerted on the overexploited marine stocks, and these stocks will grow again, allowing the ecosystem to recover. Modern processing technologies, such as enzyme treatment and fermentation, have improved the digestive power of oilseeds and nutrients, overcoming earlier limitations such as non-nutritional factors (Gatlin et al., 2007). Studies have shown that partial or complete replacement of fishmeal with processed soybeans can maintain or increase fish performance without compromising product quality or consumer health benefits.

In addition, innovations such as genetically modified oilseeds producing long-chain omega-3 fatty acids have been found only in fish oils. This makes it possible for aquafeeds to be free of fish oils. This substitution fulfills the ecological dimension, stabilizes the supply chains for feed components, and reduces the carbon footprint of fishmeal production and its marine harvesting. Furthermore, the large-scale adoption of oilseed-based feeds is expected to drastically reduce the input demand of wild fish, reduce pressure on marine ecosystems, and contribute to the global goals of biodiversity conservation and sustainable food systems. Changes in plant-based feed inputs, such as oilseeds, are an important step toward decoupling seafood production from marine abuse and generating a circular and environmentally conscious blue economy, as aquaculture burgeons to meet the increasing demand for protein. Shifts in plant-based aquafeeds support global food security and marine protection by reducing fishing pressure on wild fish and promoting durable aquaculture methods. By separating aquaculture from the exploitation of marine biodiversity, oilseed-based aquafeeds provide a revolutionary solution that contributes significantly to ocean protection and global food security (Kumar et al., 2019, 2022). Aquafeeds made of oilseeds are therefore important for making changes in the more economically and environmentally durable field of aquaculture.

11. Challenges and future perspectives in oilseed based aquafeeds

Shifts to oilse-based aquafeeds in aquaculture are driven by the increasing costs, scarcity, and environmental factors of traditional fishmeal and fish oil, which have been the major protein and lipid sources in aquafeeds for centuries. Oilseeds, such as soybean, cotton seed, sunflower, and canola (rapeseed) are potential alternatives because they are abundantly available, rich in protein, and relatively cheaper. However, the most significant among them are those challenges that affect their proper utilization in aquafeeds, including nutritional, physical, and environmental challenges. Anti-nutritional factors (ANFs), such as phytic acid, glucosinolates, tannins, protease inhibitors, lectins, saponins, and others, are a major concern because they can weaken the ability of fish to digest proteins, decrease their ability to absorb nutrients, and cause inflammation in their digestive systems. A supplement is also necessary to address the dietary needs of carnivorous animals, such as salmonids and seabass, since amino acid imbalances, particularly in oilseeds, frequently lack key amino acids, such as methionine and lysine (Kumar et al., 2012; Francis et al., 2001). Another important issue is the change in fatty acid profiles in fish flesh when plant oils replace fish oils; beneficial long-chain omega-3 fatty acids such as EPA and DHA in fish reduce the level of omega −3 fatty acids, which can affect consumer health and affect fish health and reproductive performance (Mu et al., 2020). Tolerance complicates the species-specific variability and digestive power of plant-based feeds and more feed formulations, requiring more sophisticated approaches in diet design (Gatlin et al., 2007). Further, future perspectives on many new strategies center on biotechnology intervention such as protein quality and genetic modification of oilseed crops to increase EPA/DHA biosynthesis routes; innovative feed processing techniques such as fermentation and enzyme treatment to reduce Anti-Nutritional Factors (ANFS) and improve coordination; and accurate nutrition models that optimize feed composition according to species, life phase, and upbringing conditions. The inclusion of functional feed additives (FuFAs), such as nucleotides, organic acids, probiotics, prebiotics, and certain amino acids, also shows the ability improves intestinal health and immunity in fish fed oilseed-based diets (Dawood and Koshio, 2020; Zhou et al., 2019; Hossain et al., 2024). In addition, the development of integrated life-cycle assessments and circular bio-economic models is important for evaluating the environmental and economic stability of oilseed-based feeds. Furthermore, a multidisciplinary approach in combination with plant breeding, feed formulation, fish physiology, and environmental evaluation is necessary to overcome current limitations and increase the stability of oilseed-based aquafeeds, especially in terms of increasing global demand and pressure on marine resources for aquatic agricultural products (Hua et al., 2019). Overall, while plant-based proteins provide long-term capacity for feed stability, overcoming their current boundaries seeks welfare, feed efficiency, and nutritional value to ensure research, policy assistance, and concrete efforts in adopting industry, maintained in future aquaculture systems.

12. Conclusion

Incorporating oilseed cakes and meals into aquafeeds is a sustainable, nutritionally viable, and economically attractive alternative to fishmeal, particularly in the face of rising feed costs and ecological concerns. These plant-derived by-products, including soybean, mustard, groundnut, sunflower, linseed, and cottonseed meal, offer a high protein content, balanced amino acids (with strategic supplementation), essential lipids, fiber, vitamins, minerals, and various bioactive compounds that support fish growth, immunity, and overall health. However, the presence of anti-nutritional factors (ANFs), such as phytic acid, glucosinolates, and gossypol, necessitates careful processing, including oil extraction, heat treatment, fermentation, and enzymatic detoxification, to enhance nutrient availability and safety. Feed formulation strategies tailored to species-specific needs are critical for optimizing digestibility and feed conversion efficiency of oilseed-based diets. Empirical studies have highlighted the potential of these ingredients to maintain or improve fish performance and health outcomes, particularly when integrated with modern feed technologies. Additionally, oilseed-based aquafeeds contribute to environmental sustainability by reducing the dependence on marine resources and lowering the ecological footprint of aquaculture operations. Although challenges such as nutritional variability, limited application in carnivorous species, and consumer perceptions remain, future advancements in breeding, biotechnology, and precision nutrition are expected to further enhance the role of oilseed meals in aquaculture. Thus, the strategic use of these alternative protein sources represents a significant step toward more sustainable, resilient, and economically viable aquaculture practices globally.

References

Adeparusi, E.O., Jimoh, W.A., 2011. Growth performance and hematological responses of *Clarias gariepinus* (Burchell, 1822) fed fermented castor seed meal-based diet. J. Agric. Sci. Technol. 1, 9–15.

Adeyemo, A.A., Jimoh, W.A., Hassan, A.T., Adeshina, I., 2020. Fermentation and enzymatic enhancement of plant protein ingredients for sustainable aquafeed production: effects on nutrient utilization and fish performance. Aquac. Fisher. 5 (5), 221–230.

Adolphe, J.L., Whiting, S.J., Juurlink, B.H.J., Thorpe, L.U., Alcorn, J., 2010. Health effects with consumption of the flax lignan secoisolariciresinol diglucoside. Br. J. Nutr. 103 (7), 929–938. https://doi.org/10.1017/S0007114509992753.

Bakke-McKellep, A.M., Penn, M.H., Salas, P.M., Refstie, S., Sperstad, S., Landsverk, T., Ringø, E., Krogdahl, A., 2007. Effects of dietary soyabean meal, inulin and oxytetracycline on intestinal microbiota and epithelial cell stress, apoptosis and proliferation in the teleost Atlantic salmon (*Salmo salar* L.). Br. J. Nutr. 97 (4), 699–713. https://doi.org/10.1017/S0007114507381397.

Baldwin, B.S., Hurburgh, C., McVay, K., 2010. Nutritional composition and feeding value of sunflower meal and cake products. Feedstuffs 82 (12), 18–23.

Banerjee, S., Ghosh, S., Das, A., Mandal, S.C., 2023. Effect of fermented linseed oil cake as a partial fishmeal replacer in carp diet: growth, nutrient utilization and muscle fatty acid composition. Aquac. Rep. 30.

Barrows, F.T., Gaylord, T.G., Stone, D.A.J., Smith, C.E., 2008. Effect of protein source and nutrient density on growth efficiency, histology, and plasma amino acid concentrations of rainbow trout (*Oncorhynchus mykiss*). Aquaculture 283, 90–97.

Bell, J.M., 1993. Factors affecting the nutritional value of canola meal: a review. Can. J. Anim. Sci. 73 (4), 689–697. https://doi.org/10.4141/cjas93-075.

Blekas, G., Tsimidou, M., Boskou, D., 1995. Contribution of α-tocopherol to olive oil stability. Food Chem. 52 (3), 289–294.

Brennan, C.S., 2009. Dietary fibre, glycaemic response, and health benefits of flaxseed and sesame bioactive compounds. Int. J. Food Sci. Technol. 44 (2), 213–219.

Burel, C., Boujard, T., Escaffre, A.M., Kaushik, S.J., Boeuf, G., Mol, K., 2000. Geyten, effects of dietary soybean meal on growth, digestibility, and hormonal status in rainbow trout. Aquaculture 190, 307–321.

Cao, L., Wang, W., Yang, C., Yang, Y., Diana, J., Yakupitiyage, A., Luo, Z., Li, D., 2007. Application of microbial phytase in fish feed. Enzym. Microb. Technol. 40 (4), 497–507. https://doi.org/10.1016/j.enzmictec.2007.01.007.

Carré, P., Rozan, P., 2004. Aqueous extraction of sunflower seed oil and protein: effect of dehulling and extraction conditions. LWT-Food Sci. Technol. 37 (5), 437–444.

Chattaraj, S., Behera, B.K., Das Mohapatra, P.K., 2024a. Effective valorization of Chlorella biomass and Brewers spent grain substituting fish meal and soybean meal in the diet of herbivorous fish Cirrhinus reba for higher growth, digestibility and sustainable cultivation. J. Appl. Phycol. 36 (1), 217–232.

Chattaraj, S., Chattaraj, M., Mitra, D., Ganguly, A., Thatoi, H., Das Mohapatra, P.K., 2025. 16S amplicon sequencing of the gastrointestinal microbiota of Cirrhinus reba and isolation of an autochthonous probiotic using culture based approaches. Syst. Microbiol. Biomanufact. 5 (1), 156–170.

Chattaraj, S., Ganguly, A., Mitra, D., et al., 2026a. Sustainable valorization of algal biomass in fish nutrition and dietetics: a critical review and scientometric study. Waste Biomass Valor. https://doi.org/10.1007/s12649-026-03619-7.

Chattaraj, S., Ganguly, A., Mitra, D., Mitra, D., Das Mohapatra, P.K., 2023. Study of intestinal bacteria of Cirrhinus reba and characterization of a new probiotic bacteria: an initiative to save the threatened species of Cirrhinus. Appl. Biochem. Microbiol. 59 (6), 833–849.

Chattaraj, S., Mitra, D., Chattaraj, M., Ganguly, A., Thatoi, H., Mohapatra, P.K.D., 2024b. Brewers' spent grain as fish feed ingredient: evaluation of bio-safety and analysis of its impact on gut bacteria of Cirrhinus reba by 16S Metagenomic sequencing. Curr. Res. Microb. Sci. 7, 100286.

Chattaraj, S., Mitra, D., Ganguly, A., Thatoi, H., Mohapatra, P.K.D., 2024c. A critical review on the biotechnological potential of Brewers' waste: challenges and future alternatives. Curr. Res. Microb. Sci. 6, 100228.

Chattaraj, S., Sahoo, S.S., Behera, S., Thatoi, H., 2026b. Sustainable aquafeed development using Spirulina: effects on growth performance of Catla catla. Algal Res. 104447.

Choct, M., Hughes, R.J., Wang, J., Bedford, M.R., Morgan, A.J., Annison, G., 2010. Increased small intestinal fermentation is partly responsible for the anti-nutritive activity of non-starch polysaccharides in chickens. Br. Poult. Sci. 40 (3), 419–426.

Colombo, S.M., Giovanni, T., 2021. Sustainable aquafeed ingredients and alternatives to fish-meal and fish oil in aquaculture: A review. Rev. Aquac. 13 (3), 1668–1695.

Daniel, N., 2018. A review on replacing fish meal in aqua feeds using plant protein sources. Int. J. Fish. Aquat. Stud. 6 (2), 164–179.

Dawood, M.A.O., Koshio, S., 2020. Application of fermentation strategy in aquafeed for sustainable aquaculture. Rev. Aquacult. 12 (2), 987–1002. https://doi.org/10.1111/raq.12368.

Devappa, R.K., Makkar, H.P., Becker, K., 2010. Jatropha toxicity—a review. J. Toxicol. Environ. Health Part B 13 (6), 476–507.

Devendra, C., 1988. Non-conventional feed resources in Asia and the Pacific. Food and Agriculture Organization of the United Nations.

Devendra, C., 2009. Small ruminant production systems in South and Southeast Asia. FAO.

Elleuch, M., Bedigian, D., Roiseux, O., Besbes, S., Blecker, C., Attia, H., 2011. Dietary fibre and fibre-rich by-products of food processing: characterisation, technological functionality and commercial applications: a review. Food Chem. 124 (2), 411–421. https://doi.org/10.1016/j.foodchem.2010.06.077.

El-Sayed, A.F.M., 1999. Alternative dietary protein sources for farmed tilapia, Oreochromis spp. Aquaculture 179 (1–4), 149–168. https://doi.org/10.1016/S0044-8486(99)00159-3.

Feedipedia, 2021. Soybean meal.

Francis, G., Kerem, Z., Makkar, H.P.S., Becker, K., 2002. The biological action of saponins in animal systems: a review. Br. J. Nutr. 88 (6), 587–605. https://doi.org/10.1079/BJN2002725.

Francis, G., Makkar, H.P., Becker, K., 2001. Antinutritional factors present in plant-derived alternate fish feed ingredients and their effects in fish. Aquaculture 199 (3–4), 197–227.

Gatlin III, D.M., Barrows, F.T., Brown, P., Dabrowski, K., Gaylord, T.G., Hardy, R.W., Wurtele, E., 2007. Expanding the utilization of sustainable plant products in aquafeeds: a review. Aquac. Res. 38 (6), 551–579.

Ghosh, K., Ray, A.K., Ringø, E., 2019. Applications of plant ingredients for tropical and subtropical freshwater finfish: possibilities and challenges. Rev. Aquac. 11 (3), 793–815.

Glencross, B.D., 2009. Exploring the nutritional demand for essential fatty acids by aquaculture species. Rev. Aquac. 1 (2), 71–124.

Glencross, B.D., Booth, M., Allan, G.L., 2007. A feed is only as good as its ingredients–a review of ingredient evaluation strategies for aquaculture feeds. Aquac. Nutr. 13 (1), 17–34.

Göhl, B., 1982. Tropical feeds: feed information summaries and nutritive values. FAO.

Hardy, R.W., 2010. Utilization of plant proteins in fish diets: effects of global demand and supplies of fishmeal. Aquac. Res. 41 (5), 770–776. https://doi.org/10.1111/j.1365-2109.2009.02349.x.

Hasan, M.R., Hecht, T., De Silva, S.S., Tacon, A.G., 2007. Study and Analysis of Feeds and Fertilizers for Sustainable Aquaculture Development.

Hernández, C., Olvera-Novoa, M.A., Hardy, R.W., Hermosillo, A.M., 2010. Evaluation of canola meal as a protein source in diets for juvenile Nile tilapia (*Oreochromis niloticus*). J. World Aquacult. Soc. 41 (1), 100–108. https://doi.org/10.1111/j.1749-7345.2009.00346.x.

Hoffman, L.C., Tshabalala, P.A., Muller, M., 2019. The effects of thermal processing on protein quality and Maillard reaction products in feed ingredients. Anim. Feed Sci. Technol. 251, 1–10.

Hong, K.J., Lee, C.H., Kim, S.W., 2004. Aspergillus oryzae GB-107 fermentation improves nutritional quality of food soybeans and feed soybean meals. J. Med. Food 7 (4), 430–435.

Hossain, A., Hossain, Md A., Rasul, Md G., Akter, T., Uz Zaman, Md F., Islam, Md R., 2023. Efficacy of using sunflower meal as an ingredient, and partial fishmeal—replacer, in practical feed formulated for stinging catfish (*Heteropneustes fossilis*). Aquac. Fish Fish. 3 (3), 237–246. https://doi.org/10.1002/aff2.109.

Hossain, M.S., Small, B.C., Kumar, V., Hardy, R., 2024. Utilization of functional feed additives to produce cost-effective, ecofriendly aquafeeds high in plant-based ingredients. Rev. Aquacult. 16 (1), 121–153. https://doi.org/10.1111/raq.12824.

Hua, K., Cobcroft, J.M., Cole, A., Condon, K., Jerry, D.R., Mangott, A., Praeger, C., Vucko, M.J., Zeng, C., Zenger, K., Strugnell, J.M., 2019. The future of aquatic protein: implications for protein sources in aquaculture diets. One Earth 1 (3), 316–329. https://doi.org/10.1016/j.oneear.2019.10.018.

Hutagalung, R.I., 1981. Utilization of feed from agricultural wastes in the tropics. World Anim. Rev. 37, 11–17.

Jayasena, V., Nasar-Abbas, S.M., 2011. Safflower (*Carthamus tinctorius* L.) meal as a protein supplement for ruminants. Anim. Feed Sci. Technol. 170 (1–2), 17–23.

Jones, L.A., 1991. Cottonseed meal in fish feeds. In: Proceedings of the 5th International Symposium on Fish Nutrition and Feeding, pp. 296–301.

Kakade, M.L., Rackis, J.J., McGhee, J.E., Puski, G., 1974. Determination of trypsin inhibitor activity of soy products: a collaborative analysis of an improved procedure. Cereal Chem. 51, 376–382.

Karr-Lilienthal, L.K., Grieshop, C.M., Spears, J.K., Fahey, G.C., Parsons, C.M., 2005. Soybean hulls and soybean cotyledon fiber: chemical composition and digestibility by dogs. J. Anim. Sci. 83 (2), 218–224.

Kaur, M., Sharma, S., Kaur, J., 2012. Nutritional evaluation of groundnut meal as a protein source in animal feed formulations. Ind. J. Anim. Nutr. 29 (3), 245–252.

Kaushik, S.J., Cravedi, J.P., Lalles, J.P., Sumpter, J., Fauconneau, B., Laroche, M., 1995. Partial or total replacement of fish meal by soybean protein on growth, protein utilization, potential estrogenic or antigenic effects, cholesterolemia and flesh quality in rainbow trout, *Oncorhynchus mykiss*. Aquaculture 133 (3–4), 257–274. https://doi.org/10.1016/0044-8486(94)00403-b.

Krogdahl, A., Penn, M., Thorsen, J., Refstie, S., Bakke, A.M., 2010. Important antinutrients in plant feedstuffs for aquaculture: an update on recent findings regarding responses in salmonids. Aquac. Res. 41 (3), 333–344. https://doi.org/10.1111/j.1365-2109.2009.02426.x.

Kumar, P., Amogha, K.R., Gowda, S.D., Hazarika, S., 2022. Enhancing the integration of sustainable plant products in aquafeeds: a comprehensive review. J. Plant Biota. 1, 17–22.

Kumar, V., Barman, D., Kumar, K., Kumar, V., Mandal, S.C., Clercq, E.D. (Eds.), 2012. Antinutritional factors in plant feedstuffs used in aquafeeds. World Aquaculture, pp. 64–68.

Kumar, V., Sinha, A.K., Makkar, H.P.S., Becker, K., 2010. Dietary roles of phytate and phytase in human nutrition: a review. Food Chem. 120 (4), 945–959. https://doi.org/10.1016/j.foodchem.2009.11.052.

Kumar, V., Sinha, A.K., Romano, N., Allen, K.M., Bowman, B.A., Thompson, K.R., Tidwell, J.H., 2019. The potential of genetically modified oilseed crops and plant-based alternatives in sustainable aquafeeds: current status and future perspectives. Rev. Aquac. 11 (4), 1103–1122.

Li, M.H., Robinson, E.H., 2006. Use of cottonseed meal in aquatic animal diets: a review. N. Am. J. Aquac. 68 (1), 14–22. https://doi.org/10.1577/A05-028.1.

Lim, C., Klesius, P.H., Webster, C.D., 2001. Nutritional value and physiological effects of plant antinutritional factors in fish diets. Aquac. Res. 32 (1), 349–360.

Lim, C., Lee, C.S., 2009. Alternative protein sources in aquaculture diets. CRC Press.

Liu, H., Wang, S., Cai, Y., Guo, X., Cao, Z., Zhang, Y., Xie, M., 2014. Effects of dietary inulin supplementation on growth performance, digestive enzyme activities, and intestinal microbiota of juvenile turbot (*Scophthalmus maximus*). Aquac. Nutr. 20 (3), 275–282. https://doi.org/10.1111/anu.12078.

Liu, K., 2011. Soybeans: Chemistry, technology and utilization. Springer.

Makkar, H.P.S., Tran, G., Heuzé, V., Giger-Reverdin, S., 2016. Fate of anti-nutritional factors in plant-based animal feed ingredients during processing. Anim. Feed Sci. Technol. 225, 1–13.

Malcorps, W., Kok, B., van't Land, M., Fritz, M., van Doren, D., Servin, K., Heijden, P.van der, Palmer, R., Auchterlonie, N., Rietkerk, M., Santos, M., Davies, S., 2019. The sustainability conundrum of fishmeal substitution by plant ingredients in shrimp feeds. Sustainability 11 (4), 1212. https://doi.org/10.3390/su11041212.

Moses, M., 1984. Mycotoxins in oilseed cake: occurrence and toxicological significance. Mycopathologia 86 (3), 147–150.

Mu, H., Wei, C., Zhang, Y., Zhou, H., Pan, Y., Chen, J., Zhang, W., Mai, K., 2020. Impacts of replacement of dietary fish oil by vegetable oils on growth performance, anti-oxidative capacity, and inflammatory response in large yellow croaker Larimichthys crocea. Fish Physiol. Biochem. 46 (1), 231–245. https://doi.org/10.1007/s10695-019-00712-8.

Mukhopadhyay, N., Ray, A.K., 1999. Effects of fermentation on the nutritive value of mustard oil cake in the diets for rohu, Labeo rohita (Hamilton), fingerlings. Aquac. Nutr. 5 (3), 169–174.

Mungkung, R., Aubin, J., Prihadi, T.H., Slembrouck, J., Van Der Werf, H.M.G., Legendre, M., 2013. Life cycle assessment for environmentally sustainable aquaculture management: a case study of combined aquaculture systems for carp and tilapia. J. Clean. Prod. 57, 249–256. https://doi.org/10.1016/j.jclepro.2013.05.029.

Nagalakshmi, D., Rama Rao, S.V., Panda, A.K., Sastry, V.R.B., Raju, M.V.L.N., 2016. Cottonseed meal in animal nutrition: a review. J. Anim. Feed Sci. Technol. 222, 1–18.

Natesan, S., Muthukrishnan, C.R., 2014. Nutritional composition of sunflower seed meal and its potential as animal feed. J. Oilseeds Res. 31 (3), 157–160.

Naylor, R.L., Hardy, R.W., Bureau, D.P., Chiu, A., Elliott, M., Farrell, A.P., Nichols, P.D., 2009. Feeding aquaculture in an era of finite resources. Proc. Nat. Acad. Sci. 106 (36), 15103–15110.

Naylor, R.L., Hardy, R.W., Buschmann, A.H., Bush, S.R., Cao, L., Klinger, D.H., Little, D.C., Lubchenco, J., Shumway, S.E., Troell, M., 2021. A 20-year retrospective review of global aquaculture. Nature 591 (7851), 551–563. https://doi.org/10.1038/s41586-021-03308-6.

Newkirk, R., 2009. Meal from canola and mustard for livestock feed. Canadian International Grains Institute.

Newkirk, R.W., Classen, H.L., Edney, M.J., 2003. Effects of prepress-solvent extraction on the nutritional value of canola meal for broiler chickens. Anim. Feed Sci. Technol. 104 (1–4), 111–119.

Ng, W.K., Liew, F.L., Ang, L.P., Wong, K.W., 2009. Potential of fermented oilseed meals in aquaculture feeds: effects on gut microbial ecology and fish growth performance. Aquac. Res. 40 (10), 1201–1210.

NRC, 2011. Nutrient requirements of fish and shrimp. The National Academies Press.

NRC, 2012. Nutrient requirements of swine, 11th ed. National Academies Press.

Nzikou, J.M., 2009. Characteristics and chemical composition of Cucurbitaceae seeds and seed oils. Res. J. Environ. Earth Sci. 1 (3), 39–45.

Olsen, R.L., Hasan, M.R., 2012. A limited supply of fishmeal: impact on future increases in global aquaculture production. Trends Food Sci. Technol. 27 (2), 120–128. https://doi.org/10.1016/j.tifs.2012.06.003.

Parker, R.W.R., Blanchard, J.L., Gardner, C., Green, B.S., Hartmann, K., Tyedmers, P.H., Watson, R.A., 2018. Fuel use and greenhouse gas emissions of world fisheries. Nat. Clim. Change 8 (4), 333–337. https://doi.org/10.1038/s41558-018-0117-x.

Piironen, V., Lindsay, D.G., Miettinen, T.A., Toivo, J., Lampi, A.M., 2000. Plant sterols: biosynthesis, biological function and their importance to human nutrition. J. Sci. Food Agric. 80 (7), 939–966. https://doi.org/10.1002/(SICI)1097-0010(20000515)80:7<939::AID-JSFA644>3.0.CO;2-C.

Rakshit, M., Ghosh, S., Das, P., 2019. Nutritional composition and feed value of soybean meal in livestock and aquaculture diets. J. Oilseed Res. 36 (2), 85–92.

Rama Rao, S.V., Raju, M.V.L.N., Panda, A.K., Reddy, M.R., 2006. Sunflower seed meal as a substitute for soybean meal in commercial broiler chicken diets. Br. Poult. Sci. 47 (5), 592–598. https://doi.org/10.1080/00071660600963511.

Ravindran, V., Bryden, W.L., Kornegay, E.T., 1995. Phytates: occurrence, bioavailability and implications in poultry nutrition. Poultr. Avian Biol. Rev. 6 (2), 125–143.

Refstie, S., Helland, S.J., Storebakken, T., Baeverfjord, G., 2005. Digestive capacity, intestinal morphology, and performance of Atlantic salmon (Salmo salar) fed diets supplemented with carbohydrate-degrading enzymes. Aquaculture 241 (1–4), 391–404.

Refstie, S., Korsøen, Ø.J., Storebakken, T., Baeverfjord, G., Lein, I., Roem, A.J., 2000. Differing nutritional responses to dietary soybean meal in rainbow trout (*Oncorhynchus mykiss*) and Atlantic salmon (*Salmo salar*). Aquaculture 190 (1–2), 49–63. https://doi.org/10.1016/S0044-8486(00)00382-3.

Refstie, S., Storebakken, T., Baeverfjord, G., Roem, A.J., 2004. Long-term protein and lipid growth of Atlantic salmon (*Salmo salar*) fed diets with partial replacement of fish meal with soybean meal. Aquaculture 221, 365–379. https://doi.org/10.1016/S0044-8486(03)00117-9.

Ringø, E., Hoseinifar, S.H., Ghosh, K., Doan, H.V., Beck, B.R., Song, S.K., 2016. Lactic acid bacteria in finfish—an update. Front. Microbiol. 7.

Robinson, E.H., Li, M.H., 2008. Replacement of soybean meal in channel catfish, ictalurus punctatus, diets with cottonseed meal and distiller's dried grains with solubles. J. World Aquacult. Soc. 39 (4), 521–527. https://doi.org/10.1111/j.1749-7345.2008.00190.x.

Rochfort, S., Panozzo, J., 2007. Phytochemicals for health, the role of pulses. J. Agric. Food Chem. 55 (20), 7981–7994. https://doi.org/10.1021/jf071704w.

Russo, R., Reggiani, R., 2015. Salt sensitivity in camelina sativa seedlings and polyamine content. Int. J. Plant Soil Sci. 8 (3), 1–7. https://doi.org/10.9734/ijpss/2015/20270.

Saini, R.K., Keum, Y.S., 2018. Omega-3 and omega-6 polyunsaturated fatty acids: dietary sources, metabolism, and significance — a review. Life Sci. 203, 255–267. https://doi.org/10.1016/j.lfs.2018.04.049.

Schafer, A., Koppe, W., Meyer-Burgdorff, K.H., Gunther, K.D., 2009. Effects of dietary phosphorus and phytase levels on performance and phosphorus excretion in rainbow trout (*Oncorhynchus mykiss*). Aquaculture 248, 231–240.

Schöne, F., Jahreis, G., Richter, G., Lange, R., 1997. Evaluation of rapeseed meals in pig nutrition. Anim. Feed Sci. Technol. 61, 87–99.

Serra, V., Pastorelli, G., Eurosia Angela Tedesco, D., Turin, L., Guerrini, A., 2024. Alternative protein sources in aquafeed: current scenario and future perspectives. Vet. Anim. Sci. 25, 100381. https://doi.org/10.1016/j.vas.2024.100381.

Siddhuraju, P., Becker, K., 2001. Preliminary nutritional evaluation of Mucuna seed meal (Mucuna pruriens var. utilis) in common carp (*Cyprinus carpio* L.): an assessment by growth performance and feed utilisation. Aquaculture 196 (1–2), 105–123. https://doi.org/10.1016/s0044-8486(00)00577-9.

Singh, R.P., Oswalt, D.L., 1995. Groundnut processing and utilization in the tropics.

Sugiura, S.H., Gabaudan, J., Dong, F.M., Hardy, R.W., 2001. Dietary microbial phytase supplementation and the utilization of phosphorus, trace minerals and protein by rainbow trout [*Oncorhynchus mykiss* (Walbaum)] fed soybean meal-based diets. Aquac. Res. 32 (7), 583–592. https://doi.org/10.1046/j.1365-2109.2001.00581.x.

Tacon, A.G.J., Metian, M., 2015. Feed matters: satisfying the feed demand of aquaculture. Rev. Fish. Sci. Aquacult. 23 (1), 1–10. https://doi.org/10.1080/23308249.2014.987209.

Tignani, M.V., Santolini, E., Secci, G., Bovo, M., Parisi, G., Barbaresi, A., 2024. Assessing environmental sustainability of substitute feeding formulas for gilthead seabream (*Sparus aurata*) using life cycle assessment. Sci. Total Environ. 954, 176689. https://doi.org/10.1016/j.scitotenv.2024.176689.

Torstensen, B.E., Espe, M., Sanden, M., Stubhaug, I., Waagbø, R., Hemre, G.-I., Fontanillas, R., Nordgarden, U., Hevrøy, E.M., Olsvik, P., Berntssen, M.H.G., 2008. Novel production of Atlantic salmon (*Salmo salar*) protein based on combined replacement of fish meal and fish oil with plant meal and vegetable oil blends. Aquaculture 285 (1–4), 193–200. https://doi.org/10.1016/j.aquaculture.2008.08.025.

Tripathi, M.K., Mishra, A.S., 2007. Glucosinolates in animal nutrition: a review. Anim. Feed Sci. Technol. 132 (1–2), 1–27. https://doi.org/10.1016/j.anifeedsci.2006.03.003.

Turchini, G.M., Torstensen, B.E., Ng, W.K., 2009. Fish oil replacement in finfish nutrition. Rev. Aquac. 1 (1), 10–57.

Tyapkova, O., Osen, R., Wagenstaller, M., Baier, B., Specht, F., Zacherl, C., 2016. Replacing fishmeal with oilseed cakes in fish feed—a study on the influence of processing parameters on the extrusion behavior and quality properties of the feed pellets. J. Food Eng. 191, 28–36. https://doi.org/10.1016/j.jfoodeng.2016.07.006.

Urán, P.A., Gonçalves, A.A., Taverne-Thiele, J.J., Schrama, J.W., Verreth, J.A., Rombout, J. H., 2008. Soybean meal-induced enteritis in common carp (*Cyprinus carpio*) and the effect of a probiotic. Aquaculture 283, 135–143.

Urayama, S., Manabe, T., Shirai, T., Ikenaga, M., 2021. Effect of plant-based protein inclusion on gut microbiota composition and intestinal health in aquaculture fish: a review. Fish Shellfish Immunol. 117, 228–237.

Zhang, C., Liu, Y.J., Tian, L.X., Yang, H.J., Liang, G.Y., Gao, W., 2007. Effects of dietary gossypol from cottonseed meal on growth performance and antioxidant enzymes activity in juvenile tilapia (*Oreochromis niloticus* × *O. aureus*). Aquac. Nutr. 13 (5), 335–340. https://doi.org/10.1111/j.1365-2095.2007.00486.x.

Zhou, Q.C., Tan, B.P., Mai, K.S., Liu, Y.J., 2006. Apparent digestibility of selected feed ingredients for juvenile cobia (*Rachycentron canadum*). Aquaculture 252, 491–498. https://doi.org/10.1016/j.aquaculture.2005.07.016.

Zhou, X.R., Li, J., Wan, X., Hua, W., Singh, S., 2019. Harnessing biotechnology for the development of new seed lipid traits in Brassica. Plant Cell Physiol. 60 (6), 1197–1204. https://doi.org/10.1093/pcp/pcz070.

Further reading

Glencross, B.D., Huyben, D., Schrama, J.W., 2020. The application of single-cell ingredients in aquaculture feeds—a review. Fishes 5 (3), 22.

Iliyasu, A., Mohamed, Z.A., 2016. Evaluating contextual factors affecting the technical efficiency of freshwater pond culture systems in Peninsular Malaysia: a two-stage DEA approach. Aquac. Rep. 3, 12–17.

Kaur, N., Sharma, S., Kaur, S., Nayyar, H., 2014. Selenium in agriculture: a nutrient or contaminant for crops? Arch. Agron. Soil Sci. 60 (12), 1593–1624.

CHAPTER

Utilization of grain and cereal by-products in fish feed formulation

4

Konstantinos V. Kotsanopoulos, Athanasios Exadactylos and Petros V. Martsikalis
Department of Ichthyology and Aquatic, Environment, School of Agricultural Sciences, University of Thessaly, Volos, Greece

Abbreviations

AAAA	Apparent amino acid availability
ADCs	Apparent digestibility coefficients
BF	Barley flour
BPC	Barley protein concentrate
BSG	Brewer's spent grain
CF	Crude fiber
CGM	Corn gluten meal
CM	Corn meal
CP	Crude protein
CTRL	Control
DDG	Distiller's dried grains
DDGS	Distillers dried grains with soluble
FM	Fish meal
LAB	Lactic acid bacteria
SKW	Samoon and Khubz waste
SPC	Soy protein concentrate
SSF	Solid state fermented
VWG	Vital wheat gluten
WG	Wheat gluten
WGM	Wheat gluten meal
YPCPA	Yeast protein concentrate potable alcohol
YPCR	Yeast protein concentrate refined
YPCU	Yeast protein concentrate unrefined

1. Introduction

Aquaculture is a highly sustainable method of protein production for human consumption. The Food and Agriculture Organization (FAO) has predicted that cultivation of fish and other aquatic animals will account for approximately 53% of the global seafood demand by 2030. Nevertheless, increase of aquaculture production

Fish Feed Formulation with Agroindustrial Wastes and Byproducts. https://doi.org/10.1016/B978-0-443-41728-3.00003-4

requires feed materials and their abundance and availability are of crucial importance for this industry. As a result, the focus of the aquaculture industry should be at identifying renewable sources of feed (including by-products of plant origin). The rapid expansion of aquaculture will necessitate the use of technological advancements, optimal use of natural resources, efficient use of space, and use of alternative highly nutritious feed ingredients (Albrektsen et al., 2022). The rapid expansion of the aquaculture industry and its capability to provide more fish and crustaceans than the fishing industry has been dependent on the availability of feed. Over 70% of aquaculture production around the globe depends on the use of external feed supply. The production of the aquaculture industry can only keep increasing with the current growth rate if the availability of nutritious feed keeps increasing accordingly, taking into account that other industries will compete for the available feed resources (Tacon and Metian, 2015). Materials of plant origin are widely used in aquaculture, with vegetable oil being a good example. It is important to mention though that the production of vegetable oil and protein requires higher amounts of energy than the aquatic materials they replace, and therefore an increase in the use of vegetable-derived products leads to increases in the energy required to produce feed. Alternative feed sources, such as microbial oils (e.g., yeast oil obtained through fermenting lignocellulosic hydrolyzate), can potentially cover some of the plant-oil demand, but the energy requirements of such production processes need to be carefully assessed (Sigtryggsson et al., 2023). It is clear that the utilization of more food production by-products and residues as feed can increase the availability of feed across all dependent industries (Sandström et al., 2022). Cereals have been used by humans for thousands of years to produce food. The modern production of food from cereals leads to the generation of high quantities of by-products. The germ and bran are by-products of the dry and wet milling processes of the brewers' spent grain (BSG) or the breadmaking and starch production industries. Such by-products are considered to be an inexpensive source of nutrients including dietary fibers, proteins, carbohydrates, sugars, minerals, and antioxidants (Chattaraj et al., 2024a,b). Frequently, such valuable by-products are used as animal feed or fertilizers or in some cases are unfortunately not used at all. The rapid increase in human population though has unavoidably led to the exploration of more sustainable, environmentally friendly uses for these materials (Skendi et al., 2020). Carbohydrates and fats are essential nutrients for aquatic animals (Fagbenro, 1999). Apparent digestibility coefficients (ADCs) can be used for estimating the amount of nutrients in feed and can be employed to select cost-effective ingredients for fish feed (Guimarães et al., 2008).

2. Cereals by-products in aquaculture

Proteins or plant origin are a very promising alternative to fish meal (FM) and are increasingly used in aquaculture. Table 4.1 presents the benefits and characteristics of cereal and grain by-products when used to partially replace fishmeal. Vital Wheat Gluten (VWG) is an example of such a material. It is a by-product of wheat

Table 4.1 Benefits and characteristics of cereal and grain by-products when used to partially replace fishmeal.

Fish species	Grain	Grain by-product	Benefits	References
Juvenile *Cyprinus carpio*	Fermented wheat	Yeast protein concentrate refined (YPCR), yeast protein concentrate potable alcohol (YPCPA) at 300 g kg^{-1}	High mineral retention, growth performance, and feed efficiency.	Omar et al. (2021)
		Distillers dried grains with solubles (150 and 300 g kg^{-1})		
	Fermented wheat bran	Polysaccharide	Higher final body weight and weight gain, increased muscle protein and lipid content, reduced moisture content in flesh, increased Bifidobacterium population in intestine.	Wang et al. (2020)
C. carpio	Wheat	Bakery (Samoon and Khubz) waste	No significant effect on growth performance, feed efficiency with up to 755 replacements of common cereal fish meals; gradua and significant ($P \leq .05$) increase in the dry matter and carbohydrate digestibility.	Najim and Al-Tameemi (2023)
Carps (*Catla catla*, *Labeo rohita* and *Cirhinus mrigala*)	Corn	Corn gluten	Significant yield in comparison to control (especially at 20% replacement of fish meal), increased protein and lipid in fish flesh.	Karim and Shoaib (2018)
Oreochromis niloticus	Wheat	Bran fermented by *Lactobacillus* sp.	Fermented bran was higher in protein, carbohydrates, and antioxidants but lower in phytic acic. Significantly higher fish body weight, weight gain, feed conversion ratio, specific growth rate, protein efficiency ratio, protein and fat compared to control (nonfermented).	Flefil et al. (2022)
		Bran fermented by *Saccharomyces cerevisiae*	Fermentation reduced fiber level and phytic acid and increased the levels of protein and amino acids; weight gain, specific growth rate, protein efficiency ratio, and feed conversion ratio plus blood indices, and intestinal and liver architecture optimal at 20% fermented grain in diet.	Mohammady et al. (2023)

Continued

Table 4.1 Benefits and characteristics of cereal and grain by-products when used to partially replace fishmeal.—*cont'd*

Fish species	Grain	Grain by-product	Benefits	References
	Malted barley	Spent grains from the brewing industry after aerobic fermentation	20% substitution of the diet for malted barley is suitable for satisfactory biological parameters and Nile tilapia fillet quality.	Cabanillas-Gámez et al. (2024)
	Barley	Dietary maltose	Efficient and effective in increasing body mass at 35% maltose pellets.	Ighwela et al. (2015)
	Corn	Phenolic compounds from corn husk extracts	Linear increase in antioxidant capacity, growth performance and feed intake.	Galeana-López et al. (2020)
		Corn gluten meal	20%, 40%, or 60% replacement of fish meal with by-product significantly increases protein efficiency while lowering costs.	Khader et al. (2025)
Atlantic salmon (*Salmo salar*)	Wheat	Extruded crude whole wheat, preextruded whole wheat, preextruded after-meal	Significantly higher starch digestibility with 45% preextruded after-meal; reduction in dry matter digestibility with preextruded whole wheat and crude whole wheat.	Arnesen and Krogdahl (1993)
	Soya	Solvent-extracted soybean meal, dehulled and solvent-extracted soybean meal, dehulled full-fat soybean meal, soybean concentrate.	Lower nutritional value than fish meal (apart from soybean concentrate), but no significant impact and significant cost savings at low replacement levels.	Olli et al. (1994)
		Dehulled, solvent-extracted soybean meal	At 20% replacement level, equivalent growth performance to fish oil with no detrimental effects.	Olli et al. (1995)
European seabass (*Dicentrarchus labrax*)	Wheat	Wheat germ	Wheat germ replacement at 16% improved protein apparent digestibility coefficients and feed utilization.	Reis et al. (2019)
	Corn	High protein distiller's dried grains (by-product of ethanol production)	30% replacement of soybean meal did not affect growth and physiological factors and enhanced certain health indices.	Goda et al. (2019)
	Barley, Pilsen malt, and corn	Solid-state fermented (*Aspergillus ibericus*) brewer's spent grain	Enhanced the nutritional quality of brewer's spent grain.	Estevão-Rodrigues et al. (2024)

	Wheat	Wheat germ	Replacement of 300 g/kg of standard meal with wheat germ gave 100% apparent digestibility coefficient—good protein source. Replacement with okara meal and brewer's yeast by-products not effective.	Campos et al. (2018)
	Rainbow trout (*Oncorhynchus mykiss*)	Hydrolyzed gluten	Crude protein fishmeal replacement (0%, 12.5%, 25%, and 50%) with heat protein affected extrusion parameters and physical quality but had no nutritional value effects—potential cost reduction.	Storebakken et al. (2015)
	Wheat	Fermented wheat grains	Replacement (100 g kg^{-1}) of nonfermented wheat grains with fermented (fungal mycelia) improved juvenile trout growth and nutritional performance.	Pascual et al. (2018)
Caspian brown trout	Barley/wheat	Barley protein concentrate and wheat gluten meal	Fish meal replacement up to 50% showed no negative effect on growth performance.	Zaretabar et al. (2021)
Sea bream (*Sparus aurata*), rainbow trout (*Oncorhyncus mykiss*)	Barley	Brewers' spent yeast and spent grain	Digestibility coefficients: 71%–88%, effectiveness of diet with 20%–30% replacement of fish meal had no detrimental effect compared to 100% fishmeal—potential cost reduction, more sustainable sources.	Nazzaro et al. (2021)
Sea bream (*S. aurata*)			Replacement of fishmeal (up to 30%) gives similar results in growth, food conversion and fillet final composition to 100% fishmeal, protein digestibility of 89%–95% (cost reduction, sustainability benefits).	Estévez et al. (2021)
	Corn	Gluten meal	Replacement of fish meal protein (up to 60%) does not affect growth energy retention and general fish performance.	Pereira and Oliva-Teles (2003)
			Replacement of anchovy meal (up to 10%) did not affect growth performance or protein efficiency.	Yiğit et al. (2012)

processing and is collected after starch is extracted. The main types of proteins composing it are gliadins and glutenins, that are also responsible for its viscoelastic properties. VWG contains low amounts of lysine but is rich in sulfur amino acids, when compared to other similar ingredients. It also contains high levels of glutamine, which promotes gut health and modulates immunity. VWG has demonstrated to allow very high nitrogen digestibility as it lacks protease inhibitors and is produced via a quite lenient process. Therefore, it generally does not have a detrimental effect on the growth performance of fish, even when used at high levels (Apper-Bossard et al., 2013). Draganovic et al. (2011) evaluated the extrusion behavior of blends that contained proteinaceous materials of plant origin (e.g., wheat gluten (WG) and soy protein concentrate (SPC), when offered as FM, and studied the physical characteristics of the extrudates. Different protein levels were trialed ranging from 50 to 450 g kg^{-1}. The moisture levels trialed were 20, 26, and 32 g/100 g (wet basis). The use of SPC led to a significant increase in extruder-related factors. It was found that the moisture content heavily affected the extrudate density and oil absorption. Increases in the amounts of WG and SPC led to ingredients with higher breaking strength. It was finally determined that low moisture leads to products with better composition. However, maximum FM replacement can be achieved when moisture levels increase. Draganovic et al. (2011, 2013) studied how VWG contents affect the structural characteristics of extruded fish feed as well as how they are correlated to physical and functional traits. Gluten and SPCs were blended with their gluten contents ranging from 0 to 200 g kg^{-1}. Oil uptake was dramatically reduced when a blend containing 200 g gluten kg^{-1} was used. Gluten contents ranging from 100 to 200 g kg^{-1} did not allow for products of acceptable quality to be produced. All pellets (100%) with gluten contents of 0–50 g gluten kg^{-1} sank, but this percentage fell to 36% when the gluten content increased to 200 g gluten kg^{-1}. However, a gluten level of 200 g kg^{-1} leads to pallets with smoother, nonporous surfaces.

Also, in a recent study, Sigtryggsson et al. (2023) described the mass-energy requirements of the production of yeast oil via the conversion of wheat straw hydrolyzate, during the production of biomethane and energy. It was calculated that the production of 1 ton of yeast oil (37 GJ), together with 21.5 GJ of biomethane (430 kg) and 6 GJ of power, necessitate 9.2 tons of wheat straw, 14.7 GJ of fossil energy, 14.6 GJ of electricity, and 13.3 GJ in process heat. The fossil primary energy required for the production of yeast oil was found to be 10%–38% lower than the amounts required for the production of rapeseed oil, and yeast oil can therefore be a more sustainable feed material.

3. Common carp

The common carp (*Cyprinus carpio*) is one of the most commonly produced fish species. The production of this freshwater fish amounted to about 3.4% (4.4 million tons) of the global production of fish and seafood in 2015. The feed used for the production of carps is based on cereals and is therefore not dependent on the availability of FM. Across Europe, the production of this fish species in polycultures is not uncommon (Karnai and Szucs, 2018). In this respect, Marković et al. (2016) comparatively assessed the use of cereal grains and compound feed (pellets and extruded) on the production capability, fish meat qualitative characteristics, the histology of the fish digestive tract, quality of water, as well as any changes in biocenosis in ponds used to cultivate carps. It was shown that when cereal grains are used, the ponds could produce between 500 and 1500 kg/ha, which amounted for half the production given when compound feed was provided. The flesh of fish fed with cereal grains was found to contain less protein and moisture and more fat in comparison to fish fed with compound feed, while extruded feed led to the highest amounts of n-3 polyunsaturated fatty acids and a better n-3/n-6 ratio. The quality of the water did not differ significantly with any of the abovementioned feed types.

3.1 The use of wheat by-products in carp feed

Omar et al. (2021) conducted a study to assess the effectiveness of six isonitrogenous (380 g kg^{-1} crude protein [CP]) and isolipidic (80 g kg^{-1}) diets in the production of juvenile carp *C. carpio*. FM was used in this study as control (CTRL) diet and the wheat fermentation-derived alternatives that were tested were a) yeast protein concentrate unrefined (YPCU), b) yeast protein concentrate refined (YPCR), and c) yeast protein concentrate potable alcohol (YPCPA) at 300 g kg^{-1} of total dietary protein, originating from a novel wheat fermentation process. Two levels of distillers dried grains with soluble (DDGS) (protein content 150 and 300 g kg^{-1}) were assessed in a separate experiment. Following an 8-week feeding period, fish fed with YPCU exhibited the most significant ($P \leq .05$) growth. It was also shown that all experimental diets allowed for significantly higher P and Mg intake than the CTRL diet. Fish provided with YPCR30 or YPCPA30 and DDGS30 appeared to have significantly different numbers of hepatocytes, although no such differences were seen in the groups of fish fed YPCU30 and DDGS15 diets with the CTRL group. Examination of the liver of fish fed the CTRL, YPCU, and YPCR diets showed no pathological alterations. It was therefore concluded that fermented yeast protein concentrate of wheat origin and DDGS can be effectively used as feed to lower the cost of production and promote a sustainable product life cycle.

Another study was carried out to assess how feeding juvenile common carp with dietary polysaccharide, produced via fermentation of wheat bran, can affect its growth, muscle composition, digestive enzymes, and intestinal microorganisms. A total of four feed mixtures containing 0 (D0), 1.0 (D1), 2.0 (D2), and 4.0 (D3) g/kg fermented polysaccharide were fed to different fish groups for a period of 8 weeks. It was shown that the D3 diet led to the fastest growth when compared to the CTRL D0 group. The amylase activity in the intestinal track and hepatopancreas of the D1 and D2 groups as well as the activities of pepsin and trypsin in the hepatopancreas of D2 and D3 groups were considerably higher than those of the CTRL group. The flesh of fish assigned to group D1 appeared to be richer in protein and fat, while contained less moisture, while the fat levels of fish assigned to group D3 were significantly lower. D1 and D2 positively affected the populations of *Bifidobacterium* spp. in the intestinal tracks of the fish but had no effect on lactic acid bacteria and *Escherichia coli*. It was concluded that, generally, dietary fermented polysaccharide supplements can improve enzyme activities in the digestive track, enhance the populations of Bifidobacteria, and improve the growth rate of common carp (Wang et al., 2020).

3.2 The use of barley and corn by-products in carp feed

In a recent study, Najim and Al-Tameemi (2023) examined the effect of corn meal (CM) and barley flour (BF) replacement with bakery (Samoon and Khubz) waste (SKW) on growth rate, feed efficiency, and chemical composition of common carp. A total of four isonitrogenous ($31.06 \pm 1.38\%$ CP) and isocaloric (361.07 ± 2.48 kcal/kg gross energy) diets were trialed, replacing 0%, 25%, 50%, and 75% of CM and BF with SKW. Although no statistically significant difference with the CTRL (0% replacement) diet in growth rate was found, a significant ($P \leq .05$) increase in the dry matter and carbohydrate digestibility of the fish was observed when the replacement percentage of the experimental diets increased. It was therefore proved that this experimental diet can be used to replace up to 75% of common cereal feeds, without adversely affecting production (Najim and Al-Tameemi, 2023). Corn gluten has also been trialed as a supplement in the feed of juvenile common carp and its effects on fish growth and flesh composition were examined. The feeds were composed of 31% CP, GE: 3.1 kal gG^1 and replaced 150, 270, and 490 g kgG^1 FM, respectively. The fish were fed for a period of 60 days at which point a significant increase in growth, fat and protein efficacy were seen in the CTRL diet (no replacement with the experimental feed). No significant benefit was therefore found with the use of these experimental diets (Jahanbakhshi et al., 2012).

Corn gluten meal (CGM) is a protein very commonly used in feed mixtures, mainly due to its high digestibility. To trial its benefits, 25%, 35%, and 45% (as CGM I, CGM II, and CGM III) of corn gluten were used for replacing 80%, 50%, and 20% of FM and the mixtures were fed to carps. It was demonstrated that the corn gluten feeds led to very considerable yields of 88.14, 83.86, and 98.03 kg, respectively, but, CGM III with 20% replacement of FM maximized the yield. From a nutritional point of view, the use of corn gluten in the mixtures increased the available protein and lipid by limiting the amount of moisture and ash in the flesh. It was therefore concluded that plant-based diets can play a significant role in the aquaculture of Indian major carps (Karim and Shoaib, 2018).

4. Tilapia

Tilapia is a name commonly used to describe numerous species of cichlid fish that inhabit mainly freshwater ecosystems and more rarely brackish water. Being an invasive fish species in many areas, its cultivation is becoming more and more popular and is now the second most farmed fish globally. The origins of these fish species are Africa and Middle East but are now cultivated in more than 90 countries (Prabu et al., 2019). In the study of Guimarães et al. (2008), an effort was made to assess the ADCs of protein, energy, and dry matter, and the apparent amino acid availability (AAAA) in corn, wheat middlings, broken rice, rice bran, and sorghum in diets of Nile tilapia. Chromic oxide was employed as an external inert marker. The digestibility of energy and dry matter were the highest observed in the case of broken rice (95.34% and 96.45%), followed by sorghum (82.37% and 87.29%), corn (67.34% and 82.21%), rice bran (57.58% and 55.59%), and wheat middlings (48.94% and 45.88%). Protein digestibility for corn was the highest measured (72.86%), while the lowest was that of sorghum (56.77%). AAAA values appeared to peak in the case of corn (70.05%). All in all, broken rice appeared to have the best characteristics among these cereal products and by-products for use in Nile tilapia feed due to its high nutrient digestibility.

4.1 The use of wheat by-products as tilapia feed

Fermentation of animal feed, when carried out with the correct microorganisms, can be very promising for improving the nutritional characteristics of feed and reducing its cost as well as for maintaining better water quality standards in aquaculture facilities. The process has shown to increase the bioavailability of micronutrients in fish feed. Wheat bran was therefore trialed as feed for tilapia (*Oreochromis niloticus*) in a biofloc system following fermentation with *Lactobacillus* sp. *Lactobacillus rhamnosus* KU985435. Higher levels of total protein, carbohydrates, and antioxidants were found in the bran following fermentation; however, the process led to a decrease in the levels of phytic acid (from 4.2 down to 0.07%). A number of critical growth factors, such as final body weight, weight gain, feed conversion ratio, specific growth rate, and protein efficiency ratio, improved considerably in fish fed fermented feed in comparison to fish fed with the CTRL diet, while the CP content increased, and the crude fat decreased in the flesh of the fish (Flefil et al., 2022).

In the study of Mohammady et al. (2023), another attempt was described to improve the nutritional quality of wheat bran and reduce its fiber content through fermenting it with *Saccharomyces cerevisiae*. The effects of the process on crude fiber (CF), CP, and amino acids as well as on the growth of Nile tilapia (*O. niloticus*) (feed intake, growth, blood indices, and intestinal and liver histological attributes) were recorded. It was found that fermentation lowered the levels of fibers from 13.02% to 9.28% and phytic acid from 4.31 to 0.08 mg g^{-1}, increased CF to 22.35% and had a positive effect on amino acids. The diets that were trialed included 0% (CTRL), 15%, 20%, and 25% fermented wheat bran and were given to fish for a period of 70 days. Fish fed with a diet containing 20% wheat bran appeared to have optimal final body weight, the highest weight gain, specific growth rate, protein efficiency ratio, and feed conversion ratio. However, a number of blood indices seemed to be improving as the percentage of bran in the diets increased, including hemoglobin, hematocrit, mean corpuscular volume, mean corpuscular hemoglobin, mean corpuscular hemoglobin concentration, white blood cells, monocytes, lymphocytes, and neutrophils. All in all, the use of this cereal by-product at a percentage of 20% in the feed improved the growth, feed intake, blood indices, as well as intestinal and liver health.

4.2 The use of barley by-products as tilapia feed

In another study, spent grains (malted barley), a by-product of the brewing industry, were aerobically fermented and fed to tilapia fish as replacements of 0 (CTRL), 10%, 20%, 30%, and 40% of their diets. Although no statistically significant effects were seen on the growth indices of the fish and their flesh composition or quality, a significant impact was seen on the final biofloc chemical composition. NH_4^+, NO_2^-, and NO_3^- were found to be at the proper levels for tilapia cultivation in all cases. It was therefore suggested that a 20% replacement of a common tilapia diet for malted barley can still maintain the biological parameters and flesh quality of tilapia meat at acceptable levels (Cabanillas-Gámez et al., 2024).

Another study was carried out for determining the effects of barley-extracted maltose on the growth and feed utilization of Nile tilapia, *O. niloticus*. A total of five experimental diets including the original CTRL diet were prepared in the form of pellets containing 0, 20, 25, 30, or 35% maltose, and were fed to the fish twice a day at a rate of 5% of body weight. It was observed that the growth, body weight increase, and weight gain increased significantly more quickly in fish provided with 35% maltose in comparison to fish fed with the CTRL diet. The feed conversion ratio in fish given the 35% maltose diet was higher than in all other diets (20%, 25%, or 30% maltose). It was therefore suggested that partial replacement of common tilapia diet with maltose can be effectively used and can lead to an improved fish body-mass (Ighwela et al., 2015).

4.3 The use of corn by-products as tilapia feed

Corn stover is a corn by-product of the harvesting process that can be used to produce extracts containing high levels of phenolics and antioxidants, potentially benefiting the health of fish. A study was therefore carried out, at which following identification of such phenolic compounds with ultra-performance liquid chromatography-electrospray ionization-quadrupole time of fly-mass spectroscopy, they were incorporated into tipalia diet and studied for their antioxidant effects. A total of three diets were trialed, containing 0 (CTRL), 100, and 200 mg of phenolic compounds. 2,2-diphenyl-1-picrylhydrazyl and oxygen radical absorbance capacity assays were employed to determine their antioxidant potential. The fish were fed for 2 weeks, monitoring the performance of the feed for its intake and fish growth. Both growth rate and feed intake showed a linear increase ($P < .05$) with the introduction of the phenolic compounds. This study demonstrated that corn stover extracts can add significant antioxidants to tilapia feed, improving growth and feed intake, while no negative effects were observed (Galeana-López et al., 2020).

In another study, an attempt was made to replace some FM for tilapias with CGM and the effects of the experimental diets on growth performance and cost efficiency were measured. For a total of 12 weeks, fish groups were fed with 20% FM (CTRL) and diets with 20%, 40%, 60%, 80%, and 100% of FM protein substituted by CGM. The 80% and 100% replacement diets were clearly not effective as the growth indicators of the fish were reduced when compared to the other experimental diets and the CTRL and a significant increase in serum levels, thyroid agents, and triglycerides was found. However, substitution of 20, 40%, or 60% of FM with CGM allowed for a significant ($P < .001$) increase in protein efficiency measures. No significant difference was found though in the growth indicators (specific growth rate, feed conversion rate, feed efficiency, and survival rate) at any of the experimental diets in comparison to the CTRL. Since CGM is less expensive than FM, it was concluded that the incorporation of CGM to replace FM in percentages of up to 60% can lower the production costs without negatively affecting the fish and the total production (Khader et al., 2025).

5. Salmon

In the last 4 decades, a dramatic decline in wild salmon populations has been noted in the North Atlantic as well as wide areas of the northeastern Pacific, while in the same areas the production of farmed salmon has increased (Ford and Myers, 2008). Salmon farming has known significant success in the past few decades, and it is widely accepted that innovation and productivity growth are the principal factors affecting this trend (Asche et al., 2013).

5.1 The use of wheat by-products as salmon feed

In the study of Arnesen and Krogdahl (1993), groups of Atlantic salmon farmed in sea water for a total of 101 days were provided with extruded diets that contained 0, 15, 30, or 45% crude whole wheat, preextruded whole wheat or preextruded after-meal. After-meal is a by-product of wheat that contains high levels of protein. In the whole wheat diets, the weight that fish gained was significantly lower in the 45% diet in comparison to the CTRL, and although not of statistical significance, the other experimental diets in the series also tended to reduce growth rates. Condition factors and the composition of the flesh did not differ significantly across the diets and the CTRL. Plasma glucose, liver glycogen, and hepatosomatic indices, protein and lipid digestibility were not affected by any of the diets. Significantly higher starch digestibility though was recorded by using the diet that contained 45% preextruded after-meal in comparison to the diets that contained 45% preextruded whole wheat or 45% crude whole wheat. It was finally determined that addition of high levels of wheat led to a reduction in dry matter digestibility in the preextruded whole wheat series and in the crude whole wheat series.

5.2 The use of soya by-products as salmon feed

In a study conducted to assess alternative diets for Atlantic salmon, four soybean products (solvent-extracted soybean meal, dehulled and solvent-extracted soybean meal, dehulled full-fat soybean meal, and soybean concentrate) were trialed. A replacement of premium-quality FM by these materials at 0 (CTRL), 14%, 28%, 42%, and 56% as soybean protein of total protein were trailed. The nutritional value of the soybean concentrate was found to be very similar to that of FM. The other three soyabean materials that were trialed appeared to have a detrimental effect on fish performance that increased with increasing the replacement levels (Olli et al., 1994).

In another study, diets, that contained dehulled, solvent-extracted soybean meal at percentages of 0 (CTRL), 20% and 40% soybean protein in total protein, were trailed as alternative feed meals of Atlantic salmon, *Salmon salar* L. It was found that replacement levels of up to 20% allowed for an acceptable growth performance which did not differ significantly from that obtained when high-quality FM was used, while the fat content of the flesh was also very similar. The diet that contained 40% soya protein had a significant detrimental effect on fish growth and a much lower fat content than that of the CTRL group. All fish groups though had similar protein, calcium, carcass zinc, plasma calcium, magnesium, and potassium levels. An increase in plasma zinc and a decrease in plasma iron levels were seen by adding the soyabean meal in the feed while no change in liver iron was observed (Olli et al., 1995).

6. Sea bass

The European sea bass (*Dicentrarchus labrax* L.) is an organism of great economic and cultural significance for European countries. Although sea bass farming traces back to just 1980s, the production of this species from aquaculture has recently exceeded the amount of wild fish caught (aquaculture >96% of the amounts produced in 2016). The species has been subjected to selective breeding programs (Vandeputte et al., 2019) and is abundant in European waters (Pérez-Ruzafa and Marcos, 2014).

6.1 The use of wheat by-products as salmon feed

Wheat germ, a flour milling by-product, has been investigated in different studies for its potential use as a sustainable ingredient in aquaculture feed. In a study trialing three diets of seabass, replacement of 0 (CTRL), 75 and 155 g/kg fishmeal was tested for a total of 18 weeks, a determination of ADCs was made, and the growth performance was assessed. Histology testing of intestines and the determination of fatty acids of muscle tissues were carried out, and the plasma was also tested. An increase in dry matter, protein and phosphorous ADCs, as well as protein efficiency ratio were observed with increasing the levels of wheat germ in the diets. At 155 g/kg replacement levels these changes were significantly ($P < .05$) higher compared to the CTRL, but the feed conversion ratio was not as high as in the 75 g/kg replacement level or the CTRL. It was therefore proved that partial replacement of FM in diets of European sea bass has no detrimental effects on growth or flesh composition, while can significantly reduce feed costs (Reis et al., 2019).

One of the by-products of the production of ethanol that can be of value for the aquaculture industry is distiller's dried grains (DDG). As this material is high in protein it was used in a trial to partially (30%, 40%, and 50%) replace soybean meal of European sea bass, *D. labrax*. The experimental diets that resulted were given to the fish over a period of 8 weeks and their effects on growth and physiology of the fish were examined. It was shown that replacement levels higher than 30% significantly increased the growth rate, feed intake efficiency, general health, and feed conversion ratio of the fish. A significant increase in hematological and serum indicators, including hemoglobin, red and white blood cells, packed cell volume and humeral immune factors, was seen with increasing the amounts of DDG in the diet. It was therefore indicated that replacement levels higher than 30% do not affect fish growth of sea bass and can improve the fish general health (Goda et al., 2019).

In a recent study, Estevão-Rodrigues et al. (2024) evaluated the use of solid-state fermented (SSF) BSG feed on different digestibility and health factors of European sea bass. The fermentation was carried out by *Aspergillus ibericus*. A total of five diets were prepared with 18% crude lipids and 45% CP that included no BSG (CTRL), 10% nonfermented BSG (10BSG), and 20% nonfermented BSG (20BSG), 10% fermented BSG (10BSG-SSF), and 20% fermented BSG (20BSG-SSF). It was shown that the fermentation had a significant effect on the nutritional content of BSG, raising its protein, lipid, cellulose, hemicellulose, and lignin levels by 21%, 49%, 30%, 34%, and 7.3%, respectively. Significant increases were also seen in antioxidant activity and total phenolics. The use of 20% BSG-SSF led to a significant increase in the digestibility of dry matter ($P < .01$), in comparison to the unfermented 20BSG feed. Certain histomorphological changes observed with the use of the 20BSG diet did not appear when the material was fermented. It was therefore concluded that SSF could be used to enhance the nutritional value of cereal-based by products, offering at the same time a more sustainable source of feed (Estevão-Rodrigues et al., 2024).

An assessment of ADCs of agro-food by-products for their potential to be used as feed for European seabass (*D. labrax*) was made in the study of Campos et al. (2018). In the diets that were trialed, an amount of 300 g/kg of a common commercial meal was replaced by one of the following materials: wheat germ, okara meal, poultry by-product meal, steam hydrolyzed and enzyme-treated feather meal, beta-lactoglobulin, and peptide fractions >3000 Da originated from brewer's yeast and fish by-products. These agricultural by-products appeared to have very good protein and energy digestibility and could be used to partly replace traditional fishmeal without detrimental effects on digestibility. Since these products are generally more cost-effective, such diets could reduce the overall running cost of the aquaculture and promote sustainability (Campos et al., 2018).

7. Trout

7.1 The use of wheat by-products as trout feed

Agricultural by-products have also been trailed with promising results in the production of trout meat. In the study of Storebakken et al. (2015), for example, hydrolyzed wheat gluten (HWG) was used to replace 0 (CTRL), 12.5%, 25%, and 50% of CP from fishmeal. At 0%, 12.5%, and 25% replacement levels, the pellets extruded where of adequate durability and capable of sinking when about 20% of water was added during the extrusion, and these diets led to increases in die pressure and specific mechanical energy and a decrease in torque. These changes intensified with increasing the levels of HWG. Increases in diametric expansion, sinking rate, and lipid losses were noted with incorporating HWG in the diets, while pellets coated with lipid appeared not to be as water stable with higher HWG levels. The viscoelasticity of HWG meant that extrusion at 50% replacement could only be performed when the added water to the extruder was not exceeding 11%. The pellets of this diet had lower diametric expansion and could retain less oil, but if not coated, they were sufficiently durable. In comparison to pellets produced with more water, the pellets of this diet were more unstable in water, appeared to have a lower diametric expansion, and sank more quickly. The different diets were trialed in rainbow trout for 56 days. No different in growth was observed compared to the CTRL and the feed conversion rate was about 0.8 g dry matter intake per g gain in all cases. A significant increase in apparent digestibility of cyst(e)in and phenylalanine and a decrease in isoleucine were recorded when the levels of HWG were increased. These experimental diets therefore could be effectively used for partial replacement of FM when essential amino acids are also provided (Storebakken et al., 2015). The effects of wheat by-products as feed for rainbow trout were also investigated by Pascual et al. (2018). In their study, these researchers used fermented wheat grains to replace nonfermented ones, with fermentation being performed by fungal mycelia (*Pleurotus ostreatus* or *Lentinus edodes*). Several growth factors, feeding performance and flesh compositional characteristics were evaluated. In all cases, the diets were composed of approximately 43% protein and 19% lipids, and the experiment was carried out by replacing 100 g kg^{-1} of unfermented wheat grains with an equal amount of fermented grains. Feed was provided twice daily for a total of 56 days. It was apparent that *P. ostreatus* and *L. edodes* led to a significant increase in fish body weight from the 28th day onward. An increase in feeding efficiency by $64.5 \pm 0.7\%$, $92.5 \pm 0.5\%$, and $84.8 \pm 1.5\%$ was seen for the CTRL, *P. ostreatus* and *L. edodes*, respectively, while an improvement in nutrient retention by $30.0 \pm 0.5\%$, $27.7 \pm 1.1\%$, and $21.0 \pm 0.1\%$, in the case of CP; $40.3 \pm 0.6\%$, $31.0 \pm 1.8\%$, and $16.1 \pm 0.7\%$, in the case of ether extract; and $16.1 \pm 0.1\%$,

14.0 ± 0.3%, and 11.6 ± 0.6%, in the case of phosphorus were seen for *P. ostreatus*, *L. edodes*, and the CTRL, respectively. The highest lipid levels (g/kg) were found in the flesh of *P. ostreatus* (81.4 ± 1.4), while *L. edodes* (63.2 ± 2.5) and the CTRL (42.3 ± 2.6) followed. The levels of liver glycogen were considerably increased in the cases of *L. edodes* and *P. ostreatus* (0.62 ± 0.10 and 0.21 ± 0.08% liver weight) in comparison to CTRL (0.05 ± 0.01% liver weight). Therefore, wheat-mycelium-containing meals appeared to boost rainbow trout growth and flesh nutritional characteristics, although its impact depends on the mushroom species chosen in each case.

7.2 The use of barley by-products as trout feed

An important study evaluating the use of barley by-products as trout feed is that of Zaretabar et al. (2021). In this study, a percentage of FM was replaced by barley protein concentrate (BPC) and wheat gluten meal (WGM), and the effects of these experimental diets on growth, flesh composition, digestibility, the activities of digestive enzymes, amino acids, and fatty acids of Caspian brown trout were monitored. A total of five diets were trialed: CTRL (0 g/kg BPC), 25BPC (165 g/kg BPC), 50BPC (330 g/kg BPC), 75BPC (495 g/kg BPC), and 100BPC (660 g/kg BPC). The addition of WG in the meals rendered them isonitrogenous. Feeding lasted for 8 weeks and it was demonstrated that replacement of up to 50% of FM with BPC did not affect fish growth. Digestion of protein and energy was significantly better in fish provided with CTRL, 25BPC, and 50BPC diets. The addition of BCP in the meals were found to negatively affect the activities of trypsin, pepsin, and aminopeptidase in a linear fashion. Similar, not significantly different levels of C20:5n3, C22:6n3, and long-chain polyunsaturated fatty acids (LC-PUFA) were found in the flesh of fish provided with CTRL and 50BPC diets. It was thus recommended that meals are prepared to contain a maximum of 330 g/kg BPC plus 40 g/kg WGM (Zaretabar et al., 2021).

In another study, two by-products of breweries, yeast and spent grain, were used as a replacement of certain percentages of fishmeal. The alternative products were added to isoproteic and isolipidic meals and were fed to sea bream (*Sparus aurata*) and rainbow trout (*Oncorhyncus mykiss*). A determination of the ADCs of the meals was carried out over a 30-day experiment. The ADCs of the two materials ranged between 75% and 88% and 71% and 88% in the cases of rainbow trout and seabream, respectively. The data obtained via this trial indicated that replacement of 20%–30% fishmeal with brewers' spent yeast and spent grain did not negatively affect the nutritional characteristics of the flesh and protein, lipid, and amino acid digestibility for both fish species were acceptable. Considering the large scale of the production of these materials in the European continent, they could offer an opportunity for use of a more sustainable, cost-effective protein source (Nazzaro et al., 2021).

8. Seabream

The aquaculture of gilthead sea bream (*S. aurata*) has evolved into a major industry in the Mediterranean Sea. The production takes place mainly in marine cages and recirculating aquaculture units (Seginer, 2016). The species' aquaculture has grown significantly over the last decades with Arabaci et al. (2010) highlighting the commercial importance of the species and noticing that its production in Europe back in 2008 was as large as 128,943 t. Some of the most important studies examining the use of grain by-products as feed replacements for the production of this species are discussed below.

8.1 The use of barley by-products as seabream feed

BSG is the main by-product of the brewery industry. The material has a considerable potential for use in aquaculture as it is very cost-effective, available in large quantities throughout the year and contains relatively high levels protein (20%–30% DM), although unrefined BSG is not commonly added in aquafeeds due to containing high amounts of lignin, cellulose, and hemicellulose (Karlsen, 2022). Estévez et al. (2021) conducted two trials to partially replace fishmeal, in isoproteic and isolipidic diets of sea bream, with yeast and spent grain. The study demonstrated that replacement of a maximum of 30% fishmeal with brewers' spent yeast and 15% spent grain did not negatively affect the growth, food conversion and the nutritional characteristics of the fish. There is therefore potential to use these by-products in aquaculture to lower production costs while increasing the sustainability of both the aquaculture and brewery processes (Estévez et al., 2021).

8.2 The use of corn by-products as seabream feed

In a study examining the use of corn by-products as substitute of FM for seabream, a trial was run for 12 weeks to assess if CGM can be used to partially replace fishmeal. According to this experiment a CTRL diet containing 100% fishmeal and four alternative diets containing 20%, 40%, 60%, and 80% corn gluten, with the balance being FM, were prepared. Care was taken to ensure the diets were isonitrogenous and isoenergetic. Examination of the fish indicated that at the end of the 12-week period, only fish fed with 80% corn gluten did not grow as much as the ones fed with the CTRL diet, which was attributed to insufficient amino acids levels in that diet. It was also observed that an improvement in feed efficiency can be achieved by replacing FM protein with up to 60% corn gluten. Analysis of the flesh composition of the fish at the end of the experiment indicated that the whole-body water and lipid contents were significantly lower and higher, respectively, in fish fed with the diets that included 60% and 80% corn gluten protein, compared to the CTRL, while no other apparent differences in body composition were noted.

The ADCs of dry matter was higher in the new formulas compared to the CTRL. It was therefore suggested that replacement of FM with corn gluten at a maximum level of 60% does not negatively affect the growth or flesh composition of gilthead sea bream (Pereira and Oliva-Teles, 2003). A significantly lower level of corn gluten was suggested though in the study of Yiğit et al. (2012). In this study, four experimental isonitrogenous and isocaloric diets that contained 0%, 10%, 20%, and 30% CGM that replaced anchovy meal were assessed as feed for sea bream. The trial was run for a total of 45 days and at the end of it, it was revealed that substitution of up to 10% of CGM had no negative effect on growth performance compared to the CTRL diet. Replacement of 20% and 30% of FM though with corn gluten negatively affected growth performance, the use of feed utilization and protein efficiency in comparison to CTRL. It is therefore apparent that all other aspects of farming, including the method of aquaculture, region, and other variables, should be taken into account when deciding on the percentage of FM that can be replaced with crop by-products.

9. Conclusion

The expansion of the aquaculture industry and the significant increase in production that is needed to cover the ever-increasing need for animal protein is heavily dependent on the availability of sustainable feed materials. It is imperative that such materials should be readily available all year round and inexpensive. Most importantly though the type of materials and their levels introduced in fish feed should be carefully studied to ensure they do not negatively affect fish production and growth and or have a detrimental effect on the nutritional characteristics of the flesh.

Grain and cereal by-products are very good candidates, but it is clear that their introduction into fish feed should be carefully examined to determine the levels that do not negatively affect production or end-product quality. A number of factors should be considered in each case including fish species, grain or cereal product, permissible levels of fishmeal replaced, environmental factors, cost, availability and sustainability of materials, etc. The by-products that can be considered can include DDG, wheat bran, corn meal, barley flour (i.e., bakery waste), corn gluten, spent grain (malted barley), cereal-extracted sugars (e.g., maltose), corn stover, soyabean products, wheat germ, and others. Further processing of these by-products (e.g., fermentation) may be necessary in some cases to enhance the digestibility of the new diets and improve the performance of the feed. It is clear though that the potential partial replacement of traditional animal-based FMs with plant-based by-products can not only allow the availability of fish feed but also promote the use of these by-products and create a new source of income for the grain- and cereal-processing industries.

References

Albrektsen, S., Kortet, R., Skov, P.V., Ytteborg, E., Gitlesen, S., Kleinegris, D., Mydland, L.T., Hansen, J.Ø., Lock, E.J., Mørkøre, T., James, P., Wang, X., Whitaker, R.D., Vang, B., Hatlen, B., Daneshvar, E., Bhatnagar, A., Jensen, L.B., Øverland, M., 2022. Future feed resources in sustainable salmonid production: a review. Rev. Aquacult. 14 (4), 1790–1812. https://doi.org/10.1111/raq.12673.

Apper-Bossard, E., Feneuil, A., Wagner, A., Respondek, F., 2013. Use of vital wheat gluten in aquaculture feeds. Aquat. Biosyst. 9 (1), 21. https://doi.org/10.1186/2046-9063-9-21.

Arabaci, M., Yilmaz, Y., Ceyhun, S.B., Erdogan, O., Dorlay, H.G., Diler, I., Akhan, S., Kocabas, M., Ozdemir, K., Koyun, H., Koncagul, S., 2010. A review on population characteristics of Gilthead Seabream (*Sparus auratd*). J. Anim. Vet. Adv. 9 (6), 976–981. https://doi.org/10.3923/javaa.2010.976.981.

Arnesen, P., Krogdahl, Å., 1993. Crude and pre-extruded products of wheat as nutrient sources in extruded diets for Atlantic salmon (*Salmo salar, L*) grown in sea water. Aquaculture 118 (1–2), 105–117. https://doi.org/10.1016/0044-8486(93)90284-6.

Asche, F., Roll, K.H., Sandvold, H.N., Sørvig, A., Zhang, D., 2013. Salmon aquaculture: larger companies and increased production. Aquacult. Econ. Manag. 17 (3), 322–339. https://doi.org/10.1080/13657305.2013.812156.

Cabanillas-Gámez, M., Gómez-Cadena, J., Galaviz, M.A., de Oca, G.A.R.M., Figueroa-Saavedra, F., López, L.M., Trejo-Escamilla, I., Barreto-Curiel, F., López-Valencia, G., 2024. Use of fermented malted barley by-product as partial feed replacement and carbon source for rearing Nile tilapia (*Oreochromis niloticus*) juveniles. J. World Aquacult. Soc. 55 (3), e13062. https://doi.org/10.1111/jwas.13062.

Campos, I., Matos, E., Aragão, C., Pintado, M., Valente, L.M.P., 2018. Apparent digestibility coefficients of processed agro-food by-products in European seabass (*Dicentrarchus labrax*) juveniles. Aquac. Nutr. 24 (4), 1274–1286. https://doi.org/10.1111/anu.12665.

Chattaraj, S., Behera, B.K., Das Mohapatra, P.K., 2024a. Effective valorization of Chlorella biomass and Brewers spent grain substituting fish meal and soybean meal in the diet of herbivorous fish Cirrhinus reba for higher growth, digestibility and sustainable cultivation. J. Appl. Phycol. 36 (1), 217–232.

Chattaraj, S., Mitra, D., Chattaraj, M., Ganguly, A., Thatoi, H., Mohapatra, P.K.D., 2024b. Brewers' spent grain as fish feed ingredient: evaluation of bio-safety and analysis of its impact on gut bacteria of Cirrhinus reba by 16S Metagenomic sequencing. Curr. Res. Microb. Sci. 7, 100286.

Draganovic, V., van der Goot, A.J., Boom, R., Jonkers, J., 2011. Assessment of the effects of fish meal, wheat gluten, soy protein concentrate and feed moisture on extruder system parameters and the technical quality of fish feed. Anim. Feed. Sci. Technol. 165 (3–4), 238–250. https://doi.org/10.1016/j.anifeedsci.2011.03.004.

Draganovic, V., van der Goot, A.J., Boom, R., Jonkers, J., 2013. Wheat gluten in extruded fish feed: effects on morphology and on physical and functional properties. Aquac. Nutr. 19 (6), 845–859. https://doi.org/10.1111/anu.12029.

Estevão-Rodrigues, T., Fernandes, H., Moutinho, S., Filipe, D., Fontinha, F., Magalhães, R., Couto, A., Ferreira, M., Gamboa, M., Castro, C., Belo, I., Salgado, J., Oliva-Teles, A., Peres, H., 2024. Effect of solid-state fermentation of Brewer's spent grain on digestibility and digestive function of European seabass (*Dicentrarchus labrax*) juveniles. Anim. Feed. Sci. Technol. 315, 116018. https://doi.org/10.1016/j.anifeedsci.2024.116018.

Estévez, A., Padrell, L., Iñarra, B., Orive, M., San Martin, D., 2021. Brewery by-products (yeast and spent grain) as protein sources in gilthead seabream (*Sparus aurata*) feeds. Aquaculture 543, 736921. https://doi.org/10.1016/j.aquaculture.2021.736921.

Fagbenro, O., 1999. Apparent digestibility of various cereal grain by-products in common carp diets. Aquac. Int. 7 (4), 277–281. https://doi.org/10.1023/a:1009285214776.

Flefil, N.S., Ezzat, A., Aboseif, A.M., Negm El Dein, A., 2022. *Lactobacillus* fermented wheat bran, as an economic fish feed ingredient, enhanced dephytinization, micronutrients bioavailability, and tilapia performance in a biofloc system. Biocatal. Agric. Biotechnol. 45, 102521. https://doi.org/10.1016/j.bcab.2022.102521.

Ford, J.S., Myers, R.A., 2008. A global assessment of salmon aquaculture impacts on wilds almonids. PLoS Biol. 6 (2), 0411–0417. https://doi.org/10.1371/journal.pbio.0060033.

Galeana-López, J.A., Hernández, C., Leyva-López, N., Esmeralda Lizárraga-Velázquez, C., Sánchez-Gutiérrez, E.Y., Heredia, J.B., 2020. Corn husk extracts as an antioxidant additive in diets for Nile tilapia (*Oreochromis niloticus*) fingerlings: effect on growth performance, feed intake and toxicity. Biotecnia 22 (2), 147–154. https://doi.org/10.18633/biotecnia.v22i2.1256.

Goda, A.A.S., Srour, T.M., Omar, E., Mansour, A.T., Baromh, M.Z., Mohamed, S.A., El-Haroun, E., Davies, S.J., 2019. Appraisal of a high protein distiller's dried grain (DDG) in diets for European sea bass, *Dicentrarchus labrax* fingerlings on growth performance, haematological status and related gut histology. Aquac. Nutr. 25 (4), 808–816. https://doi.org/10.1111/anu.12898.

Guimarães, I.G., Pezzato, L.E., Barros, M.M., Tachibana, L., 2008. Nutrient digestibility of cereal grain products and by-products in extruded diets for Nile tilapia. J. World Aquacult. Soc. 39 (6), 781–789. https://doi.org/10.1111/j.1749-7345.2008.00214.x.

Ighwela, K.A., Ahmad, A.B., Abol-Munafi, A.B., 2015. Effect of dietary maltose on growth and feed utilization of Nile tilapia (*O. niloticus*) fingerlings. Res. J. Recent Sci. 4 (IVC-2015), 1–5.

Jahanbakhshi, A., Ramazi, F.G., Soudagar, M., 2012. Effects of dietary corn gluten supplementation on body composition and growth performance in common carp (*Cyprinus carpio*) juvenile. Glob. Vet. 9 (1), 85–88.

Karim, A., Shoaib, M., 2018. Influence of corn gluten meal on growth parameters and carcass composition of Indian major carps (*Catla catla*, *Labeo rohita* and *Cirhinus mrigala*). Turk. J. Fish Aquat. Sci. 19 (1), 1–6. https://doi.org/10.4194/1303-2712-v19_1_01.

Karlsen, F., 2022. Evaluation of Chemical and Bilogical Refinement Strategies for Transforming Brewer's Spent Grain Into a High-quality Protein Source for Fish Feed. DTU Aqua.

Karnai, L., Szucs, I., 2018. Outlooks and perspectives of the common carp production. Ann. Pol. Assoc. Agric. Aribus. Econ. XX (1), 64–72. https://doi.org/10.5604/01.3001.0011.7230.

Khader, M., Shehata, S., Ebrahim, M., Al-Marakby, K., Zaki, E., Naiel, M., 2025. Effect of replacement of fish meal by corn by product meal on growth performance for Nile tilapia (*Oreochromis niloticus*). Egypt J. Vet. Sci. 56 (2), 321–334. https://doi.org/10.21608/ejvs.2024.267728.1825.

Marković, Z., Stanković, M., Rašković, B., Dulić, Z., Živić, I., Poleksić, V., 2016. Comparative analysis of using cereal grains and compound feed in semi-intensive common carp pond production. Aquac. Int. 24 (6), 1699–1723. https://doi.org/10.1007/s10499-016-0076-z.

Mohammady, E.Y., Aboseif, A.M., Soaudy, M.R., Ramadan, E.A., Hassaan, M.S., 2023. Appraisal of fermented wheat bran by *Saccharomyces cerevisiae* on growth, feed utilization, blood indices, intestinal and liver histology of Nile tilapia, *Oreochromis niloticus.* Aquaculture 575. https://doi.org/10.1016/j.aquaculture.2023.739755.

Najim, S.M., Al-Tameemi, R.A., 2023. The evaluation of bakery waste as a replacement for corn meal and barley flour in the diets of the common carp (*Cyprinus carpio L.*) fingerlings. Egypt. J. Aquat. Biol. Fish. 27 (2), 709–721. https://doi.org/10.21608/ejabf.2023.297432.

Nazzaro, J., San Martin, D., Perez-Vendrell, A.M., Padrell, L., Iñarra, B., Orive, M., Estévez, A., 2021. Apparent digestibility coefficients of brewer's by-products used in feeds for rainbow trout (*Oncorhynchus mykiss*) and gilthead seabream (*Sparus aurata*). Aquaculture 530, 735796. https://doi.org/10.1016/j.aquaculture.2020.735796.

Olli, J.J., Krogdahl, Å., van den Ingh, T.S.G.A.M., Brattås, L.E., 1994. Nutritive value of four soybean products in diets for Atlantic salmon (*Salmo salar L.*). Acta. Agric. Scand. Sect. A Anim. Sci. 44 (1), 50–60. https://doi.org/10.1080/09064709409410181.

Olli, J.J., Krogdahl, Å., Våbenø, A., 1995. Dehulled solvent-extracted soybean meal as a protein source in diets for Atlantic salmon, *Salmo salar L.* Aquac. Res. 26 (3), 167–174. https://doi.org/10.1111/j.1365-2109.1995.tb00899.x.

Omar, S.S., Anwar, A.Y., El-Haroun, E.R., Davies, S.J., 2021. Evaluation of protein enriched co-products originating from wheat fermentation in diets of common carp *Cyprinus carpio* to examine effects on growth response, mineral retention, haematological status and intestinal integrity. Aquac. Nutr. 27 (5), 1336–1351. https://doi.org/10.1111/anu.13272.

Pascual, M.M., Hualde, J.P., Bianchi, V.A., Moreno, P., Castro, J.M., Luquet, C.M., 2018. Wheat grains fermented by fungal mycelia (*Pleurotus ostreatus* or *Lentinus edodes*) as alternative feed ingredients for juvenile rainbow trout (*Oncorhynchus mykiss*). Aquac. Int. 26 (6), 1339–1352. https://doi.org/10.1007/s10499-018-0286-7.

Pereira, T.G., Oliva-Teles, A., 2003. Evaluation of corn gluten meal as a protein source in diets for gilthead sea bream (*Sparus aurata L.*) juveniles. Aquac. Res. 34 (13), 1111–1117. https://doi.org/10.1046/j.1365-2109.2003.00909.x.

Pérez-Ruzafa, Á., Marcos, C., 2014. Ecology and distribution of *Dicentrarchus labrax* (Linnaeus 1758). In: Biology of European Sea Bass. CRC Press, pp. 3–33. https://doi.org/10.1201/b16043.

Prabu, E., Rajagopalsamy, C.B.T., Ahilan, B., Jegan Michael Andro Jeevagan, I., Renuhadevi, M., 2019. Tilapia – an excellent candidate species for world aquaculture: a review. Annu. Res. Rev. Biol. 31 (3), 1–14. https://doi.org/10.9734/arrb/2019/v31i330052.

Reis, B., Marques, A., Campos, I., Dias, E., Valente, L.M.P., 2019. Wheat germ as an alternative ingredient to a fair average quality fishmeal in diets for European seabass. Aquac. Nutr. 25 (4), 932–945. https://doi.org/10.1111/anu.12912.

Sandström, V., Chrysafi, A., Lamminen, M., Troell, M., Jalava, M., Piipponen, J., Siebert, S., van Hal, O., Virkki, V., Kummu, M., 2022. Food system by-products upcycled in livestock and aquaculture feeds can increase global food supply. Nat. Food 3 (9), 729–740. https://doi.org/10.1038/s43016-022-00589-6.

Seginer, I., 2016. Growth models of gilthead sea bream (*Sparus aurata L.*) for aquaculture: a review. Aquac. Eng. 70, 15–32. https://doi.org/10.1016/j.aquaeng.2015.12.001.

Sigtryggsson, C., Karlsson Potter, H., Passoth, V., Hansson, P.A., 2023. From straw to salmon: a technical design and energy balance for production of yeast oil for fish feed from wheat straw. Biotechnol. Biofuels. Bioprod. 16 (1), 140. https://doi.org/10.1186/s13068-023-02392-2.

Skendi, A., Zinoviadou, K.G., Papageorgiou, M., Rocha, J.M., 2020. Advances on the valorisation and functionalization of by products and wastes from cereal-based processing industry. Foods 9 (9), 1243. https://doi.org/10.3390/foods9091243.

Storebakken, T., Zhang, Y., Ma, J., Øverland, M., Mydland, L.T., Kraugerud, O.F., Apper, E., Feneuil, A., 2015. Feed technological and nutritional properties of hydrolyzed wheat gluten when used as a main source of protein in extruded diets for rainbow trout (*Oncorhynchus mykiss*). Aquaculture 448, 214–218. https://doi.org/10.1016/j.aquaculture.2015.05.029.

Tacon, A.G.J., Metian, M., 2015. Feed matters: satisfying the feed demand of aquaculture. Rev. Fish Sci. Aquacult. 23 (1), 1–10. https://doi.org/10.1080/23308249.2014.987209.

Vandeputte, M., Gagnaire, P.-A., Allal, F., 2019. The European sea bass: a key marine fish model in the wild and in aquaculture. Anim. Genet. 50 (3), 195–206. https://doi.org/10.1111/age.12779.

Wang, R.F., An, X.P., Wang, Y., Qi, J.W., Zhang, J., Liu, Y.H., Weng, M.Q., Yang, Y.P., Gao, A.Q., 2020. Effects of polysaccharide from fermented wheat bran on growth performance, muscle composition, digestive enzyme activities and intestinal microbiota in juvenile common carp. Aquac. Nutr. 26 (4), 1096–1107. https://doi.org/10.1111/anu.13067.

Yiğit, M., Bulut, M., Ergün, S., Güroy, D., Karga, M., Kesbiç, O.S., Yılmaz, S., Acar, Ü., Güroy, B., 2012. Utilization of corn gluten meal as a protein source in diets for gilthead sea bream (*Sparus aurata L.*) juveniles. J. FisheriesSciences.com 6 (1), 63–73. https://doi.org/10.3153/jfscom.2012008.

Zaretabar, A., Ouraji, H., Abedian Kenari, A., Yeganeh, S., Esmaeili, N., Keramat Amirkolaee, A., 2021. One step toward aquaculture sustainability of a carnivorous species: fish meal replacement with barley protein concentrate plus wheat gluten meal in Caspian brown trout (*Salmo trutta caspius*). Aquac. Rep. 20. https://doi.org/10.1016/j.aqrep.2021.100714.

CHAPTER

5 Rendered industry by-products and insect products for the development of aquafeeds

Ayusman Behera, Sukadev Das and Gargee Mohanty

P.G. Department of Zoology, Maharaja Sriram Chandra Bhanja Deo University, Baripada, Odisha, India

1. Introduction

Over the past few decades, the global aquaculture industry has witnessed a golden period due to a surge in the demand for seafood and the decline of wild fish stocks. According to the Food and Agriculture Organization (The State of World, 2022), the aquaculture sector is providing over half of the fish consumed worldwide, making it a key element of global food security. As the sector widens, the necessity for high-quality aquafeed constituents, more specifically the protein and lipid sources, has increased. Traditionally, fishmeal and fish oil have been used as main ingredients in the formulation of aquafeed because of their nutritional profile and balanced amino acid and fatty acid profile (Chattaraj et al., 2024a, 2026a). However, currently the aquafeed industry is encountering problems in getting the marine-derived ingredients because of overfishing and fluctuations in stocks of wild fish and edible aquatic species. Such components are also causing environmental concern. Thus, the present challenges emphasize the urgent need for sustainable and cost-effective protein and fat sources for formulating aquafeeds.

2. Need for an alternative protein and fat source in aquafeed

Aquafeeds are formulated for meeting the nutritional needs of various cultured and farmed aquatic species that include finfish, crustaceans, shrimps, prawns, molluscs, oysters, snails, lobsters, etc. Among all nutrients, proteins and lipids are the main ingredients in the aquafeed, as these nutrients directly make the edible part of the species. Protein is the most critical nutrient among all, as it contributes to the growth, maintenance, and reproduction of aquatic animals (Tacon and Metian, 2008). Similarly, fish oil—a co-product of fishmeal production—has also declined and is now a challenge to obtain for producing aquafeed. Fish oil is a valuable source of long-

Fish Feed Formulation with Agroindustrial Wastes and Byproducts. https://doi.org/10.1016/B978-0-443-41728-3.00013-7

chain omega-3 polyunsaturated fatty acids, particularly EPA (eicosapentaenoic acid) and DHA (docosahexaenoic acid), when derived from marine fish. These fatty acids are important for cardiovascular health in humans. In aquaculture, fish oil remains an essential lipid source, particularly for carnivorous species.

Conventionally, fishmeal and fish oil have been the preferred protein and lipid sources in aquafeed production because of their high digestibility, balanced amino acid and fatty acid profile, and palatability (Naylor et al., 2021). However, the use and production of fishmeal and fish oil have stagnated at approximately 4–6 million tons per year due to their economic and ecological limitations (Stadtlander, 2021). Furthermore, both fishmeal and fish oil are expensive and increasingly subject to price instability linked to the availability of wild fish and underutilized fish stocks. Hence, searching for alternate protein and lipid sources for aquafeed formulation and production without compromising feed quality and the health of aquatic species is the present-day call and challenge. The primary goals are thus to identify protein and lipid sources that are nutritionally adequate, economically viable, environmentally sustainable, and acceptable in aquafeeds for different aquatic species.

Currently, the alternative protein and lipids are sourced from terrestrial plants, animal by-products, and insect by-products. These include soybean meal and corn meal, poultry by-products such as bone and blood meal, feather meal, insects, and microbial biomass (Henry et al., 2015; Makkar and Ankers, 2014). Each of these options provides distinct advantages and limitations regarding their nutrient profile, digestibility, palatability, availability, and environmental footprints.

A number of market drivers are accelerating the shift toward the alternative protein and lipid options other than fishmeal and fish oil in formulating aquafeeds. These drivers reflect wider apprehensions related to sustainability, circular economy, and cost-effectiveness.

Among all three, sustainability has emerged as the foremost driver behind the development and incorporation of alternative protein and lipids in aquafeeds. Such alternatives will not only reduce the overexploitation of wild fish caught to produce fishmeal and fish oil in making aquafeed but will also protect the marine ecosystem and its biodiversity (Naylor et al., 2021; Chattaraj et al., 2025). Furthermore, aquaculture itself is encountering sustainability challenges which include nutrient pollution, emission of greenhouse gases, degradation of habitats, etc. In such situations, animal by-products and insect by-products and other organic sources which are considered to be the nonprofitable products in animal-based industries and insect-utilizing insects have been considered to be potent protein and lipid sources for feed formulation for aquatic farmed animals. These are potential sources for circular resource recycling due to their low water and land use (Henry et al., 2015).

In addition, efforts to integrate principles of circular bioeconomy where waste streams are valorized into valuable feed ingredients are getting accelerated. Rendering industry by-products such as meat, bone, and blood meals-derived protein and lipids, considered to be the organic waste of the respective industries, contribute to closing nutrient loops and refining resource efficiency (Tacon and Metian, 2015). Likewise, insect farming can convert organic waste, such as food scraps or manure, into nutrient-rich biomass. This new approach not only diverts waste from landfills but also produces sustainable protein and lipids with a low environmental footprint (Makkar and Ankers, 2014). Involving such ideas of circular bioeconomy in aquafeed production aids in achieving sustainable developmental goals by promoting responsible consumption, reducing environmental pollution, and contributing to a resilient food production system in the aquaculture sector.

Incorporating the animal industry by-products and insect waste also helps in cost-cutting in the production of aquafeed. The use of such ingredients helps to attain stable feed prices. For instance, rendered animal by-products and insect by-products generally cost less per unit of crude protein than the fishmeal and fish oil (Mau) (Lu et al., 2022). Putting such sources in aquafeed often leads to formulating and processing more nutritionally balanced feeds while managing the costs effectively, making it a profitable business in the aquaculture sector.

This chapter thus aims to explore the incorporation of such valuable waste in production, making a more nutritionally potent, profitable product, and making a promise to achieve the sustainable developmental goals.

3. The rendered—Products division in aquafeeds

3.1 Definition of rendered products and rendering process

The rendering process refers to the critical industrial process which involves the transformation of animal and insect by-products through thermal and mechanical applications into required products such as aquafeeds in our case. In this process, the animal by-products, such as blood, bone, skin, feathers, offal, and other nonedible or waste tissues, are turned into stable, protein- and fat-rich products to be further used in the formulation and manufacturing of aquafeeds.

The mechanical and thermal processes typically involve grinding the raw materials followed by cooking them at high temperatures between 115 and 145°C, thereafter separating fat, and drying the proteinaceous material to yield concentrated aquafeed (Meeker, 2006; Environmental, 2023). Further processing required the steps of stabilization and decontamination and detoxification of the products (if they are perishable products), effectively converting the waste and nonedible animal or insect products into valuable feed ingredients.

4. Role of the rendering industry in processing animal by-products

The rendering industry plays a two-fold role in managing environmental waste and enhancing economic efficiency. This is achieved through recycling of abattoir and meat processing waste; rendering diverts millions of tons of organic waste from landfills annually and significantly declines methane and other greenhouse gas emissions (Ogunji et al., 2020). In aquafeed manufacturing, rendered products have become increasingly attractive, as these are nutritionally and profitably replacements for fish meal and fish oil, whose availability has been presently restricted by overfishing and have ecological constraints (Naylor et al., 2021).

The core products derived from rendering animal by-products include blood and bone meal, meat, poultry by-products, and feather meal. Each product differs in terms of its nutritional profile, digestibility, functional characteristics, and utility in aquafeed formulations.

5. Nutrient composition of rendered products

Rendered animal by-products are rich in amino acids, small peptides, and even proteins, with essential nutrients making them excellent fish meal substitutes. Among these, meat and bone meals **(MBM)** are the commonly used rendered products in aquafeeds. The protein content of MBM typically oscillates between 48% and 52%, with a remarkable mineral content because of incorporated bone meal, especially calcium (Ca) and phosphorus (P) (Hu et al., 2008; Jobling, 2012; Tan et al., 2005). The MBM also provides 8%–12% fat and a high ash content. A good ash content is required for formulating a diet for animal feed, as low ash content may result in a supply of more minerals than the required amount. Likewise, the blood meal, which is procured from dried animal blood, consists of high protein amounts. The percentage of proteins present in the blood meal ranges between 80% and 90%. Protein from blood meal is generally found to be rich in lysine, making it a complementary ingredient in feeds where plant proteins may be a limiting factor in the supply of this amino acid (Makkar and Ankers, 2014). The use of cow blood meal in tilapia diets demonstrated high blood meal protein content and successful partial replacement of fishmeal (Ogunji et al., 2020). However, blood meal is found to have low fat and ash content and therefore needs to be amalgamated with other protein sources to obtain a balanced amino acid profile.

Similarly, poultry by-product meal (PBM) is derived from parts of poultry carcasses, organs, trimmings, and frames. However, feathers are excluded. PBM offers a more balanced nutritional profile. It contributes about 60% to 65% of animal protein, 10%–15% of fat, and 15% to 20% of ash content. It is particularly valued for its high palatability and digestibility, especially in carnivorous fish species (Glencross, 2016; Glencross et al., 2023; Simon et al., 2019).

Presently, feather meal obtained from the hydrolysis of feathers is also added in the preparation of aquafeed. Though simple thermal treatment or other mechanical techniques do not yield many amino acids or proteins, when these feathers are hydrolyzed, the keratin protein, which is present predominantly in feathers, yields around 80%–90% of crude proteins. Feather meal is notably high in sulfur-containing amino acids, particularly cysteine, although it remains limited in lysine and methionine (Grazziotin et al., 2006). A summarized nutritional comparison of these rendered meals is presented in Table 5.1.

Table 5.1 Nutritional composition of major rendered meals in aquafeeds.

Meal type	Crude protein (%)	Fat (%)	Ash (%)	Key nutritional traits	Source
Meat and bone meal (MBM)	48–52	8–12	30–35	High in Ca and P, moderate lysine	Jobling (2012); Environmental (2023)
Blood meal	80–90	<2	<10	Extremely high in lysine, low in minerals	Makkar and Ankers (2014); Ogunji et al. (2020)
Poultry by-product meal (PBM)	60–65	10–15	15–20	Balanced amino acid profile, palatable	Glencross et al. (2023)
Feather meal	80–90	<10	5–10	High in cysteine, needs hydrolysis	Grazziotin et al. (2006)

6. Functional properties of rendered products

Besides their nutritional contents, the rendered products also possess various functional properties that influence their utility in aquafeeds. A great amount of digestibility, palatability, and pellet-binding characteristics make them more likely to become feed for aquatic animals.

The digestibility property of any feed is a crucial factor, especially for aquaculture species, as these species usually require more feed and nutrients. Both blood meal and PBM typically show a higher rate of digestibility (nearly 85%–90%) when properly processed.

MBM, however, exhibits a comparatively lower digestibility rate as compared to other meals because of their higher ash content and variability. Both ash content and variability in the quality of MBM depend on the source of procurement and processing conditions (Tacon and Metian, 2008). Correspondingly, feather meal's digestibility and palatability depend largely on the process of hydrolysis and thereafter its processing. If the process of hydrolysis is done with proper handling and processing with precision, then the high-quality protein yield reaches up to 87%, though in some fish species, the digestibility problem is noted with feather protein (Henry et al., 2015).

Palatability of rendered meals is reported to be higher for carnivores and omnivorous fish species. Fish species such as seabass, rainbow trout, and catfish readily consume diets containing PMB or MBM as their source of nutrients. In some cases, it has been found that direct incorporation of MBM in aquafeed leaves a distinct aroma which is not readily accepted by some aquatic species (Ogunji et al., 2020). Similarly, feather meal impacts the palatability if not hydrolyzed properly.

Properties like pellet durability and water stability are also essential in aquaculture systems to minimize nutrient leaching and ensure feed intake before disintegration. Rendered meals, more specifically MBM and PBM, contain proteins like gelatine and collagen in them. These two proteins (collagen and gelatine) act as natural binders and play a key role in making pellets. The properties of cohesion as well as reduction in water disintegration make them good binders in feed formulation (Glencross et al., 2023). However, blood meal, due to its fine particulate structure and lack of binding capacity, is usually included in smaller proportions and combined with other ingredients for optimal pellet structure. Table 5.2 shows the functional and economic properties of rendered animal by-products in aquafeeds.

7. The rendered animal by-products in aquafeeds formulation

7.1 Utilization of different rendered animal by-products in diet formulation of key aquaculture species

Every species are different from each other, and hence their nutritional demands vary. Rendered animal by-products have been proven to be used in feed formulation effectively across a wide range of aquatic species. Their effective utilization and proportion of each variety of rendered animal by-product vary from species to species because of their differences in digestive physiology, nutritional demands, and feeding behavior. Several carnivorous fish species, such as *Salmo salar* (salmon), require highly digestible protein in their feed. Research findings showed that no adverse effect was observed in the health and performance of salmon when the diet of salmon was formulated with around 60% PBM in place of fishmeal (Glencross et al., 2023). In addition to the newly formulated meal, it was supplemented with methionine and lysine. Contrarywise, when blood meal and MBM were added to the formulated diet in place of fishmeal, the portion added was within the range of 10%–30% because of their potential issues with palatability and higher ash content.

In the case of *Oreochromis niloticus* (tilapia), it was found that being an omnivorous species, it is highly tolerable and adaptable to a variety of animal-based proteins. Hence, in formulating feed for tilapia, about 50% fishmeal could be able to get replaced with blood meal. The newly formulated feed was found beneficial for the species, as it showed no adverse effect and did not produce any negative impact on its growth pattern and performance (Kirimi et al., 2017). Correspondingly, inclusion of 20%–40% of MBM and PBM showed promising outcomes in the diet of tilapia, while inclusion of 25% hydrolyzed feather meal also showed positive effects on growth and feed intake by the same species (Poolsawat et al., 2021).

While experimenting with the feed formulation for *Clarias gariepinus*, it was found that when 50% of fishmeal was replaced with hydrolyzed feather meal proteins in combination with blood meal and PBM, no negative impact was observed in the growth and health of the catfish (Ogunji et al., 2008). MBM used at a rate of 30% in the feed for catfish fingerlings proved effective. Though the shrimps are more sensitive toward the dietary changes, they could tolerate up to 25%–30% inclusion of PMB in their diet in place of fishmeal.

Table 5.2 Functional and economic properties of rendered animal by-products in aquafeeds.

Parameter	Fishmeal	MBM	Blood meal	PBM	Feather meal (hydrolyzed)	Source
Digestibility (%)	90–95	80–85	85–90	85–90	85–87	Glencross et al. (2023)
Palatability	Excellent	Good	Variable	High	Moderate	Henry et al. (2015)
Pellet binding	Excellent	Good	Poor	Good	Limited	Tessenderlo (2023)
Environmental impact	High	Low	Low	Low	Low	Environmental (2023; Ogunji et al. (2020)
Cost per kg protein (USD)	3.5–4.0	1.5–2.0	1.0–1.5	1.8–2.2	1.0–1.3	Future (2024)

8. Impact of rendered animal by-products in aquafeed on the growth performance and feed conversion ratios

Formulated aquafeeds with rendered animal by-products, when properly processed, and nutritionally balanced, can maintain or enhance growth rates and feed conversion ratios (FCRs). In formulating a diet for salmon, the inclusion of PBM and BM resulted in getting FCRs of 1.1–1.2, which was comparable to a fishmeal-l-formulated diet, particularly when amino acid profiles were corrected (Glencross, 2016). Correspondingly, the FCRs of the tilapia diet consisting of 30%–40% blood meal substitution were found to range between 1.4 and 1.6 and showed similar weight gains to fishmeal-fed groups.

Diets of catfish formulated with hydrolyzed feather meal in combination with PBM and BM were found to have FCRs ranging between 1.5 and 1.8, exhibiting the utilization of protein hydrolysates efficiently by the fish species (OGUNJI et al., 2008). In the case of shrimps, formulated diets incorporating 20%–25% of PBM exhibited satisfactory growth and FCR values when essential amino acids were supplemented in the required amount (Henry et al., 2015).

9. Palatability tests and adaptations

Rendered meals, though, offer excellent nutritional properties, yet the palatability often limits their inclusion in feed specific to a strong undesirable odor that occurs because of lysine content from the animal by-products. This odor thus reduces the intake of diet by fastidious species like salmon and sea bass. Similarly, improper hydrolysis of feathers impacts the feed texture and acceptance by the aquatic animals.

In order to combat such limitations in feed formulation, feed manufacturers make several modifications. They carry out all types of hydrolysis (acid, alkali, and enzyme) and then process for their neat isolation and utilization in feed making to enhance properties of palatability and digestibility. Furthermore, different soothing flavors like krill meal and synthetic amino acid blends have been found to get blended with the diets for their ready acceptance by the aquatic species. Palatability and moisture content of feed are also increased through coating the formulated feed with fish oil fats. In species such as rainbow trout and European seabass, inclusion of less than 15% blood meal along with flavors such as krill extract showed a significant rise in the intake of feed by these species (Glencross et al., 2023).

10. Blending of both animal- and plant-based protein in aquafeed formulation

Rendered proteins obtained from animal by-products, when blended with plant-based ingredients, allow for a balanced diet that meets the nutritional requirements of the aquatic animal and imparts fewer negative effects on their growth or performance as compared to feed formulated with single ingredients. A good formulation needs a good strategy to be adapted for aquafeed production. A typical formulation strategy might include 15%–20% PMB to supplement lysine requirements, 5%–10% used feather meal with sulfur-containing amino acids like cysteine, and the remainder of the protein matrix can be fulfilled by adding plant-based proteins such as soybean meal, corn gluten meal, and synthetic amino acids such as threonine and methionine. Such a type of formulation may support more than one requirement in aquafeeds. On one hand, the formulated feed with protein >85% digestion ability with an enhanced nitrogen retention feature with a cost reduction up to 30% compared to fishmeal-laden formulated feed. In addition, this will also help in safeguarding variability in raw material quality, which is especially valuable in multi-phase feeding systems, where nutrient demands vary significantly across developmental stages.

The blending of rendered animal by-products along with plant-based products serves as a sustainable, cost-effective source for aquafeed. Such a formulation also minimizes the over-reliance on any one resource to be used as a protein source for aquafeed and makes use of underutilized proteins from the rendering and processing industries. This holistic approach to feed formulation using the rendered industrial animal by-products aligns well with the sustainable aquaculture practices and the principles of a circular bioeconomy.

11. Limitations and challenges of using rendered animal by-products in aquafeed

In spite of the lot of nutritional and economic advantages of using rendered animal by-products in aquafeeds, several limitations and challenges must be talked about to ensure their safe and effective utilization in feed formulation and replacing fishmeal. The challenges being put forward in the inclusion and use of rendered animal by-products range from regulatory and safety concerns to processing inconsistencies and palatability and odor issues. Each one of these factors can influence the overall viability of rendered ingredients in the commercial manufacture of aquafeed.

12. Regulatory limits

One of the foremost barriers to the extensive use of rendered animal by-products in aquafeeds is the regulatory limits, more specifically in the region of the European Union. In this regard, one stringent regulation has been imposed on the use of processed animal proteins (PAP) in animal feed to avoid the contamination of animal feed with transmissible spongiform encephalopathies (TSEs), particularly bovine spongiform encephalopathy. For many years, the European Union imposed a complete ban on the use of ruminant-derived PAPs in feed for all food-producing animals. While these rules have been revised to allow the use of nonruminant PAPs (processed proteins obtained from poultry and pigs) in aquafeeds under strict traceability and processing controls, regulatory sprints still continue (European Commission, 2021).

The rules and regulations on the use of nonconventional animal proteins in feed formulation vary significantly in their regulatory frameworks, which complicates the international trade of rendered ingredients. In countries where rules and regulations for the incorporation of animal proteins in animal feed are less developed or absent, concerns over product quality, traceability, and consumer confidence can arise. Consequently, harmonizing regulations and promoting scientifically informed policies and regulations are essential for incorporating and expanding the use of rendered animal by-products globally.

13. Pathogen and contaminant risks

Another major challenge and limit in the use of rendered animal by-products is the potential presence of pathogens and chemical toxicants. Although modern rendering processes involve high temperature and pressure, which helps in the elimination of most microbial risks, the quality of raw material used and their handling prior to processing can still introduce biological risks. Improper storage and transport of animal tissues may lead to contamination with *Salmonella* spp., *Clostridium* spp., or other pathogenic strains of *Escherichia coli* (Henry et al., 2022).

Chemical toxicants such as heavy metals, pesticide residues, and dioxins could be able to make their way to the aquafeed through the PAPs, especially when the raw materials are procured from the industrial zones or poorly regulated facilities. These contaminants not only pose risks to animal health but also play a major role in bioaccumulation and biomagnification phenomena and ultimately affect human consumers through food chains and food webs. To lessen such problems and risks, proper and stringent quality assurance protocols must be adopted throughout the rendering and feed manufacturing processes and industries. Regulatory standards must be laid down, and quality control units for testing and monitoring the presence of microbial contamination and chemical residues in feed prepared with rendered animal proteins must be established. Certification systems like Hazard Analysis and Critical Control Points and ISO standards can play a crucial role in ensuring the safety of rendered ingredients used in aquaculture.

14. Challenges with sensory and acceptability

Aquafeeds generated from rendered meals do fulfil all the nutritional requirements of the species to which they are fed, yet the sensory characteristics of these feeds, which include the color, texture, and odor of the feed, affect the acceptability and palatability. Some high-value and sensitive aquaculture species, like salmon and trout, reject the feed made up of blood meal, as it carries a strong metallic odor and dark coloration. Similarly, mixing of improperly hydrolyzed feather meal results in gritty textures and reduced palatability of feed by aquatic species. To overcome such challenges, feeds are made palatable with the use of taste enhancers like krill meal, squid extracts, or proprietary flavoring agents. In some cases, lipid coatings are also put to mask the undesirable odors and to improve texture. As the sensory attributes of feed not only reduce the feed intake by the aquatic species but also adversely affect their feed conversion efficiency, especially if fish or shrimps reject pellets or consume them hesitantly.

15. Processing variability and nutritional inconsistency

One of the inherent challenges with rendered animal by-products is the variability in nutrient content due to differences in raw materials, rendering methods, and batch-to-batch consistency. For instance, MBM produced from a mix of bone and soft tissues will differ significantly in protein, ash, calcium, and phosphorus content compared to MBM sourced primarily from muscle trimmings. Similarly, feather meal quality is directly influenced by the degree and method of hydrolysis applied (Grazziotin et al., 2006).

Such variability complicates formulation efforts for feed manufacturers. Inconsistent amino acid profiles, digestibility, and mineral content require precise formulation adjustments and sometimes limit the inclusion rate of rendered products in commercial feeds. This challenge becomes even more critical when formulating diets for species with narrow nutritional requirements or in life stages demanding precise nutrient delivery.

Standardizing rendering practices and applying real-time nutrient analysis technologies can help minimize variability. Moreover, the integration of feed formulation software and near-infrared spectroscopy (NIRS) tools can improve ingredient evaluation and batch standardization. These technologies, when combined with robust supplier quality programs, can ensure more consistent use of rendered products in aquafeeds.

16. Summary of challenges

To better illustrate the primary limitations associated with the use of rendered animal by-products in aquaculture, Table 5.3 provides a concise overview.

17. Common insect species used in manufacturing of aquafeed

The inclusion of insect-derived by-products, either in whole form or defatted larvae meal, insect oils, and frass, the residual substrate and excrement, has emerged as an

Table 5.3 Inclusion levels and performance metrics of rendered meals in key aquaculture species.

Species	Ingredient (s)	Max inclusion level (%)	Reported FCR	Source
Salmon (*Salmo salar*)	PBM	60	1.1–1.2	Glencross et al. (2023)
Tilapia (*O. niloticus*)	BM, MBM, FeM	50, 40, 25	1.4–1.6	Kirimi et al. (2017)
Catfish (*C. gariepinus*)	BM, PBM, FeM	50 combined	1.5–1.8	Ogunji et al. (2008)
Shrimp (*L. vannamei*)	PBM, BM	30, 10	~1.6	Henry et al. (2022)

Table 5.4 Summary of limitations and mitigation strategies for rendered products in aquafeeds.

Limitation	Description	Mitigation strategy
Regulatory restrictions	Regional bans or limitations on PAPs due to TSE concerns	Policy revision, product traceability, compliance with international standards
Pathogen and contaminant risks	Microbial contamination, heavy metals, and toxins	Strict HACCP protocols, raw material screening, quality certifications
Sensory and acceptability concerns	Unpleasant odour, taste, or appearance; consumer skepticism	Use of palatants, lipid coatings, consumer education, and transparency
Processing variability	Inconsistent nutrient profiles and digestibility across batches	NIRS testing, feed formulation software, sourcing from certified renderers

ecological and nutritionally compelling alternative. These insect by-products are rich in high-quality protein, balanced amino acids, essential lipids, and micronutrients. They thus help in producing biomass which aids in simultaneous valorization of organic waste streams. This action of these insects supports circular agriculture and sustainability (Table 5.4).

18. Black soldier fly (*Hermetia illucens*)

18.1 Lifecycle and rearing conditions

The black soldier fly (BSF) has a relatively short and competent lifecycle of approximately 45 days from egg to adult, thriving at temperatures between 24 and 36°C and humidity around 60% and 80% (Chia et al., 2020; Meneguz et al., 2018). Larvae undergo six instar stages before pupating and emerging as adults. This short lifecycle allows quick biomass turnover, which is an essential property for scalable feed production.

19. Substrates and nutritional profile

BSF larvae can feed upon any kind of organic matter. They efficiently process organic side-streams, including food waste, agricultural by-products, and organic manure (Chia et al., 2020; Lalander et al., 2019; Meneguz et al., 2018). The macronutrient composition of the substrate influences the nutritional profile of the insect larvae meal. The type of protein, fat, carbohydrates, and fiber found in the diet of the insect larvae strongly influences both larval growth and nutrient composition of the larvae and adult (Chia et al., 2020; Meneguz et al., 2018). Protein levels in larvae range between 25% and 60% dry matter, while fat content varies widely from 7% to 58%, heavily influenced by diet (Chia et al., 2020; Oonincx et al., 2015).

BSF naturally accumulates medium-chain fatty acids such as lauric acid (C12:0), which provide antimicrobial properties and support gut health in fish (Lalander et al., 2019; Oonincx et al., 2015). By modifying substrates such as mixing fish waste or algae, it has been reported that the larvae are enriched with long-chain omega-3 fatty acids (EPA, DHA) by up to 40% (Rodrigues et al., 2022; Chattaraj et al., 2026b), significantly improving their value to be used for manufacturing marine aquafeeds.

20. Production scale and efficiency

BSF larvae outpace many livestock species with low FCRs. The FCRs of this insect larvae range from 1.4 to 2.6 and have an efficient nutrient conversion (Oonincx et al., 2015). Industrial scalability is feasible with smart rearing systems equipped with environmental sensors and automated aeration (e.g., "smart bin" systems) that have demonstrated adult-equivalent yields to conventional methods with lower labor and optimized conditions (Avila et al., 2022).

Bioaccumulation studies reveal that larvae bear more nutrients relative to their substrates, with average crude protein around 40.8% and fat around 28.6% on a dry basis, although these levels can increase further in defatted meals (>60% protein) through mechanical or solvent extraction processes (Chia et al., 2020; Oonincx et al., 2015).

21. By-products and circular benefits

BSF frass contains minerals like nitrogen, potassium, and phosphorus, chitin, and residual nutrients, offering dual use as a soil amendment and plant fertilizer. Their larvae reduce waste volumes by up to 50%, promoting their role in circular waste-to-feed systems (Chia et al., 2020; Meneguz et al., 2018).

22. Mealworms (*Tenebrio molitor*)

22.1 Lifecycle and rearing

Tenebrio molitor, commonly called mealworms, complete their life cycle in roughly 100–150 days under temperatures of 25–30°C, with moderate humidity (van Broekhoven et al., 2015). They pass through 15–20 larval instars. They grow well on grain-based substrates and are easily cultured in bins. Such natural adjustments make them ideal for modular and midscale production (Park et al., 2014; van Broekhoven et al., 2015).

23. Efficiency and diet

Mealworms claim to have a robust feed-to-growth efficiency. It produces about 50% protein and 20%–30% lipids when fed cereal by-products. A balanced protein-to-carbohydrate ratio (approximately 1:1) is critical, with at least 15% protein ensuring optimal development. A dietary lipid content of more than 3% has often proven to be detrimental (Davis, 1978; van Broekhoven et al., 2015). They have a fatty acid composition rich in oleic and palmitic acids and are relatively consistent.

24. Aquafeed applications

Mealworms offer a high percentage of essential amino acid content (e.g., lysine, methionine) comparable to fishmeal. Research reports on such aquaculture trials reveal successful substitution of these worm meals up to 50%–100% without any adverse effects on fish growth or feed efficiency (van Broekhoven et al., 2015). European Union approvals and consistent results are accelerating their adoption.

25. Housefly larvae (*Musca domestica*)

25.1 Lifecycle and rapid rearing

Houseflies (*Musca domestica*) complete their development from egg to prepupa stage within 5–7 days and so offer the fastest cycle of providing insect protein sources. They rapidly consume organic wastes such as cattle manure, kitchen waste, and food scraps, making them ideal candidates for producing low-cost insect proteins in regions where conducive environmental conditions prevail in rapid turnover.

26. Nutritional value

The nutritional profile of housefly larvae meal has been reported to have a high protein content ranging from 40% to 60% and lipids between 9% and 26%. In addition, chitin and antimicrobial peptide content in housefly larvae meal confer functional benefits and so support fish health and offer disease resistance.5

27. Aquaculture trials

A number of studies have been carried out on aquatic species, especially on tilapia and catfish. Studies have demonstrated comparable or improved growth performance when fishmeal is partially replaced by housefly larvae meal, emphasizing their potential as cost-effective feed ingredients in aquafeed markets.

28. Nutritional profile of insect by-products in aquafeed

Insect meal, as we have discussed above, is rich in nutrients and can be replaced with fish meal and fish oil. Insect meal offers high-quality protein, vital lipids, bioactive compounds, and chitin, all of which support and impart positively toward fish growth, immunity, and feed efficiency.

29. Protein and fat levels

Insect meals consistently provide major nutritional value, especially protein and fats. The larvae meal of the BSF typically contains about 47% of crude protein and 8%–36% of fat on a dry matter basis, whereas defatted BSF meal gives about 48% of protein (Saputra and Lee, 2023). These protein levels surpass the protein content of fishmeal, which is conventionally used for aquafeed formulation and production. Most remarkably, it has been observed that defatted insect meals, by removing the lipid fraction from them, deliver an even higher percentage of protein, amounting to about 60%–70% (Saputra and Lee, 2023). Though these meals contain a higher percentage of protein and fat, they are still highly digested by the aquatic species. Full-fat BSF meals show protein digestibility of 81%–97%, and defatted forms maintain a similar course as compared to or even better than that of fishmeal-containing feeds (Protein, 2020). Correspondingly, larvae meals of mealworms and houseflies provide around 50%–55% protein and 20%–30% fat content, making them suitable as complementary protein sources (Aragão et al., 2022). Similarly, housefly larvae meal contains 40%–60% and 9%–26% of protein and fat, respectively, with research finding validating their use as partial substitutes for fishmeal in species like tilapia and catfish (Riekkinen et al., 2022). Thus, reports suggest that insect meals offer excellent nutrient profiles which are comparable to fishmeal with the added management flexibility that comes from defatting or blending fats and lipids into other formulations adapted for aquafeeds.

30. Amino acids and bioactive lipid contents in insect by-products

Insect by-products or insect meals not only provide crude protein and lipids but also deliver valuable bioactive lipids and essential amino acids. Lauric acid (C12:O) is highly abundant in the larvae of the BSF. These larvae carry almost 59% of lauric acid among other saturated fatty acids (Alfiko et al., 2022; Rossi et al., 2025). Lauric acid has been reported to have antimicrobial activities that help in improving the gut health of fish, help in inhibiting pathogens, and regulate gut inflammation, making it a promising functional lipid in aquafeeds (Hsieh et al., 2024). Fish trials have shown that incorporating black soldier oil rich in lauric acid surges growth, rate of feed conversion, and immune resilience (Rossi et al., 2025). Apart from lauric acid, the insect meal of the BSF also contains essential amino acids like methionine and lysine. Methionine is present approximately at a rate of 2.1 g/16 g nitrogen, while lysine is around 6.6 g per nitrogen (Saputra and Lee, 2023).

Likewise, mealworms and housefly by-products contain almost similar levels of lauric acid and essential amino acids as the products of BSF carry (Luzardo-Álvarez et al., 2023). Experiments on fish feeding confirm that these amino acids are bioavailable and support healthy growth when insect meals are included in aqua feeds (Hsieh et al., 2024).

31. Chitin: nutritional challenge and immunostimulant

Insect by-products that are used for feed formulation also contain chitin, a polysaccharide found in the exoskeleton of insects. Chitin gives both positive and negative properties to the feed formulated with insect meal. Negative properties of chitin include its indigestible property. The indigestibility property of any feed leads to availability and absorption of nutrients because of unreceptive access to the enzymes. It was reported that when diets containing chitin between 4% and 10% were fed to rainbow trout, the rate of protein digestion, as well as the growth rate got down (Regan et al., 2025). Apart from this, nitrogen of chitin shows up in crude protein assays, sometimes inflating the protein content of the diet (Regan et al., 2025).

Chitin also bears positive properties, and this includes its functionality as an immunostimulant. Experimental studies have revealed that low-level inclusion (around 1%–2%) of chitin in feed can stimulate macrophages, enhance lysosome activity, and promote beneficial gut microbiota, which in turn increases pathogen resistance in fish and shrimp (Hasan et al., 2023; Regan et al., 2025). Studies on sea bass and sea bream showed increased gut microbial diversity and production of short-chain fatty acids in the presence of chitin (Regan et al., 2025). Aquaculture scientists are on their way to explore the ways to treat chitin with acid, base, or even enzymes like chitinase to process and defat it so that the digestibility of it can be enhanced, and this could provide immunity along with proper gut health (Hasan et al., 2023).

32. Effect of substrate on nutrient composition

Another most valuable property of insect feedstocks is their nutritional plasticity. Nutritional plasticity refers to the ability to tailor nutrient profiles through substrate choice.

Alternating the standard growth medium with different kinds of waste streams significantly imparts larvae composition. For BSFs, feeding substrates enriched with fish offal and microalgae enhances the synthesis of long-chain omega-3 fatty acids such as eicosapentaenoic acid (EPA) and docohexaenoic acid (DHA) by a surge of 40%, making them more suitable for marine species (Cruz-Romero et al., 2021; Chattaraj et al., 2024b). On the contrary, vegetable-based or high-fibre-based feeds produce lean larvae with a higher protein-to-lipid ratio (Riekkinen et al., 2022). Therefore, substrate choice enables producers to target nutritional thoughtfulness to address the nutrient needs of specific species. Like BSFs, mealworms also respond to the protein and fat content of their substrates on which they are reared. Larvae raised on grain and on brewery leftovers were found to synthesize around 54% protein, while larvae raised on a high-fat or fibrous diet exhibited slower growth and lower protein digestibility (Bordiean et al., 2022; Lu et al., 2022). Feed scientists have also correlated nutrient richness of substrate with increased micronutrient levels, such as vitamins and minerals in insect meals (Regan et al., 2025). Such substrate dependency nutrient control offers feed formulators a valuable tool for adjusting insect farming input, which may result in shaping the protein content and quality, fatty acid content and quality, and availability of micronutrients in feed to cater to the needs of specific requirements of aquatic species.

33. Challenges and limitations of insect by-products in aquafeed

33.1 Economic limitations

Insect by-products, though, provide nutritional and functional advantages; insect-based ingredients remain significantly more expensive as compared to conventional protein sources, posing a critical barrier to widespread adoption in aquaculture. In Europe, BSF meals cost approximately 3.50–7 Euros/kg, which is nearly double or triple the cost of fishmeal in this region (Mulazzani et al., 2021). The hike in price might be due to the fact that the current production of insect by-products is limited, though the number of insect farms has increased to a significant number. Approximately 3000 t per farm are supplying insect meal, but this figure is far below the quantities of meals needed to achieve cost-effective economies of scale (Rossi et al., 2025). So, unless production volumes expand exponentially and technology becomes more reliable and stronger, insect meals are likely to remain niche ingredients in high-value feeds rather than mainstream aquaculture.

34. Technical constraints

Converting live insects into high-quality, stable feed constituents fetches several technical complexities. Crucial and important steps of feed making, such as drying, defatting, and oil extraction, need to safeguard the properties of essential nutrients and functional compounds like lauric acid. Overheating often degrades proteins and bioactive compounds, undermining the very benefits of utilizing insect meals in aquafeed composition (Mau) (Lu et al., 2022). In addition, the presence of chitin in insect meal creates problems in digestibility and so requires enzyme supplementation or de-chitinization using chitinase enzyme treatment or through other chemical, physical, or mechanical methods, which again puts a burden on process expense and complexity (Hasan et al., 2023). Small- and medium-scale insect meal producers often lack access to optimized equipment, making it difficult to produce consistent, high-quality products critical for feed industry confidence.

35. Legislative and regulatory challenges

Regulatory frameworks to use insect meal as feed ingredients remain fragmented and thoughtful, further impeding its growth. In the European Union, only seven insect species are officially approved to be used in aquafeed production, and strict rules forbid feeding insects on catering, slaughterhouse, or manure.

Regulatory frameworks for insect feed ingredients remain fragmented and cautious, further hampering growth. In the EU, only seven insect species are authorized for use in the feed industry (The State, 2022). Asia and North America also lack coherent standards, resulting in fragmented market access and barriers to cross-border trades (Mau) (Lu et al., 2022). Such regulatory ambiguity discourages investment in large-scale infrastructure setup and limits outreach to producers who are deprived of clarity on acceptable substrates, safety protocols, standard operating protocols, and labeling requirements. Without unified guidelines and standards, scaling insect-based aquafeed formulation and production will continue to be an uphill battle.

36. Social acceptance and consumer acuity

Consumer acuity remarkably influences market acceptance of insect-fed fish and seafood. Research studies in Europe reveal that consumers often perceive that fish fed with insect-based diets have poor taste, bear potential health risks, and have general unfamiliarity. This is despite positive results obtained from sensory panel taste, which showed minimal or no detectable differences in taste or quality as compared to conventionally fed fish (Wendin and Nyberg, 2021). However, many consumers appreciate the beneficial factors for using insect meal, such as sustainability, but factors like neophobia, repulsion, and insufficient information on insect meal benefits reduced willingness to pay perks beyond conventional prices for aquafeed (Ferrer Llagostera et al., 2019). Research reports from Spain reveal that environmental messaging helps in boosting acceptance. However, problems with people on taste concern and inadequate information regarding the benefits of insect feed still persist (Ferrer Llagostera et al., 2019). Hence, successful strategies to improve perception include targeted education campaigns, proper and informative labelling, sustainability certifications from competent organizations and praising words for taste, which are the keys to social acceptance of insect-based aquafeed.

37. Pathways to overcoming challenges

To expose the full potential of insect by-products in aquafeed production, a multifaceted approach is of utmost need. Looking from an economic point of view, investment in large-scale procurement and production of insect meal and further in production of feed required establishment of automated production units, which further would help in driving down the prices by at least 30% by 2030, thinning the gap with fishmeal (Mulazzani et al., 2021). On the technical aspect, research input is required on lowering the cost of the de-chitinization process regarding standardized protocols for drying and defatting. Likewise, regulatory progress is equally vital to develop evidence-based, globally synchronized standards on substrate use, processing processes, and labeling information so as to surge the confidence and investment of consumers.

38. Sustainability and environmental profile of conventional and unconventional aquafeed ingredients

Life cycle assessments (LCAs) suggest having a strong framework to evaluate the environmental profile and efficiencies of different unconventional ingredients used for aquafeed formulation and production.

Fishmeal is the conventional ingredient in aquafeed, but it has its own environmental setbacks. The production of fishmeal typically results in between 7 and 10 kg of CO_2 equivalent per kilogram of protein while simultaneously contributing to an elevated pressure on wild fish stocks (Maiolo et al., 2020). However, industrial advancement, especially in sourcing more sustainable fish species and improvising feed conversion, has achieved almost 20% lower greenhouse gas emissions since 2018 (MacLeod et al., 2020). In spite of such achievements, dependency on wild-caught fish still remains a significant ecological issue.

Consider rendered animal by-products like poultry or MBM products and blood meal as ingredients of aquafeed; though unconventional, they offer a more sustainable alternative since they use waste from slaughterhouses rather than new agricultural produce. Reports reveal that rendered animal by-products often carry a lower carbon footprint as compared to fishmeal because of their indirect utilization of otherwise discarded biomass (Devendra and Sevilla, 2002; Sandström et al., 2022). In the case of rendered animal by-products, the use of land and water footprints is also minimal, as the resource claims for rendering are centralized and comparatively low when equated to primary agricultural products.

In the case of insect meal production, it has been found that not all insect production is equally sustainable. When larvae are reared on conventional feed substrates like soyabean or corn, or processed with fossil fuels, the amount of carbon emissions increases significantly, sometimes far exceeding the carbon footprint of fishmeal. Evidence of LCA studies claims that insect systems reared on high-value feed may outpace the carbon footprints of plant produce like soya and corn and even fish.

Therefore, production methods and energy sources for the inclusion of conventional and unconventional ingredients are crucial determinants in aquafeed formulation and production.

References

Alfiko, Y., Xie, D., Astuti, R.T., Wong, J., Wang, L., 2022. Insects as a feed ingredient for fish culture: status and trends. Aquac. Fish. 7 (2), 166–178. https://doi.org/10.1016/j.aaf.2021.10.004.

Aragão, C., Gonçalves, A.T., Costas, B., Azeredo, R., Xavier, M.J., Engrola, S., 2022. Alternative proteins for fish diets: implications beyond growth. Animals 12 (9). https://doi.org/10.3390/ani12091211.

Avila, K.U., Campbell, M., Mauck, K., Gebiola, M., Karydis, K., 2022. Development and testing of a smart bin toward automated rearing of black soldier fly larvae. In: IEEE International Conference on Automation Science and Engineering. IEEE Computer Society, United States, pp. 1238–1243. https://doi.org/10.1109/CASE49997.2022.9926430.

Bordiean, A., Krzyżaniak, M., Aljewicz, M., J. Stolarski, M., 2022. Influence of different diets on growth and nutritional composition of yellow mealworm. Foods 11 (19), 3075. https://doi.org/10.3390/foods11193075.

Chattaraj, S., Behera, B.K., Das Mohapatra, P.K., 2024b. Effective valorization of Chlorella biomass and Brewers spent grain substituting fish meal and soybean meal in the diet of herbivorous fish Cirrhinus reba for higher growth, digestibility and sustainable cultivation. J. Appl. Phycol. 36 (1), 217–232.

Chattaraj, S., Chattaraj, M., Mitra, D., Ganguly, A., Thatoi, H., Das Mohapatra, P.K., 2025. 16S amplicon sequencing of the gastrointestinal microbiota of Cirrhinus reba and isolation of an autochthonous probiotic using culture based approaches. Syst. Microbiol. Biomanufact. 5 (1), 156–170.

Chattaraj, S., Ganguly, A., Mitra, D., et al., 2026b. Sustainable valorization of algal biomass in fish nutrition and dietetics: a critical review and scientometric study. Waste Biomass Valor. https://doi.org/10.1007/s12649-026-03619-7.

Chattaraj, S., Mitra, D., Chattaraj, M., Ganguly, A., Thatoi, H., Mohapatra, P.K.D., 2024a. Brewers' spent grain as fish feed ingredient: evaluation of bio-safety and analysis of its impact on gut bacteria of Cirrhinus reba by 16S Metagenomic sequencing. Curr. Res. Microb. Sci. 7, 100286.

Chattaraj, S., Sahoo, S.S., Behera, S., Thatoi, H., 2026a. Sustainable aquafeed development using Spirulina: effects on growth performance of Catla catla. Algal Res. 104447.

Chia, S.Y., Tanga, C.M., Osuga, I.M., Cheseto, X., Ekesi, S., Dicke, M., van Loon, J.J.A., 2020. Nutritional composition of black soldier fly larvae feeding on agro-industrial by-products. Entomol. Exp. Appl. 168 (6–7), 472–481. https://doi.org/10.1111/eea.12940.

Cruz-Romero, M.C., O'flynn, C.C., Troy, D., Mullen, A.M., Kerry, J.P., 2021. The use of potassium chloride and tapioca starch to enhance the flavour and texture of phosphate- and sodium-reduced low fat breakfast sausages manufactured using high pressure-treated meat. Foods 11 (1), 17. https://doi.org/10.3390/FOODS11010017.

Davis, G.R.F., 1978. Growth response of larvae of *Tenebrio molitor* L. to concentrations of dietary amino acids. J. Stored. Prod. Res. 14 (1), 69–71. https://doi.org/10.1016/0022-474x(78)90022-x.

Devendra, C., Sevilla, C.C., 2002. Availability and use of feed resources in crop–animal systems in Asia. Agric. Syst. 71 (1–2), 59–73. https://doi.org/10.1016/s0308-521x(01)00036-1.

Environmental Sustainability of Rendering, 2023. Available from: https://nara.org/sustainability/environmental-sustainability/.

Ferrer Llagostera, P., Kallas, Z., Reig, L., Amores de Gea, D., 2019. The use of insect meal as a sustainable feeding alternative in aquaculture: current situation, Spanish consumers' perceptions and willingness to pay. J. Clean. Prod. 229, 10–21. https://doi.org/10.1016/j.jclepro.2019.05.012.

Glencross, B., 2016. Understanding the nutritional and biological constraints of ingredients to optimize their application in aquaculture feeds. In: Aquafeed formulation. Elsevier Inc., Australia, pp. 33–73. https://doi.org/10.1016/B978-0-12-800873-7.00003-8.

Glencross, B., Fracalossi, D.M., Hua, K., Izquierdo, M., Mai, K., Øverland, M., Robb, D., Roubach, R., Schrama, J., Small, B., Tacon, A., Valente, L.M.P., Viana, M.T., Xie, S., Yakupityage, A., 2023. Harvesting the benefits of nutritional research to address global challenges in the 21st century. J. World. Aquacult. Soc. 54 (2), 343–363. https://doi.org/10.1111/jwas.12948.

Grazziotin, A., Pimentel, F.A., de Jong, E.V., Brandelli, A., 2006. Nutritional improvement of feather protein by treatment with microbial keratinase. Anim. Feed Sci. Technol. 126 (1–2), 135–144. https://doi.org/10.1016/j.anifeedsci.2005.06.002.

Hasan, I., Gai, F., Cirrincione, S., Rimoldi, S., Saroglia, G., Terova, G., 2023. Chitinase and insect meal in aquaculture nutrition: a comprehensive overview of the latest achievements. Fishes 8 (12), 607. https://doi.org/10.3390/fishes8120607.

Henry, M., Gasco, L., Piccolo, G., Fountoulaki, E., 2015. Review on the use of insects in the diet of farmed fish: past and future. Anim. Feed Sci. Technol. 203 (1), 1–22. https://doi.org/10.1016/J.ANIFEEDSCI.2015.03.001.

Henry, M.A., Golomazou, E., Asimaki, A., Psofakis, P., Fountoulaki, E., Mente, E., Rumbos, C.I., Athanassiou, C.G., Karapanagiotidis, I.T., 2022. Partial dietary fishmeal replacement with full-fat or defatted superworm (Zophobas morio) larvae meals modulates the innate immune system of gilthead seabream, *Sparus aurata*. Aquac. Rep. 27, 101347. https://doi.org/10.1016/j.aqrep.2022.101347.

Hsieh, Y.K., Hsieh, J.W., Hu, W.C., Tseng, Y.C., 2024. AIoT-based shrimp larvae counting system using scaled multilayer feature fusion network. IEEE Internet Things J. 11 (22), 36438–36451. https://doi.org/10.1109/JIOT.2024.3410539.

Hu, M., Wang, Y., Wang, Q., Zhao, M., Xiong, B., Qian, X., Yujiang, Z., Luo, Z., 2008. Replacement of fish meal by rendered animal protein ingredients with lysine and methionine supplementation to practical diets for gibel carp, *Carassius auratus* gibelio. Aquaculture 275 (1–4), 260–265. https://doi.org/10.1016/j.aquaculture.2008.01.005.

Jobling, M., 2012. National Research Council (NRC): nutrient requirements of fish and shrimp. Aquac. Int. 20 (3), 601–602. https://doi.org/10.1007/s10499-011-9480-6.

Kirimi, J.G., Musalia, L.M., Munguti, J.M., 2017. Effect of replacing fish meal with blood meal on chemical composition of supplement for Nile Tilapia (*Oreochromis Niloticus*). East Afr. Agric. For. J. 82 (1), 1–9. https://doi.org/10.1080/00128325.2016.1158898.

Lalander, C., Diener, S., Zurbrügg, C., Vinnerås, B., 2019. Effects of feedstock on larval development and process efficiency in waste treatment with black soldier fly (*Hermetia illucens*). J. Clean. Prod. 208, 211–219. https://doi.org/10.1016/j.jclepro.2018.10.017.

Lu, S., Taethaisong, N., Meethip, W., Surakhunthod, J., Sinpru, B., Sroichak, T., Archa, P., Thongpea, S., Paengkoum, S., Purba, R.A.P., Paengkoum, P., 2022. Nutritional composition of black soldier fly larvae (*Hermetia illucens* L.) and its potential uses as alternative protein sources in animal diets: a review. Insects 13 (9). https://doi.org/10.3390/insects13090831.

MacLeod, M.J., Hasan, M.R., Robb, D.H.F., Mamun-Ur-Rashid, M., 2020. Quantifying greenhouse gas emissions from global aquaculture. Sci. Rep. 10 (1). https://doi.org/10.1038/s41598-020-68231-8.

Maiolo, S., Parisi, G., Biondi, N., Lunelli, F., Tibaldi, E., Pastres, R., 2020. Fishmeal partial substitution within aquafeed formulations: life cycle assessment of four alternative protein sources. Int. J. Life Cycle Assess. 25 (8), 1455–1471. https://doi.org/10.1007/s11367-020-01759-z.

Makkar, H.P.S., Ankers, P., 2014. Towards sustainable animal diets: a survey-based study. Anim. Feed Sci. Technol. 198, 309–322. https://doi.org/10.1016/j.anifeedsci.2014.09.018.

Meeker, D., 2006. Essential Rendering, All about the Animal By-Products Industry. Available from. https://scholar.google.com/scholar?hl=en&as_sdt=0%2C5&q=Meeker%2C+D.L.+and+Hamilton%2C+C.R.%2C+2006.+Essential+rendering.+All+about+the+animal+by-products+industry.&btnG=.

Meneguz, M., Schiavone, A., Gai, F., Dama, A., Lussiana, C., Renna, M., Gasco, L., 2018. Effect of rearing substrate on growth performance, waste reduction efficiency and chemical composition of black soldier fly (*Hermetia illucens*) larvae. J. Sci. Food Agric. 98 (15), 5776–5784. https://doi.org/10.1002/jsfa.9127.

Mulazzani, L., Madau, F.A., Pulina, P., Malorgio, G., 2021. Acceptance of insect meal in aquaculture feeding: a stakeholder analysis for the Italian supply chains of trout and seabass. J. World Aquacult. Soc. 52 (2), 378–394. https://doi.org/10.1111/jwas.12766.

Naylor, R.L., Hardy, R.W., Buschmann, A.H., Bush, S.R., Cao, L., Klinger, D.H., Little, D.C., Lubchenco, J., Shumway, S.E., Troell, M., 2021. A 20-year retrospective review of global aquaculture. Nature 591 (7851), 551–563. https://doi.org/10.1038/s41586-021-03308-6.

Ogunji, J.O., Kloas, W., Wirth, M., Schulz, C., Rennert, B., 2008. Housefly maggot meal (Magmeal) as a protein source for *Oreochromis niloticus* (Linn.). Asian Fish Sci. 21 (3). https://doi.org/10.33997/j.afs.2008.21.3.006.

Ogunji, J.O., Iheanacho, S.C., Abe, G.A., Ikeh, O.R., 2020. Assessing effects of substituting dietary fish meal with boiled donkey and cow blood meal on growth performance and digestive enzyme activities of *Clarias gariepinus* juvenile. J. World Aquacult. Soc. 51 (4), 1066–1079. https://doi.org/10.1111/jwas.12716.

Oonincx, D.G.A.B., Van Broekhoven, S., Van Huis, A., Van Loon, J.J.A., 2015. Feed conversion, survival and development, and composition of four insect species on diets composed of food by-products. PLoS One 10 (12). https://doi.org/10.1371/journal.pone.0144601.

Park, J.B., Choi, W.H., Kim, S.H., Jin, H.J., Han, Y.S., Lee, Y.S., Kim, N.J., 2014. Developmental characteristics of *Tenebrio molitor* larvae (*Coleoptera: Tenebrionidae*) in different instars. Int. J. Ind. Entomol. 28 (1), 5–9. https://doi.org/10.7852/ijie.2014.28.1.5.

Poolsawat, L., Yang, H., Sun, Y.F., Li, X.Q., Liang, G.Y., Leng, X.J., 2021. Effect of replacing fish meal with enzymatic feather meal on growth and feed utilization of tilapia (*Oreochromis niloticus* × *O. aureus*). Anim. Feed Sci. Technol. 274. https://doi.org/10.1016/j.anifeedsci.2021.114895.

Protein, 2020. Protein sources in aquaculture feed: quality and nutrition. Available from: https://www.burdockgroup.com/protein-sources-in-aquaculture-feed-quality-and-nutrition.

Regan, N., Francis, P.M., Fredrick, O., Farida, M., Gordian, R.M., 2025. Coastal mariculture: status, challenges, and future perspectives of milkfish (Chanos chanos) farming in Tanzania. J. Fish. Sci. 24 (1), 1–17. https://doi.org/10.22092/ijfs.2024.132008.

Riekkinen, K., Väkeväinen, K., Korhonen, J., 2022. The effect of substrate on the nutrient content and fatty acid composition of edible insects. Insects 13 (7), 590. https://doi.org/10.3390/INSECTS13070590/S1.

Rodrigues, D.P., Ameixa, O.M.C.C., Vázquez, J.A., Calado, R., 2022. Improving the lipid profile of black soldier fly (*Hermetia illucens*) larvae for marine aquafeeds: current state of knowledge. Sustainability (Switzerland) 14 (11). https://doi.org/10.3390/su14116472.

Rossi, G., Psarianos, M., Ojha, S., Schlüter, O.K., 2025. Review: insects as a novel feed ingredient: processing technologies, quality and safety considerations. Animal 19, 101495. https://doi.org/10.1016/j.animal.2025.101495.

Sandström, V., Chrysafi, A., Lamminen, M., Troell, M., Jalava, M., Piipponen, J., Siebert, S., Hal, O.van, Virkki, V., Kummu, M., 2022. Food system by-products upcycled in livestock and aquaculture feeds can increase global food supply. Nat. Food 3 (9), 729–740. https://doi.org/10.1038/s43016-022-00589-6.

Saputra, I., Lee, Y.N., 2023. Nutrition composition of commercial full-fat and defatted black soldier fly larvae meal (*Hermetia illucens*) as a potential protein resource for aquafeeds. Biodiversitas J. Biol. Divers. 24 (9). https://doi.org/10.13057/biodiv/d240930.

Simon, C.J., Salini, M.J., Irvin, S., Blyth, D., Bourne, N., Smullen, R., 2019. The effect of poultry protein concentrate and phosphorus supplementation on growth, digestibility and nutrient retention efficiency in barramundi *Lates calcarifer*. Aquaculture 498, 305–314. https://doi.org/10.1016/j.aquaculture.2018.08.069.

Stadtlander, T., 2021. Aquaculture feeding–problematic, but not without alternatives. Rural Times 21 (4), 31–34. https://orgprints.org/42887/.

Tacon, A.G.J., Metian, M., 2008. Global overview on the use of fish meal and fish oil in industrially compounded aquafeeds: trends and future prospects. Aquaculture 285 (1–4), 146–158. https://doi.org/10.1016/j.aquaculture.2008.08.015.

Tacon, A.G.J., Metian, M., 2015. Feed matters: satisfying the feed demand of aquaculture. Rev. Fish. Sci. Aquacult. 23 (1), 1–10. https://doi.org/10.1080/23308249.2014.987209.

Tan, B., Mai, K., Zheng, S., Zhou, Q., Liu, L., Yu, Y., 2005. Replacement of fish meal by meat and bone meal in practical diets for the white shrimp Litopenaeus vannamai (Boone). Aquac. Res. 36 (5), 439–444. https://doi.org/10.1111/j.1365-2109.2005.01223.x.

The State of World, 2022. The State of World Fisheries and Aquaculture 2022. https://doi.org/10.4060/CC0461EN.

van Broekhoven, S., Oonincx, D.G.A.B., van Huis, A., van Loon, J.J.A., 2015. Growth performance and feed conversion efficiency of three edible mealworm species (Coleoptera: Tenebrionidae) on diets composed of organic by-products. J. Insect Physiol. 73, 1–10. https://doi.org/10.1016/j.jinsphys.2014.12.005.

Wendin, K.M., Nyberg, M.E., 2021. Factors influencing consumer perception and acceptability of insect-based foods. Curr. Opin. Food Sci. 40, 67–71. https://doi.org/10.1016/j.cofs.2021.01.007.

CHAPTER

6 Processing techniques for nonconventional ingredients and agricultural byproducts in aquafeed production

Pratap Mukhopadhyay[1] **and Urna Banerjee**[2]
[1]*ICAR-CIFA, Bhubaneswar, Odisha, India;* [2]*Kolkata, West Bengal, India*

1. Introduction

Aquaculture is not only a viable rural enterprise in most parts of India but also an important component of blue economy and our food production system as well. Its contribution toward provision of human food having very high biological value, creation of livelihood options, and conservation of endangered aquatic bioresources are becoming significant. Many of the technical aspects of aquaculture including year round seed production in captivity, essential pond management measures, nutrition and dietetics, and fish health monitoring have been more or less standardized and all these together contributed to a 10-fold increase in production over the past decades. The potential of Indian fisheries today (world's second best with about 18 mmt. annually of which aquaculture being of the order of 12 mmt) in terms of changing socioeconomic scenario is enormous. The key drivers are the resources themselves in terms of both water and fish diversity. Freshwater aquaculture is the main stay in India in terms of total yield as well as catering to the Indian food basket (Chattaraj et al., 2024a; Chattaraj and Das Mohapatra, 2024). Contributing to over 60% of the total fish production which is based mainly on cyprinids consisting of three major carp species and three species of exotic carps. The sector has been growing further by way of horizontal expansion and higher productivity per unit area with the available technology backup, time-tested traditional knowledge base of the elderly farming community, financial investment/credit support, entrepreneurship development, and strategic planning. Such rapid stride in its development has also simultaneously brought several issues and problems to the forefront like inconsiderate use of inputs including feeds in fish husbandry (affecting water and soil quality) aimed at obtaining higher yield per unit water area. Thus, major challenge remains on how to ensure a long-term sustainable aquaculture development. From a nutritional perspective the key factors influencing the productivity are quality feed supplying essential nutrients and energy and the right kind of feeding strategy. India

Fish Feed Formulation with Agroindustrial Wastes and Byproducts. https://doi.org/10.1016/B978-0-443-41728-3.00012-5

has a broad variety of agro/horti-based products and byproducts from more than 330 mmt of agricultural/horticultural produce annually. These include cereal grains, pulses, oilseeds, legumes, tuber crops, plantation crops, millets, to mention a few. Given the media attention on environmentally friendly culture system development based on ecological principals that improves overall efficiency of resource use. There is a definite need to increase the utilization efficiency of such inputs for economic and ecological reasons or rather in simple terms, a farming system approach that takes care of fish production through adoption of innovative management approach without ignoring environmental factors (Chattaraj et al., 2026a).

2. Aquaculture practices: Ideal versus field experiences

Ideal practice: Semiintensive culture (with three major carps viz., catla, rohu, and mrigal along with three minor carp species viz., silver carp, common carp, and grass carp) begins with larval rearing in nursery ponds. Fish larvae from the hatchery are stocked at 10 million pcs/ha and attain a body weight of 130 mg during 15 days rearing period. At this stage, they readily accept formulated feeds immediately after yolk sac absorption.

The feed mixture generally consists of rice polish/rice bran, groundnut oilcake, roasted soyabean meal with occasional use of fish meal and a mixture of fish/vegetable oil and a vitamin-mineral premix. This reduces the complete dependence on live feed during early larval stages. Fry are then transferred to rearing ponds where they are stocked at 100,000 fry/ha for about 45 days during which they grow to fingerlings of around 15 g each. In the fry rearing stage, an almost similar type of feed as in nursery phase is given, generally by broadcasting, either in powdered or granular form spread over the pond surface following a rigid schedule.

During the subsequent period, 5% of the required feed is broadcast and the rest is given in the form of a dough ball placed in a bamboo basket/tray suspended from a long bamboo pole. These are sometimes placed within the three layers of pond water—upper, middle, and bottom for easy access by the different species in three zones, viz., surface feeders (catla, silver carp), column feeder (rohu and grass carp), and bottom feeders (mrigal and common carp). Fingerlings are then stocked into grow-out production at a density of 5000–10,000 fingerlings/ha. The supplementary feed consists of rice bran, groundnut oilcake, roasted soyabean meal, a probiotic and a vitamin-mineral supplement in the form of spaghetti type pellets spread on a bamboo basket. These feeding baskets/trays are placed at three to four locations in the pond and are monitored closely for voluntary feed intake (VFI). Feed amounts are adjusted periodically. As a standard practice, feeding is done once in the morning and once in the afternoon (Chattaraj et al., 2023, 2024b).

The three-tier system for nursery, rearing, and grow-out production is the recommended practice. Fish production of 3–10 mt/ha/yr may be achieved depending on the levels of inputs such as feed and organic manure. There is an ample evidence that the use of simple processing techniques such as grinding to reduce feed particle size and compacting (pellets) can improve feed efficiency. Adequate aeration in culture pond contributes to better production and water quality.

Field experience: However small farmers who contribute the bulk of production are not always able to follow the package of practices generally followed in the farm described above. They are constrained by availability of sufficient number of ponds and financial capacity. Many small farmers procure fish seed from local vendors and stock their ponds (often without prior acclimatization to the new environment) with fry (25 mm) or fish larvae (10 mm) instead of fingerlings. This is also due to the lack of nursery and rearing facilities as well as to the lack of sufficient fingerlings. In doing so, they have the tendency to overstock ponds not only in terms of numbers but also in terms of proportions of species. The small farmers also neglect to follow prestocking and poststocking pond management procedures, involving intermittent liming and fertilization.

The most neglected practice is in the feeds and feeding. A survey indicated that farmers use nine major ingredients and live feed types. Ingredients are rice bran, groundnut oil cake, cottonseed meal, sunflower meal, soyabean meal, mustard oil cake, wheat bran, common salt, and mineral mixture. The feed types are rice bran only; rice bran and cotton seed meal; rice bran and groundnut oil cake; rice bran and sunflower meal; and rice bran and mustard oil cake. In general, the tendency is to use a mixture of more rice bran (sometimes with husk only) and less oil cake, presumably due to costs. The feeding method is broadcasting throughout the year, irrespective of summer or winter months. They are reluctant to use a hand pelletizer although it is not expensive. This results in feed wastage and deterioration of water and sediment quality. It is also common to feed fish when convenient for farmers. More awareness on the merits of feeding fish to satiation in tune with biological rhythms and an appropriate feed form after processing can help improve production.

However, some progressive farmers, particularly those in Andhra Pradesh and West Bengal, are well aware of the development of technologies. They have a strong mindset and entrepreneurship approach and have become capable in transforming freshwater aquaculture into a major economic activity in the region. These farmers follow a unique practice of raising stunted carp yearlings for stocking in grow out ponds. Some farmers who cannot afford to have their own ponds for such purpose, procure fingerlings from nursery farms which raise fry to larger fish varying from 50 to 100 g for a period of 9–10 months with a restricted feeding regime. After this period of nutritional restriction (or under nutrition), a number of compensatory responses are invoked at the start of re-alimentation in grow-out ponds. This results in hyperfagia, rapid weight gain, better feed conversion efficiency compared to that seen in continuously fed fish or the normal ones. These farmers thus take advantage of compensatory or catch up growth phenomenon following a period of food deprivation. Another approach is to stock two carp species, rohu and catla instead of three carp species at a stocking density of 10,000 fish/ha with rohu as the main species. They will then deliberately introduce fingerlings of the snakehead species *Channa straitus* at a stocking density of 500 fish/ha. These serve as scavengers and control aquatic insects and never compete with carp either for food or for basic environmental factors. With regard to feeding the fish in culture ponds, most Andhra Pradesh farmers use farm-made feeds instead of industrially produced compounded feeds. This also follow the method of traditional demand feeding with perforated nylon bags half immersed and tied to poles.

3. Use of plant ingredients

There is a large number of terrestrial agricultural, horticultural feed resources, locally grown grasses, selected aquatic macrophytes and quite a few nonconventional plant-derived sources available in India that can be used in aquafeeds. Proximate composition of some of these components has been depicted in a previous communication (Mukhopadhyay and Mitra, 2007). The digestibility values of nutrients/energy have been established for some in carps. Carps have a long intestine which allows for digestion of a variety of feedstuffs and they also have a high amylase activity. The lack of some essential amino acid in certain plants protein sources can be compensated by using a combination of different plant ingredients. Several studies have shown that plant protein sources could be successfully utilized even by the species generally considered as carnivores if such sources can be processed well (Table 6.1).

However, the main concern is the presence of some antinutritional factors (ANFs) in some oilseeds, legumes, pulses, etc. (Table 6.2). This is the limiting factor in their potential use in fish feeds. ANFs which are endogenously produced within

Table 6.1 Presence of different (ANFs) in plant-based feedstuffs.

Plant-derived nutrient source	Antinutrients present
Soyabean meal	Protease inhibitors, lectins, phytic aids, saponins, phytoestrogens, antivitamins, allergens
Rapeseed meal	Protease inhibitors, glucosinolates, phytic acid, tannins
Mustard oil cake	Glucosinolates, tannins
Sunflower oil cake	Protease inhibitors, saponins, arginase inhibitor
Cottonseed meal	Phytic acid, phytoestrogens, gossypols antivitamins, cyclopropenoic acid
Sesame meal	Phytic acid, protease inhibitors
Lupin seed meal	Protease inhibitors, saponins, phytoestrogens, alkaloids
Pea seed meal	Protease inhibitors, lectins, tannins, cyanogens, phytic acid, saponins, antivitamins
Leuceana leaf meal	Mimosine
Alfalfa leaf meal	Protease inhibitors, saponins, phytoestrogens, antivitamins

Table 6.2 Quality criteria for final acceptance/prorate deduction/rejection.

Parameter	Acceptable	Deduction	Rejection
Mustard cake (expeller)			
Moisture	10% max.	10%–11% (1% unit)	Above 11.0%
CP	35% min.	32%–35% (1% unit)	Below 32.0%
Ether extract	7% min.	5%–7% (3% unit)	Below 5.0%
Crude fiber	9% max.	9%–11% (1% unit)	Above 11.0%
Acid insoluble ash	2.5% max.	2.5%–3.5% (1% unit)	Above 3.5%
Groundnut cake (expeller)			
Moisture	9% max.	9%–10% (1% unit)	Above 10%
CP	41% min.	38%–41% (1% unit)	Below 38.0%
Ether extract	7% min.	7%–7% (3% unit)	Below 5%
Crude fiber	12% max.	12%–14% (1% unit)	Above 14.0%
Acid insoluble ash	2.5% max.	2.5%–3.5% (1% unit)	Above 3.5%
Rice bran (de-oiled)			
Moisture	12% max.	13%–15% (1% unit)	Above 12.0
CP	15% min.	13%–15% (1% unit)	Below 13.0%
Crude fiber	13% max.	12%–13% (1% unit)	Above 15.0%
Acid insoluble ash	6%	6%–7% (1% unit)	Above 7%
Bajra/maize grain (yellow-red)			
Moisture	12% max.	12%–13% (1% unit)	Above 13.0%
Common salt			
Moisture	1%	1%–2% (1% unit)	Above 2.0%

the plant tissues as a part of their defense mechanism these compounds are generally classified under the group allelochemicals. They are generally not lethal but diminish nutrient bioavailability to the ingesting animals (Chattaraj et al., 2025). In fish species like cyprinids which are stomach less, such diminishing effects on nutrient availability become remarkable unlike higher vertebrates specially ruminants where they are degraded within the rumen. Several methods have been used to identify the ANFs in plant-derived feedstuff, their mode of action, established and amelioration processes, as well to get rid of these factors and have been applied successfully which can increase the nutritive value of these ingredients thereby increasing the bioavailability of nutrients.

Uncooked seeds of many legumes such as soybean, grass pea, horse gram, black gram, oilseeds like rapeseed, mustard groundnut, sunflower, sesame, linseed, and palm kernel contains ANFs such as trypsin inhibitors, tannins, lectin, saponin, phytate, etc. Roasting or heat treatment will not only inactivate most of these but will also increase nutrient digestibility. However, excessive heat is undesirable because of the Millard reaction which may not allow bioavailability of basic amino acids like lysine. One of the key factors that determine the final quality of the aquafeeds is the adoption of appropriate processing technology along with working out right combination factors in processing that leads to feed production of desired quality attributes. Generally, the ingredient processing steps are in the order grinding, sieving, mixing, and palletization (feed mixture gets compacted either using a pelletizer or extruder. The pellet from the later processing normally floats or sinks slowly while for the former the pellets are generally of sinking type).

Several processing methods such as germination, soaking in water (even dry heat treatment, fermentation, cooking are used to inactivate or reduce undesirable allelochemicals from the ingredients). Fermentation of legumes may be used to improve nutritive value and decrease certain ANFs such as phytate and protease inhibitors. Extrusion is another means of processing to improve the nutritive value of legumes primarily. This method is a means of reducing the level of heat-labile ANFs. Thermal processing by extrusion and micronizing (infrared heat) is positively correlated to digestibility.

The irradiation technology is a promising alternative or additional processing technique for the purpose of reducing both heat stable and heat labile ANFs. Irradiation levels up to 10 kGy which are permissible for food irradiation seem to be effective in inactivating ANFs like protease inhibitor, lectins, and phytate in particular. Approximately two-third of total phosphorus in oilseed meals grains is present in phytate, an indigestible form of phosphorus which binds protein, basic amino acids, and cations like Zn, Fe, Ca, Mn. Phytase is an enzyme specific to phytate hydrolysis. This enzyme is absent in the digestive tract of fish and even if it is, the amount is normally too small to digest dietary phytase to fish feeds (prior to palletization) containing oilseed meal or legumes and their by-products meal, discharge of phosphorus into the aquatic environment can be reduced. Commercial phytase is currently available as a feed supplement for fish and other animal feeds. With the use of phytase, the exogenous supply of P and other minerals such as Ca, Mg, Zn, Mn, and Fe may be reduced or possibly unnecessary. Moreover, by the virtue of the chelating process, supplemental phytase may have protein-sparing effect in feeds by releasing phytate bound protein.

Plant feedstuffs may contain nonstarch polysaccharides (NSPs) such as cellulose, xylans, and mannans found in cell wall material that reduce the nutritive value of feeds. Intestinal enzymes to digest these carbohydrates are not produced in most fish species. By supplementing NSP degrading enzymes, fiber containing plant materials can be effectively utilized in fish. Dietary use of such enzymes will not only improve nutrient utilization but also minimize waste excretion in water.

To improve digestibility of these nutrients, bioconversion of solid-state fermentation of the plant ingredients with suitable microbial culture and optimizing the fermentation condition may lead to predigestion. This will enhance nutrient utilization from potential feed ingredients. While considering the effectiveness of enzyme applications, one must also take into account the findings that have shown no significant impact on enzyme application (Mukhopadhyay and Mitra, 2007).

3.1 Nonconventional ingredients and agricultural byproducts in fish feed production

In a broader sense, the ingredients that lack any conventional source of origin is termed as "non-conventional." Plant-based ingredients, therefore, are those whose procurance is strictly restricted to the plants only and are mainly recognized as multiple, e.g., cereal grains, pulses, oil-seeds, legumes, seeds, leaves, edible flowers. It had been depicted that, at about eight nonconventional feed resources for their explicit use in fish nutrition, as for example, Cassava leaves, *Moringa olifera* leaf, Moringa steno leaf, Taro leaf, Pigeon pea leaves, rice bran, sugarcane filter cake, brewer's spent grain, etc., algae (*Spirulina* sp., *Chlorella vulgaris*, *Tisochrysis lutea*, *Tetraselmis suecica*, *Nannochloropsis* sp.), both micro and macro (seaweeds) and seeds rich in oils and protein from the plant species of the Brassicaceae Family (*Brassica napus*—rape, *Brassica carinata*—ethiopian mustard, *Camelina sativa*—false flax), sunflower (*Helianthus annuus*), lin (*Linum usitatissimum*), coconut (*Cocos nucifera*), cotton (*Gossypium arboreum*), soyabean (*Glycine max*) meal, peanuts, corn flour, sweet potato (*Ipomoea batatas*), banana (*Ipomoea batatas*), salvinia (*Salvinia molesta*), papaw (*Carica papaya*), Gliricidia (*Gliricidia sepium*), habarala (*Colocasia esculenta*), jack seeds (*Artocarpus heterophyllus*) , wheat gluten meal, corn gluten meal, horse-bean, defatted rubber seed meal , fruit wastes like discarded peels, algal macrophytes, duckweeds, algae, some specified terrestrial plants viz., *Leucaena leucocephala* (Mukhopadhyay and Jena), etc., are further addendums suggested for their inclusion in fish feeds. Furthermore, in the section of algae due to an impressive content of C.P. levels in the plant-based algal sources including a multiple beneficial both physiologic and immunologic impacts on fishes, several yeast species can also be deliberately included in fish feeds, as for example—*Saccharomyces cerevisiae*, *Candida utilis*, and *Kluyveromyces marxianus* (Chattaraj et al., 2024a, 2026b).

4. Processing techniques

In simple terms, "processing" is referred to all those series or chained set of activities performed from "land" to farm in order to upscale the qualitative status of the feed ingredients and make it a perfect fit to comply with nutritional demand of the fishes and confirm their healthy sustenance in an uninterrupted manner. Primarily, rather theoretically though such techniques are split in two phases—"land to lab" and "lab to field," in a true sense, the processing implies its actual scientific endeavors in the later ones where besides the very basic steps like grinding, mixing (following Pearson-square method) and pelleting them in multi-faced forms, thermal treatment, ANFs' removal, microbial fermentation, and enzyme actions still imprints a major footprints in valorizing the nutritional upgradation of the feed ingredients to an accelerated level. Anthropogenic causes of climate change including aquatic pollution and accumulation of lethal toxicants in the global water bodies, as nowadays a fatal threat bio-life sustenance, monitoring their remnants in the animal-based ingredients in order to ensure their optimal bio-safety standards are also to be considered as a stringent processing protocol along with the nutritional parameters as well.

4.1 Physical

Definitionally, any amendments done to the physical structures of the ingredients availing the mechanical interventions is termed as "physical treatment." In this, the key methods are pulverization, thermal action, etc. The principal objectives of physical treatments may be three—i. amending the surface area of the feeds for the easy accessibility of chemical processing to onset, ii. favoring a better chance of the nutritional components to come in contact with either of the chemical processing agents, and iii. Optimizing nutritional functionality in fish physiology to the fullest. Extrusion, in this case, renders as the simplest rather the foremost step of such processing methods in which the process is carried out employing two types of extruders—single screw and twin-screw extruder, followed by the successive steps like pulverization (grinding), heat treatment, enzymatic treatment, fermentation using microbial intervention and all alike, as adhered in the feed processing technique prior their application in the culture sites. Application of heat treatments to the ingredients is however exclusively prevalent in the extruding barrels of the extruder systems used for re-structuring the feeds shape, and is liberated solely due to friction faced by the ingredients while passing through the barrels. Moisture is the key driver in deteriorating the quality of fish feeds invisibly. Removal of moisture through dehydration of the feed ingredients while its journey through the extruders in tandem with its rotational dynamics are also an important step of feed processing technique (SØRENSEN, 2012). However, further impressive results could be drawn via implementation of remodulations in the screw types, understanding the potential of the interactive dynamics of the feed ingredients with the engineered modulations and lastly by allowing a restricted lapse for the consequent condensation of the feeds so as to prevent the onset of the chemical reactions, the example of which can delivered from "starch gelatinization" which is still a common problem in the fish feeds occurred by the absorbance of moisture in within and henceforth causing the feed particle to swell accordingly hampering their attractability and acceptability in the culture ponds simultaneously.

4.2 Chemical

4.2.1 Fermentation

Fermentation is nothing but increasing the shelf-life of any biological (perishable) commodity intervening microbiological entities or their enzymatic extracts naturally or any synthetic nonmicrobial components that are purely lab-based in nature. In other words, it is a simple process, in which decomposition of the complex molecular structures of the substrates (complex feed substances) are performed under a collaborative conjugate of microbes and enzymes thereby improvising proximate values (or interchangeably can be regarded as "nutritional value") of the feeds multifold. Microbes here normally considers the feeds as a substrate for their growth that are pretreated beforehand (Wang et al., 2024). "Growth" as is regarded as continuous process both in somatic or in length-weight relationship point toward withdrawal of the feeds by the microbes via external digestion mechanisms—in which such symbiotic partnership glimpsed between the feed and the microbial community helps to enhance the protein contents of the feeds in which the proteinaceous compounds liberated on the feeds may are granted into account for elevating the overall nutritional profile of the feeds. Natural fermentation of the feed ingredients for improvising their shelf-life, deactivating some selective enzymes, and boosting immune response in the fed fishes employs microbial clusters viz., yeasts, filamentous fungi, and bacteria to generate energy from carbohydrate substrate through a series of glycolytic pathway (Siddik et al., 2024). Types of such fermentation techniques are based upon the nature of ingredients used—solid state fermentation (SSF) and submerged fermentation (SF)—both of which include the environmental factors during the execution of fermentation reactions that are an effective impactor of governing the success of such microbial technique, including its qualitative completion. Initial pH and optimizing temperature here, comes first in the list of environmental influencers that are to be perfectly tuned aligning with the microbe substrate, organism, production time (Neves et al., 2024), and technique (Dawood and Koshio, 2020) and lastly the density of the sample upon which it acts (Neves et al., 2024), etc., for keeping postqualitative traits of the enacted samples unchanged as before in respect to its every single aspects—texture, color, odor, all of which was observed to remain unaltered in the experiment of Jannathulla and Dayal (2022), in which it was successfully resulted that fermentation had a negligible effect on the physical characterizations of the plant-based proteins except a selective few. This however coincides with the investigation of Ang et al. (2021), in tandem with the increase in protein, NFE, and lipid (fatty acid) in fermented plant-based feed ingredients (as, for example, the macroalgal species for their exaggerated application in fish nutrition owing to their fantastic biofiltration and nutritional profile), fish growth tends to fate a remarkable effect.

4.3 Enzymatic application

Any biologically originated compound that preholds proteinaceous nature and are capable of interacting in a substrate specific mechanism is regarded as "enzyme." Though distinguished separately, such classification may also be submerged into the fermentation technique by the virtue of their methodological overview as enzymes can be classified as either of the two types viz., i. endogenous, i.e., the ones, which are obtained from biological sources (entities, often) and ii. exogeneous, i.e., the ones which are purely based upon market availability and is purely synthetic or artificial in nature, and in such cases the first one seldom, meagres with the microbial fermentation techniques leaving the later ones to be emphasized only in regard to the "enzymatic applications." Shortest nomenclature of these enzymes can however be performed by suffixing an "*ase*" to the substrates upon which it acts. In abidance to this, although three kinds of enzymes, proteases, amylase, lipase, are widely referred, yet, in biochemistry classification of enzyme enlists its six overall types—oxidoreductase, transferases, hydrolases, lyases, isomerases, transferases, and ligases. Depending upon the source of these enzymes from where it originates, it can be classified as microbe-liberated ones and substrates specific ones among which the best known team players in violating the suppressive roles of ANFs are phytase extracted *Bacillus subtilis*, pectinase from *Aspergillus niger*, cellulase from *Trichoderma harzianum*, xylanase from *Trichoderma longibrachiatum*, α-galactosidase from *Trichoderma reesei* (Li et al., 2020) in reporting better growth, feed conversion ratio, protein efficiency ratio, and mineral deposition in several experimental fishes like juvenile *Nile tilapia* and *Oreochromis niloticus*, especially when fed with phytase (Ghosh and Mukhopadhyay, 2006). In connection to the diversification of the intestinal biome for better utilization of oligosaccharide (stachyose and raffinose) indigestible ingredients, enzymes (especially microbe derived) deserve a special mention here. In addition to endogenous ones, some exogenous ones for striking the deleterious effects of the starch derived components in fish nutrition viz., poor growth, lowered digestion, malnourishment, are also known to include as feed additives for a satisfactory increase growth. Their inclusion protocols either in encapsulation or a surplus elementary addendum in the fish diets however remains finely lined with multiple considerations—water temperature, rearing conditions of the fishes, their particular growth stages (Liang et al., 2022), a synchronous relation with the balance between the two viz., secretory rates of endogenous enzymes versus application (inclusion) rates of endogenous enzymes per kg. or gram of feed.

4.4 Removal of ANFs

Definitionally, ANFs are those which suppress the activity of nutrients or reduce their bioavailability thereby hampering fish growth in ultimate runs. More clearly, these components can be integral substances of a feedstuff and may result from natural contamination or may be accidental contaminants derived from human sources, leading to compromisation of diet quality or exceedance of the legal restrictions either. List of such antinutrients (which may be interchangeably termed as ANFs) are as follows.

4.4.1 Trypsin inhibitors

Raw soyabeans contain some crystalline globular proteins that act as trypsin inhibitors by forming irreversible complexes with trypsin and can simultaneously be inactivated via heat processing; although excessive heat application can reduce the amino acid content, especially lysine. The growth inhibition of underheated soyabean products for fingerling channel catfish and rainbow trout has been presumed to be due to interfere with protein utilization. Several studies, however, with common carp, channel catfish, and rainbow have indicated that other ANFs in the soyabean may also be responsible for the suppressive actions like decreased intestinal activity in channel catfish. Other fish species, viz., channel catfish or carps, however, are more sensitive to such impacts as confirmed by subsequent experimental trials where the channel catfish fed with 35% C.P. diet showed higher tolerance to soyabean meal in comparison to those fed with 25% of the same (Nutrient Requirement of Fish, 1993).

The mechanism whereby the trypsin inhibitors cause growth inhibition has been the object of much study, but still a satisfactory answer had not been derived. There is a little doubt that hypertrophy of the pancreas constitutes one of the primary physiological effects produced by the inhibitor, an effect which is accompanied by a stimulation of its secretory activity. This results in an endogenous loss of the pancreatic enzyme trypsin and chymotrypsin, which are rich in the sulfur containing amino acids and finally exerts its optimal effects on fish physiology deleteriously.

The presence of naturally occurring trypsin inhibitor in soyabean and other legume plant seeds had long been known among which some exerts toxicity while the others, antinutritional effects. The method of measuring trypsin inhibitor activity (TIA) in soyabean products are mainly colorimetric and is based on (Kunitz, 1947) original procedure in which trypsin substrate was availed. A synthetic substrate, benzol-DL-arginine-*p*-nitroanilide hydrochloride (BAPA) was introduced in the trypsin assay. Kakade et al. (1969), recommended BAPA as a substrate for the TIA assay after evaluating casein versus BAPA. However, later on questions on reliability on the original procedure of Kakade et al. (1969) led to a collaborative study organized by American Association of Cereal Chemist and American Oil Chemist Society. A modified method, i.e., an AACC method was described as a result of this collaborative study and was universally adopted as well (Liu and Markakis, 1989).

In general, the extent to which trypsin inhibitors are destroyed is a function of the temperature, duration of heating, particle size, and moisture condition, variables which must be closely controlled in the processing of soyabean meal in order to obtain a product having the maximum nutritious value. Destruction of over 95% of the TIA atmospheric steaming in the presence of 5% to 19% moisture is accompanied by a marked improvement in the nutritive value of protein.

4.4.2 Phytic acid (Phytates)

Approximately 70% of the phosphorus and in soyabean meal and many other feedstuffs of plant origin is in the form of phytate and its availability to fish is negligible. Phytates act as strong chelators and form protein-phytic acid complexes that may reduce the bioavailability of proteins and minerals viz., Zn, Mn, Cu, Mo, Ca, Mg and Fe apart from depressing the absorption of nutrients due to destruction of intestinal pyloric caeca region. Relevant publication however postulated that this reduction in mineral bioavailability needs the additional mineral supplementation in soyabean based diets as compared with those based on fish meal. Phytates in conjugation with high concentrations of Ca, causes Zn deficiency in chinook salmon despite its feeding by Zn enriched feeds. Addition of 0.5% phytic acid to chemically defined diets fed to rainbow trouts results in 10% growth reduction and FCR, asides exhibiting no effects on Zn absorption, concluding that growth retarding effects on phytic acids is absolutely related to reduced protein availability in fishes (Nutrient Requirements of Fish, 1993). Removal of the outer layer of the seeds through milling or fermentation process carried out by the application of yeast or lactic acid bacteria however may lead to reduction of phytic acid contents and superior growth in fishes.

4.4.3 Tannins

Tannins are water soluble phenolic compounds of plants with a molecular weight of >500 and with the ability to precipitate gelatin and other proteins from aqueous solutions. Hydrolyzed tannins (HT) and non-condensed tannins (CT) are two groups of these compounds which may be differentiated by their structure and reactivity toward hydrolytic reagents. The HTs consists of a carbohydrate moiety in which hydroxyl groups are esterified to gallic acid or *m*- digallic acid (Gallotannins) or hydroxydiphenic acid. These types of tannins are readily hydrolyzed by acids, bases or by enzymes, e.g., *Penicillium* tannase and are seen to exhibit their anti-nutritional effects through diminishing the dry matter digestion and later on the permeability of the gut walls via interacting with outer cellular layer of the gut (D'Mello and Devendra, 1995). Several experimental protocols are however explicitly available for estimating these antinutrients in order restrict their deleterious effects on fish physiology. One of these include overnight extraction of samples with methanol at room temperature followed by addition of aliquot of extracts to a solution of vanillin and hydrochloric acid and a final reading of the resultant color in calorimeter or spectrophotometer at 500–525 nm (Burns, 1971).

Removal or inactivation of the antinutrients which is the immediate step after their estimation is normally carried out via simple steps viz., wilting, field-drying, freeze-drying, dehulling of the seeds, autoclaving or treatment with alkali, treatment of the tannin-containing samples with respective oxidizing agents or even supplementing them with tannin-complexing agents, polyethylene, glycol etc so as to mitigate their effects in fishes. Caution however, should be exercised in incorporating seeds and agro-industrial by-products containing high amount of tannins.

4.4.4 Cyanogens

A wide variety of plants contains glycosides capable of releasing prussic acid or hydrogen cyanide (HCN) upon hydrolysis, the process of which is known as cyanogenesis. The entire plant or any part may exhibit this property. The glycosides usually co-occur with beta-glycosidases that cleave sugars from the glycoside and hydroxynitrile lyases that decompose the resulting cyanohydrins. The glycosides occur in vacuoles in the plant cell and the enzymes are found in the cytosol. Damage to the plant cell during mastication results in the enzyme and glycoside coming together and produce HCN, which when exceeds 20 mg/100 g may pose deleterious effect or serious animal health problems (D'Mello and Devendra, 1995).

4.4.5 Saponins

Saponins are another class of glycosides containing a polycyclic aglycone moiety of either C27 steroid or C30 triterpenoid (collectively termed as sapogenins) attached to a carbohydrate. They are widely distributed in plant kingdom and in essential aquafeed ingredients viz., soyabean and groundnut, and have a characteristic bitter taste and foaming properties (D'Mello and Devendra, 1995). They are also known to reduce the feeding intensity (VFI) in fishes through conducting an irritating effect on the mouth and throat membranes. Retardation of growth are however the ultimate outcomes.

Removal or overcome of the adverse effects of saponins from the ingredients may be done by washing the samples in water and/or conducting heat treatments, although it had been shown that the hemolytic activity of the soyabean saponins remains unaffected by heat treatment necessary to produce the optimum nutritional value of soyabean meal.

4.4.6 Mimosine

One of the most distinguishing features of tropical legumes in comparison with their temperate counterparts is the ubiquitous occurrence of a diverse array of nonprotein amino acids in sufficient concentration to precipitate adverse effects in farm animals. Mimosine is the prime example of such amino acid occurring as it does in the leaves and seeds of *Leucaena leucocephala*, where they may be regarded as a structural analogue of the nutritionally important amino acid tyrosine and its neurotransmitter products, dopamine and noradrenaline found in the brain. Concentration of mimosine in leaves ranges from 10–25 g/kg (D.M.) with even higher quantities of upto 145 g/kg D.M. in the seeds which can immediately be revaluated by multiple method can be employed viz., colorimetric assay, ion-exchange and high performance liquid chromatography (D'Mello and Devendra, 1995).

4.4.7 Gossypol

Gossypols are normally found in the pigment glands of cotton and may account for as much as 2.4% of seed weight in certain varieties. Free Gossypol is tolerated at varying amounts by different fish species, but excessive concentrations may depress growth rates and cause damage to various organ tissues. Gossypols had also been identified as a carcinogenic with aflatoxin B in rainbow trout. Similarly, Roehm et al. (1967), reported adverse effects of free gossypol on growth of rainbow trout fed with concentrations of 1000 mg/kg of feed or higher, but none at 250 mg/kg of feed. It was found that although growth did not occur at concentrations lower than 290 mg/kg of feed, histopathological changes were still noted at 95 mg/kg of feed and included thickening of glomerular basement membrane of the kidney and necrosis and ceroid deposition in the liver (Wood and Yasutake, 1956), further noted similar histopathological alterations in rainbow trout. Coming to the carp species, growth inhibitions were prominent in fingerling channel catfishes when fed at 900 mg/kg of feed but a portion of the depression was possibly due to a dietary lysine and gossypol. However, in case of solvent extracted cottonseed meal containing free gossypol @400–800 mg/kg, would not provide toxic concentration of free gossypol, conditioned, the concentration of available lysine in the feeds should be scrutinized when cottonseed meal replaces soyabean meal in the fish dietetics. Blue tilapia moreover, in this regard, have shown tolerance to free gossypol concentrations at as high as 1800 mg/kg of feed with no apparent growth depressions as documented experimentally (Nutrient Requirements of Fish, 1993).

4.4.8 Glucosinolates

Glucosinolates are found naturally in oilseed crops such as rapeseed. The compounds are not themselves harmful, but upon enzyme hydrolysis the products release thiocyanate ion, isothiocyanate, goitrin and nitrites, all functioning as potent antithyroid agents. Thiocyanate ions inhibit the uptake of iodine by the thyroid while isocyanates and nitrites presumably are precursors to thiocyanate ion. The effects of these compounds can be reversed with additional iodine supplementation in diet. Goitrin is the most potent glucosinolate which inhibits the ability of thyroid to bind iodine. The effects of goitrin cannot be reversed with dietary iodine.

Glucosinolate concentration content of meal from traditional rapeseed ranges from 3%–8%, but selective breeding has led to the low glucosinolate varieties of rapeseed, i.e., canola, which provide meal with less than 0.2 mg/g glucosinolates. Additionally it was showed that feeding rainbow trout traditional rapeseed meal caused thyroid hyperplasia and reduced plasma thyroxine concentration. Heat treatment inactivates the enzyme myrosinase, which hydrolyzes the glucosinolates to their toxic by-products. Extraction of rapeseed meal with water reduced the glucosinolate content and led to improved growth of rainbow trout. Treatment of rapeseed however failed to eliminate all growth suppressing factors for common carp (Nutrient Requirements of Fish, 1993).

4.4.9 Erucic acid

A constituent of rapeseed oil, is a 22-carbon MUFA and may constitute 20%–55% oil. Diets with erucic acid containing rapeseed oil are normally cardiotoxic in nature and causes lipid accumulation followed by necrosis of heart muscle fibers. Unpublished works of Parker and Hendricks has shown that inclusion of erucic acid in the diet of coho salmon at 3%–6% led to mortalities and pathologies of skin, gill, kidney and heart. Selective breeding program have developed canola that emphasized low erucic acid as well as low glucosinolates. No erucic acid related pathologies have been associated with the inclusion of rapeseed meal in natural ingredient fish feeds (Nutrient Requirements of Fish, 1993).

4.4.10 Alkaloids

The pyrrolizidine alkaloids are toxins found in several plant families, but not in any that are commonly used as feedstuffs in fish feeds. Many of these plants are grown in conjugation with soyabean and cotton, and therefore may find their way into fish feeds as contaminants of soyabean or cottonseed meal. The compounds are metabolized to toxic pyrroles (McLean, 1970) by mixed function oxidases in the liver. Pyrrolizidine alkaloids at 100 mg/kg of feed causes severe growth depression and mortality in rainbow trout as evidenced experimentally; and at concentrations of 2 mg/kg of feeds these toxins causes hepatic lesions including necrosis, megalocytis, fiber tissue scarring, and occlusion of the hepatic veins which are again proved scientifically. Liver lesions and mortalities are also noted after 6 months of removal of the alkaloids from the feeds (Nutrient Requirements of Fish, 1993).

4.4.11 Thiaminase

Many species of fish mainly freshwater ones, contains thiaminase that hydrolyzes thiamin in feed preparation. Thiaminase is destroyed by heat treatment and also by the acidification used to produce fish silage. Thiaminase is a relatively slow-acting enzyme therefore feeding raw fish combined with dry mash within hours of preparation is not a high risk. An alternative strategy is to feed raw fish and dry mash containing thiamine separately.

4.4.12 Histamine and Gizzerosine

Histamine, 4-(2-aminoethyl) imidazole, is a primary amine arising from the decarboxylation of the amino acid, L-histidine. Histamine production in fish muscle can be controlled by low storage temperatures, which often help to reduce enzymatic and the most microbial activities. However, if onboard refrigeration is inadequate, many of the fish captured could undergo significant deterioration prior to processing into fish meal. During the processing of fish meal, microbial and enzymatic degradation can take place rapidly at warm temperatures to produce high concentrations of histamine. Gizzerosine, an active compound, on the other hand is formed through the condensation of histamine with the epsilon amino group of lysine during high-temperatures processing raw fish into fish meal and are observed to land up with development of gastric abnormalities in the fed fishes, for example, juvenile rainbow trout. However, feed intake or gastric function remains un-impaired as observed experimentally.

4.4.13 Phytotoxins

Many species of marine algae that are capable of producing toxins including paralytic shellfish toxin (PSP), diarrheic shellfish toxin (DSP) and domoic acid, which causes amnesic shellfish poisoning (ASP). Shellfish specifically bivalve molluscs, concentrate these toxins which are the principal vectors through which they are transmitted possibly to farmed fish. One of the notable route of such transmition is like such; Sardines and presumably other fish utilizing algae as food consume algae containing domoic acid without any sign of toxicity. Further, accumulation of specific toxic amino acids like domoic acids may be triggered in the fish meal based upon their harvest for the production of aquafeed ingredients. However, such contention may be nullified through observing nullified effects on rainbow trout especially in exhibiting toxicity even after consuming 50 μg of domoic acid per kg body weight (Table 6.3).

Table 6.3 Microbial enzymes having potential in aquafeed application.

Microbial enzymes	Advantages
Amylase	Utilization of starch and complex polysaccharides
Arbinase, pectinase	Degradation of antinutritional factors
Cellulase	Degradation of cellulose in plant feed stuffs
Keratinase	Degradation of feather keratin for production of feather meal
Lipase	Emulsification and esterification of oils (fats)
Phytase	Degradation of phytase, increment of protein and mineral availability, reduction of fecal phosphorus loss
Protease	Utilization of animal and plant proteins
Tannase	Elimination of tannin
Xylanase	Hydrolysis of nonstarch polysaccharides like pentosans
α-Galactosidase	Acts as alpha-galactosides to reduce antinutritional factors
β-Glucosidase	Acts on beta-glucans and NSPs to unlock proteins

4.4.14 Microbial toxins

Molds that grow on feed ingredients and on prepared feeds are an important factor for causing toxication in fish. In particular aflatoxins which are produced by the mold *Aspergillus flavis* cause serious health problems in fish at much lower intake levels than other terrestrial animals and are of rising concern in aquaculture due to high carcinogenic properties and frequent occurrence in mainly South-east stretches. Cottonseed meal, peanut meal, and corn products are the most problematic feed ingredients with respect to aflatoxins while rice, wheat, barley and oats are next to it. Prolonged intake of very low aflatoxin (<1 ppb) causes liver cancer after 1 year in rainbow trout, the most sensitive vertebrate to aflatoxin intake, while the common effects include impaired blood clotting, sensitivity to bruising, decreased immunogenic response, increased mortality, injury to the lining of internal anatomy (viz., stomach, intestine, spleen, spleen, heart and kidney). Acute toxicity in rainbow trout is observed when fishes are fed diets containing 0.8–1.9 μg aflatoxin per gm of feed, depending upon their types. Differences in sensitivity exist among strains of rainbow trout, among trout species, among other salmonids, among other species of fish. Coho salmon and channel catfish, for example, are much resistant to aflatoxin exposure than rainbow trout. Other mold toxins of concern in fish feeding include ochratoxin, slaframine, citrinin, vomitoxin and T-2 toxin found in cereal grains. Of these, vomitoxins has been evaluated in fish with dietary levels of 20 μg toxin/gm feed or higher causing feed refusal in trout, concluding that the feeding rates should be optimized with distinct removal techniques of mycotoxins in the feeds in order to ensure feed acceptability and bioavailability both (Antinutritional factors by Dong, F. M., Hardy, R.W. and Higgs, D. A.).

Under the group of mycotoxins, another one which gains relevant attention zearalenone and tricothescenes, the former of which are estrogenic metabolites causing reproductive problems in farmed animals, including reduced growth rates, decreased RBC formation, widespread hemorrhage, poor blood clotting, impaired immune system and death (Lovell, 1984).

4.5 Enzyme supplementation of fish feeds

Though not inter-wined directly, enzyme supplementation, especially the digestive ones in fish feeds coincides with its direct function in alleviating the nutritional supremacy of the aquafeeds which obviously is the fundamental basis of "ANF deactivation," although such idea is not new in the field of aquaculture anyway. Typically, these enzymes are classified in two types: intestinal (driven from fish intestine) and microbial (secreated by microbe community) enzymes, where it was seen that enzyme supplemented feeds induced only a small positive effect on common carp (Dabrowski et al., 1979; Dabrowski and Glogowski, 1977a, 1977b). Alpha-amylase in contrary upon inclusion in minimal rates (0.2%) resulted in better growth and increased FCR value in fish (Tomassion et al. 1982). *Salmo salar* oppositely, exhibited nil effects on the consumption-growth relationships when studied by three rations of a control diet or feed utilization efficiency (Carter et al., 1992). But, moreover, use of enzyme extracts in the intestinal tissue of fishes finds some limitations besides its positive impacts on fish physiology and most deliberate one is its "heat labile" nature which makes them very difficult to keep viable after processing of the feed ingredients during feed preparation. In addition, purification of enzymes from intestinal extracts is not only hard to scale but also expensive.

In all higher animals, digestion of food materials occur by dual action of enzyme of the consumer itself and also contributed by inhabitant microorganism. A good number of evidences have the positive role of intestinal microbial community in supplementing in digestive enzyme. Enzymes derived from microbial source are the practice of new decade and is gaining popularity with time. It is easy to scale up and is increasingly being used as additives in animal husbandry (Dvorak, 2000a, b; Murillo et al., 2000; Pixnos et al., 2000a; Lopez-Soto et al., 2000) as well asides aquaculture. Bottcher (1985), reported that the addition of brewery enzyme to a cyprinoid starter feed resulted in a 1.2–1.4-fold increase in the percentual daily biomass, growth rate of carp fry. Zhong et al. (1994) described the use of microbial enzyme in preparation of prawn feed where the individual weight of prawn fed with feeds supplemented with 1 ppt of feed enzyme preparation agent increased by 12.3% and the per unit yield raised to 13.87%. The effects of dietary supplementation of thermostable bacterial alpha-amylase was equally found effective in producing growth enhancement and improved body composition of rohu (*Labeo rohita*) fingerlings (Ghosh et al., 2001).

From the invitro study of enzyme producing capacity it is evident that the bacterial flora present in the gastrointestinal tract of fish are potent producers of proteolytic enzymes and may also produce cellulose moderately (Saha and Ray, 1998;

Ghosh et al., 2002). However, there is probably less information on amylase production by the intestinal microflora in fish (Sugita et al., 1996). Seeto et al. (1996), documented the role of bacterial enzymes in fermentation of carbohydrates in marine herbivores fishes, *Odax exanomelas* and *Carinodus lophodont.* Tang et al. (1994) reported that there are stable microflora in common carp, *Cyprinus carpio* intestine and bacterial amylase plays a promoting effect on growth on starch digestion by host.

Increased mortality of the hatchlings/spawns mainly occur when they start feeding. The intestinal tract of fish larvae is much more shorter and simply organized than that of adults (Stroband and Dabrowski, 1979) and that may be correlated with relatively low production of digestive enzymes. The larvae therefore simply do not have the necessary enzymes or the amount of enzymes to digest the feed at the optimum level. Supplementation of exogenous enzyme extracted from microorganisms, or enzyme producing microorganisms could therefore be practiced to aid in digestion and this would potentially lead to increased nutritionally efficiency and survival of fish larvae in turn.

5. Conclusion

There is a large amount of scientific literature on the benefits of aquaculture and merit of fish eating on human health (supply of protein of high biological value, provision of long chain n3 PUFAs, critical vitamin cyanocobalamin and essential mineral elements including calcium phosphorus, zinc and iodine). Aquaculture thus has a major role in providing bases for better human health. It is now known that one can tailor aquaculture product quality through application of suitable processing techniques, incorporation of locally available agri-horti cultural products and byproducts as well as following nutritional principles in aquafeed production. It is vital that recommended processing techniques be adopted for their production. This will have high relevance in raising the nutrient and energy digestibility and decreasing the emission of metabolites for development of sustainable culture system. Adoption of appropriate processing technique also contributes to water stability of the feed pellets and even increase their shelf life. Over and above these adoption of aquatic biosecurity methods, stringent fish seed certification, integrated water management measures are priority areas now especially, when there is a paradigm shift toward protein based preferential food production system aiming at least carbon and water footprint in order to fulfill the commitments envisaged in the Sustainable Development Goals by 2030.

References

Ang, C.Y., Yong, A.S.K., Azad, S.A., Lim, L.S., Zuldin, W.H., Lal, M.T.M., 2021. Valorization of macroalgae through fermentation for aquafeed production: a review. Fermentation 7 (4). https://doi.org/10.3390/fermentation7040304.

Antinutritional factors by Dong, F. M., Hardy, R.W. and Higgs, D. A.

Bottcher, U., 1985. Untersuchungen Zur Aufsucht der Brut Von Karpfen (*Cyprinus carpio* L.) and Silberkarprfen (*Hopyphthalmichthys molitrix* Val.) mit veriierten Chargen des Cyprinidstarterfulters (SF-A). Fisch. Forsch. 23 (2), 7–12.

Burns, R.E., 1971. Method for estimation of tannin in grain sorghum 1. Agron. J. 63 (3), 511–512. https://doi.org/10.2134/agronj1971.00021962006300030050x.

Carter, C.G., Houlihan, D.F., McCarthy, I.D., 1992. Feed utilization efficiencies of Atlantic salmon (*Salmo salar* L.) parr: effects of a single supplementary enzyme. Comp. Biochem. Physiol. 101, 396–374.

Chattaraj, S., Behera, B.K., Das Mohapatra, P.K., 2024. Effective valorization of Chlorella biomass and Brewers spent grain substituting fish meal and soybean meal in the diet of herbivorous fish *Cirrhinus reba* for higher growth, digestibility and sustainable cultivation. J. Appl. Phycol. 36 (1), 217–232.

Chattaraj, S., Chattaraj, M., Mitra, D., Ganguly, A., Thatoi, H., Das Mohapatra, P.K., 2025. 16S amplicon sequencing of the gastrointestinal microbiota of *Cirrhinus reba* and isolation of an autochthonous probiotic using culture based approaches. Syst. Microbiol. Biomanuf. 5 (1), 156–170.

Chattaraj, S., Das Mohapatra, P.K., 2024a. Immunostimulatory and antagonistic potential of the methanolic extract of Oedogonium intermedium SCB in *Cirrhinus reba* challenged with *Aeromonas hydrophila*. Aquac. Int. 32 (3), 2523–2548.

Chattaraj, S., Ganguly, A., Mitra, D., et al., 2026a. Sustainable valorization of algal biomass in fish nutrition and dietetics: A critical review and scientometric study. Waste Biomass Valor. https://doi.org/10.1007/s12649-026-03619-7.

Chattaraj, S., Ganguly, A., Mitra, D., Mitra, D., Das Mohapatra, P.K., 2023. Study of intestinal bacteria of *Cirrhinus reba* and characterization of a new probiotic bacteria: an initiative to save the threatened species of *Cirrhinus*. Appl. Biochem. Microbiol. 59 (6), 833–849.

Chattaraj, S., Mitra, D., Chattaraj, M., Ganguly, A., Thatoi, H., Mohapatra, P.K.D., 2024b. Brewers' spent grain as fish feed ingredient: evaluation of bio-safety and analysis of its impact on gut bacteria of *Cirrhinus reba* by 16S Metagenomic sequencing. Curr. Res. Microbial Sci. 7, 100286.

Chattaraj, S., Sahoo, S.S., Behera, S., Thatoi, H., 2026b. Sustainable aquafeed development using Spirulina: effects on growth performance of Catla catla. Algal Res. 104447.

Dabrowski, K., Glogowski, J., 1977a. Studies on the role of exogenous proteolytic enzymes in the digestive process in fish. Hydrobiologia 54, 129–134.

Dabrowski, K., Glogowski, J., 1977b. A study on the application of proteolytic enzymes to fish food. Aquaculture 12, 349–360.

Dabrowski, H., Grudniewski, H., Dabrowski, K., 1979. Artificial diets for common carp: effect of the addition of enzyme extracts. Prog. Fish-Cult. 41 (4), 196–200.

Dawood, M.A.O., Koshio, S., 2020. Application of fermentation strategy in aquafeed for sustainable aquaculture. Rev. Aquacult. 12 (2), 987–1002. https://doi.org/10.1111/raq.12368.

D'Mello, J.F., Devendra, C. (Eds.), 1995. Tropical legumes in animal nutrition. CAB International, pp. 95–133.

Dvorak, R.A., 2000a. Effects of yeast/enzyme supplement on the performance of newly-arrived feeder steers. J. Anim. Sci. 78.

Dvorak, R.A., 2000b. Efficacy of an enzyme/live yeast culture product with and without monesin/trypsin in high grain diets for yearling steers. J. Anim. Sci. 78.

Ghosh, K., Mukhopadhyay, P.K., 2006. Application of enzymes in aqua feeds. Aqua Feeds Formul. Beyond. 3 (4), 7–10.

Ghosh, K., Chakraborty, K., Sen, S.K., Ray, A.K., 2001. Effects of thermostable bacterial α-amylase on growth and feed utilization in Rohu, Labeo rohita (Hamilton), fingerlings. Isr. J. Aquac. Bamidgeh 53 (3–4), 101–109.

Ghosh, K., Sen, S.K., Ray, A.K., 2002. Characterization of bacilli isolated from gut of rohu, *Labeo rohita*, fingerling and its significance in digestion. J. Appl. Aquacult. 12 (3), 33–42.

Jannathulla, R., Dayal, J.S., 2022. Effect of fungal fermentation on proximate composition, physical and functional properties of five different plant proteins used in aquafeed formulation. Indian J. Fish. 69 (3), 161–166. https://doi.org/10.21077/ijf.2022.69.3.120851-20.

Kakade, M.I., Simons, N., Liener, I.E., 1969. An evaluation of natural versus synthetic substances for measuring the antitryptic activity of soyabean samples. Cereal Chem. 46, 518–526.

Kunitz, M., 1947. Crystalline soybean trypsin inhibitor. J. Gen. Physiol. 30 (4), 291–310. https://doi.org/10.1085/jgp.30.4.291.

Li, S., Ding, G., Song, F., Sang, C., Wang, A., Chen, N., 2020. Comparison of dehulled, fermented and enzyme-treated soybean meal in diets for largemouth bass, *Micropterus salmoides*: effects on growth performance, feed utilization, immune response and intestinal morphology. Anim. Feed Sci. Technol. 267, 114548. https://doi.org/10.1016/j.anifeedsci.2020.114548.

Liang, Q., Yuan, M., Xu, L., Lio, E., Zhang, F., Mou, H., Secundo, F., 2022. Application of enzymes as a feed additive in aquaculture. Mar. Life Sci. Technol. 4 (2), 208–221. https://doi.org/10.1007/s42995-022-00128-z.

Liu, K., Markakis, P., 1989. An improved colorimetric method for determining antitryptic activity in soyabean products. Cereal Chem 66, 415–422.

Lopez-Soto, M.A., Plascencia, A., Zinn, R.A., 2000. Interaction of maceration and Fibrolytic enzyme supplementation on digestion of rice straw in Holstein cows. J. Anim. Sci. 78, 109.

Lovell, T., 1984. Microbial toxins in fish feeds. Fish feed & nutrition. Aquacult. Mag. 10 (6A), 34–36.

McLean, E.K., 1970. The toxic actions of pyrrolizidine (senecio) alkaloids. Pharmacol. Rev. 22 (4), 429–483. https://doi.org/10.1016/s0031-6997(25)06884-x.

Mukhopadhyay, P.K., Mitra, G., 2007. Plant derived feedstuffs for freshwater aquaculture in India. Aqua Cul. Asia Pac 4 (2), 19–21.

Murillo, M., Castro, H.L., Sanchez, J.F., Vasquez, M.S., Cuz, J., Alvaez, E.G., Zinn, R.A., 2000. Interaction of forage level and Fibrolytic enzymes on digestive function in cattle. J. Anim. Sci. 78.

Neves, N.O.D.S., De Dea Lindner, J., Stockhausen, L., Delziovo, F.R., Bender, M., Serzedello, L., Cipriani, L.A., Ha, N., Skoronski, E., Gisbert, E., Sanahuja, I., Perez Fabregat, T.E.H., 2024. Fermentation of plant-based feeds with lactobacillus acidophilus improves the survival and intestinal health of juvenile Nile tilapia (*Oreochromis niloticus*) reared in a biofloc system. Animals 614 (2). https://doi.org/10.3390/ani14020332.

Nutrient Requirement of Fish, 1993. Committee on animal nutrition board on agriculture national research council. National Academy Press, Washington, D.C, pp. 38–42.

Pinos, R.J., Gonzalez, S., Mendosa, G., Cobo, M., Bacena, R., Hernandez, A., Martinez, A., Ortega, M., Hoyos, G., Jacques, K., 2000. Effect of fibrolytic enzyme supplement (Fibrozyne) on intake and apparent digestibility of alfalfa and ryegrass fed to lambs. J. Dairy Sci. 29, 907–912.

Roehm, J.M., Lee, D.J., Sinhhuber, R.O., 1967. Accumulation and elimination of dietary gossypol in the organs of rainbow trout. J. Nutr. 92, 425–428.

Saha, A.K., Ray, A.K., 1998. Cellulase activity in rohu fingerlings. Aquac. Int. 6 (4), 281–291. https://doi.org/10.1023/A:1009210929594.

Seeto, G.S., Veivers, P.C., Clements, K.D., Slaytor, M., 1996. Carbohydrate utilisation by microbial symbionts in the marine herbivorous fishes *Odax cyanomelas* and *Crinodus lophodon*. J. Comp. Physiol. B 165 (7), 571–579. https://doi.org/10.1007/bf00387519.

Siddik, M.A.B., Julien, B.B., Islam, S.M.M., Francis, D.S., 2024. Fermentation in aquafeed processing: achieving sustainability in feeds for global aquaculture production. Rev. Aquacult. 16 (3), 1244–1265. https://doi.org/10.1111/raq.12894.

Sørensen, M., 2012. A review of the effects of ingredient composition and processing conditions on the physical qualities of extruded high-energy fish feed as measured by prevailing methods. Aquac. Nutr. 18 (3), 233–248. https://doi.org/10.1111/j.1365-2095.2011.00924.x.

Stroband, H., Dabrowski, K., 1979. Morphological and physiological aspects of the digestive system and feeding in freshwater fish larvae. Actes du Colluque CNERNA 355–374.

Sugita, H., Kawasaki, J., Kumazawa, J., Deguchi, Y., 1996. Production of amylase by the intestinal bacteria of Japanese coastal animals. Lett. Appl. Microbiol. 23 (3), 174–178. https://doi.org/10.1111/j.1472-765x.1996.tb00058.x.

Tomassion, H., Anjelov, A., Paukehieva, T., Kadjinikolova, L., Gadjeva, M., Danova, L., 1982. Testing for new recipes for combined pellets for consumer trout. Proc. Freshwater. Fish. Res. Inst. Plovid. 16, 112–126.

Tang, F., Xiaoyan, Z., Xingzhong, Z., 1994. The influences of common carp intestinal bacteria and its amylases on the host digestion. J. Fish. China. Shuichan. Xuebao. 18, 177–182.

Wang, Q., Qi, Z., Fu, W., Pan, M., Ren, X., Zhang, X., Rao, Z., 2024. Research and prospects of enzymatic hydrolysis and microbial fermentation technologies in protein raw materials for aquatic feed. Fermentation 10 (12), 648. https://doi.org/10.3390/fermentation 10120648.

Wood, E.M., Yasutake, W.T., 1956. Ceroid in fish. Am. J. Pathol. 32, 591–603.

Zhong, J., Wang, J., Ma, M., 1994. On application of feed enzyme preparation agent to make prawn's diet. Shandong Fish. Qilu Yuye. 11 (3), 30–32.

CHAPTER

Valorization of brewery byproducts to be utilized as fish feed ingredients

7

Arindam Ganguly[1], Suman Das[1], Riyan Kaibarta[1], Debasis Mitra[2], Sourav Chattaraj[3] and Koushik Guchhait[4]

[1]*Department of Microbiology, Bankura Sammilani College, Bankura, West Bengal, India;* [2]*Department of Microbiology, Graphic Era (Deemed to be University), Dehradun, Uttarakhand, India;* [3]*Centre for Industrial Biotechnology Research, School of Pharmaceutical Sciences, Siksha 'O' Anusandhan Deemed to be University, Bhubaneswar, Odisha, India;* [4]*Department of Biochemistry, Medical College of Wisconsin, Milwaukee, WI, United States*

1. Introduction

Aquaculture is emerging as a dynamic and rapidly growing sector that plays a crucial role in ensuring food security for a large populace (Ganguly et al., 2018, 2018b). It supports livelihood and creates employment opportunities through the cultivation of aquatic organisms (Ganguly et al., 2017, 2018, 2024). Small indigenous fish species are often promoted to meet the nutritional demands of adolescent girls and individuals with health concerns (Ganguly et al., 2025a, 2025b). Trade expansion, technological advancements, and adoption of cost-effective practices fueled the growth of aquaculture industries (Chattaraj et al., 2022; Ganguly et al., 2022). The application of bio-compatible autochthonous probiotics not only protects against virulent pathogens but also enhances fish productivity (Ganguly et al., 2019, 2025c). However, the continued dependence on traditional protein sources such as fish meal and soybean meal in fish feed formulation poses two key challenges: elevated production costs and environmental concerns. In response, there is a growing global emphasis on developing sustainable and affordable alternative protein sources for aquaculture feed.

The global food and beverage (F&B) industry has experienced remarkable growth over the past two decades, driven by continuous innovation in processes, products, and services. Beer is one of the most widely consumed alcoholic beverages worldwide. In the year 2018, global consumption of beer across 170 major countries and regions stood at approximately 1.8879 billion hectolitres. However, it is often accompanied by the generation of massive amounts of waste, in both solid and liquid forms, which poses a significant environmental and economic challenge to the brewing industry (Nad et al., 2024; Ganguly et al., 2025f). The modern world is grappling with significant challenges in managing waste effectively (Chattaraj et al., 2025b). The waste is generated in every production cycle, starting from malt mixing, boiling,

Fish Feed Formulation with Agroindustrial Wastes and Byproducts. https://doi.org/10.1016/B978-0-443-41728-3.00014-9

fermentation, filtering, etc.; the end product is characterized by different waste processing technologies, which in turn depend on the volumes, compositions, and possible value attached to these residues (Kerby and Vriesekoop, 2017). Advances in technology most certainly change the brewing industry, as most approaches in cost-effectiveness employ saving waste and by-products throughout the brewing process (Kumar et al., 2025). Generally, waste from breweries falls into three major categories: Brewers' Spent Grain (BSG), Brewers' Spent Yeast (BSY), and Brewers' Spent Hops (BSH) (Chattaraj et al., 2024a). Most of them are hot trub formed as by-products blocked in the brewing process, but their production rate is so low that it is considered insignificant (Chattaraj et al., 2024a). Brewers' spent grain (BSG) is the main by-product of the brewing industry, amounting to almost 85% of the total weight of brewing by-products (Mussatto et al., 2006). The biomass generated during the stage of mashing is produced in high amounts, and the fiber, protein, and residual starch make up most of its composition. Due to its varied nutritional benefits, BSG has good prospects for valorization into animal feeds, fuels, and functional foods. Brewer's spent yeast (BSY) is the second main waste by-product of brewing, while its disposal presents an environmental challenge. Researchers have identified BSY as a possible substitute for fish meal in aquaculture and a complementary feed ingredient for livestock, such as pigs and ruminants (Oliva-Teles and Gonçalves, 2001; Ozório et al., 2012; Stewart and Priest, 2006). Brewers' yeast is rich in proteins, vitamins, and minerals, which makes it a boon as a nutritional supplement to help growth and health in farmed fish and other animals (Ovie and Eze, 2014). The brewery industry, up to this point in time, has utilized colossal raw materials, huge amounts of water, and energy. It is no wonder that such enormous amounts of raw materials were used when it produced 190 million cubic meters of beer worldwide in 2022 alone (Report, 2023). Consequently, this implies that huge amounts of liquid and solid by-products are generated. Therefore, efficient management and valorization of the by-products are necessary to ensure environmental mitigation strategies and sustainable development in the brewing industry. Increased protein levels in fish after feeding represent one of the advantages of using Brewers' Spent Yeast (BSY) in fish feeds (Oliva-Teles and Gonçalves, 2001). Being a good source of protein, essential amino acids, and bioactive substances that stimulate growth and nutrition in fish, BSY could be a possible substitute for the conventional protein sources presently used in aquafeeds. Brewer's spent grain (BSGs) represents the options regarded by the studies to transform brewery waste into real gains for aquaculture while fulfilling its circular economy functions (Sampathkumar et al., 2023). So, by using these nutrient-rich by-products, brewing and aquaculture will also work together in reducing waste, improving sustainability, and decreasing feed production costs; thus, contributing to a more resource-efficient and environmentally sustainable food system. Plant protein industrial food waste sources offer considerable prospects for future replacement as fishmeal for aquaculture purposes, according to many investigations; however, challenges and disadvantages remain with plant protein sources. Low digestibility due to antinutritional factors and imbalanced amino acids would impair their effectiveness. Next to fishmeal, plant protein sources could even impose negative effects on fish performance (Sampathkumar et al., 2023). Alternative protein sources with better nutritional composition and higher sustainability in aquafeed production should, therefore, be considered, placing more emphasis on brewery waste. BSGs also improve feed digestibility in European sea bass (Fernandes et al., 2021). BSG's high fiber content and prebiotic properties can support gut health and nutrient absorption, thus defining its role as a functional ingredient in aquafeeds.

2. Types of brewery byproducts

In the process of brewing beer, the primary by-products are brewers' spent grains—brewers' spent hops/hot trub, and residual brewers' yeast (dos Santos Mathias et al., 2014). They slightly differ from one another in such properties (composition, moisture content, and ash content). Even though the chemical content of brewery waste varies with the nature and quality of ingredients used and the stages of the brewing process performed, its nutrient content is typically rich in nutrients. Thus, they are always a good source of nutrients (Rachwał et al., 2020).

1. **Brewers' spent grains:** Brewer's spent grains represent the principal solid waste of the entire 80%–85%, the greatest by-product of the brewing process (Mussatto, 2014; Mussatto et al., 2006). Brewers' Spent Grains dry matter production is estimated at 38.6 million tons annually on a worldwide level (Mussatto, 2014). It consists of cellulose, hemicellulose, lignin, proteins, lipids, portion of an ash fraction, and soluble compounds as well which are present in the background such as glucose, maltose and malt oligomers (Akermann et al., 2020; Mussatto, 2014). Hordein, glutelin, albumin, and globulin are major proteins present in BSG. Hordeins are the most prevalent storage proteins (prolamins), which are subdivided as B (30–50 kDa), C (55–80 kDa) and D hordeins (95 kDa) based on their molecular weights (Ikram et al., 2017). BSG has been studied as a field of research for enhancing the nutritional value of BSG and its utilization in animal feeding including aqua nutrition. With fishmeal and agricultural products like carp (*Cyprinus carpio* L.) (Morgan et al., 2013) and Nile tilapia (*Oreochromis niloticus*) (Tidwell et al., 2023) and omnivorous fish species, BSG has found its use in aquaculture fish diet only. Due to the fairly high percentage of fiber (50% DM) and just a few of the intermediates available (20%–30% DM) protein sources per meal, the application of BSG in the carnivorous fishes' diets was restricted (Nazzaro et al., 2021; San Martin et al., 2020). The BSG is very precious because it has the ability to produce a large quantity of useful products like ethanol (Parchami et al., 2022), proanthocyanidine phenolic compounds (Leite et al., 2019), lactic acid (Mussatto, 2014) and enzymes (Liguori et al., 2015). The monosaccharides-the most abundant in BSG are glucose, xylose, and arabinose. Once again, BSG contains some minerals, vitamins and amino acids. BSG has mostly essential amino acids such as methionine, phenylalanine, lysine, tryptophan, and histidine, with

nonphenylalanine amino acids such as alanine, glycine, proline, and serine present (Mitri et al., 2022).

2. **Brewers' spent yeast:** Brewers' spent yeast (BSY) is the other major by-product of brewing, and the by-product holds promise as a substitute fish meal in aquaculture fish feeds, pigs, and ruminant feeds (Estévez et al., 2021; Stewart and Priest, 2006). Seabass body composition after partial fishmeal replacement with BSY, and the findings indicated seabass improved protein level after feeding on the replaced diet (Oliva-Teles and Gonçalves, 2001). Spent brewer's yeast (SBY) (or excess yeast or residual yeast) is among the major by-products of the brewing process for beer (Puligundla et al., 2020). BSY is composed of proteins, nucleotides, minerals, and vitamins like vitamin B (predominantly B1, B2, B3, B6, and B8), vitamin C, and vitamin D as a supplement to animal nutrition (Demirgul et al., 2022; Gokulakrishnan et al., 2022; Podpora et al., 2016). Hence, not only is BSY rich in amino acids such as leucine, lysine, tyrosine, arginine, cysteine, histidine, isoleucine, methionine, phenylalanine, threonine, tryptophan, and valine, but it is also a good source of high-quality protein (Fărcas et al., 2017). Hence, different other molecules with bioactive properties such as β-glucans, mannan oligosaccharides also occur in significant quantity in this by-product (Ferreira et al., 2010). Brewers' residual yeast contains high moisture content, i.e., approximately 74%–86%. Spent yeast's mineral residue (ash) content is 2%–8.5% and dry matter content is 10%–16% as cited in the source (Mathias et al., 2015). SBY is accepted as generally recognized as safe (GRAS) for ingestion by humans (Ferreira et al., 2010). Nevertheless, due to high concentrations (6%–15%) of nucleic acids in them, SBY application as a source of protein for human nutrition is limited (Podpora et al., 2016). It is well known that high concentrations of nucleic acids in human diet may lead to increased levels of uric acid in the blood and consequently may lead to hyperuricemia (Podpora et al., 2016). No growth performance and survival rate adverse effect on giant freshwater prawn was observed after 90 days of feeding with different levels of inclusion of BSY (maximum 60% inclusion) (Nguyen et al., 2019). Brewers' Spent Yeast not only serves as an economic source of nitrogen, but it was also rich in feeding material with high level of B-complex vitamins and minerals (Ferreira et al., 2010).

3. **Brewers' spent hops:** In comparison to BSH, almost 85% of the hops utilized are fed into beer by-products (Kerby and Vriesekoop, 2017). BSH consists of polyphenols, isomerized acids adsorbed onto the solid trub residues and insoluble hop constituents, and condensation compounds of the wort proteins (Nigam and Pandey, 2009). It usually produces 0.2–0.4 kg of BSH waste per hectoliter of beer generated, and it contains 80%–90% water. BSH contains lipids, amino acids, and precipitated proteins and could be used in the manufacture of high-value products. The main ways to dispose of BSH include reuse as compost or fertilizer (because it contains high nitrogen) or mixed with BSG as animal feed. The BSH essential oils were also investigated for application in insect repellents (Kanagachandran and Jayaratne, 2006; Kerby and Vriesekoop,

2017). Besides, valorization of brewing by-products (BSG and BSH) would be realized in the recovery of valuable components such as proteins, fibers, polysaccharides, flavor compounds and phytochemicals useful as functional ingredients in beverage and food preparation, dietary supplements and cosmetic formulations (Fărcas et al., 2017).

3. Nutritional value of brewery by-products

The by-products of brewing have a very complicated chemical composition varying according to the variety of substrate used, harvesting and malting time, and mixing time (Lynch et al., 2016).

1. **Chemical composition of brewers' spent grains:** Over 80%–85% of the Brewers' Spent Grains (BSG) comprise barley malt grain husks with an admixture of some of the pericarp and seed coat layers of these grains (Table 7.1). These are usually high in fiber content, primarily consisting of cellulose, hemicellulose, noncellulosic polysaccharides, and lignin. There are a few minerals, with perhaps some protein and lipid. The husk is rich in silica and contains the major part of the polyphenolic compounds of the barley grain (Macleod, 1979; Mussatto, 2014). Hemicellulose and cellulose are sugar fractions that contain sugars, the richest of which are xylose, arabinose and glucose in BSG (Mussatto, 2014). BSG is also abundant in essential and nonessential amino acids (Ikram et al., 2017). The essential amino acids make up 30% of the protein content, with lysine being the highest at 14.3%, then leucine 6.12%, phenylalanine 4.64%, isoleucine 3.31%, threonine 0.71% and tryptophan 0.14%. Nonessential amino acids in BSG (70% of the total protein) are composed mainly of histidine (on 26.27%) and glutamic acid (16.59%), along with small quantities of aspartic acid (4.81%), valine (4.61%), arginine (4.51%), alanine (4.12%), serine (3.77%), tyrosine (2.57%), glycine (1.74%), asparagine (1.47%), γ-amino butyric acid (0.26%) and glutamine (0.07%) (Waters et al., 2012). Triacylglycerols (TAG) (55%–67% total lipids), free fatty acids (FFA) (18%–30%), diacylglycerols (DAG) (7.7%–5.7%), monoacylglycerols (MAG) (1.7%), phospholipids (PL) (9.1%) and steroid compounds (SC) (hydrocarbons, ketones, free sterols, sterol esters and sterol glycosides) (5%) were the most abundant lipid classes found in the total lipids of BSG (del Río et al., 2013; Niemi et al., 2012). The presence of a high level of free fatty acid in BSG has been largely due to the endogenous lipase that has the ability to release the free fatty acid from triacylglycerol and polar lipids during mashing and malting (del Río et al., 2013; Niemi et al., 2012). BSG also possesses waxes, essential oils, and tannins (Ikram et al., 2017). The most significant phenolic compounds in BSG are hydroxycinnamic acids (HCAs), with the predominance of ferulic (FA), p-coumaric (PCA), sinapic (SA), syringic (SyA), and caffeic acids (CA), and hydroxybenzoic acids (HBAs) contained in extremely small amounts (Faulds et al., 2002; Fărcas et al., 2015; Szwajgier et al., 2010) (Fauld). The

Table 7.1 Chemical composition of brewer's spent grains (BSG).

Components	% of dry weight	References
		Muthusamy (2014)
Hemicellulose	20–25	Xiros and Christakopoulos (2012); Steiner et al. (2015)
Proteins	19–30	Xiros et al. (2008)
Lignin	12–28	
Lipids	10	
Cellulose	12–25	
Ash	2–5	
Starch	2.7	
Phenolics	2	
Minerals	**Value** (ppm)	
Phosphorus	2000	Khidzir et al. (2010)
Magnesium	690	
Calcium	1040	
Copper	nd	Mussatto (2014)
Potassium	nd	
Iron	nd	
Manganese	nd	
Vitamins	**Value** (ppm)	
Pantothenic acid	8.5	Mussatto et al. (2006) Huige (1995)
Pyridoxine	0.7	
Biotin	0.1	
Choline	1800	
Riboflavin	1.5	
Folic acid	0.2	
Thiamine	0.7	
Niacin	44	

Nd= Not detected

major minerals in the spent grains of brewers are phosphorus (6000 mg/kg), calcium (3600 mg/kg), and magnesium (1900 mg/kg) (Ikram et al., 2017; Lynch et al., 2016), accompanied by traces of iron and even less of copper, potassium, and manganese (Mussatto, 2014). Vitamins present in BSG are biotin (0.1 ppm), niacin (44 ppm), choline (1800 ppm), folic acid (0.2 ppm), pantothenic acid (8.5 ppm), riboflavin (1.5 ppm), thiamine (0.7 ppm), and pyridoxine (0.7 ppm) (Ikram et al., 2017; McCarthy et al., 2013; Mussatto et al., 2006).

1. **Chemical composition of brewers' spent yeast:** The major chemical constituents of Brewers' Spent Yeasts (BSY) are carbohydrates, proteins, free amino acids, ash, vitamins, and fatty acids (Fărcas et al., 2017). Brewer's spent yeast is

a low-cost by-product of high protein value (Jaeger et al., 2020). The residual dry matter consisted of approximately 64.1% protein (dry weight), which could make it stable under storage conditions (Vieira et al., 2016). Valine, isoleucine, threonine, lysine, tyrosine, leucine, arginine, cysteine, histidine, methionine, phenylalanine and tryptophan are the major amino acids found in brewer's spent proteins of yeast (Fărcas et al., 2017). The limiting ones in restriction to the nutritive value of yeast proteins are usually sulfur amino acids (methionine and cysteine) (FAO/WHO, 1990). In addition, an elevated level (34%) of flavor-enhancing amino acids was identified as well (Table 7.2), i.e., glutamic acid, aspartic acid, glycine, and alanine (Fărcas et al., 2017). Approximately 83% of BSY polysaccharides have been found to be insoluble in terms of the total content of polysaccharides (Pinto et al., 2015). The BSY polysaccharides

Table 7.2 Chemical composition of brewer's spent yeast (BSY).

Components	(g/100g dry weight)	References
α-amino nitrogen	3.79 ± 0.23	Vieira et al. (2016)
Ash	14.0 ± 0.2	
Carbohydrates	12.9 ± 0.1	
Protein	64.1 ± 0.2	
RNA	4.00 ± 0.16	
Fat	1.32 ± 0.04	
Minerals	**(mg/100g dry weight)**	
Calcium	27.1 ± 0.40	
Sodium	1228 ± 22	
Magnesium	273 ± 2.31	
Iron	1.76 ± 0.03	
Potassium	9148 ± 69	
Chromium	0.019 ± 0.000	
Zinc	11.9 ± 0.29	
Manganese	0.564 ± 0.013	
Copper	0.364 ± 0.001	
Cobalt	0.030 ± 0.001	
Selenium	0.030 ± 0.000	
Molybdenum	0.003 ± 0.000	
Vitamins	**(mg/100g dry weight)**	
Pyridoxine	55.1 ± 2.5	
Riboflavin	Nq	
Nicotinic acid	77.2 ± 1.1	
Folic acid	3.01 ± 0.02	
Cyanocobalamin	Nq	

Nq = Not quantified

consist primarily of β-glucans, mannoproteins, and glycogen. The cell wall constitutes 25%–50% of volume and 15%–30% dry weight of *Saccharomyces cerevisiae* (Lipke and Ovalle, 1998). The cell wall consists of 50%–60% β-glucan, 35%–40% mannoproteins, 1%–3% chitin, and 1%–23% glycogen (Pinto et al., 2015). Glycogen is an α-glucan that is present in two forms in the yeast cell wall–as a soluble form and an insoluble form (Arvindekar and Patil, 2002). The-glucan of the cell wall is predominantly made up of β-(1,3) glucan as the framework (comprising about 85% of yeast's glucan content) and a small amount of β-(1,6) glucan (about 3% of glucan content) (Manners et al., 1973). Thirteen phenolic compounds isolated by HPLC from the spent brewer yeast extracts were reported with the following order of elution: gallic acid, protocatechuic acid, (±)-catechin, p-coumaric acid, caffeic acid, vanillic acid, chlorogenic acid, (−)-epicatechin, syringic acid, ferulic acid, rutin, isoquercetin, and cinnamic acid (Vieira et al., 2016). The free fraction thus contained two compounds quantified-gallic acid and (±)-catechin-and six compounds quantified in the bound fraction, the predominant one being (+)-catechin (24.6 mg/100 g dw), followed by gallic acid (2.11 mg/100 g dw), cinnamic acid (1.24 mg/100 g dw), p-coumaric acid (10.3 mg/100 g dw), protocatechuic acid (13.1 mg/100 g dw), ferulic acid (9.22 mg/100 g dw), as reported by Vieira et al. (2016). BSY was reported to be an excellent source of B vitamins with high mean vitamin B3 (77.2 mg/100 g), vitamin B6 (55.1 mg/100 g), and vitamin B9 (3.01 mg/100 g) content. The mean vitamin B3 content of 77.2 mg/100 g dw exceeded that reported by Pinto et al. (2015) for lyophilized brewer's yeast residue remaining after alkaline treatment without cell wall removal (0.79 mg/100 g). Also, the mean vitamin B6 (pyridoxine) contents of 55.1 mg/100 g dw were much greater than the contents of Pinto et al. (2015), 9.99 mg/100 g for dried brewer's yeast waste. As for vitamin B9 (folic acid), the mean content detected (3.01 mg/100 g dw) was also greater than the contents described by Pinto et al. (2015) (0.25 mg/100 g). These results indicate that the operations applied for the extraction of yeast extracts significantly influence their vitamin profile, but the yeast wall removal improved the level of vitamins B3, B6, and B9 (Vieira et al., 2016). The content of the macro-minerals (sodium, potassium, calcium, and magnesium) and trace elements (chromium, iron, manganese, copper, cobalt, molybdenum, selenium, and zinc) of brewer's spent yeast extracts (Vieira et al., 2016).

1. **Chemical composition of brewers' spent hops:** Spent hops have proteins, sugars, and free fatty acids (Daenicke et al., 1991). Spent hops are a rich source of the biosynthetic pathways of the essential oils of Sesquiterpene hydrocarbons, nonterpene derivatives, Oxygenated monoterpenes, Monoterpene hydrocarbons, and Oxygenated sesquiterpenes (Karlović et al., 2020). Proteins present in spent hops are very numerous (Table 7.3). This is due to the fact that proteins boil and are precipitated during boiling wort, and precipitated proteins are removed along with exhausted hops just before fermentation (Mathias et al.,

Table 7.3 Chemical composition of brewer spent hops (BSHs).

Components		Value (%)	References
Essential oil	Oxygenated sesquiterpenes	8	Stewart and Priest (2006) Hardwick (1995)
	Nonterpene derivatives	18	
	Sesquiterpene hydrocarbons	37	
	Monoterpene hydrocarbons	27	
	Oxygenated monoterpenes	4	
Proteins		22–23	
Ash		6.65	
Lipids		4.5	
Nitrogen free extract		40	
Crude fiber		23–26	

2015). The presence of Ca^{2+} allows for the creation of insoluble complexes between proteins and soluble hop constituents and peptides, which influences the final chemical makeup of exhausted hops (Karlović et al., 2020).

4. Processing techniques for brewery by-products

Brewery byproducts, such as brewer's spent grain, brewer's spent yeast, along with wastewater, offer good, green replacements for fishmeal in aquaculture; they cut down waste, reuse nutrients and fit with circular bioeconomy plans. But these byproducts hold much water, contain antinutritional factors like lignin and phytic acid, plus show natural changes in what they contain—these qualities stop direct use in fish food. Processing becomes necessary to improve their food value, raise their digestibility, and make them safe for fish. Some major methods include thermal drying, solid state fermentation, enzymatic hydrolysis, microbial protein fermentation, also microwave assisted pretreatment. These methods together raise how much protein fish get and reduce harm to the environment. Difficulties remain, particularly rising energy costs as well as complex rules. Drying is the main step for BSG and BSY, which involves drying because these materials hold a lot of moisture, about 80%–85%, so they tend to spoil from microbes. Drying lowers the moisture to less than 10%. This allows storage and pelleting for fish feed. Hot air drying at 60–80°C sees wide use because of its scalability and ability to restrict contamination by fungi, like Aspergillus, which produces mycotoxins (Lynch et al., 2016). As an example, dried BSG helps the diet of rainbow trout (*Oncorhynchus mykiss*) because it gives a stable, nutrient-rich feed (Heuzé et al., 2017). BSY often undergoes deactivation by boiling or machine drying, which stops fermentation and, as a result, a single-cell protein source with 49% protein is obtained (Gokulakrishnan et al., 2023). But drying uses a lot of energy. This raises costs and potentially takes away from the economic benefits of low-cost byproducts (Heuzé et al., 2017). Sun-drying costs less, but it depends on the weather. It also risks quality that varies in humid regions (Lynch et al., 2016). Solid-state fermentation (SSF) improves BSG's nutritional make-up by increasing the protein amount and lowering ANFs like cellulose and phytic acid, which limit a fish's ability to digest food. SSF uses organisms such as *Aspergillus ibericus*, which break down complex polysaccharides, and that makes amino acids more available. As an example, BSG treated with SSF, at 10%, improved growth and feed efficiency in European seabass (*Dicentrarchus labrax*) (Estevão-Rodrigues et al., 2025). SSF also enriches BSY with bioactive compounds, like β-glucans, which help in fish immunity (Gokulakrishnan et al., 2023). SSF needs careful handling of temperature and humidity, which is very challenging at a large size. The costs of SSF may stop small fish farms from using it. Different types of microbial strain as well as varied fermentation results, also make it harder to set standard procedures (Sarker et al., 2024). Enzymatic hydrolysis raises the digestibility of BSG. It uses enzymes, such as cellulases and proteases, to break down lignin and nonstarch polysaccharides. This process increases the bio accessibility of protein, turning BSG fit for carnivorous fish. For example, fish like gilthead seabream (Estevão-Rodrigues et al., 2025). Hydrolysis also pulls out bioactive peptides with antioxidant properties, which might improve fish health (Sarker et al., 2024). But the price of enzymes and special equipment limits accessibility, especially in developing areas (Heuzé et al., 2017). The process also creates waste, which needs proper disposal to reduce harm to the environment (Sodhi et al., 2022). Even with these problems, enzymatic hydrolysis shows promise for making good feed items from brewery waste. From brewery wastewater and stillage, producing microbial biomass offers a new way to reuse liquid byproducts. Timber Fish Technologies, for example, grows microbial biomass and invertebrates from brewery wastewater. This creates feed ingredients rich in nutrients. MiAlgae also changes distillery byproducts into algae, which is rich in omega-3. This reduces dependency on fish oil (Sustainable, 2024). These methods match the circular bioeconomy because they reuse waste streams, as shown in studies on valorizing food industry by-products (Ganguly et al., 2025d, 2025e). But these methods need complex bioreactors and a lot of technical skill, which limits their use in places with poor infrastructure (Gokulakrishnan et al., 2023). The nutrients in microbial biomass also vary, so quality control must be strict (Sarker et al., 2024). Microwave pretreatment is a new way to improve BSG's nutritional content; it reduces NSPs and improves feed effectiveness. Tests with gilthead seabream showed that the fish that ate microwave-treated BSG grew 0.69% faster and their immune system also improved (Estevão-Rodrigues et al., 2025). This method takes less time than enzymatic hydrolysis. It also uses less energy than hot air drying, but its scalability is limited by equipment costs and optimization needs (Martínez-Antequera et al., 2025). The method can make food easier to digest, and because of this, it offers a good choice for making feed in a sustainable way.

5. Growth performance

The incorporation of Brewer's Spent Yeasts and Brewer's Spent Grains as feedstuff improves the growth, nutritional content, fecundity, and overall health of livestock and fish (Chattaraj et al., 2024a). The use of BSGs in aquaculture effectively curtails the overuse of soybean meal in fish feed manufacturing (Table 7.4). The substitution of soybean meal with BSGs in the diet of *Pangasianodon hypophthalmus* and *Cirrhinus reba* not only significantly improves growth parameters, but also reduces the cost of fish feed by 27.56% and 8.55%, respectively (Chattaraj and Das Mohapatra, 2024; Jayant et al., 2018; Kaur and Saxena, 2004). Adding BSG to the diet of carps (*Cirrhina mrigala*, *Labeo rohita*, and *Catla catla*) resulted in enhanced growth, accompanied by enhanced intake of feed and better byproduct utilization. BSG is also nutritious as Jayant et al., (2018) indicated its 20.71% nitrogen free extracts, 2.24% crude fat, 13.75% crude fiber, 47.26% crude protein, and 6.83% ash. Fernandes et al. (2021) showed that BSG has the potential to enhance feed digestibility for European seabass. However, the high fiber content of BSG poses challenges for its inclusion in fish diets, which is one of the problems in aquaculture feed formulation, even though maximum substitution of soybean meal at 100% and 50% by BSG has been documented by Hassan et al. (2016) and Jayant et al. (2018), respectively. BSG diets showed better growth performance than the control diets, which included wheat meal and yellow corn flour meal. The BSG meal was safe with no toxicity observed in the carp *Cirrhinus reba* (Chattaraj et al., 2024b). *Cirrhinus reba,* which received BSG meal, showed significantly higher total protein, albumin, and globulin as compared to the fish on control diets. Histopathological studies on the intestine of *Cirrhinus reba* on BSG meal did not show any significant change in the tunica muscularis, villi, microvilli, goblet cells, or lamina propria as compared to control (Chattaraj et al., 2024b). The aminotransferase (ALT) and Aspartate aminotransferase (AST) of the control group were significantly higher than those of the group supplemented with the BSG diet (Chattaraj et al., 2024b). *Aspergillus ibericus* boosted its protein level in Solid-state fermentation (SSF) of BSG while simultaneously reducing the cellulose, hemicellulose, lignin, and lipid components of the fermented product (Filipe et al., 2023). The fermentation of Brewer's Spent Grain with *Rhizopus* sp. and *Rhizopus oligosporus* or resulted in a decrease of lipid, fiber, sugar, and neutral detergent fiber while increasing protein (Cooray and Chen, 2018; Ibarruri et al., 2019). In two commercially valued fish species, gilthead seabream (*Sparus aurata*) and rainbow trout (*Oncorhynchus mykiss*), 20% of fishmeal was substituted with BSG, which preserved protein digestibility while decreasing lipid digestibility (Nazzaro et al., 2021). Substitution of wheat constituents (wheat gluten, wheat starch, and soybean meal) with BSG up to 15% did not affect protein and lipid digestibility in rainbow trout and gilthead seabream (Estévez et al., 2022). While a high fiber diet can negatively affect fish intestinal histomorphology, it has been shown that the European seabass has a high tolerance to dietary fiber, with up to 16% fiber inclusion having no impact on gut morphology (Bonvini et al., 2018). The increase in goblet cells in the anterior intestine of fish fed with fermented

Table 7.4 Growth performance of fish fed brewers' meal.

Sl. No.	Meal	Fish	Growth performance	Hematological/serum biochemical parameters	References
1.	Brewers' spent grains (BSG)	*Cirrhinus reba*	BSG meal-fed group exhibited superior growth performance compared to the control meal.	The BSG incorporated diet group demonstrated nonsignificantly greater concentration of total protein, albumin, and globulin compared to the control group. Alanine Aminotransferase and Aspartate aminotransferase were higher in the control group compared to the BSG incorporated diet group.	Chattaraj et al. (2024b)
2.	Hydrolyzed spent grain (HSG), Hydrolyzed spent yeast (HSY), Dried spent yeast (DSY) and Dried spent grain (DSG)	*Oncorhynchus mykiss*	*Oncorhynchus mykiss* on spent yeast (DSY and HSY) exhibits the greatest growth.	N.d	Estévez et al. (2022)
3.	Brewers' spent grains (BSG)	*Dicentrarchus labrax*	BSG improves feed utilization and growth of *Dicentrarchus labrax.*	N.d	Fernandes et al. (2022)
4.	Brewers' spent yeast (BSY)	*Labeo rohita*	The growth characteristics of *Labeo rohita* fed on various levels of BSY.	Total RBC was much greater in fish feed with BSY. The control diet based on fishmeal has the least amount of RBC. Fish fed with 30% fishmeal replacement with BSY have a much reduced count of WBC. The total hemoglobin content was much greater in fish fed with BSY.	Pradhan et al. (2024)
5.	Brewers' spent grains (BSG)	*Cirrhinus reba*	Fish receiving (Chlorella meal + Soybean meal) and (Chlorella meal + BSG) diet possess an important growth parameter enhancement, whereas fish reared on (Fish meal + BSG) diet experienced nonsignificant growth changes of the overall growth toward the control group (Fish meal as well as Soybean meal).	The supplementation with *Chlorella* sp. as an alternative to fish meal significantly increased the protein levels in C. reba's serum compared to the group (Fish meal + Soybean meal). Albumin content was considerably increased in groups (Chlorella meal + Soybean meal) and (Chlorella meal + BSG). The levels of total protein, albumin, and globulin were highest in group (Chlorella meal + Soybean meal), while the lowest levels were in group (Fish meal + BSG).	Chattaraj et al. (2024c)

6.	Brewers' spent grains (BSG)	*Oreochromis niloticus*	The digestibility coefficients were high, referring to the good absorption of the ingredient being tested.	n.d	Glowka et al. (2025)
7.	Brewers' spent grains (BSG)	*Dicentrarchus labrax*	High concentrations of fiber-rich materials (BSG) can be detrimental to fish growth; however, this phenomenon was not found in the current investigation. *Dicentrarchus labrax* possesses a high tolerance activity toward fibers.	The growth of goblet cells was noted in the anterior intestine of fish consuming a diet with BSG. *Dicentrarchus labrax* consuming the BSG diet exhibited reduced total protease and trypsin activities compared to those consuming the control diet.	Estevão-Rodrigues et al. (2025)

Nd= Not detected.

BSG and nonfermented BSG can be associated with a rise in dietary fiber (Machado et al., 2013). The feed efficiency and growth indicators of climbing perch fingerlings, such as the final body weight, weight gain per fingerling, average weight gain per day, specific growth rate, efficiency of protein conversion into weight, and the value of protein produced showed marked increase along with the increased incorporation rate of brewer's spent yeast (Gokulakrishnan et al., 2022). There was a significant rise in serum total protein, albumin, and globulin of climbing perch fingerlings with the raised levels of incorporation of BSY. Hemoglobin, as well as RBC, WBC, and differential leukocyte counts including eosinophils and lymphocytes, showed the above response to increasing BSY levels (Gokulakrishnan et al., 2022). With increasing BSY levels, there were significant increases in climbing perch fingerlings' α-amylase activity and total protease activity (Gokulakrishnan et al., 2022). Much higher protein value was obtained from the muscle of fish fed with Hydrolyzed Spent Yeast at 20% (Estévez et al., 2022). No significant growth performance differences with the partial replacement of 50% of fishmeal protein by brewer's yeast, the highest concentration tested in this study corresponding to an addition of 55% brewer's yeast to the diet, on a dry matter basis, have been reported by Oliva-Teles and Gonçalves (2001).

6. Immunomodulatory effect on fish

Brewer's spent yeast (BSY) meal is used in aquaculture for the stimulation of fish immune systems, gut microbiota modulation, and growth promotion (Mahdy et al., 2022). BSY contains special polymers (chitin, mannoproteins, β-glucans) in its cell wall that can trigger an innate immune response (Gokulakrishnan et al., 2022), suppress bacterial infection, and reinforce disease resistance in certain fish species, such as Nile tilapia (El-Boshy et al., 2010), salmonids (Tukmechi and Bandboni, 2014), Japanese seabass (*Lateolabrax japonicas*) (Yu et al., 2014), and common carp (Dawood et al., 2020a). Pathogen-associated molecular pattern (PAMP) molecules include β-glucan, Mannan oligosaccharide (MOS), peptidoglycan, and nucleic acid. β-glucan activates macrophages and dendritic cells on the surface of the cells to activate innate as well as acquired immunity, including T-helper cell (Aderem and Ulevitch, 2000; Bricknell and Dalmo, 2005). MOS is a beta-glucan from the yeast cell wall and may stimulate the liver to release mannose-binding protein that binds to bacteria's capsule and trigger a cascade reaction of the complement, a form of the innate immune response (Abu-Elala et al., 2018). Peptidoglycan and nucleic acids have the ability to strengthen the host immune system as well as modulate immune-related genes and intestinal microbial weapons by various physiologically active substances (Dawood et al., 2020b). The lysozyme activity has been increased when the beer yeast cell wall fractions were used as an additive feed (Welker et al., 2012; Abu-Elala et al., 2013; Torrecillas et al., 2014). β-glucan or chitosan from *Saccharomyces cerevisiae*, Toll-like receptors (TLRs) of macrophages are always activated by it, it is assumed that phagocyte function was stimulated by the brewer's yeast by-product cell wall polysaccharides (Beutler, 2004). Cytokines produced in immune-related cells such as macrophages, dendritic cells, and are involved in growth, differentiation, and activation of the immune response. β-glucan or MOS derived from brewer's yeast is expected to show increased immune-associated cytokine expression in carp, with an increase in innate immune activity (Faulds et al., 2002; White et al., 2002). Immunomodulatory polysaccharides also increased macrophage function through the generation of reactive oxygen species (ROS) (Ferreira et al., 2015). Macrophages use highly active oxygen generated in the cytoplasm, these highly active or reactive oxygen species (ROS) are catabolized by the enzyme Superoxide Dismutase (SOD) (Cavinato et al., 2020). SOD gene expression was greatly increased when brewer's yeast was used as feed (Yang et al., 2023). Brewer's yeast products work better as prebiotics, probiotics, and immunostimulants than immunizations and antibiotics to reduce fish mortality (Sultana et al., 2024; Chattaraj et al., 2025a) (Fig. 7.1).

7. Economic and environmental benefits

Using brewery by-products, like brewer's spent grain (BSG) and spent yeast, as fish-feeding ingredients brings economic plus environmental advantages. BSG and brewer's yeast contain a lot of protein. They are abundant as well as costing less than common fishmeal and soybean meal. This helps lower the cost of aquafeed, which is a big expense for fish farmers (Gokulakrishnan et al., 2023). Research shows that up to half of the fishmeal in shrimp as well as fish diets can be changed for protein-rich BSG or yeast, which does not hurt animal growth or feed conversion efficiency. The cost of producing these feeds is much less than the price of fishmeal. That helps the brewery and aquaculture businesses stay financially sound (Estévez et al., 2022). Brewery waste valorization further supports local economies by creating value-added supply chains, reducing breweries' waste management costs, and aligning both sectors with circular bioeconomy principles (Bonato et al., 2022). In terms of the environmental impact, the use of brewery side-streams in aquafeed provides a noticeable reduction in organic waste going to landfill and greenhouse gas emissions, reducing emissions by 6% or greater than if standard feeds are used (Fonseca et al., 2024; Iñarra et al., 2022) (Arantes da). Each kilogram of BSG diverted from a landfill prevents the estimated release of up to 3 kg of CO^{2-} equivalent emissions (D'Ascenzo et al., 2024; Newman et al., 2025). Several life-cycle assessment studies have concluded that formulated feeds containing brewery by-products take up less agricultural land and water, decrease the risk of eutrophication, and avoid species depletion of marine fish stocks—therefore protecting biodiversity and lowering the aquaculture sector's environmental footprint (Fonseca et al., 2024; Karlsen and Skov, 2022) (Arantes da). Furthermore, innovations such as fermentation will have a positive impact in terms of the nutritional value, digestibility of BSG for the fish, fish health, and resilience, while minimizing nutrient leaching into aquatic ecosystems (Fernandes et al., 2022). The synergy between breweries

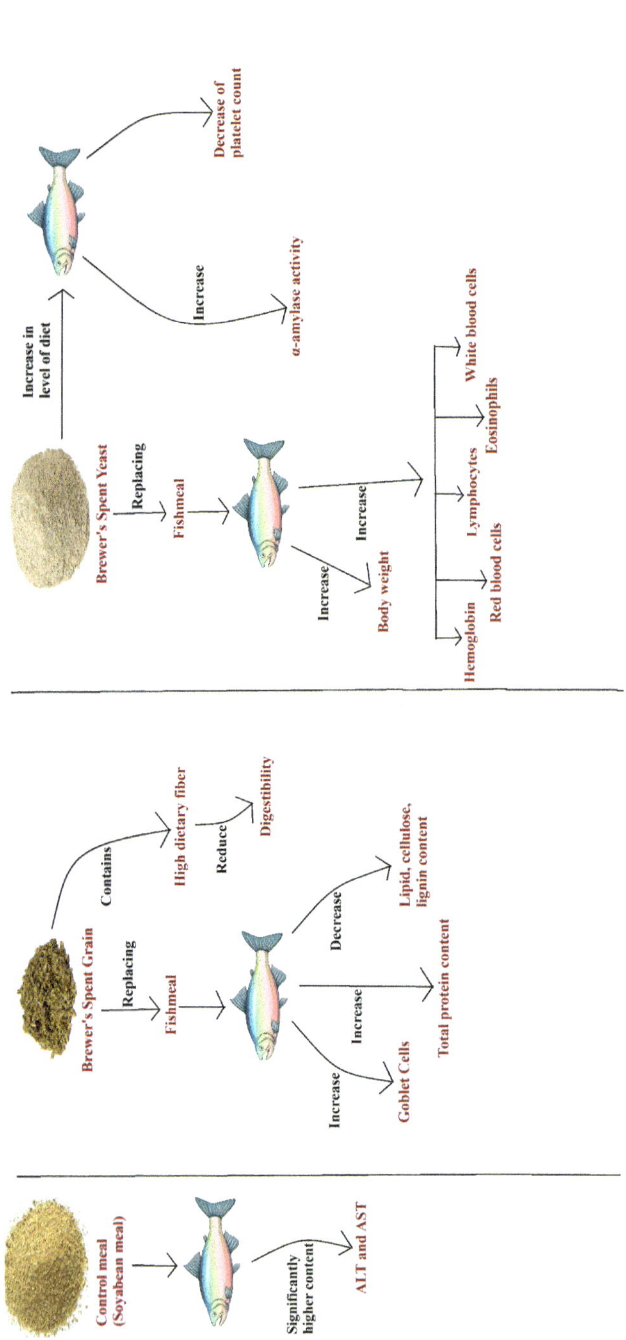

FIGURE 7.1

Comparative analysis of brewer's spent meal with cornmeal on growth and hematological properties of fish.

and fish farms thus advances sustainability, lowers aquafeed costs, cuts greenhouse gas emissions, and supports a circular economy. The synergy between breweries and fish farms thus supports sustainability, lowers aquafeed costs, cuts greenhouse gas emissions, and supports a circular economy.

8. Future direction

The use of by-products from breweries in aquaculture feed represents an exciting opportunity for sustainable food production, and much potential for the future. Given the increasing demand for aquaculture products against a backdrop of growing pressure on traditional feed sources, the valorization of rich residues such as brewer's spent grain (BSG) and brewer's spent yeast (BSY) is a practical and environmental imperative. In the future, the processing of these by-products via biotechnology, including enzymatic hydrolysis, microbial fermentation, and protein extraction, will ensure that these products are nutritionally suited and digestible for a variety of aquatic species. Specifically, the application of new combinations of microbes may help to neutralize antinutritional factors, increase amino acid accessibility, and extend shelf life, factors that represent a necessary first step for uptake in the feed industry. Collaboration among all aspects of industry and policy support will be critical to sustainably scaling these innovations. New public-private partnerships between breweries, feed producers, and aquaculture companies will create the foundation for regional supply chains to achieve by-product coordination, processing, and distribution. Funding for locally based decentralized valorization units in proximity to breweries would reduce transport costs and ensure that feed was sourced and produced locally, even in rural and coastal areas. Standardizing protocols for safety, quality assurance, and nutrient composition is crucial to establishing trust among industry and regulators that will facilitate faster integration in the market. Research institutions and policymakers will also need to group together to develop robust frameworks for the use of brewery by-products as part of certified feed products, while limiting variabilities and inclusion of contaminants. The possibility of coupling brewery waste valorization with other sustainable feed alternatives, such as insect protein or algae cultivation systems, to inform a hybrid feed with an optimal nutritional profile, minimize environmental consequences, and be economically viable, is also being developed (Chattaraj et al., 2026a, 2026b). Digital tools such as life cycle assessment models, nutrient optimization tools, and blockchain traceability could continue to facilitate the brewery system of aquafeeds through enhanced transparency and efficiency, and consumer confidence. To support circular economy principles, even the downstream products created from aquaculture systems could be reintroduced into the circular economy, e.g., biocomposting, or energy capture, which would complete a closed-loop model that minimizes waste and creates value at all levels.

Valuing brewery by-products stimulates a range of global challenges from an ecological and socio-economic perspective, such as reducing greenhouse gas emissions, addressing waste disposal issues, conserving ocean biodiversity, and bettering the world's food security. For many developing countries, where feed ingredients tend to be imported or expensive, using agro-industrial waste as a low-cost local opportunity can reduce environmental footprints, create economic independence, and increase clean resources for the world. Climate change is forcing more environmental priorities sooner rather than later, and using brewery waste for animal feed is a straightforward opportunity that is immediate in importance but has long-term significance in sustainability. In conclusion, brewery waste valorization, specifically as fish feed, is more than just a scientific advancement it represents a real shift toward rethinking industrial ecosystems as collaborative sustainable systems. With a strong level of research, technological solutions, policy and political support, and stakeholder involvement, this networked model provides an opportunity to turn waste into wealth, reduce ecological footprints, and create resilient food systems of the future.

References

Abu-Elala, N., Marzouk, M., Moustafa, M., 2013. Use of different *Saccharomyces cerevisiae* biotic forms as immune-modulator and growth promoter for *Oreochromis niloticus* challenged with some fish pathogens. Int. J. Vet. Sci. Med. 1 (1), 21–29. https://doi.org/10.1016/j.ijvsm.2013.05.001.

Abu-Elala, N.M., Younis, N.A., AbuBakr, H.O., Ragaa, N.M., Borges, L.L., Bonato, M.A., 2018. Efficacy of dietary yeast cell wall supplementation on the nutrition and immune response of Nile tilapia. Egypt J. Aquat. Res. 44 (4), 333–341. https://doi.org/10.1016/j.ejar.2018.11.001.

Aderem, A., Ulevitch, R.J., 2000. Toll-like receptors in the induction of the innate immune response. Nature 406 (6797), 782–787. https://doi.org/10.1038/35021228.

Akermann, A., Weiermüller, J., Christmann, J., Guirande, L., Glaser, G., Knaus, A., Ulber, R., 2020. Brewers' spent grain liquor as a feedstock for lactate production with *Lactobacillus delbrueckii* subsp. *lactis*. Eng. Life Sci. 20 (5–6), 168–180. https://doi.org/10.1002/elsc.201900143.

Arvindekar, A.U., Patil, N.B., 2002. Glycogen - a covalently linked component of the cell wall in *Saccharomyces cerevisiae*. Yeast 19 (2), 131–139. https://doi.org/10.1002/yea.802.

Beutler, B., 2004. Inferences, questions and possibilities in Toll-like receptor signalling. Nature 430 (6996), 257–263. https://doi.org/10.1038/nature02761.

Bonato, S.V., Augusto de Jesus Pacheco, D., Schwengber ten Caten, C., Caro, D., 2022. The missing link of circularity in small breweries' value chains: unveiling strategies for waste management and biomass valorization. J. Clean. Prod. 336. https://doi.org/10.1016/j.jclepro.2021.130275.

Bonvini, E., Bonaldo, A., Parma, L., Mandrioli, L., Sirri, R., Grandi, M., Fontanillas, R., Viroli, C., Gatta, P.P., 2018. Feeding European sea bass with increasing dietary fibre levels: impact on growth, blood biochemistry, gut histology, gut evacuation. Aquaculture 494, 1–9. https://doi.org/10.1016/j.aquaculture.2018.05.017.

Bricknell, I., Dalmo, R., 2005. The use of immunostimulants in fish larval aquaculture. Fish Shellfish Immunol. 19 (5), 457–472. https://doi.org/10.1016/j.fsi.2005.03.008.

Cavinato, L., Genise, E., Luly, F.R., Domenico, E.G.D., Del Porto, P., Ascenzioni, F., 2020. Escaping the phagocytic oxidative burst: the role of SODB in the survival of *Pseudomonas aeruginosa* within macrophages. Front. Microbiol. 11. https://doi.org/10.3389/fmicb.2020.00326.

Chattaraj, S., Das Mohapatra, P.K., 2024. Immunostimulatory and antagonistic potential of the methanolic extract of *Oedogonium intermedium* SCB in *Cirrhinus reba* challenged with *Aeromonas hydrophila*. Aquac. Int. 32 (3), 2523–2548. https://doi.org/10.1007/s10499-023-01283-5.

Chattaraj, S., Ganguly, A., Mandal, A., Das Mohapatra, P.K., 2022. A review of the role of probiotics for the control of viral diseases in aquaculture. Aquac. Int. 30 (5), 2513–2539. https://doi.org/10.1007/s10499-022-00915-6.

Chattaraj, S., Mitra, D., Ganguly, A., Thatoi, H., Mohapatra, P.K.D., 2024a. A critical review on the biotechnological potential of brewers' waste: challenges and future alternatives. Curr. Res. Microb. Sci. 6, 100228.

Chattaraj, S., Mitra, D., Chattaraj, M., Ganguly, A., Thatoi, H., Mohapatra, P.K.D., 2024b. Brewers' spent grain as fish feed ingredient: evaluation of bio-safety and analysis of its impact on gut bacteria of *Cirrhinus reba* by 16S Metagenomic sequencing. Curr. Res. Microb. Sci. 7, 100286.

Chattaraj, S., Behera, B.K., Das Mohapatra, P.K., 2024c. Effective valorization of *Chlorella* biomass and brewers spent grain substituting fish meal and soybean meal in the diet of herbivorous fish *Cirrhinus reba* for higher growth, digestibility and sustainable cultivation. J. Appl. Phycol. 36 (1), 217–232.

Chattaraj, S., Chattaraj, M., Mitra, D., Ganguly, A., Thatoi, H., Das Mohapatra, P.K., 2025a. 16S amplicon sequencing of the gastrointestinal microbiota of *Cirrhinus reba* and isolation of an autochthonous probiotic using culture based approaches. Syst. Microbiol. Biomanufact. 5 (1), 156–170.

Chattaraj, M., Goswami, A., Ganguly, A., Chattaraj, S., 2025b. Considerations of policy and morality in waste management AI technologies for enhancing recycling processes. IGI Global, India, pp. 233–266.

Chattaraj, S., Sahoo, S.S., Behera, S., Thatoi, H., 2026a. Sustainable aquafeed development using Spirulina: Effects on growth performance of *Catla catla*. Algal Res. 104447.

Chattaraj, S., Ganguly, A., Mitra, D., et al., 2026b. Sustainable valorization of algal biomass in fish nutrition and dietetics: A critical review and scientometric study. Waste Biomass Valor. https://doi.org/10.1007/s12649-026-03619-7.

Cooray, S.T., Chen, W.N., 2018. Valorization of brewer's spent grain using fungi solid-state fermentation to enhance nutritional value. J. Funct. Foods 42, 85–94. https://doi.org/10.1016/j.jff.2017.12.027.

Daenicke, R., Rohr, K., Engling, F.P., 1991. Influence of brewers' spent hops silage in diets for dairy cows on digestion and performance variables. CABI Digital Library, pp. 539–554.

D'Ascenzo, F., Vinci, G., Maddaloni, L., Ruggeri, M., Savastano, M., 2024. Application of life cycle assessment in beer production: systematic review. Beverages 10 (3), 86.

Dawood, M.A., Magouz, F.I., Essa, M., Mansour, M., 2020a. Impact of yeast fermented poultry by-product meal on growth, digestive enzyme activities, intestinal morphometry and immune response traits of common carp (*Cyprinus carpio*). Ann. Anim. Sci. 20 (3), 939–959.

Dawood, M.A., Moustafa, E.M., Elbialy, Z.I., Farrag, F., Lolo, E.E., Abdel-Daim, H.A., Van Doan, H., 2020b. Lactobacillus plantarum L-137 and/or β-glucan impacted the histopathological, antioxidant, immune-related genes and resistance of Nile tilapia (*Oreochromis niloticus*) against Aeromonas hydrophila. Res. Vet. Sci. 130, 212–221.

del Río, J.C., Prinsen, P., Gutiérrez, A., 2013. Chemical composition of lipids in brewer's spent grain: a promising source of valuable phytochemicals. J. Cereal. Sci. 58 (2), 248–254. https://doi.org/10.1016/j.jcs.2013.07.001.

Demirgul, F., Simsek, O., Sagdic, O., 2022. Amino acid, mineral, vitamin B contents and bioactivities of extracts of yeasts isolated from sourdough. Food Biosci. 50, 102040. https://doi.org/10.1016/j.fbio.2022.102040.

dos Santos Mathias, T.R., de Mello, P.P.M., Sérvulo, E.F.C., 2014. Solid wastes in brewing process: a review. J. Brew. Distill. 5 (1), 1–19.

El-Boshy, M.E., El-Ashram, A.M., AbdelHamid, F.M., Gadalla, H.A., 2010. Immunomodulatory effect of dietary *Saccharomyces cerevisiae*, β-glucan and laminaran in mercuric chloride treated Nile tilapia (*Oreochromis niloticus*) and experimentally infected with Aeromonas hydrophila. Fish Shellfish Immunol. 28 (5–6), 802–808. https://doi.org/10.1016/j.fsi.2010.01.017.

Estevão-Rodrigues, T., Fernandes, H., Moutinho, S., Ferreira, M., Castro, C., Belo, I., Peres, H., 2025. Effect of solid- fermented brewer's spent grain on growth, metabolism, and oxidative status of European seabass (*Dicentrarchus labrax*). Fishes 10 (2), 49.

Estévez, A., Padrell, L., Iñarra, B., Orive, M., San Martin, D., 2021. Brewery by-products (yeast and spent grain) as protein sources in gilthead seabream (*Sparus aurata*) feeds. Aquaculture 543, 736921. https://doi.org/10.1016/j.aquaculture.2021.736921.

Estévez, A., Padrell, L., Iñarra, B., Orive, M., San Martin, D., 2022. Brewery by-products (yeast and spent grain) as protein sources in rainbow trout (*Oncorhynchus mykiss*) feeds. Front. Mar. Sci. 9, 862020.

Faulds, C., Sancho, A., Bartolomé, B., 2002. Mono- and dimeric ferulic acid release from brewer's spent grain by fungal feruloyl esterases. Appl. Microbiol. Biotechnol. 60 (4), 489–494. https://doi.org/10.1007/s00253-002-1140-3.

FAOWHO, 1990. Energy and protein requirements. Report of a joint FAO/WHO/UNU expert consultation technical report. FAO/WHO and United Nations University, Geneva, pp. 116–129.

Fărcas, A.C., Socaci, S.A., Dulf, F.V., Tofană, M., Mudura, E., Diaconeasa, Z., 2015. Volatile profile, fatty acids composition and total phenolics content of brewers' spent grain by-product with potential use in the development of new functional foods. J. Cereal. Sci. 64, 34–42. https://doi.org/10.1016/j.jcs.2015.04.003.

Fărcas, A.C., Socaci, S.A., Mudura, E., Dulf, F.V., Vodnar, D.C., Tofană, M., Salanță, L.C., 2017. Exploitation of brewing industry wastes to produce functional ingredients. Brewing Technology 137–156.

Fernandes, H., Moyano, F., Castro, C., Salgado, J., Martínez, F., Aznar, M., Fernandes, N., Ferreira, P., Gonçalves, M., Belo, I., Oliva-Teles, A., Peres, H., 2021. Solid-state fermented brewer's spent grain enzymatic extract increases in vitro and in vivo feed digestibility in European seabass. Sci. Rep. 11 (1). https://doi.org/10.1038/s41598-021-02393-x.

Fernandes, S., Ferreira, V., Fernandes, C., 2022. Belo, Valorization of brewer's spent grain using biological treatments and its application in feeds for European seabass (*Dicentrarchus labrax*). Front. Bioeng. Biotechnol. 10.

Ferreira, V.O., Pinho, O., Vieira, E., Tavarela, J.G., 2010. Brewer's *Saccharomyces* yeast biomass: characteristics and potential applications. Trends Food Sci. Technol. 21 (2), 77–84. https://doi.org/10.1016/j.tifs.2009.10.008.

Ferreira, S.S., Passos, C.P., Madureira, P., Vilanova, M., Coimbra, M.A., 2015. Structure-function relationships of immunostimulatory polysaccharides: a review. Carbohydr. Polym. 132, 378–396. https://doi.org/10.1016/j.carbpol.2015.05.079.

Filipe, D., Dias, M., Magalhães, R., Fernandes, H., Salgado, J., Belo, I., Oliva-Teles, A., Peres, H., 2023. Solid-state fermentation of distiller's dried grains with solubles improves digestibility for European seabass (*Dicentrarchus labrax*) juveniles. Fishes 8 (2), 90. https://doi.org/10.3390/fishes8020090.

Fonseca, Y.A.d., Fernandes, A.R.A.C., Gurgel, L.V.A., Baêta, B.E.L., 2024. Comparative life cycle assessment of early-stage technological layouts for brewers' spent grain upcycling: a sustainable approach for adding value to waste. J. Water Process Eng. 66. https://doi.org/10.1016/j.jwpe.2024.105904.

Ganguly, A., Mandal, A., Khan, M.A., Dutta, T.K., Raha, S., Das Mohapatra, P.K., 2017. Study of physico-chemical parameters, planktonic diversity and bacterial load of *Clarias batrachus* cultivation pond at Bankura, WB, India. Int. J. Biol. Sci. 6 (12), 23–34.

Ganguly, A., Banerjee, A., Mandal, A., Das Mohapatra, P.K., 2018. Probiotic-based cultivation of *Clarias batrachus*: importance and future perspective. Acta Biol. Szeged. 62 (2), 158–168. https://doi.org/10.14232/abs.2018.2.158-168.

Ganguly, A., Banerjee, A., Mandal, A., Das Mohapatra, P.K., 2018b. Study of indigenous freshwater fish diversity of Bankura (West Bengal), India with special reference to *Clarias batrachus*. J. Appl. Nat. Sci. 10 (4), 1162–1172. https://doi.org/10.31018/jans.v10i4.1892.

Ganguly, A., Banerjee, A., Mandal, A., Das Mohapatra, P.K., 2019. Study of bile salt hydrolase in Lysinibacillus sphaericus: a potent fish probiotic and its in silico structure prediction for catalytic interaction. Rom. Arch. Microbiol. Immunol. 78, 81–90.

Ganguly, A., Konar, U., Kundu, A., Chatterjee, S., Nandi, S., Guin, R.K., Duari, M., Mandal, A., Das Mohapatra, P.K., 2022. Traditional fishing gears of Bankura district, WB, India: some uniqueness in fish catching. Not. Sci. Biol. 14 (1). https://doi.org/10.15835/nsb14111132.

Ganguly, A., Chattaraj, S., Ganguly, M., Chattaraj, M., Banerjee, A., Mandal, A., Ali Khan, M., Sen, S.K., Das Mohapatra, P.K., 2024. Effect of three probiotic *Bacillus* strains supplemented feeds on growth, carcass composition and blood parameters of *Clarias magur* (Hamilton). J. Appl. Aquacult. 37 (1), 19–44. https://doi.org/10.1080/10454438.2024.2330499.

Ganguly, A., Konar, R., Kirtania, D.K., Ganguly, M., Chattaraj, M., Mitra, D., Chattaraj, S., 2025. Employing technological innovation for eco-friendly waste disposal at a global scale. In: AI Technologies for Enhancing Recycling Processes. IGI Global, pp. 135–152. https://doi.org/10.4018/979-8-3693-7282-1.ch008.

Ganguly, M., Jana, S., Ganguly, A., Chattaraj, M., Midya, D.K., 2025a. Menstrual hygiene management among the rural school-going adolescent girls with special emphasis to the Kora tribe in West Bengal, India. Discover Public Health 22, 265. https://doi.org/10.1186/s12982-025-00663-x.

Ganguly, M., Ganguly, A., Chattaraj, S., Midya, D.K., 2025b. A review on menstrual health in adolescent girls emphasizing multi-omics and machine learning strategies for preventing reproductive tract infections. Discover Public Health 22, 105. https://doi.org/10.1186/s12982-025-00485-x.

Ganguly, A., Kirtania, D.K., Chhetri, M., Chattaraj, S., 2025c. Sustainable development through probiotics: a scientometric analysis of Indian research and innovation. Discover Sustainability 6, 455. https://doi.org/10.1007/s43621-025-01263-0.

Ganguly, A., Kaibarta, R., Das, S., Kundu, R., Paramanik, S., Mohapatra, S., Mukherjee, S.K., Mitra, D., Chattaraj, S., 2025d. Biodegradation and valorisation of plastic based food packets: a microbial solution for sustainability and circular economy. Discover Sustainability 6 (447). https://doi.org/10.1007/s43621-025-01257-y.

Ganguly, A., Konar, R., Karuppiah, P., Eswaranpillai, U., Ganguly, M., Chattaraj, M., Mitra, D., Chattaraj, S., 2025e. Sustainable valorization of food waste through bacterial biofilm: a pathway towards environmental safety and economic viability. In: Biofilm Applications to Revolutionize Food Technology (Chapter 13). Springer Nature. https://doi.org/10.1007/978-3-031-85205-3_13.

Glowka, R.P., Weingartner, M., Rogério Boscolo, W., 2025. Determination of apparent digestibility coefficients of brewery waste for Nile tilapia. Observ. Econ. Latinoam. 23 (2), e8951. https://doi.org/10.55905/oelv23n2-046.

Gokulakrishnan, M., Kumar, R., Pillai, B.R., Nanda, S., Bhuyan, S.K., Kumari, R., Debbarma, J., Ferosekhan, S., Siddaiah, G.M., Sundaray, J.K., 2022. Dietary brewer's spent yeast enhances growth, hematological parameters, and innate immune responses at reducing fishmeal concentration in the diet of climbing perch. Front. Nutr. 9. https://doi.org/10.3389/fnut.2022.982572.

Gokulakrishnan, M., Kumar, R., Ferosekhan, S., Siddaiah, G.M., Nanda, S., Pillai, B.R., Swain, S.K., 2023. Bio-utilization of brewery waste (Brewer's spent yeast) in global aquafeed production and its efficiency in replacing fishmeal: from a sustainability viewpoint. Aquaculture 565. https://doi.org/10.1016/j.aquaculture.2022.739161.

Hardwick, W.A., 1995. Commercial and economic aspects, pp. 1–35.

Hassan, M.A., Aftabuddin, M., Meena, D.K., Mishal, P., Gupta, S.D., 2016. Effective utilization of distiller's grain soluble—an agro-industrial waste in the feed of cage-reared minor carp Labeo bata in a tropical reservoir, India. Environ. Sci. Pollut. Control Ser. 23 (16), 16090–16095. https://doi.org/10.1007/s11356-016-6732-z.

Heuzé, V., Tran, G., Sauvant, D., Lebas, F., 2017. Brewers grains. Available from: https://feedipedia.org/node/74.

Huige, N.J., 1995. Brewery by-products and effluents. In: Food science and technology. Marcel Dekker, New York, p. 501.

Ibarruri, J., Cebrián, M., Hernández, I., 2019. Solid state fermentation of brewer's spent grain using Rhizopus sp. to enhance nutritional value. Waste Biomass Valoriz. 10 (12), 3687–3700. https://doi.org/10.1007/s12649-019-00654-5.

Ikram, S., Huang, L.Y., Zhang, H., Wang, J., Yin, M., 2017. Composition and nutrient value proposition of brewers spent grain. J. Food Sci. 82 (10), 2232–2242. https://doi.org/10.1111/1750-3841.13794.

Iñarra, B., San Martin, D., Ramos, S., Cidad, M., Estévez, A., Fenollosa, R., Martinez, J.M., Ferdinando, A., De Smet, A.M., Zufía, J., 2022. Ecodesign of new circular economy scheme for brewer's side streams. Sustain. Chem. Pharm. 28, 100727. https://doi.org/10.1016/j.scp.2022.100727.

Jaeger, A., Arendt, E.K., Zannini, E., Sahin, A.W., 2020. Brewer's spent yeast (BSY), an underutilized brewing by-product. Fermentation 6 (4). https://doi.org/10.3390/fermentation6040123.

Jayant, M., Hassan, M.A., Srivastava, P.P., Meena, D.K., Kumar, P., Kumar, A., Wagde, M.S., 2018. Brewer's spent grains (BSGs) as feedstuff for striped catfish, Pangasianodon hypophthalmus fingerlings: an approach to transform waste into wealth. J. Clean. Prod. 199, 716–722. https://doi.org/10.1016/j.jclepro.2018.07.213.

Kanagachandran, K., Jayaratne, R., 2006. Utilization potential of brewery waste water sludge as an organic fertilizer. J. Inst. Brew. 112 (2), 92–96. https://doi.org/10.1002/j.2050-0416.2006.tb00236.x.

Karlović, A., Jurić, A., Ćorić, N., Habschied, K., Krstanović, V., Mastanjević, K., 2020. By-products in the malting and brewing industries—Re-usage possibilities. Fermentation 6 (3), 82. https://doi.org/10.3390/fermentation6030082.

Karlsen, F., Skov, P.V., 2022. Review-Potentials and limitations of utilising brewer's spent grain as a protein source in aquaculture feeds. J. Clean. Prod. 357, 131986.

Kaur, V.I., Saxena, P.K., 2004. Incorporation of brewery waste in supplementary feed and its impact on growth in some carps. Bioresour. Technol. 91 (1), 101–104. https://doi.org/10.1016/S0960-8524(03)00073-7.

Kerby, C., Vriesekoop, F., 2017. An overview of the utilisation of brewery by-products as generated by British craft breweries. Beverages 3 (2), 24. https://doi.org/10.3390/beverages3020024.

Khidzir, K.M., Abdullah, N., Agamuthu, P., 2010. Brewery spent grain: chemical characteristics and utilization as an enzyme substrate. Malays. J. Sci. 29 (1), 41–51.

Kumar, R., Gupta, G., Hussain, A., Rani, A., Thapliyal, A., Gunsola, D., Chattaraj, S., Ganguly, A., Panneerselvam, P., Guerra-Sierra, B.E., Mitra, D., 2025. Pioneering zero-waste technologies utilization and its framework on sustainable management: international, national and state level. Discov. Appl. Sci. 7 (3). https://doi.org/10.1007/s42452-025-06693-z.

Leite, P., Silva, C., Salgado, J.M., Belo, I., 2019. Simultaneous production of lignocellulolytic enzymes and extraction of antioxidant compounds by solid-state fermentation of agro-industrial wastes. Ind. Crop. Prod. 137, 315–322. https://doi.org/10.1016/j.indcrop.2019.04.044.

Liguori, R., Soccol, C.R., de Souza Vandenberghe, L.P., Woiciechowski, A.L., Faraco, V., 2015. Second generation ethanol production from brewers' spent grain. Energies 8 (4), 2575–2586. https://doi.org/10.3390/en8042575.

Lipke, P.N., Ovalle, R., 1998. Cell wall architecture in yeast: new structure and new challenges. J. Bacteriol. 180 (15), 3735–3740. https://doi.org/10.1128/jb.180.15.3735-3740.1998.

Lynch, K.M., Steffen, E.J., Arendt, E.K., 2016. Brewers' spent grain: a review with an emphasis on food and health. J. Inst. Brew. 122 (4), 553–568. https://doi.org/10.1002/jib.363.

Machado, M.R.F., Souza, H.d.O., de Souza, V.L., de Azevedo, A., Goitein, R., Nobre, A.D., 2013. Morphological and anatomical characterization of the digestive tract of Centropomus parallelus and C. undecimalis. Acta Sci., Biol. Sci. 35 (4), 467–474. https://doi.org/10.4025/actascibiolsci.v35i4.14352.

Macleod, A.M., 1979. The physiology of malting. Brewing Science 1, 145–232.

Mahdy, M.A., Jamal, M.T., Al-Harb, M., Al-Mur, B.A., Haque, M.F., 2022. Use of yeasts in aquaculture nutrition and immunostimulation: a review. J. Appl. Biol. Biotechnol. 10 (5), 59–65. https://doi.org/10.7324/JABB.2022.100507.

Manners, D.J., Masson, A.J., Patterson, J.C., 1973. The structure of a β (1$\rightarrow$3) d glucan from yeast cell walls. Biochem. J. 135 (1), 19–30. https://doi.org/10.1042/bj1350019.

Martínez-Antequera, F.P., Simó-Mirabet, P., de las Heras, V., Román, M., Mancera, J.M., Martos-Sitcha, J.A., Moyano, F.J., 2025. Effect of two different pretreatments of brewers spent grain used as feed ingredient on nutritional, immunological, and metabolical parameters in gilthead seabream (*Sparus aurata*). Biology 14 (6). https://doi.org/10.3390/biology14060585.

Mathias, T.R.D.S., Alexandre, V.M.F., Cammarota, M.C., de Mello, P.P.M., Sérvulo, E.F.C., 2015. Characterization and determination of brewer's solid wastes composition. J. Inst. Brew. 121 (3), 400–404.

McCarthy, A.L., O'Callaghan, Y.C., Piggott, C.O., FitzGerald, R.J., O'Brien, N.M., 2013. Ireland Brewers' spent grain; Bioactivity of phenolic component, its role in animal nutrition and potential for incorporation in functional foods: A review. Proc. Nutr. Soc. 72 (1), 117–125. https://doi.org/10.1017/S0029665112002820.

Mitri, S., Salameh, S.J., Khelfa, A., Leonard, E., Maroun, R.G., Louka, N., Koubaa, M., 2022. Valorization of brewers' spent grains: pretreatments and fermentation, a review. Fermentation 8 (2). https://doi.org/10.3390/fermentation8020050.

Morgan, D.K.J., Verbeek, C.J.R., Rosentrater, K.A., Hicks, B.J., 2013. The palatability of flavoured novel floating pellets made with brewer's spent grain to captive carp. N. Z. J. Zool. 40 (2), 170–174. https://doi.org/10.1080/03014223.2012.719912.

Mussatto, S.I., 2014. Brewer's spent grain: a valuable feedstock for industrial applications. J. Sci. Food Agric. 94 (7), 1264–1275. https://doi.org/10.1002/jsfa.6486.

Mussatto, S.I., Dragone, G., Roberto, I.C., 2006. Brewers' spent grain: generation, characteristics and potential applications. J. Cereal. Sci. 43 (1), 1–14. https://doi.org/10.1016/j.jcs.2005.06.001.

Muthusamy, N., 2014. Chemical composition of brewers spent grain. Int. J. Sci. Environ. Technol. 3, 2109–2112.

Nad, S., Konar, U., Chattaraj, S., Ganguly, A., 2024. Valorization of feather waste by microbial enzymatic activity: bioconversion, production and application, 2935. Springer Science and Business Media LLC, pp. 337–363. https://doi.org/10.1007/978-3-031-58025-3_15.

Nazzaro, J., San Martin, D., Perez-Vendrell, A.M., Padrell, L., Iñarra, B., Orive, M., Estévez, A., 2021. Apparent digestibility coefficients of brewer's by-products used in feeds for rainbow trout (*Oncorhynchus mykiss*) and gilthead seabream (*Sparus aurata*). Aquaculture 530, 735796. https://doi.org/10.1016/j.aquaculture.2020.735796.

Newman, E., Nitin, N., Spang, E., Fox, G., 2025. Carbon footprint assessment on the viability of utilizing brewer's spent grain to produce biochar. Appl. Sci. 15 (10), 5525. https://doi.org/10.3390/app15105525.

Nguyen, N.H.Y., Trinh, L.T., Chau, D.T., Baruah, K., Lundh, T., Kiessling, A., 2019. Spent brewer's yeast as a replacement for fishmeal in diets for giant freshwater prawn (Macrobrachium rosenbergii), reared in either clear water or a biofloc environment. Aquac. Nutr. 25 (4), 970–979. https://doi.org/10.1111/anu.12915.

Niemi, P., Tamminen, T., Smeds, A., Viljanen, K., Ohra-aho, T., Holopainen-Mantila, U., Faulds, C.B., Poutanen, K., Buchert, J., 2012. Characterization of lipids and lignans in brewer's spent grain and its enzymatically extracted fraction. J. Agric. Food Chem. 60 (39), 9910–9917. https://doi.org/10.1021/jf302684x.

Nigam, P.S.N., Pandey, A. (Eds.), 2009. Biotechnology for agro-industrial residues utilisation: utilisation of agro-residues. Springer Science & Business Media.

Oliva-Teles, A., Gonçalves, P., 2001. Partial replacement of fishmeal by brewers yeast (*Saccaromyces cerevisae*) in diets for sea bass (*Dicentrarchus labrax*) juveniles. Aquaculture 202 (3–4), 269–278. https://doi.org/10.1016/s0044-8486(01)00777-3.

Ovie, S.O., Eze, S.S., 2014. Utilization of *Saccharomyces cerevisiae* in thepartial replacement of fishmeal in *Clarias gariepinus* diets. Int. J. Adv. Agric. Res. 2, 83–88.

Ozório, R.O.A., Portz, L., Borghesi, R., Cyrino, J.E.P., 2012. Effects ofdietary yeast (*Saccharomyces cerevisia*) supplementation in practical diets of tilapia (*Oreochromis niloticus*). Animals 2 (1), 16–24.

Parchami, M., Agnihotri, S., Taherzadeh, M.J., 2022. Aqueous ethanol organosolv process for the valorization of Brewer's spent grain (BSG). Bioresour. Technol. 362, 127764. https://doi.org/10.1016/j.biortech.2022.127764.

Pinto, M., Coelho, E., Nunes, A., Brandão, T., Coimbra, M.A., 2015. Valuation of brewers spent yeast polysaccharides: a structural characterization approach. Carbohydr. Polym. 116, 215–222. https://doi.org/10.1016/j.carbpol.2014.03.010.

Podpora, B., Świderski, F., Sadowska, A., Rakowska, R., Wasiak-Zys, G., 2016. Spent brewer's yeast extracts as a new component of functional food. Czech J. Food Sci. 34 (6), 554–563. https://doi.org/10.17221/419/2015-CJFS.

Pradhan, D., Swain, H.S., Upadhyay, A., Sahu, B., Nanda, S., Patra, S.K., Samantaray, K., Mohanta, K.N., Giri, S.S., 2024. On valorization of brewer's yeast as an environmentally sustainable fishmeal replacement in Labeo rohita nutrition: insight to growth attributes, digestive enzyme activities and haemato-biochemical indices. Waste Biomass Valoriz. 15 (6), 3503–3517. https://doi.org/10.1007/s12649-023-02289-z.

Puligundla, P., Mok, C., Park, S., 2020. Advances in the valorization of spent brewer's yeast. Innov. Food Sci. Emerg. Technol. 62, 102350. https://doi.org/10.1016/j.ifset.2020.102350.

Rachwał, K., Waśko, A., Gustaw, K., Polak-Berecka, M., 2020. Utilization of brewery wastes in food industry. PeerJ 8, e9427. https://doi.org/10.7717/peerj.9427.

Report 2022/2023. (2023).

Sampathkumar, K., Yu, H., Loo, S.C.J., 2023. Valorisation of industrial food waste into sustainable aquaculture feeds. Future Foods 7. https://doi.org/10.1016/j.fufo.2023.100240.

San Martin, D., Orive, M., Iñarra, B., Castelo, J., Estévez, A., Nazzaro, J., Iloro, I., Elortza, F., Zufía, J., 2020. Brewers' spent yeast and grain protein hydrolysates as second-generation feedstuff for aquaculture feed. Waste Biomass Valoriz. 11 (10), 5307–5320. https://doi.org/10.1007/s12649-020-01145-8.

Sarker, A., Ahmmed, R., Ahsan, S.M., Rana, J., Ghosh, M.K., Nandi, R., 2024. A comprehensive review of food waste valorization for the sustainable management of global food waste. Sustain. Food Technol. 2 (1), 48–69.

Singh Sodhi, A., Sharma, N., Bhatia, S., Verma, A., Soni, S., Batra, N., 2022. Insights on sustainable approaches for production and applications of value added products. Chemosphere 286 (Pt 1), 131623. https://doi.org/10.1016/j.chemosphere.2021.131623.

Steiner, J., Procopio, S., Becker, T., 2015. Brewer's spent grain: source of value-added polysaccharides for the food industry in reference to the health claims. Eur. Food Res. Technol. 241 (3), 303–315. https://doi.org/10.1007/s00217-015-2461-7.

Stewart, G.G., Priest, F.G., 2006. Brewery by-products and effluents. Informa UK Limited, pp. 670–729. https://doi.org/10.1201/9781420015171-22.

Sultana, S., Biró, J., Kucska, B., Hancz, C., 2024. Factors affecting yeast digestibility and immunostimulation in aquatic animals. Animals 14 (19), 2851. https://doi.org/10.3390/ani14192851.

Sustainable omega-3, 2024. Sustainable omega-3 production from distillery byproducts. Available from: https://www.mialgae.com/.

Szwajgier, D., Waśko, A., Targoński, Z., Niedźwiadek, M., Bancarzewska, M., 2010. The use of a novel ferulic acid esterase from lactobacillus acidophilus K1 for the release of phenolic acids from brewer's spent grain. J. Inst. Brew. 116 (3), 293–303. https://doi.org/10.1002/j.2050-0416.2010.tb00434.x.

Tidwell, J.H., Coyle, S.D., Rossi, W., Rucker, K., 2023. Evaluation of brewers spent grains with different levels of exogenous enzymes on the production performance and body composition of Nile tilapia (*Oreochromis niloticus*) and channel catfish (*Ictalurus punctatus*). J. Appl. Aquacult. 35 (2), 257–272. https://doi.org/10.1080/10454438.2021.1956669.

Torrecillas, S., Montero, D., Izquierdo, M., 2014. Improved health and growth of fish fed mannan oligosaccharides: potential mode of action. Fish Shellfish Immunol. 36 (2), 525–544. https://doi.org/10.1016/j.fsi.2013.12.029.

Tukmechi, A., Bandboni, M., 2014. Effects of *Saccharomyces cerevisiae* supplementation on immune response, hematological parameters, body composition and disease resistance in rainbow trout, *Oncorhynchus mykiss* (Walbaum, 1792). J. Appl. Ichthyol. 30 (1), 55–61. https://doi.org/10.1111/jai.12314.

Vieira, E.F., Carvalho, J., Pinto, E., Cunha, S., Almeida, A.A., Ferreira, V.O., 2016. Nutritive value, antioxidant activity and phenolic compounds profile of brewer's spent yeast extract. J. Food Compos. Anal. 52, 44–51. https://doi.org/10.1016/j.jfca.2016.07.006.

Waters, D.M., Jacob, F., Titze, J., Arendt, E.K., Zannini, E., 2012. Fibre, protein and mineral fortification of wheat bread through milled and fermented brewer's spent grain enrichment. Eur. Food Res. Technol. 235 (5), 767–778. https://doi.org/10.1007/s00217-012-1805-9.

Welker, T.L., Lim, C., Yildirim-Aksoy, M., Klesius, P.H., 2012. Effect of short-term feeding duration of diets containing commercial whole-cell yeast or yeast subcomponents on immune function and disease resistance in channel catfish, *Ictalurus punctatus*. J. Anim. Physiol. Anim. Nutr. 96 (2), 159–171. https://doi.org/10.1111/j.1439-0396.2011.01127.x.

White, L.A., Newman, M.C., Cromwell, G.L., Lindemann, M.D., 2002. Brewers dried yeast as a source of mannan oligosaccharides for weanling pigs. J. Anim. Sci. 80 (10), 2619–2628.

Xiros, C., Paul, C., 2012. Biotechnological potential of brewers spent grain and its recent applications. Waste Biomass Valoriz. 3 (2), 213–232. https://doi.org/10.1007/s12649-012-9108-8.

Xiros, C., Topakas, E., Katapodis, P., Christakopoulos, P., 2008. Hydrolysis and fermentation of brewer's spent grain by Neurospora crassa. Bioresour. Technol. 99 (13), 5427–5435. https://doi.org/10.1016/j.biortech.2007.11.010.

Yang, E.C., Choi, J.H., Jung, S.M., Jang, T.W., Kim, J.H., Hwang, Y.J., Choi, S., 2023. Effects of lactobacillus fermented brewer's yeast by-products on growth performance, innate immunity and antibacterial activity in Carp, *Cyprinus carpio*. J. Fish Pathol. 36 (2), 323–336.

Yu, H.H., Han, F., Xue, M., Wang, J., Tacon, P., Zheng, Y.H., Wu, X.F., Zhang, Y.J., 2014. Efficacy and tolerance of yeast cell wall as an immunostimulant in the diet of Japanese seabass (*Lateolabrax japonicus*). Aquaculture 432, 217–224. https://doi.org/10.1016/j.aquaculture.2014.04.043.

CHAPTER

8 Immunostimulatory potential of brewers spent yeast and other agroindustrial byproducts in fish and shellfish

Arindam Ganguly[1], Suman Das[1], Riyan Kaibarta[1], Mousumi Ganguly[2] and Sourav Chattaraj[3]

[1]*Department of Microbiology, Bankura Sammilani College, Bankura, West Bengal, India;* [2]*Department of Nutrition, Bankura Sammilani College, Bankura, West Bengal, India;* [3]*Centre for Industrial Biotechnology Research, School of Pharmaceutical Sciences, Siksha 'O' Anusandhan Deemed to be University, Bhubaneswar, Odisha, India*

1. Introduction

One of the goals to be achieved in the future, is the use of sustainable protein ingredients in the production of aquafeed, for instance, by-products and waste biomass coming from other production processes, in order to support the development of aquaculture in the future (Chattaraj et al., 2025; Nad et al., 2024). Agricultural-based businesses have a significant amount of waste that is generated in the areas of collection, storage, transport, and processing of raw materials. Waste biomass is primarily composed of organic matter, and it constitutes a reservoir of bioactive compounds. Bioactive compounds may induce a physiological effect on a living organism (Chattaraj et al., 2026a,b; Leyva-López et al., 2020). Single-cell biomass, in contrast with other plant and animal proteins, Brewer's Spent Yeast or BSY (it is an agro-industrial waste biomass of *Saccharomyces* sp. from breweries) containing bioactive compounds (mannoproteins, chitin, β-glucans) on its cell wall. Some bioactive compounds (terpenes, glucosinolates, phenolic compounds, and saponins) are derived from agro-industrial waste. These bioactive compounds can induce immunostimulatory effects, growth performance, development, and simultaneous disease resistance in fishes (Gokulakrishnan et al., 2022; Leyva-López et al., 2020; Sultana et al., 2024). Immunostimulatory polysaccharides enhanced macrophage function by the production of reactive oxygen species, enhancing the differentiation and proliferation (Ferreira et al., 2015). β-glucans used in farm animals to improve feed conversion ratios (FCR), antibiotics substitutes in the digestive system of fish, eliminate opportunistic bacteria, improve intestinal function and morphology, and promote both acquired immunity and innate functions (Ganguly

Fish Feed Formulation with Agroindustrial Wastes and Byproducts. https://doi.org/10.1016/B978-0-443-41728-3.00017-4

et al., 2017; Tungland, 2018). Candida utilis yeast or Torula yeast, which is available from different agro-industrial by-products, regulates rainbow trout immune responses (Sheikhzadeh et al., 2012). BSY application as probiotics to fish feed that have the capability to enhance fish health by improving immune response, digestion of feed, disease management, stress resistance, and growth rate (Leyva-López et al., 2020). Furthermore, probiotics form barriers against entering pathogens, induce the secretion of intestinal mucus, and improve the growth of microvilli (Ahmadifar et al., 2019). Yeast in fermented poultry byproduct meal in the diet of *Cyprinus carpio* at the level of 15%–20% improved growth, immune response, and activities of digestive enzymes of the fish (Dawood et al., 2020a). Yeast products are considered more potent as antibiotics to lower fish mortality and immunostimulants compared to immunizations (Leyva-López et al., 2020). Brewers' spent grains (BSG) is a high-protein waste product which is obtained during the brewing process. BSG inclusion in fish diets has resulted in the enhanced growth rate and nutrient digestibility of different fish species (Chattaraj et al., 2024a). Goblet cells in the anterior intestine of fish fed BSG-containing diets were significantly increased (Estevao-Rodrigues et al., 2024). Apple and pear pomaces are products of the fruit processing and juice industry, consisting of peels, seeds, stems, and pomace. Energy source from apple by-products (peels, pomace, seeds) as a feed supplement for common carp caused the immune response, feed utilization, and growth performance to be enhanced (Hoseinifar et al., 2021). Grape pomace, a by-product of the wine industry, consists of the seeds and peels of the fruit, and the by-product can be potentially used as a feed ingredient for the aquaculture sector (Quagliardi et al., 2024). Another set of experiments revealed that a dietary supplement with an increased level of grape seed proanthocyanidin resulted in the Nile tilapia's immune response, growth performance, feed consumption, and welfare being improved (Yang et al., 2023; Zhai et al., 2014). The olive oil industry produces large quantities of by-products (including olive oil pomace, olive leafs and olive pomace) with applicability for the aquafeed industry. Pumpkin skins and seeds are parts that are thrown away as waste in the process of pumpkin (Lourenço et al., 2025). Nile Tilapia diet demonstrated that by partially replacing soybean meal with pumpkin seed by 40%, fish growth performance, feed conversion ratio, and immune and oxidative responses were improved significantly (Mounes et al., 2024). The banana industry generated wastes, namely peel, leaves, and pseudostems, that are rich in some nutrients like protein, vitamins, carbohydrates, and dietary fiber (Kumari et al., 2023). Banana peel was used to supplement the diets of Nile tilapia with good results in terms of nutrient digestibility, growth performance, and feeding efficiency (Yossa et al., 2022). The use of brewer's spent yeast (BSY) and other agricultural wastes in aquaculture diets for fish and shellfish has great promise for sustainable aquaculture. Wastes are rich in proteins, vitamins, minerals, and bioactive molecules like β-glucans and chitin. Besides that, these substances also have the ability to change the immune response, making the fish more resistant to diseases and thus lowering the need for fishmeal and antibiotics (Konstantinidis et al., 2022; Sultana et al., 2024). BSY possesses a variety of immunostimulatory functions; for instance, the number of leukocytes is elevated, and the hematological status is improved as a result of the stimulation of innate immunity and gut health in different fish species (Gokulakrishnan et al., 2022). Agro-industrial wastes like fruit peels and grape pomace not only contribute to the improved immunity of the organism, but also its antioxidative status due to the phytonutrients that are naturally present (Habotta et al., 2022; Wei et al., 2023). As a result of the insertion of these substances in aqua-feed, not only the circular bioeconomy gets strengthened, but also food production becomes more sustainable through the bioconversion of waste materials into a high-value functional feed product (Ganguly et al., 2025a, 2025b; Konstantinidis et al., 2022).

2. Brewer's spent yeast

Spent yeast from brewers is the second largest by-product of the brewing industry (Chattaraj et al., 2024b). This residue slurry contributes to up to a maximum of 15% total by-products produced during the brewing process (Rachwał et al., 2020). In contrast to Brewers' spent grains, which are largely husks of the grains, *Saccharomyces cerevisiae* constitutes the bulk of BSY (Sampathkumar et al., 2023). The richest source of yeast biomass is the bottom fermentation beer production process with two stages of fermentation (first at 8–15°C and second at −1 to 4°C) (Skendi et al., 2018). In the first stage of fermentation, the heavy multiplication of yeast is going on, and thus, after fermentation and maturation, the excess of yeast (spent yeast) is available and should be removed from the beer (Gilliland, 1962). From the second fermentation, which is considerably longer, there is sediment made up of leftover yeast cells that can also be recycled. Even though during brewing the yeast biomass is utilized 4–6 times (largely), taken from a fermentation batch to another, enormous world beer production creates massive amounts of BSY waste (Skendi et al., 2018). Brewer's spent yeast as animal feed has also been taken into account in aquaculture (Jaeger et al., 2020). The use of BSY in dietary supplementation was found to enhance the digestive abilities of white seabream and meager fish species (Castro et al., 2013; Chattaraj et al., 2024c).

3. Brewer's spent yeast production

Brewer's spent yeast (BSY) is a large-scale by-product of brewing, occurring at the stage of fermentation in breweries using strains of *Saccharomyces cerevisiae* and related species. After the wort is cooled and loaded into fermenters, yeast inoculation and commencement of fermentation occur. Fermenting the sugar using yeast, over a period of several days, produces ethanol and CO_2 while the yeast grows at very high rates until fermentation is complete. In this phase, a surplus of yeast weight, known as spent yeast, accumulates at the bottom of the fermenter (Avramia and Amariei, 2021; Konstantinidis et al., 2022). Yeast recovery can be bottom-cropping (suitable for lager brewing) or top-cropping (suitable for ales), but modern breweries predominantly employ cylindro-conical fermenters for economically harvesting yeast from the bottom of the vessel with less contamination and maximum yield (Huige, 2006). After the recovery, the suspended cells, trub, residual beer, and brewery waste are removed from the unwanted solids in the yeast slurry. Several washing and rinsing operations are carried out to eliminate excess hop particles, fermentation residues, and protein precipitates in order to recover the spent yeast for reuse (Oliveira et al., 2022). To dehydrate the yeast mass and eliminate excessive water, breweries employ centrifugation, which is also less expensive in terms of transport and storage. BSY is unstable as it contains water and carryover fermentation substrates, and stabilization is therefore required. BSY may be pasteurized to inactivate enzymes and avoid spoilage, or dried (drum drying, spray drying, or freeze-drying) to yield flakes or powders of yeast for storage and other feed and food uses (Avramia and Amariei, 2021; Huige, 2006). In order to meet the requirements of certain applications, breweries can treat BSY by autolysis, an operation of controlled heating (usually 45–60°C for 16–30 h) which takes advantage of the naturally occurring enzymes in yeast that break down cell walls and thus release proteins, peptides, B-vitamins, and other bioactive nutrients into the solution (Oliveira et al., 2022). Autolyzed BSY is typically used in the production of yeast extract, a main ingredient for flavorings and nutritional products. Increased valorization of BSY involves additional lysis, enzymatic hydrolysis, or even mechanical or chemical breakage to produce high-value products. One group of such is β-glucans and mannans, which may be used for pharmaceutical, nutraceutical, or functional feed purposes, along with nucleotides, glutathione, and other antioxidants (Avramia and Amariei, 2021; Ciobanu et al., 2024). The nutritional value of BSY (roughly 40%–49% crude protein, essential amino acids, B-complex vitamins, minerals) makes it a promising raw material in aquafeed and animal feed, thus helping to reduce the reliance on fishmeal and supporting sustainable food systems (Konstantinidis et al., 2022; Oliveira et al., 2022). During the process, sanitary processing and premature stabilization are of great concern to avoid microbial growth and deterioration, particularly for food, feed, or biotechnology applications. As circular bioeconomy rises to the forefront, brewery industries now maximize BSY valorization and reuse waste yeast as valuable raw material for feed, food, agriculture, and industrial biotechnology (Avramia and Amariei, 2021; Oliveira et al., 2022).

4. Composition and bioactive components of BSY

The principal chemical constituents of BSY are illustrated by carbohydrates, proteins, free amino acids, ash, fatty acids, and vitamins (Fărcas et al., 2017). Brews spent yeast contains about 83% of total polysaccharides and a considerable amount of BSY polysaccharides that are insoluble in nature (Jaeger et al., 2020). BSY (*Saccharomyces cerevisiae*) cell wall consists of 1%–3% chitin, 1%–23% glycogen,

50%–60% β-glucan, and 35%–40% mannoproteins (Pinto et al., 2015). The glucan part of the cell wall is a β-(1,3) glucan backbone for the most part (about 85% of yeast glucan component), with a very small amount of a β-(1,6) glucan (around 3% of the glucan component). β-glucans have been linked to a wide range of health benefits, such as antidiabetic, prebiotic, and immune-modulating effects, as well as antiinflammatory and anticancer (Jaeger et al., 2020). The latter being a result of the formation of a covalent bond between glycogen and cell wall β-(1,3) glucan through a β-(1,6) linkage. Glycogen, an-glucan, exists in two forms within the yeast cell wall: a soluble and an insoluble one (Jaeger et al., 2020). Brewer's spent yeast is a by-product with a high protein content. This product contains 45%–60% protein, and the FDA has given it the GRAS (generally regarded as safe) status (Jaeger et al., 2020). The main amino acids in the proteins of brewers' spent yeast are threonine, valine, leucine, arginine, isoleucine, lysine, methionine, cysteine, tyrosine, tryptophan, histidine, and phenylalanine (Fărcas et al., 2017). A high content (34%) of flavor-enhancing amino acids (alanine, glutamic acid, glycine, and aspartic acid) was also found, which presents the potential of BSY or BSY extracts (Jaeger et al., 2020). The mannoproteins of brewer's spent yeast (*Saccharomyces cerevisiae*) consist primarily of mannose (89%–96%), in branched short α-(1-2,6)-linked chain structures, and also contain xylose, glucose, and galactose residues attached to the protein moiety (Skendi et al., 2018). The mannoproteins, a family of glycoproteins, also have potential applications as food industry emulsifiers (Cameron et al., 1988). One more reason to consider BSY is that it has been ranked as an ideal source of B vitamins and, as a result, it presents high average concentrations of vitamin B3 (72 mg/100 g), B6 (55.1 mg/100 g), and B9 (3.01 mg/100 g). The content of B vitamins in the spent brew yeast extract was determined by Vieira et al. (2016) using High-Performance Liquid Chromatography (HPLC). Through HPLC, the authors identified and quantified the 13 phenolic compounds (ferulic acid, (±)-catechin, gallic acid, chlorogenic acid, rutin, p-coumaric acid, vanillic acid, protocatechuic acid, isoquercetin, caffeic acid, cinnamic acid, (−)-epicatechin, and syringic acid) separated from BSY. The free fraction thus had only two compounds quantified, gallic acid and (±)-catechin, six compounds being quantified in the bound fraction. (+)-catechin was the compound that was found in the highest concentration, followed by p-coumaric acid, gallic acid, protocatechuic acid, ferulic acid, and cinnamic acid (Vieira et al., 2016). The macro-minerals (calcium, sodium, potassium, and magnesium) and trace elements (zinc, chromium, copper, iron, manganese, selenium, cobalt, and molybdenum) composition of brewer's spent yeast extracts (Vieira et al., 2016). Analyses have shown that brewer's residual yeast is characterized by high carbon content, with a content ranging between 45% and 47% dry matter of yeast. The carbon/nitrogen ratio of the residue is 5.1–5.8 (Mathias et al., 2015). BSY is also rich in polyphenolic compounds (Podpora et al., 2016). Brewers' residual yeast contains a high moisture content, that is about 74%–86%. According to the source, spent yeast's mineral residue (ash) content is 2%–8.5% and its dry matter content is 10%–16% (Mathias et al., 2015). Glutathione is a compound found in fungi, animals, and plants with significant antioxidant and other metabolic properties. It is a peptide within the body, made up of glutamate, cysteine, and an unspecified amino acid of a low molecular weight, maybe serine or glycine. In yeasts (*Saccharomyces cerevisiae*), glutathione is generated through the breakdown of glucose to pyruvate, and then the pyruvate and Acetyl-CoA are fed into the citric acid cycle (Jaeger et al., 2020).

5. Immunostimulatory mechanisms of BSY in fish and shellfish

Spent yeast (*Saccharomyces cerevisiae*) from the brewer is a sustainable novel component of aquafeed as it can have positive impacts on immunostimulation and nutrition of various aquaculture species (Mahdy et al., 2022). *Saccharomyces cerevisiae* can be used as an immunostimulant for captive-reared fish that can take advantage of spent yeast's capability to stimulate growth and resistance to disease. Various studies have been carried out to establish the impacts of these immunostimulants on intestinal morphology, growth, and general health in fish (Pratiwy and Kharima, 2022). Immunostimulants are a class of biological and synthetic substances that enhance both nonspecific (innate) immunity (phagocytic, lysozyme, respiratory burst activity, complement, bactericidal activity, myeloperoxidase, total haemocytes, phenoloxidase, and nitric oxide) and specific (adaptive immunity) (agglutination titer and antibody) against the infectious diseases in various species of fish and shellfish (Srivastava and Pandey, 2015). The administration of immunostimulants can raise the level of resistance to a point where it eliminates infections caused by opportunistic pathogens and hence results in enhanced performance, increased growth, and lowered mortality during the production cycle (Raa, 1996). Brewers' spent yeast contains bioactive molecules (β-glucans, mannoproteins, chitin) in its cell wall that can trigger an innate immune response (Gokulakrishnan et al., 2022), suppress bacterial infection, and activate disease resistance in certain species of fish, such as common carp (Dawood et al., 2020a), salmonids (Tukmechi and Bandboni, 2014), Nile tilapia (El-Boshy et al., 2010), and Japanese seabass (*Lateolabrax japonicas*) (Yu et al., 2014). In fish and shellfish, the natural response has been thought of as critical element in resisting pathogens because of adaptive immune system shortcomings, their poor antibody repertoire and lymphocyte memory, maturity and slow expansion (Srivastava and Pandey, 2015). The innate immune system is activated and stimulated by foreign polysaccharide particles, called pathogen associated molecular patterns (PAMPs) molecules such as peptidoglycan, β-glucan, nucleic acid, Mannan oligosaccharide, which are recognized and bound by pattern recognition proteins (PRPs), and these molecules such as β-glucan activates cell surface resident dendritic cells and macrophages to induce innate immunity, activates prophenoloxidase (proPO) system as well as adaptive immunity which involves T-helper cells (Aderem and Ulevitch, 2000; Bricknell and Dalmo, 2005; Cerenius et al., 2008; Janeway and Medzhitov, 2002). Brewers' yeasts are a rich source of β-glucan, which is highly recognized in fishes and other vertebrates for immunostimulation (Rodrigues et al., 2020). β-glucan or

Mannan oligosaccharide of brewer's yeast will most probably possess high immune-related cytokine expression in carp, resulting in an increase in innate immune response (White et al., 2002). Mannan oligosaccharide isolated from yeast cell walls have the ability to trigger the liver to secrete mannose-binding protein that binds to the bacterial capsule and triggers a cascade reaction of the complement, one of the inbuilt defense mechanisms (Abu-Elala et al., 2018). In this mechanism of immunostimulation, β-glucan and active metabolites derived from β-glucan, like short-chain fatty acids, function as PAMP upon entry of spent yeasts into the body of fish. Other receptors, including C-type lectin receptor, Toll-like receptors (TLRs), and complement receptor have been reported to bind β-glucans as pattern recognition receptor (PRRs) (DoñateJimeno, 2009; Rodrigues et al., 2020). TLRs, along with lactosylceramide and scavenger receptor, are able to bind yeast ligands, and can be defined as β-glucan receptor for triggering signaling by the MyD88 pathway, which, subsequently, initiates further release of cytokines for conferring antifungal immunity (DoñateJimeno, 2009; Rodrigues et al., 2020). Binding of β-glucans or short-chain fatty acids to known or new PRRs has been described to activate the immune system of fish toward the promotion of phagocytosis, reactive oxygen species (ROS) production, leukotriene, interleukin, and cytokine production, and other immune reactions related to Th1 immunity (DoñateJimeno, 2009; Haque, 2017; Pogue et al., 2021; Rodrigues et al., 2020).. In this process, expert antigen-presenting cells like macrophages are stimulated by β-glucan to phagocytose and process pathogens and present antigens more effectively through autophagy and phagocytosis, which, subsequently, trigger other immune cells (DoñateJimeno, 2009; Haque, 2017; Rodrigues et al., 2020). Activation of the membrane-associated NADPH oxidase upon stimulation of the phagocytic cell membrane triggers an enhanced consumption of oxygen and the generation of reactive oxygen species (ROS), which have been found to be toxic to bacterial pathogens (Srivastava and Pandey, 2015). Leukotriene from activated macrophages has been reported to induce antiinfection immunity. Interleukin and cytokines from activated macrophages have been found to stimulate T cells, B cells, NK cells, and other macrophages for more effective defense against pathogens (Mahdy et al., 2022). Chitosan from *Saccharomyces cerevisiae*, consistently activates toll-like receptors (TLRs) of macrophages, it is assumed that phagocyte function was stimulated by brewer's yeast by-products' cell wall polysaccharides (Beutler, 2004). Peptidoglycan, nucleic acids have the capability to strengthen the immune system of the host as well as control immune-related genes and intestinal microbial guns by various physiologically active materials (Dawood et al., 2020b). Lysozyme in fish is produced in the liver and extra hepatic tissues and participates in a wide array of defense mechanisms, including immune response, bacteriolysis, antimicrobial, opsonization as well as limited antiviral and antineoplastic activity (Srivastava and Pandey, 2015). In aquatic invertebrates, nitric oxide biological functions were associated with immune defense, hemocyte aggregation, feeding, metamorphosis, environmental stress, regulation of blood pressure, symbiosis, and swimming (Palumbo, 2005). Adaptive immunity, leading to the generation of memory cells and membrane-bound receptors like T-cell receptors and immunoglobulin (Ig) enabling the rapid and effective clearance of the specific fish pathogens. Immunoglobulin (antibodies) in the form of either a B-lymphocyte receptor or secreted in plasma. A tetrameric IgM exists in teleost fish. Specific pathogen recognition and initiation of adaptive immune response is done by the lymphocytes, the B- and T-cells (Srivastava and Pandey, 2015; Swain, 2006). B-cells are involved in humoral response, whereas the T-cells are engaged in the cell-mediated response (Randelli et al., 2008). Spontaneous hemolytic activity due to the alternative complement pathway is generally high in fish serum (Lange et al., 2001). Fishes, both from sea and freshwater, possess the alternative complement pathway and classical complement pathway of activation (Holland and Lambris, 2002). The alternative complement activity independent of antibody is highly visible in fish serum; this pathway plays a significant role in fish compared to mammals (Srivastava and Pandey, 2015). Fig. 10.1 illustrates the immunomodulatory effects of brewer's spent grain in fish and shellfish. Bioactive components such as β-glucans, short-chain fatty acids, and mannan oligosaccharides activate macrophages via specific receptors, enhancing phagocytosis and cytokine secretion. This activation stimulates T cells, natural killer cells, interferon-γ production, and leukotriene-mediated inflammation, while improving immune gene expression, complement activation, and overall host defense against bacterial infections.

6. Applications of BSY in aquaculture nutrition

Aquaculture contributes significantly to livelihoods and employment generation through the cultivation of various aquatic organisms (Ganguly et al., 2024). In recent years, small indigenous fish species have been increasingly promoted to address the Fig. 8.1 nutritional needs of adolescent girls and individuals with specific health conditions (Ganguly et al., 2025c,d). The global aquaculture industry aims to develop sustainable feed formulations that reduce environmental impact while ensuring optimal growth and health performance in cultured species (Ganguly et al., 2018a,b,c). Traditional aquafeeds are heavily dependent on fishmeal, a finite and expensive marine resource. As a result, researchers and feed manufacturers are seeking alternative protein sources that are cost-effective, nutritionally balanced, and environmentally sustainable. The utilization of BSY represents not only an effective method of waste valorization but also a critical step toward a circular bioeconomy. It as a high-value ingredient capable of replacing a substantial proportion of fishmeal without compromising nutritional quality (Leyva-López et al., 2020).

7. Nutritional superiority and biochemical profile

Brewer's Spent Yeast is widely recognized as a high-protein byproduct with Generally Regarded As Safe (GRAS) status. The protein content of BSY typically ranges from 45% to 60% on a dry weight basis, with some concentrated sources reporting up to 64% (Jaeger et al., 2020). The protein is rich in essential amino acids such as

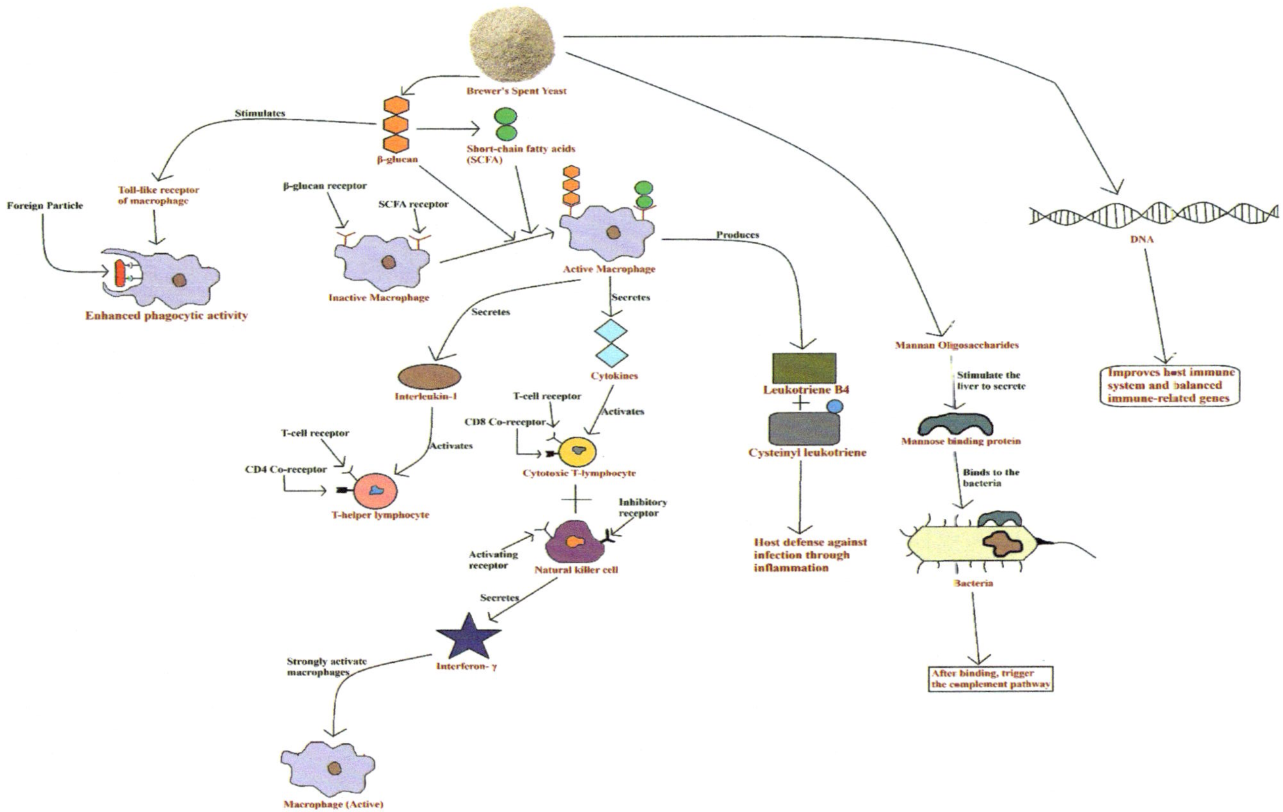

FIGURE 8.1

Immunomodulatory effect of brewer's spent grain on fish and shellfish.

lysine, methionine, threonine, and arginine. This superior amino acid profile makes BSY an effective replacement for fishmeal in aquafeeds. Gokulakrishnan et al. (2022) reported that BSY could replace up to 75% of fishmeal in the diet of *Anabas testudineus* (climbing perch) without compromising growth performance or feed utilization efficiency. In addition to macro-nutrients, BSY contains functional microcomponents such as nucleotides, B-complex vitamins, and trace minerals that play key roles in metabolic regulation, cell proliferation, and immune defense (Vieira et al., 2016). These compounds also contribute to improved stress tolerance and disease resistance in fish and crustaceans. The yeast cell wall components (β-glucans and mannan oligosaccharides) further enhance gut health and immune modulation, providing dual nutritional and functional benefits (Sharma, 2025).

8. Enhancing bioavailability through advanced processing

While BSY's nutritional potential is significant, its digestibility is limited by the rigid yeast cell wall, which encapsulates valuable proteins, peptides, and micronutrients (Ciobanu et al., 2023). Therefore, processing techniques that disrupt the cell wall are essential to maximize nutrient availability. Conventional methods such as autolysis and acid hydrolysis have been explored; however, each presents limitations. Enzymatic hydrolysis has emerged as the most efficient and environmentally friendly method for BSY valorization. It involves using specific proteolytic enzymes to break down the yeast cell wall, releasing intracellular proteins and generating bioactive peptides. This process significantly improves digestibility, palatability, and overall nutrient utilization (Ciobanu et al., 2023). Enzymatically hydrolyzed BSY has been shown to enhance feed conversion ratios (FCR) and promote faster growth in finfish and crustaceans, making it a high-value feed additive rather than a low-cost supplement.

Feed processing methods also play a crucial role in nutrient retention and feed acceptance. Therefore, optimizing both chemical and physical processing is necessary to maintain nutrient quality and ensure desirable organoleptic properties in BSY-based feeds.

9. Functional roles in digestive modulation and growth

BSY also exerts functional effects on the digestive physiology of aquatic animals. The inclusion of yeast-based components in the diet stimulates the secretion of endogenous digestive enzymes, thereby enhancing the breakdown and absorption of nutrients. Studies have shown that BSY supplementation (1%–2%) in *Diplodussargus* (white seabream) increased amylase and protease activities, while in *Argyrosomusregius* (meager), it specifically elevated amylase activity (Castro et al., 2013). These findings indicate that BSY acts as a digestive modulator, optimizing nutrient utilization efficiency even at low inclusion levels. The improved enzyme activity and digestibility translate into better growth performance and feed efficiency. In climbing perch, 50%–75% of the dietary substitution of fishmeal with BSY resulted improved growth performance and nutrient retention (Gokulakrishnan et al., 2022). These outcomes confirm BSY's potential to support efficient aquaculture production while minimizing reliance on marine-derived proteins.

10. Other agro-industrial by-products as immunostimulants

Agro-Industrial by-products also serve as key sources of minerals, pigments, vitamins, and other high phytogenic properties compounds (such as flavonoids and phenolic acids), antimicrobial compounds, immune system stimulants, antioxidants, and enzymes in digestion. Agro-Industrial by-products have a high nutritional value, they possess high levels of simple and complex carbohydrates and high fiber (predominantly in the pomaces, leaves, and peels). Such material, such as plant-derived substances, immunoglobulins, and lectins, is one of the immunomodulatory substances found in agro-industrial by-products that not only increase their resistance to diseases and pathogens but also modulate the immune system of aquatic animals (Lourenço et al., 2025). The use of bio-compatible autochthonous probiotics offers dual benefits by protecting fish against virulent pathogens and simultaneously enhancing their overall productivity (Chattaraj et al., 2022; Ganguly et al., 2019; Ganguly et al., 2025e). Some by-products contain probiotic microorganisms or have a prebiotic effect, which facilitates the development of a healthy gut flora in fish and, thus, their immune system gets strengthened (Ringø et al., 2010).

1. **Olive oil industry:** Olive oil production produces significant quantities of by-products of importance to aquafeed industry, i.e., the olive leaves, olive pomace, and olive oil pomace (Difonzo et al., 2021; Hazreen-Nita et al., 2022). The olive pomace consists, predominantly, of carbohydrates (such as fiber) followed by protein and fat. The olive pomace oil consists of the lipid part of this by-product and contains a high proportion of oleic acid, accompanied by palmitic, linoleic, and stearic acids (Nunes et al., 2018). The olive oil pomace appears to have the most potential as a (partial) fish oil substitute (Nasopoulou and Zabetakis, 2012). The olive pomace oil was effectively utilized to substitute fish oil with inclusion levels of up to 8% in rainbow trout and gilthead seabream diets, demonstrating that apart from the positive responses in terms of feed utilization and growth performance, inclusion of these sources of fatty acids significantly enhanced the lipid quality of fish fillet (Lourenço et al., 2025).
2. **Pea processing industry:** Peas have been referred to as an excellent source of proteins of good quality that comprise a significant amount of essential amino acids such as lysine, leucine, phenylalanine, threonine, methionine, and tryptophan (Nasir et al., 2024; Kumari and Deka, 2021). Not only that, but pea pods, which have been considered waste, are still packed with proteins and, in addition, they contain fiber and storage carbohydrates. The substances are

mostly Mg, Ca, and K, with the Na/K ratio of 0.07, while the rest of the minerals (Cu, Fe, Zn) are in traces (Lourenço et al., 2025). Pea peel meal was used to substitute fish meal at the levels of 15% and 25% in aquafeed for Nile tilapia (Desouky et al., 2023).

3. **Wine industry:** Grape pomace, which is a by-product of the wine sector, consists of the peels and seeds as well as the stems. Grape by-product can be utilized as a feed aquaculture ingredient and is rich in lipids, proteins, and starch (Quagliardi et al., 2024). Meanwhile, the pulp of grapes also carries polyphenols and flavonoids (Lourenço et al., 2025; Quagliardi et al., 2024). In their review article, they reported that by-products of grapes constitute a class of good phytonutrients that could be used as supplements in aquafeed, thereby promoting the growth and health of fish. Grape seeds are concentrated with polyphenolic compounds and proanthocyanidins having one or more health-promoting properties like antiinflammatory, antioxidant, and toxicity resistance-enhancing (Mehrinakhi et al., 2021). Yang et al., (2023) experimented on the influence of grape seed extract on the growth, immune response, and resistance to the disease of the common carp. They tested four levels of inclusion (0, 10, 20, and 30 g/kg) in a basal diet with 23% DM crude protein content in a feeding trial. Their findings showed that the inclusion of 20 g/kg or more led to an up-regulation of immune activity, growth performance, feed efficiency, and fish resistance to pathogenic infections. It is worth mentioning that grape seed oil was reported to be a good candidate for the partial replacement of fish oil in a dressing of rainbow trout; thus, the antioxidative power, the growth rate, and the fatty acid content of rainbow trout were enhanced (Mahmoodi et al., 2023).
4. **Tomato processing industry:** Tomato processing by-products are seeds, peels, and fibrous fractions, commonly known as tomato bagasse or pomace. Tomato byproducts comprise various valuable compounds such as carotenoids (in the form of β-carotene and lycopene) and phenolic compounds identified as enhancer activities (Lourenço et al., 2025). Dried tomato pomace as a feed ingredient for the common carp (Amirkolaie et al., 2015). Kesbiç et al. (2022) revised the common carp diets by replacing 0.5%, 1%, 2% and 5% of corn starch with dried tomato pomace. Results showed that the inclusion of this carotenoid-rich industrial byproduct in the feed improved the health of the fish.
5. **Apple and pear processing industry:** Pomaces, peels, seeds, and stems of apple and pear are products of the fruit processing and juice sector. Apple and pear pomace contain rich amounts of carbohydrate (sugars and fiber) and low amounts of protein and fats (Lourenço et al., 2025). Apple pomace has higher contents of Na, K, Mg; P, Ca, and still retains trace amounts of Mn, Fe, Cu, and Zn (Skinner et al., 2018). Pear pomace contains minerals such as, in order of decreasing amount Na, P, K, Fe, Mg, and Ca (Lomba-Viana et al., 2022). Pectin, a carbohydrate found in fruit peels, is a potential candidate for dietary supplements. Supplementing feed with 2%–4% apple peel-purified pectin in

common carp led to enhancement in growth performance, feed utilization, immune and oxidative status (Hoseinifar et al., 2021).

6. **Banana processing industry:** Banana by-products such as leaves, overripe fruits, blossoms, peels, and pseudostems contain a wealth of nutrients such as dietary fiber, vitamins, protein, and carbohydrates (Kumari et al., 2023). The waste products of bananas have a lot of phytochemicals in them that include phenolic compounds, carotenoids, vitamin C, vitamin E, and vitamin B complex. Besides potassium, which is the major mineral in banana by-products, there are also manganese (Mn), calcium (Ca), and phosphorus (P) (Lourenço et al., 2025). To enhance the growth of Nile tilapia, banana peel was used with a positive effect on the overall growth, feed efficiency, and nutrient digestibility. The addition of banana peel at 30% level in diets resulted in the enhancement of growth performance and feed utilization (Yossa et al., 2022).
7. **Pumpkin processing industry:** Pumpkin peels and seeds that are normally thrown away during pumpkin processing are still loaded with nutrients. Peels consist of a high proportion of fiber, followed by protein and, to a lesser extent, by lipids, whereas the seeds consist primarily of lipids and proteins (Lourenço et al., 2025). Pumpkin peels are also a good source of vitamin C, β-carotene, minerals (i.e., Fe and P), and phenolic compounds. The richest minerals present in seeds are Zn and K, supplemented by Na, Fe, and Ca (Hussain et al., 2022). Pumpkin seeds also have significant levels of tocopherol and phenolics, especially flavonoids (Hussain et al., 2021, 2022). Pumpkin seed cake as a replacement for soybean meal in Nile tilapia diets, with evidence that replacing 40% soybean meal enhanced growth performance of fish, feed conversion ratio, and immune and oxidative stress responses (Mounes et al., 2024).

11. Bioactive compounds from agro-industrial byproducts

A number of value-added products and bioactive compounds with the potential to be applied in the aquafeed sector are produced by agro-industrial by-products (Dawood et al., 2022b). Any plant part can be defined as a bioactive compound source (Veneziani et al., 2017). Various bioactive compounds of fruit processing by-products are crucial for the development and health performances of aquatic animals, i.e., saponins, phenolic compounds, glucosinolates, terpenes, and dietary fiber (Leyva-López et al., 2020).

1. **Terpenes:** Terpenes are products of isoprene units. Terpenes are grouped based on the number of condensed isoprene units (Kirby and Keasling, 2009). Terpenes are further grouped into essential oils and carotenoids that are distinguished by their antioxidant activity (Hassoun and Coban, 2017). The antimicrobial activity of essential oils is connected with their hydrophobicity. This characteristic enables essential oils to permeate the cell wall and cytoplasmic membrane and disrupt the structure of its constituents. Apart from that,

essential oils can inhibit energy regulation and structural component synthesis enzymes (Hyldgaard et al., 2012).

2. **Saponins:** Saponins are amphipathic molecules consisting of sugar residues attached to a network of polycyclic rings by glycosidic linkages (Yang et al., 2014). Saponins occur in plant products like agave or legumes (Sidana et al., 2016). Saponins possess immunostimulatory activities (Sarikahya et al., 2018). The action of saponins is increased against Grampositive microorganisms, whereas Gramnegative bacteria are resistant, perhaps because they contain the double lipid membrane (Arabski et al., 2012). Saponins have microbiota modulating effect, which is connected with their antimicrobial potential. Also, saponins have the ability to break down the cell membrane and hence movement of extracellular and intracellular contents is facilitated (Jacob et al., 1991).
3. **Phenolic compounds:** Phenolic compounds are one of the secondary metabolites in plants that are usually present in the form of glycosides or esters (Vermerris and Nicholson, 2008). The main structure of phenolic compounds consists of an aromatic ring with one or more hydroxyl groups attached. Phenolic compounds have been found to be endowed with a wide range of biological effects, namely antiparasitic, immunostimulants, anticancer, antiviral, antioxidants, antihypertensives, antiinflammatory, and microbiota modulators (Leyva-López et al., 2020). Phenolic compounds can perform their bioactivity through the interaction with the components of the cell membrane, enzymes, transcription factors, and receptors. Phenolic compounds bring antioxidant protection to the membrane by the interaction of their hydrophilic and hydrophobic regions with the polar head and nonpolar chain of lipid bilayer (Fraga et al., 2010). Phenolic compound has indirect antioxidant and antiinflammatory effects through Nrf2 activation and inhibition of NF-κB nuclear translocation, respectively (Hussain et al., 2017).
4. **Dietary fiber:** Dietary fiber is made up of carbohydrate polymers of three or more units that are resistant to the action of the body's digestive enzymes and, therefore, cannot be hydrolyzed or absorbed by the small intestine (Makki et al., 2018). Fiber is categorized into insoluble and soluble fiber. Soluble fibers such as β-glucans, fructooligosaccharides, galactooligosaccharides, and some pectins are digested by intestinal microbiota (Deehan et al., 2017). In aquaculture, β-glucans are an essential source of energy for the immune system (Medina-Gali et al., 2018). The immunostimulatory properties of β-glucan depend on their interaction and identification with receptors on the membrane (for example, dectin-1, and CR3) (Jin et al., 2018).
5. **Glucosinolates:** The glucosinolates are glycosides structured by β-D-glucopyranose residue attached to a (Z)-N-hydroximinosulfate ester by sulfur bridges and an amino acid derivative radical. They are present in all Brassica family species, including canola, broccoli, arugula, and mustard (Bischo, 2016). Glucosinolates and the degradation products thereof (isothiocyanates) exhibit antibacterial activity, antioxidant, and anticancer. As a matter of fact, these compounds act as indirect antioxidants because they have the ability to regulate

the activity of enzymes involved in the metabolism of foreign substances (Phase I and Phase II), which is the very first step of long-term antioxidant reactions (Vig et al., 2009). The release of the killing agents from the glucosinolate metabolism has been linked to the situation where the intracellular enzyme systems of one of the targeted bacteria, which are responsible for energy-generating ATP synthesis, are affected, thus resulting in bacterial cessation (Leyva-López et al., 2020).

12. Future directions

The future of BSY in aquaculture nutrition is focused on sustainability aspects in aquaculture, as well as the valorization of waste for the creation of value in the circular bioeconomy. Indeed, with the growth in global aquaculture production, anticipated to be over 200 million metric tons by 2030, the aquaculture industry must grow without damaging the environment. The major source of fishmeal and fish oil for aquaculture from wild fish is coming under pressure due to environmental concerns and high prices. BSY, particularly from the brewing industry, provides a potential solution in converting low-value waste to a precious feed resource. This goes with circular economy thinking, which is that what one sector calls waste becomes another's resource. Breweries produce large volumes of spent yeast yearly, which are directly landfilled or incinerated, or are used for low-value products such as bedding for animals. Instead, the use of such biomasses for aquacultural sources of feed will result in cost reduction and environmental harm limitation, less reliance on marine proteins, a decrease of the fish-in-fish-out (FIFO) ratio, as well as alleviated stress on overfished populations. Integration has almost all the possibilities. BSY may dominate the future of aquaculture in IMTA, where the yeast-based bioflocs serve as a principal source of on-site feed while cleaning the water. In biofloc systems, BSY-derived carbon and nitrogen are converted by bacteria into microbial protein easily available for fish and shrimp. In addition, this protein production on site enhances feed efficiency while reducing water changes and releasing waste. Further, the β-glucans and mannoproteins in BSY assist the formation of floc by microbial groups, allowing them to be robust and prevent nutrient loss. The role of BSY in such systems may allow for practically waste-free, closed-loop designs. Pilot studies with tilapia and shrimp have shown that, in terms of growth performance, BSY contribution at the range of 30%–50% can keep fish growing as well as those fed standard feeds, whereas the external feed input may be lowered by 20%–40%. In order to scale this model, there is then a need for common procedures of preparing BSY, using the microbial additives, and administering the real-time carbon-nitrogen ratio to avoid ammonia spikes. One of the other significant improvements in animal nutrition is that feeds are now built to specifications according to the immune and body requirements of the species. This immune booster is strong through BSY. On the whole, however, the active parts of BSY—β-1,3/1,6-glucans, mannoproteins, nucleotides, and peptides—show differing activity with different aquatic species. For example, the fish species such as trout have been found to respond well to the downregulation of proinflammatory cytokines when using glucans, whereas shrimp gained from more hemocyte production working with phenoloxidase. Conversely, they get better digestive health and digestive enzymes release because of nucleotides and short-chain peptides from BSY. The future will be developing precise formulations where BSY could be subjected through enzymes or microbes to get particular components to that species and life stage. For instance, the case with young fish aspects of feeds could focus on easy-to-digest hydrolysates that are high in free amino acids and nucleotides, which will help in the development of early immunity in risky periods. Meanwhile, grow-out diets may focus on whole cell wall parts, which will maintain that long-term defense to pathogens. Improving directed fermentation using cocultures of bacteria could actually break down nucleic acids but preserve or augment the contributions by β-glucan, therefore rendering possible the designer BSY with better use. Integrating BSY with precision aquaculture affects nutrient management, where real-time health checks with biosensors and machine learning can modify feed according to bodily needs. For instance, an animal tracked acoustically that has high cortisol levels or lower feed intake would receive more BSY to promote stress resistance. Digestive issues could prescribe feed withdrawal to avoid uric acid buildup. Using big data, AI models should make excellent disease risk predictions, allowing BSY to prime immunity prior to pathogen invasion. These systems will reduce fish death by 25%–35% in high-stress periods in aquaculture systems. The combination of digital technologies with BSY renders the feed a very useful tool in improving health with very little wastage. The speed of adoption will depend totally on the viability of BSY to scale up and become cost-competitive. Its efficacy has been proven by research; however, logistics, regulations, and cost pose big challenges under commercialization. Also, BSY is dependent on brewing strain, type of fermentation, and debittering, and hence standardization and good control are very important. Automating ways to maintain quality will be important with the incorporation of the work in the feed mill. Indications suggest that BSY could, at a very low price in the long run, be produced on a large scale, thus saving on feed costs. Partnerships, farms, and policy will help reduce the investment risk. Advancing technology will enhance nutrient use efficiency and produce even more products, thus improving economic value. This way, BSY may nominally capture a share of the aquafeed protein market, replacing fishmeal and achieving long-term sustainability for aquaculture.

13. Conclusion

BSY represents a radical change in aquaculture nutrition and is a great example of how agro-industrial by-products can be turned into cheap, environmentally friendly immunostimulants and protein sources. BSY, with its protein content varying between 45% and 55% and being completely supplemented with all essential amino acids, nucleotides, β-glucans, and mannoproteins, allows a very high replacement of fishmeal doses, growth, and product quality not being compromised, from 30% in carnivorous salmonids to more than 80% in omnivorous tilapia and carp. Growth rates, feed conversion ratios, and survival rates are consistently reported for the inclusion levels between 10% and 31% and range from 1.5% to 2.3% per day, 1.2–1.8%, and over 90%, respectively, even as culture conditions are difficult. Fish fed with BSY in biofloc systems benefit from better protein synthesis from the microbial community, improved water quality, and a lower necessity for external feed, thus initiating a positive nutrition-environment interaction cycle. Brewer's spent yeast can be considered a multifunctional feed ingredient, which basically helps productivity and, at the same time, animal welfare, because of its ability to repair the intestinal mucosa, increase the activity of digestive enzymes, and regulate the expression of immune genes. One of the most notable features of BSY is its immunostimulatory property, which becomes very useful in a time when disease pressure is increasing, and there are restrictions from the regulators on the use of antibiotics. BSY, which is able to do so by activating pattern recognition receptors, enhancing phagocytic activity, and also upregulating those immune effectors lysozyme, complement proteins, and acute-phase reactants, provides strong protection against bacterial pathogens like *Vibrio* sp., Aeromonas sp., and Streptococcus sp. In shrimp, it elevates hemocyte counts and phenoloxidase activity that lead to better survival in White Spot Syndrome Virus or Early Mortality Syndrome situations. In finfish, it alleviates the inflammatory damage that takes place during stress periods, thus the quality as well as the value of the fillet is preserved. These phenomena, which are the main reasons for a radical reduction in the use of antibiotics, are the most important ones among the effects that collectively account for less than 70%–80% of the disease management costs in intensive farming. BSY, by allowing the replacement of prophylactic drugs with natural immunomodulators, plays a dual role: on the one hand, it dampens the menace of antimicrobial resistance that is a global threat, and on the other hand, it responds to the consumers' preference for clean-label and sustainably produced seafood. After BSY, a larger suite of agro-industrial by-products—spent grains, rice bran hydrolysates, fruit pomace, distillers dried grains, and poultry by-product meals—are going through the same process; thus, collectively, they are forming a diversified, resilient feed matrix. These alternatives make the price volatilities of traditional commodities less vulnerable, facilitate nutrient recycling, and considerably reduce the environmental footprint of aquafeed production. In combination, they significantly lower the fish-in-fish-out (FIFO) ratio, which is around 0.5:1 or even less, as opposed to 0.8:1 in traditional systems, thus helping to conserve marine biodiversity and ensure protein availability for the future. The carbon footprint of feed production is reduced accordingly with BSY-based diets, which could be the reason for the saving of 1.5–2.0 kg CO_2-equivalent per kg fish produced. This eco-friendly benefit is also realized in terms of water and land use: the reuse of brewery waste not only stops the emission of methane from landfills but also reduces the area of agricultural land needed for soy or corn growing. The financial consequences are of similar magnitude. At 30%–50% fishmeal substitution,

BSY cuts feed costs by 10%–20%, thereby increasing the profit margins of farmers and making aquaculture products more affordable to consumers. In biofloc-integrated systems, the overall production costs may be lowered by 25%–35% due to the savings in water, energy, and labor. These cost savings have a great impact, especially on the regions that are still developing—Asia, Africa, and Latin America—where aquaculture of smallholders is prevalent, and the availability of premium ingredients is quite limited. By using brewery by-products that are locally available, these producers become more competitive; thus, rural employment in waste collection, processing, and feed manufacturing gets promoted. The social equity side, therefore, being a core part of the BSY story, is the element that makes high-performance nutrition accessible to the less fortunate, thus giving them the opportunity to become part of the blue economy. From an environmental point of view, BSY represents a regenerative aquaculture model, which is a system that lessens damage and, at the same time, helps to bring back the environmental balance. By implementing closed loops for nutrients, lowering the chances of eutrophication, and also lessening the need for wild fisheries, BSY is, first of all, a tool for the marine ecosystems' restoration and secondly for biodiversity hotspots' preservation. Where global warming is concerned, the ability of BSY to remain stable under different temperature and pH levels makes it a resilient source of food for aquatic organisms that live in warming waters. Moreover, by supplying the heat shock protein through induction and the antioxidant enzyme through activation, BSY, as a climate-adaptive feed solution, plays a role in the alleviation of environmental stress. As weather extremes cause disruption in supply chains, the local and waste-derived nature of BSY makes the supply chain resilient, thus it is capable of providing a buffer against global shocks. Its partnership with agro-industrial waste and its allies in the realm of aquaculture nutrition signifies a paradigmatic change: going from linear and extractive models to circular and restorative ones. The said waste becomes a resource, vulnerability turns into resilience, and, finally, environmental liability converts into an ecological asset. The changeover is not simply technical but philosophical in nature. In fact, feed utilization as a loss-making department is replaced by that of a sustainability leverage tool. To materialize this dream, it is necessary to have a coordinated approach. That means the commitment to supply the processing plants with the necessary infrastructure, the regulatory harmonization for the new ingredients, farmer training in functional feed use, and consumer knowledge of the merit of upcycled products. The full acceptance of BSY will make it possible not only to prolong the growth of aquaculture but also to raise the latter as a net-positive contributor to global food systems, planetary health, and intergenerational equity. BSY is an example of a far-reaching possibility: a demonstration that innovation, if it is based on circularity and inclusivity, can feed the world without costing the Earth. BSY is a demonstration of the viability of a far-reaching opportunity: that innovation, if firmly grounded in circularity and inclusiveness, could feed the world without putting the Earth in danger.

References

Abu-Elala, N.M., Younis, N.A., AbuBakr, H.O., Ragaa, N.M., Borges, L.L., Bonato, M.A., 2018. Efficacy of dietary yeast cell wall supplementation on the nutrition and immune response of Nile tilapia. Egypt. J. Aquat. Res 44 (4), 333–341. https://doi.org/10.1016/j.cjar.2018.11.001.

Aderem, A., Ulevitch, R.J., 2000. Toll-like receptors in the induction of the innate immune response. Nature 406 (6797), 782–787. https://doi.org/10.1038/35021228.

Ahmadifar, E., Shahriari Moghadam, M., Dawood, M.A.O., Hoseinifar, S.H., 2019. *Lactobacillus* fermentum and/or ferulic acid improved the immune responses, antioxidative defence and resistance against Aeromonas hydrophila in common carp (*Cyprinus carpio*) fingerlings. Fish Shellfish Immunol 94, 916–923. https://doi.org/10.1016/j.fsi.2019.10.019.

Amirkolaie, A.K., Dadashi, F., Ouraji, H., Khalili, K.J., 2015. The potential of tomato pomace as a feed ingredient in common carp (*Cyprinus carpio* L.) diet. J. Anim. Feed Sci 24 (2), 153–159. https://doi.org/10.22358/jafs/65642/2015.

Arabski, M., Węgierek-Ciuk, A., Czerwonka, G., Lankoff, A., Kaca, W., 2012. Effects of Saponins against clinical *E. coli* strains and eukaryotic cell line. J. Biomed. Biotechnol 2012, 1–6. https://doi.org/10.1155/2012/286216.

Avramia, I., Amariei, S., 2021. Spent Brewer's yeast as a source of insoluble β-glucans. Int. J. Mol. Sci. 22 (2), 825. https://doi.org/10.3390/ijms22020825.

Beutler, B., 2004. Inferences, questions and possibilities in Toll-like receptor signalling. Nature 430 (6996), 257–263. https://doi.org/10.1038/nature02761.

Bischo, K.L., 2016. Chapter 40—glucosinolates. In: Gupta, R.C. (Ed.), Nutraceuticals. Academic Press, Boston, MA, USA, pp. 551–554.

Bricknell, I., Dalmo, R., 2005. The use of immunostimulants in fish larval aquaculture. Fish Shellfish Immunol 19 (5), 457–472. https://doi.org/10.1016/j.fsi.2005.03.008.

Cameron, D.R., Cooper, D.G., Neufeld, R.J., 1988. The mannoprotein of *Saccharomyces cerevisiae* is an effective bioemulsifier. Appl. Environ. Microbiol 54 (6), 1420–1425. https://doi.org/10.1128/aem.54.6.1420-1425.1988.

Castro, C., Pérez-Jiménez, A., Coutinho, F., Pousão-Ferreira, P., Brandão, T.M., Oliva-Teles, A., Peres, H., 2013. Digestive enzymes of meagre (Argyrosomusregius) and white seabream (*Diplodussargus*): effects of dietary brewer's spent yeast supplementation. Aquaculture 416–417, 322–327. https://doi.org/10.1016/j.aquaculture.2013.09.042.

Cerenius, L., Lee, B.L., Söderhäll, K., 2008. The proPO-system: pros and cons for its role in invertebrate immunity. Trends Immunol 29 (6), 263–271. https://doi.org/10.1016/j.it.2008.02.009.

Chattaraj, S., Chattaraj, M., Mitra, D., Ganguly, A., Thatoi, H., Das Mohapatra, P.K., 2025. 16S amplicon sequencing of the gastrointestinal microbiota of Cirrhinus reba and isolation of an autochthonous probiotic using culture based approaches. Syst. Microbiol. Biomanufact. 5 (1), 156–170.

Chattaraj, S., Ganguly, A., Mandal, A., Das Mohapatra, P.K., 2022. A review of the role of probiotics for the control of viral diseases in aquaculture. Aquac. Int 30 (5), 2513–2539. https://doi.org/10.1007/s10499-022-00915-6.

Chattaraj, S., Behera, B.K., Das Mohapatra, P.K., 2024a. Effective valorization of *Chlorella* biomass and Brewers spent grain substituting fish meal and soybean meal in the diet of herbivorous fish *Cirrhinus reba* for higher growth, digestibility and sustainable cultivation. J. Appl. Phycol. 36 (1), 217–232. https://doi.org/10.1007/s10811-023-03106-9.

Chattaraj, S., Mitra, D., Ganguly, A., Thatoi, H., Mohapatra, P.K.D., 2024b. A critical review on the biotechnological potential of Brewers' waste: challenges and future alternatives. Curr. Res. Microb. Sci. 6, 100228. https://doi.org/10.1016/j.crmicr.2024.100228.

Chattaraj, S., Mitra, D., Chattaraj, M., Ganguly, A., Thatoi, H., Mohapatra, P.K.D., 2024c. Brewers' spent grain as fish feed ingredient: evaluation of bio-safety and analysis of its impact on gut bacteria of *Cirrhinusreba* by 16S Metagenomic sequencing. Curr. Res. Microb. Sci. 7, 100286. https://doi.org/10.1016/j.crmicr.2024.100286.

Chattaraj, S., Ganguly, A., Mitra, D., et al., 2026b. Sustainable valorization of algal biomass in fish nutrition and dietetics: a critical review and scientometric study. Waste Biomass Valor. https://doi.org/10.1007/s12649-026-03619-7.

Ciobanu, L.T., Constantinescu-Aruxandei, D., Tritean, N., Lupu, C., Negrilă, R.N., Farcasanu, I.C., Oancea, F., 2023. Valorization of spent brewer's yeast bioactive components via an optimized ultrasonication process. Fermentation 9 (11), 952. https://doi.org/10.3390/fermentation9110952.

Chattaraj, S., Sahoo, S.S., Behera, S., Thatoi, H., 2026a. Sustainable aquafeed development using Spirulina: effects on growth performance of Catla catla. Algal Res. 104447.

Ciobanu, L.T., Constantinescu-Aruxandei, D., Farcasanu, I.C., Oancea, F., 2024. Spent brewer's yeast lysis enables a best out of waste approach in the beer industry. Int. J. Mol. Sci. 25 (23), 12655. https://doi.org/10.3390/ijms252312655.

Dawood, M.A., Magouz, F.I., Essa, M., Mansour, M., 2020a. Impact of yeast fermented poultry by-product meal on growth, digestive enzyme activities, intestinal morphometry and immune response traits of common carp (*Cyprinus carpio*). Ann. Anim. Sci. 20 (3), 939–959. https://doi.org/10.2478/aoas-2020-0021.

Dawood, M.A., Moustafa, E.M., Elbialy, Z.I., Farrag, F., Lolo, E.E., Abdel-Daim, H.A., Van Doan, H., 2020b. Lactobacillus plantarum L-137 and/or β-glucan impacted the histopathological, antioxidant, immune-related genes and resistance of Nile tilapia (*Oreochromis-niloticus*) against *Aeromonashydrophila*. Res. Vet. Sci. 130, 212–221. https://doi.org/10.1016/j.rvsc.2020.03.019.

Dawood, M.A.O., Habotta, O.A.E., Elsabagh, M., Azra, M.N., Van Doan, H., Kari, Z.A., Sewilam, H., 2022. Fruit processing by-products in the aquafeed industry: a feasible strategy for aquaculture sustainability. Rev. Aquacult 14 (4), 1945–1965. https://doi.org/10.1111/raq.12680.

Deehan, E.C., Duar, R.M., Armet, A.M., Perez-Muñoz, M.E., Jin, M., Walter, J., 2017. Modulation of the gastrointestinal microbiome with nondigestible fermentable carbohydrates to improve human health. Microbiol. Spectr 5 (5). https://doi.org/10.1128/microbiolspec.BAD-0019-2017.

Desouky, A.M., Hwihy, H.M., Shaban, W.M., Azab, A.M., 2023. Evaluating of pea peels meal as a fishmeal alternative in formulated diet ingredients of *Oreochromis niloticus*. Egypt. J. Aquat. Biol. Fish 27 (6), 725–737. https://doi.org/10.21608/ejabf.2023.330193.

Difonzo, G., Troilo, M., Squeo, G., Pasqualone, A., Caponio, F., 2021. Functional compounds from olive pomace to obtain high-added value foods – a review. J. Sci. Food Agric 101 (1), 15–26. https://doi.org/10.1002/jsfa.10478.

DoñateJimeno, C., 2009. A transcriptomic approach toward understanding PAMP-driven macrophage activation and dietary immunostimulation in fish.

El-Boshy, M.E., El-Ashram, A.M., AbdelHamid, F.M., Gadalla, H.A., 2010. Immunomodulatory effect of dietary *Saccharomyces cerevisiae*, β-glucan and laminaran in mercuric chloride treated Nile tilapia (*Oreochromis niloticus*) and experimentally infected with Aeromonas hydrophila. Fish Shellfish Immunol. 28 (5–6), 802–808. https://doi.org/10.1016/j.fsi.2010.01.017.

Estevao-Rodrigues, T., Fernandes, H., Moutinho, S., Filipe, D., Fontinha, F., Magalhães, R., Peres, H., 2024. Effect of solid-state fermentation of brewer's spent grain on digestibility and digestive function of European seabass (*Dicentrarchuslabrax*) juveniles. Anim. Feed Sci. Technol. 315, 116018.

Fărcas, A.C., Socaci, S.A., Mudura, E., Dulf, F.V., Vodnar, D.C., Tofană, M., Claudia Salanță, L., 2017. Exploitation of Brewing Industry Wastes to Produce Functional Ingredients. InTech. https://doi.org/10.5772/intechopen.69231.

Ferreira, S.S., Passos, C.P., Madureira, P., Vilanova, M., Coimbra, M.A., 2015. Structure-function relationships of immunostimulatory polysaccharides: a review. Carbohydr. Polym. 132, 378–396. https://doi.org/10.1016/j.carbpol.2015.05.079.

Fraga, C.G., Galleano, M., Verstraeten, S.V., Oteiza, P.I., 2010. Basic biochemical mechanisms behind the health benefits of polyphenols. Mol. Aspect. Med. 31 (6), 435–445. https://doi.org/10.1016/j.mam.2010.09.006.

Ganguly, A., Banerjee, A., Mandal, A., Das Mohapatra, P.K., 2018a. Probiotic-based cultivation of *Clarias batrachus*: importance and future perspective. Acta Biol. Szeged. 62 (2), 158–168. https://doi.org/10.14232/abs.2018.2.158-168.

Ganguly, A., Banerjee, A., Mandal, A., Das Mohapatra, P.K., 2018b. Study of indigenous freshwater fish diversity of Bankura (West Bengal), India with special reference to *Clariasbatrachus*. J. Appl. Nat. Sci. 10 (4), 1162–1172. https://doi.org/10.31018/jans.v10i4.1892.

Ganguly, A., Banerjee, A., Mandal, A., Das Mohapatra, P.K., 2018c. Isolation and characterization of bacteria from the intestine of *Clariasbatrachus* for probiotic organism. Proc. Zool. Soc. 72, 411–419. https://doi.org/10.1007/s12595-018-0283-x.

Ganguly, A., Banerjee, A., Mandal, A., Das Mohapatra, P.K., 2019. Study of bile salt hydrolase in Lysinibacillussphaericus: a potent fish probiotic and itsin silico structure prediction for catalytic interaction. Rom. Arch. Microbiol. Immunol. 78, 81–90.

Ganguly, A., Chattaraj, S., Ganguly, M., Chattaraj, M., Banerjee, A., Mandal, A., Ali Khan, M., Sen, S.K., Das Mohapatra, P.K., 2024. Effect of three probiotic *Bacillus* strains supplemented feeds on growth, carcass composition and blood parameters of *Clariasmagur* (Hamilton). J. Appl. Aquacult. 37 (1), 19–44. https://doi.org/10.1080/10454438.2024.2330499.

Ganguly, A., Kaibarta, R., Das, S., Kundu, R., Paramanik, S., Mohapatra, S., Mukherjee, S.K., Mitra, D., Chattaraj, S., 2025a. Biodegradation and valorisation of plastic based food packets: a microbial solution for sustainability and circular economy. Discover Sustain. 6 (447). https://doi.org/10.1007/s43621-025-01257-y.

Ganguly, A., Konar, R., Karuppiah, P., Eswaranpillai, U., Ganguly, M., Chattaraj, M., Mitra, D., Chattaraj, S., 2025b. Sustainable valorization of food waste through bacterial biofilm: a pathway towards environmental safety and economic viability. In: *Biofilm Applications to Revolutionize Food Technology* (Chapter 13). Springer Nature. https://doi.org/10.1007/978-3-031-85205-3_13.

Ganguly, M., Jana, S., Ganguly, A., Chattaraj, M., Midya, D.K., 2025c. Menstrual hygiene management among the rural school-going adolescent girls with special emphasis to the Kora tribe in West Bengal, India. Discover Public Health 22, 265. https://doi.org/10.1186/s12982-025-00663-x.

Ganguly, M., Ganguly, A., Chattaraj, S., Midya, D.K., 2025d. A review on menstrual health in adolescent girls emphasizing multi-omics and machine learning strategies for preventing reproductive tract infections. Discover Public Health 22, 105. https://doi.org/10.1186/s12982-025-00485-x.

Ganguly, A., Kirtania, D.K., Chhetri, M., Chattaraj, S., 2025e. Sustainable development through probiotics: a scientometric analysis of Indian research and innovation. Discover Sustain. 6, 455. https://doi.org/10.1007/s43621-025-01263-0.

Ganguly, A., Mandal, A., Khan, M.A., Dutta, T.K., Raha, S., Das Mohapatra, P.K., 2017. Study of physico-chemical parameters, planktonic diversity and bacterial load of Clarias-batrachus cultivation pond at Bankura, WB, India. Int. J. Biol. Sci. 6 (12), 23–34.

Gilliland, R.B., 1962. Yeast reproduction during fermentation. J. Inst. Brew. 68 (3), 271–275. https://doi.org/10.1002/j.2050-0416.1962.tb01863.x.

Gokulakrishnan, M., Kumar, R., Pillai, B.R., Nanda, S., Bhuyan, S.K., Kumari, R., Debbarma, J., Ferosekhan, S., Siddaiah, G.M., Sundaray, J.K., 2022. Dietary brewer's spent yeast enhances growth, hematological parameters, and innate immune responses at reducing fishmeal concentration in the diet of climbing perch, Anabas testudineus fingerlings. Front. Nutr. 9. https://doi.org/10.3389/fnut.2022.982572.

Habotta, O.A., Dawood, M.A.O., Abdul Kari, Z., Tapingkae, W., Van Doan, H., 2022. Antioxidative and immunostimulant potential of fruit derived biomolecules in aquaculture. Fish Shellfish Immunol. 130, 317–322. https://doi.org/10.1016/j.fsi.2022.09.029.

Holland, M.C.H., Lambris, J.D., 2002. The complement system in teleosts. Fish Shellfish Immunol. 12 (5), 399–420.

Haque, M.F., 2017. Autophagy-mediated antigen presentation and its importance in adoptive immunotherapy. Int. J. Pharm. Pharmaceut. Res. 10 (2), 45–59.

Hassoun, A., Çoban, Ö.E., 2017. Essential oils for antimicrobial and antioxidant applications in fish and other seafood products. Trends Food Sci. Technol. 68, 26–36.

Hazreen-Nita, M.K., Abdul Kari, Z., Mat, K., Rusli, N.D., Mohamad Sukri, S.A., Che Harun, H., Lee, S.W., Rahman, M.M., Norazmi-Lokman, N.H., Nur-Nazifah, M., Firdaus-Nawi, M., Dawood, M.A.O., 2022. Olive oil by-products in aquafeeds: opportunities and challenges. Aquac. Rep. 22. https://doi.org/10.1016/j.aqrep.2021.100998.

Hoseinifar, S.H., Jahazi, M.A., Mohseni, R., Yousefi, M., Bayani, M., Mazandarani, M., Van Doan, H., El-Haroun, E.R., 2021. Dietary apple peel-derived pectin improved growth performance, antioxidant enzymes and immune response in common carp, *Cyprinus carpio* (Linnaeus, 1758). Aquaculture 535. https://doi.org/10.1016/j.aquaculture.2020.736311.

Huige, N.J., 2006. Brewery by-products and effluents. In: Handbook of Brewing. CRC Press, pp. 670–729.

Hussain, T., Tan, B., Liu, G., Murtaza, G., Rahu, N., Saleem, M., Yin, Y., Venditti, A., 2017. Modulatory mechanism of polyphenols and Nrf2 signaling pathway in LPS challenged pregnancy disorders. Oxid. Med. Cell. Longev. 2017 (1). https://doi.org/10.1155/2017/8254289.

Hussain, A., Kausar, T., Din, A., Murtaza, M.A., Jamil, M.A., Noreen, S., Rehman, H.u., Shabbir, H., Ramzan, M.A., 2021. Determination of total phenolic, flavonoid, carotenoid, and mineral contents in peel, flesh, and seeds of pumpkin (Cucurbita maxima). J. Food Process. Preserv. 45 (6). https://doi.org/10.1111/jfpp.15542.

Hussain, A., Kausar, T., Sehar, S., Sarwar, A., Ashraf, A.H., Jamil, M.A., Noreen, S., Rafique, A., Iftikhar, K., Quddoos, M.Y., Aslam, J., Majeed, M.A., 2022. A Comprehensive review of functional ingredients, especially bioactive compounds present in pumpkin peel, flesh and seeds, and their health benefits. Food Chem. Adv. 1. https://doi.org/10.1016/j.focha.2022.100067.

Hyldgaard, M., Mygind, T., Meyer, R.L., 2012. Essential oils in food preservation: mode of action, synergies, and interactions with food matrix components. Front. Microbiol. 3, 12.

Jacob, M.C., Favre, M., Bensa, J.C., 1991. Membrane cell permeabilisation with saponin and multiparametric analysis by flow cytometry. Cytometry 12 (6), 550 558.

Jaeger, A., Arendt, E.K., Zannini, E., Sahin, A.W., 2020. Brewer's spent yeast (BSY), an underutilized brewing by-product. Fermentation 6 (4), 123. https://doi.org/10.3390/fermentation6040123.

Janeway Jr., C.A., Medzhitov, R., 2002. Innate immune recognition. Annu. Rev. Immunol. 20 (1), 197–216.

Jin, Y., Li, P., Wang, F., 2018. β-glucans as potential immunoadjuvants: a review on the adjuvanticity, structure-activity relationship and receptor recognition properties. Vaccine 36 (35), 5235–5244.

Kesbiç, O.S., Acar, Ü., Hassaan, M.S., Yılmaz, S., Guerrera, M.C., Fazio, F., 2022. Effects of tomato paste by-product extract on growth performance and blood parameters in common carp (*Cyprinus carpio*). Animals 12 (23), 3387.

Kirby, J., Keasling, J.D., 2009. Biosynthesis of plant isoprenoids: Perspectives for microbial engineering. Annu. Rev. Plant Biol. 60, 335–355. https://doi.org/10.1146/annurev.arplant.043008.091955.

Konstantinidis, I., Firmino, J.P., Ruiz, A., Iñarra, B., San Martin, D., Estévez, A., Fernandes, J.M.O., Gisbert, E., 2022. Brewer's spent dry yeast modulates immunity in gilthead sea bream (*Sparus aurata*). Front. Mar. Sci. 9. https://doi.org/10.3389/fmars.2022.857592.

Kumari, T., Deka, S.C., 2021. Potential health benefits of garden pea seeds and pods: a review. Legume Sci. 3 (2). https://doi.org/10.1002/leg3.82.

Kumari, P., Gaur, S.S., Tiwari, R.K., 2023. Banana and its by-products: a comprehensive review on its nutritional composition and pharmacological benefits. EFood 4 (5), e110.

Lange, S., Guđmundsdottir, B.K., Magnadottir, B., 2001. Humoral immune parameters of cultured Atlantic halibut (*Hippoglossushippoglossus* L.). Fish Shellfish Immunol. 11 (6), 523–535.

Leyva-López, N., Lizárraga-Velázquez, C.E., Hernández, C., Sánchez-Gutiérrez, E.Y., 2020. Exploitation of agro-industrial waste as potential source of bioactive compounds for aquaculture. Foods 9 (7). https://doi.org/10.3390/foods9070843.

Lomba-Viana, X., Raymundo, A., Prista, C., Alegria, M.J., Sousa, I., 2022. Clean label "Rocha" pear (*Pyrus communis* L.) snack containing juice by-products and euglena gracilis microalgae. Front. Nutr. 9. https://doi.org/10.3389/fnut.2022.825999.

Lourenço, S., Neves, M., Gonçalves, E.M., Roseiro, C., Pombo, A., Raimundo, D., Pinheiro, J., 2025. Application of fruit and vegetable processing by-products as ingredients in aquafeed. Discov. Food 5 (1). https://doi.org/10.1007/s44187-025-00309-6.

Mahdy, M.A., Jamal, M.T., Al-Harb, M., Al-Mur, B.A., Haque, M.F., 2022. Use of yeasts in aquaculture nutrition and immunostimulation: a review. J. Appl. Biol. Biotechnol. 10 (5), 59–65. https://doi.org/10.7324/JABB.2022.100507.

Mahmoodi, B., Aberoumand, A., Ziaei-nejad, S., 2023. Effects of diets containing grape pomace on the growth, nutrition indices, and the quality traits of common carp (*Cyprinus carpio*). dried tomato pomace as an ingredient for the common carp feed Food Sci. Nutr. 11 (10), 6660–6669.

Makki, K., Deehan, E.C., Walter, J., Bäckhed, F., 2018. The impact of dietary fiber on gut microbiota in host health and disease. Cell Host Microbe 23 (6), 705–715. https://doi.org/10.1016/j.chom.2018.05.012.

Mathias, T.R.D.S., Alexandre, V.M.F., Cammarota, M.C., de Mello, P.P.M., Sérvulo, E.F.C., 2015. Characterization and determination of brewer's solid wastes composition. J. Inst. Brew. 121 (3), 400–404.

Medina-Gali, R.M., Ortega-Villaizan, M.d.M., Mercado, L., Novoa, B., Coll, J., Perez, L., 2018. Beta-glucan enhances the response to SVCV infection in zebrafish. Dev. Comp. Immunol. 84, 307–314. https://doi.org/10.1016/j.dci.2018.02.019.

Mehrinakhi, Z., Ahmadifar, E., Sheikhzadeh, N., Moghadam, M.S., Dawood, M.A.O., 2021. Extract of grape seed enhances the growth performance, humoral and mucosal immunity, and resistance of common carp (cyprinus carpio) against aeromonas hydrophila. Ann. Anim. Sci. 21 (1), 217–232. https://doi.org/10.2478/aoas-2020-0049.

Mounes, H.A.M., Abd-El Azeem, Z.M.A., Abd El-Bary, D.A., Al-Sagheer, A.A., Abd-Elhakim, Y.M., Hassan, B.A., Sadek, S.S., Ahmed, K.M., 2024. Effect of substituting soybean meal in *Oreochromis niloticus* diets with pumpkin (Cucurbita maxima) seed cake on water quality, growth, antioxidant capacity, immunity, and carcass composition. Animals 14 (2). https://doi.org/10.3390/ani14020195.

Nad, S., Konar, U., Chattaraj, S., Ganguly, A., 2024. Valorization of feather waste by microbial enzymatic activity: bioconversion, production and application. In: Agro-waste to Microbe Assisted Value Added Product: Challenges and Future Prospects, pp. 337–363. https://doi.org/10.1007/978-3-031-58025-3_15.

Nasir, G., Zaidi, S., Tabassum, N., Asfaq, 2024. A review on nutritional composition, health benefits and potential applications of by-products from pea processing. Biomass Convers. Biorefinery 14 (10), 10829–10842. https://doi.org/10.1007/s13399-022-03324-0.

Nasopoulou, C., Zabetakis, I., 2012. Benefits of fish oil replacement by plant originated oils in compounded fish feeds. A review. Lwt 47 (2), 217–224.

Nunes, M.A., Costa, A.S., Bessada, S., Santos, J., Puga, H., Alves, R.C., Oliveira, M.B.P., 2018. Olive pomace as a valuable source of bioactive compounds: a study regarding its lipid-and water-soluble components. Sci. Total Environ. 644, 229–236.

Oliveira, A.S., Odila Pereira, J., Ferreira, C., Faustino, M., Durão, J., Pereira, A.M., Oliveira, C.M., Pintado, M.E., Carvalho, A.P., 2022. Spent yeast valorization for food applications: effect of different extraction methodologies. Foods 11 (24). https://doi.org/10.3390/foods11244002.

Palumbo, A., 2005. Nitric oxide in marine invertebrates: a comparative perspective. Comp. Biochem. Physiol. Mol. Integr. Physiol. 142 (2), 241–248. https://doi.org/10.1016/j.cbpb.2005.05.043.

Pinto, M., Coelho, E., Nunes, A., Brandão, T., Coimbra, M.A., 2015. Valuation of brewers spent yeast polysaccharides: a structural characterization approach. Carbohydr. Polym. 116, 215–222. https://doi.org/10.1016/j.carbpol.2014.03.010.

Podpora, B., Świderski, F., Sadowska, A., Rakowska, R., Wasiak-Zys, G., 2016. Spent brewer's yeast extracts as a new component of functional food. Czech J. Food Sci. 34 (6), 554–563. https://doi.org/10.17221/419/2015-CJFS.

Pogue, R., Murphy, E.J., Fehrenbach, G.W., Rezoagli, E., Rowan, N.J., 2021. Exploiting immunomodulatory properties of β-glucans derived from natural products for improving health and sustainability in aquaculture-farmed organisms: Concise review of existing knowledge, innovation and future opportunities. Curr. Opin. Environ. Sci. Health 21. https://doi.org/10.1016/j.coesh.2021.100248.

Pratiwy, F.M., Kharima, Z.P., 2021. The use of bread yeast (*Saccharomyces cerevisiae*) in increasing growth, disease resistance, and as immunostimulating for various kinds of fish: a review. Asian J. Fish. Aquat. Res. 64–70. https://doi.org/10.9734/ajfar/2022/v20i6517.

Quagliardi, M., Frapiccini, E., Marini, M., Panfili, M., Santanatoglia, A., Kouamo Nguefang, M.L., Roncarati, A., Vittori, S., Borsetta, G., 2024. Use of grape by-products in aquaculture: new frontiers for a circular economy application. Heliyon 10 (5). https://doi.org/10.1016/j.heliyon.2024.e27443.

Raa, J., 1996. The use of immunostimulatory substances in fish and shellfish farming. Rev. Fish. Sci. 4 (3), 229–288. https://doi.org/10.1080/10641269609388587.

Rachwał, K., Waśko, A., Gustaw, K., Polak-Berecka, M., 1996. Utilization of brewery wastes in food industry. PeerJ 8, e9427. https://doi.org/10.7717/peerj.9427.

Randelli, E., Buonocore, F., Scapigliati, G., 2008. Cell markers and determinants in fish immunology. Fish Shellfish Immunol. 25 (4), 326–340.

Ringø, E., Olsen, R.E., Gifstad, T.Ø., Dalmo, R.A., Amlund, H., Hemre, G.I., Bakke, A.M., 2010. Prebiotics in aquaculture: a review. Aquac. Nutr. 16 (2), 117–136.

Rodrigues, M.V., Zanuzzo, F.S., Koch, J.F.A., de Oliveira, C.A.F., Sima, P., Vetvicka, V., 2020. Development of fish immunity and the role of β-glucan in immune responses. Molecules 25 (22). https://doi.org/10.3390/MOLECULES25225378.

Sampathkumar, K., Yu, H., Loo, S.C.J., 2023. Valorisation of industrial food waste into sustainable aquaculture feeds. Future Foods 7, 100240.

Sarikahya, N.B., Nalbantsoy, A., Top, H., Gokturk, R.S., Sumbul, H., Kirmizigul, S., 2018. Immunomodulatory, hemolytic and cytotoxic activity potentials of triterpenoid saponins from eight Cephalaria species. Phytomedicine 38, 135–144. https://doi.org/10.1016/j.phymed.2017.11.009.

Sharma, N., 2025. Exploring the potential of mannan oligosaccharides in enhancing animal growth, immunity, and overall health: a review. Carbohydr. Polym. Technol. Appl 9, 100603. https://doi.org/10.1016/j.carpta.2024.100603.

Sheikhzadeh, N., Heidarieh, M., Karimi Pashaki, A., Nofouzi, K., Farshbafi, M.A., Akbari, M., 2012. Hilyses®, fermented *Saccharomyces cerevisiae*, enhances the growth performance and skin non-specific immune parameters in rainbow trout (*Oncorhynchus mykiss*). Fish Shellfish Immunol. 32 (6), 1083–1087. https://doi.org/10.1016/j.fsi.2012.03.003.

Sidana, J., Singh, B., Sharma, O.P., 2016. Saponins of agave: Chemistry and bioactivity. Phytochemistry 130, 22–46.

Skendi, A., Harasym, J., Galanakis, C.M., 2018. Recovery of high added-value compounds from brewing and distillate processing by-products. In: Sustainable Recovery and Reutilization of Cereal Processing By-Products. Woodhead Publishing, pp. 189–225.

Skinner, R.C., Gigliotti, J.C., Ku, K.M., Tou, J.C., 2018. A comprehensive analysis of the composition, health benefits, and safety of apple pomace. Nutr. Rev. 76 (12), 893–909. https://doi.org/10.1093/nutrit/nuy033.

Srivastava, P.K., Pandey, A.K., 2015. Role of immunostimulants in immune responses of fish and shellfish. Biochem. Cell. Arch. 15 (1), 47–73.

Sultana, S., Biró, J., Kucska, B., Hancz, C., 2024. Factors affecting yeast digestibility and immunostimulation in aquatic animals. Animals 14 (19), 2851. https://doi.org/10.3390/ani14192851.

Swain, P., 2006. Species defence mechanisms of fish: the basis of antibody production and cell-mediated immune response. In: Fish Shellfish Immunology: An Introduction.

Tukmechi, A., Bandboni, M., 2014. Effects of *Saccharomyces cerevisiae* supplementation on immune response, hematological parameters, body composition and disease resistance in rainbow trout, *Oncorhynchus mykiss* (Walbaum, 1792). J. Appl. Ichthyol. 30 (1), 55–61.
Tungland, B., 2018. Human microbiota in health and disease: from pathogenesis to therapy human microbiota in health and disease: from pathogenesis to therapy. Elsevier, undefined, pp. 1–680. https://doi.org/10.1016/C2017-0-01893-1.
Veneziani, G., Novelli, E., Esposto, S., Taticchi, A., Servili, M., 2017. Applications of recovered bioactive compounds in food products. Elsevier BV, pp. 231–253. https://doi.org/10.1016/b978-0-12-805314-0.00011-x.
Vermerris, W., Nicholson, R., 2008. Families of phenolic compounds and means of classification. In: Phenolic Compound Biochemistry. Springer Netherlands, Dordrecht, pp. 1–34.
Vieira, E.F., Carvalho, J., Pinto, E., Cunha, S., Almeida, A.A., Ferreira, I.M., 2016. Nutritive value, antioxidant activity and phenolic compounds profile of brewer's spent yeast extract. J. Food Compos. Anal. 52, 44–51. https://doi.org/10.1016/j.jfca.2016.07.006.
Vig, A.P., Rampal, G., Singh Thind, T., Arora, S., 2009. Bio-protective effects of glucosinolates – a review. LWT - Food Sci. Technol. (Lebensmittel-Wissenschaft -Technol.) 42 (10), 1561–1572. https://doi.org/10.1016/j.lwt.2009.05.023.
Wei, L.S., Kari, Z.A., Kabir, M.A., Khoo, M.I., Azra, M.N., Wee, W., 2023. Promoting growth and health of african catfish, Clariasgariepinus, through dietary novel supplement, ginger, zingiberofficinalerosc, leaf powder. Aquacult. Studies 24 (4).
White, L.A., Newman, M.C., Cromwell, G.L., Lindemann, M.D., 2002. Brewers dried yeast as a source of mannan oligosaccharides for weanling pigs. J. Anim. Sci. 80 (10), 2619–2628. https://doi.org/10.1093/ansci/80.10.2619.
Yang, Y., Laval, S., Yu, B., 2014. Chemical synthesis of saponins. Adv. Carbohydr. Chem. Biochem. 71, 137–226.
Yang, H., Li, Y., Wang, G., Xie, J., Kaneko, G., Yu, E., 2023. Dietary grape seed proanthocyanidin extract improved the chemical composition, antioxidant capacity, myofiber growth and flesh quality of Nile tilapia muscle. Aquac. Rep. 33, 101878. https://doi.org/10.1016/j.aqrep.2023.101878.
Yossa, R., Ahmad Fatan, N., Kumari, J., Schrama, J.W., 2022. Apparent digestibility coefficients of banana peel, cassava peel, cocoa husk, copra waste, and sugarcane bagasse in the GIFT strain of Nile tilapia (*Oreochromis niloticus*). J. Appl. Aquacult. 34 (3), 734–754.
Yu, H.H., Han, F., Xue, M., Wang, J., Tacon, P., Zheng, Y.H., Wu, X.F., Zhang, Y.J., 2014. Efficacy and tolerance of yeast cell wall as an immunostimulant in the diet of Japanese seabass (Lateolabrax japonicus). Aquaculture 432, 217–224. https://doi.org/10.1016/j.aquaculture.2014.04.043.
Zhai, S.W., Lu, J.J., Chen, X.H., 2014. Effects of dietary grape seed proanthocyanidins on growth performance, some serum biochemical parameters and body composition of tilapia (*Oreochromis niloticus*) fingerlings. Ital. J. Anim. Sci. 13 (3), 3357.

CHAPTER

Effective valorization of microalgal biomass for utilization in fish feed

9

Arpan Mondal, Sovan Sengupta and Harisankar Dey

Applied Phycology Laboratory, Post Graduate Department of Botany, Ramakrishna Mission Vivekananda Centenary College, Kolkata, West Bengal, India

1. Introduction

Fish serves as a major protein source for billions of people as they are easily digestible and cheap in price. The aquaculture industry has experienced significant growth over recent decades with the increasing demand for seafood (Ruben et al., 2025). Traditional feed ingredients, such as fish oil and fishmeal, derived from wild-caught fish, are both economically and environmentally unsustainable. As these resources deplete, there is an urgent need to explore alternative feed ingredients that can support sustainable aquaculture practices (Ganguly et al., 2025).

Fish oil and fishmeal have been the cornerstone of aquafeed due to their high nutritional value (Majluf et al., 2024). Nevertheless, their production relies heavily on wild fish stocks, which are under considerable pressure due to overfishing. Additionally, the use of terrestrial crops like soybean meal poses its own set of issues, including land use change, deforestation, and the competition with human food crops. These concerns necessitate the exploration of alternative, sustainable feed sources to ensure the future viability of aquaculture (Chattaraj and Das Mohapatra, 2024).

Microalgae, microscopic photosynthetic organisms found in aquatic environments, have emerged as a promising alternative. These organisms are renowned for their ability to produce high-quality biomass rich in essential nutrients, making them an ideal candidate for fish feed. The valorization of microalgal biomass entails converting this raw material into a form suitable for aquafeed, leveraging its nutritional benefits while addressing sustainability concerns (Al-Muhtaseb et al., 2022).

This chapter aims to provide a comprehensive analysis of the effective valorization of microalgal biomass for utilization in fish feed (Fig. 9.1). It will delve into the nutritional profile of microalgae, their digestibility and nutrient absorption in fish, the environmental and economic benefits, practical applications through case studies, and the challenges and future prospects in this field. By synthesizing current research, this article will highlight the potential of microalgal biomass to revolutionize aquaculture feed practices and contribute to more sustainable fish farming.

Fish Feed Formulation with Agroindustrial Wastes and Byproducts. https://doi.org/10.1016/B978-0-443-41728-3.00008-3

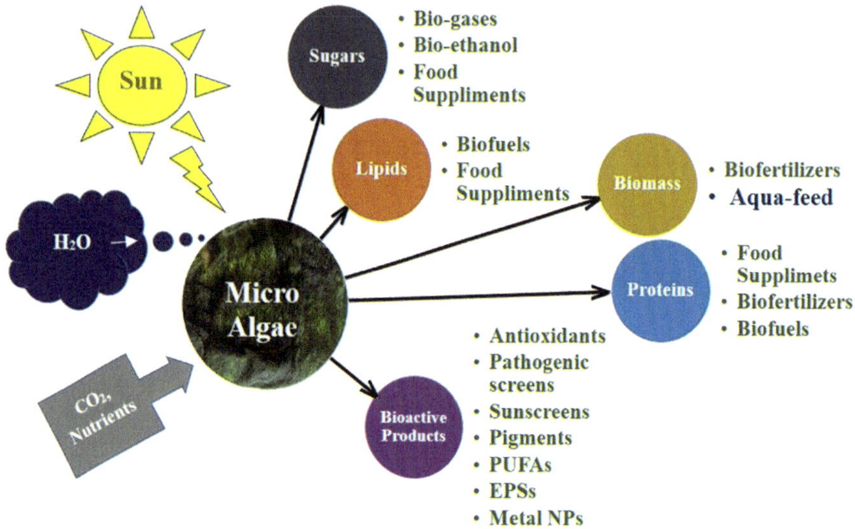

FIGURE 9.1 Utilizations of microalgal products in various fields.

Microalgae-derived bioproducts with various industrial applications.

2. Nutritional profile of microalgal biomass

2.1 Composition of microalgal biomass

Mcroalgae being a diversified group of photosynthetic microorganisms, known for their nutritional profile, makes them highly suitable for use in fish feed. The primary constituents of microalgal biomass include proteins, lipids, carbohydrates, amino acids, and fatty acids, each of which plays a crucial role in the diet of fish (Chattaraj et al., 2026a).

Protein is an important part of fish diet for fish growth and development, providing the necessary amino acids that are the building blocks of muscle tissue and other vital organs. The microalgae are known for their high protein content, ranging from 20% to 70% of their dry weight, depending on the species and cultivation conditions. Microalgae provide a complete profile of essential amino acids required by fish. Essential amino acids are those that cannot be synthesized by the fish and must be obtained through their diet. These include lysine, methionine, and histidine, among others. The balanced amino acid profile of microalgae contributes to efficient protein synthesis and muscle growth in fish.

The lipid content of microalgae varies between 1% and 40% of their dry weight. Importantly, microalgal lipids are rich in polyunsaturated fatty acids (PUFAs), including omega-3 fatty acids such as eicosapentaenoic acid (EPA) and docosahexaenoic acid (DHA). These fatty acids are vital for maintaining fish health, improving immune response, and enhancing the quality of fish fillets (Chattaraj et al., 2026b).

Microalgae also contain significant amounts of carbohydrates, which can constitute 10%–50% of their dry weight. Carbohydrates serve as an energy source for fish, supporting their metabolic activities and overall vitality.

Phenolic acids, flavonoids, lignans, and stilbenes are examples of the phenolic compounds that are present in a wide variety of plant foods. The antioxidant qualities of microalgal extracts increase with the concentration of phenolic acids and phenologlucinol. Although microalgae may not have as many phenolic compounds as other plants, their chemical structure—such as the amount of hydroxyl groups or their whereabouts in the aromatic cycle—directly affects their antioxidant efficacy. *Spirulina platensis* had the highest concentration of gallic acid chlorogenic acid that makes them a good anti-oxidant containing supplementary product.

Chlorella, well known as a super food for its high nutrient values. *Chlorella* has the protein content of approximately 45%–60% of its dry weight (Table 9.1) (Dimopoulou et al., 2025; Chattaraj et al., 2024), lipid content of around 20%–30% of its dry weight and carbohydrate content of about 10%–20% of its dry weight.

Spirulina is globally consumed for its high health benefits, as it is rich in several nutrients. It serves as a high protein source with a protein content of 60%–70% of its dry weight (Milledge et al. "The next big superfood could be green and slimy," 2011). Besides high protein content it contains a considerable amount of lipid (around 7%–12% of its dry weight) and carbohydrate (around 15%–20% of its dry weight) also that makes it a very good food supplement for not only human but also for all other live stocks.

2.2 Comparison with traditional feed ingredients

One of the primary advantages of microalgal biomass over traditional feed ingredients is its sustainability. Traditional feed ingredients like fishmeal and fish oil rely on wild-caught fish, leading to overfishing and depletion of marine resources. In contrast, microalgae can be cultivated in controlled environments using non-arable land and minimal freshwater resources (Ali et al., 2025). The production of fishmeal and fish oil often involves species like anchovies and sardines, which are harvested in large quantities, disrupting marine ecosystems. Microalgae such as *Nannochloropsis* can be grown in photobioreactors, offering a renewable and eco-friendly alternative.

Microalgal biomass is known for its nutrient density, providing a comprehensive profile of proteins, lipids, and essential nutrients (Sarker, 2023). This makes it an excellent substitute for traditional feed ingredients that may lack certain nutritional components. *Spirulina (Arthrospira platensis)* contains around 60%–70% protein content, which is significantly higher than soybean meal (typically 40%–50%) (Chattaraj et al., 2026b). Additionally, *Nannochloropsis* is rich in omega-3 fatty acids like EPA, which are crucial for fish health and are typically sourced from fish oil (Bernaerts et al., 2018).

Table 9.1 Carbohydrate, protein and fat content of different microalgal species.

Name of the species	Carbohydrates % (w/w)	Protein % (w/w)	Fat % (w/w)	References
Botryococcus braunii	19–31	39–40	25–34	Tibbetts et al. (2015)
Chaetoceros calcitrans	27	36	15	The next big superfood could be green and slimy (2011)
Chlamydomonas reinhardtii	17	48	21	The next big superfood could be green and slimy (2011)
Chlorella vulgaris	12–17	51–58	14–22	Becker and E.W. (2007)
Dunaliella salina	32	57	6	The next big superfood (2011)
Euglena gracilis	14–18	39–61	22–38	The next big superfood (2011)
Haematococcus pluvialis	27	48	15	Bleakley and Hayes (2017)
Isochrysis galbana	10–17	50–56	12–14	The next big superfood (2011)
Nannochloropsis granulate	27–36	18–34	24–28	Tibbetts et al. (2015)
Phaeodactylum tricornutum	16.8	34.8	16.1	Brown (1991)
Porphyridium cruentum	40–57	28–39	9–14	The next big superfood (2011)
Prymnesium parvum	25–33	28–45	22–38	Krimpen et al. (2013)
Scenedesmus dimorphus	21–52	8–18	16–40	The next big superfood (2011)
Scenedesmus quadricauda	21–52	47	1.9	Krimpen et al. (2013)
Skeletonema costatum	4.6	25	10	Brown (1991)
Spirogona sp.	33–64	6–20	11–21	Krimpen et al. (2013)
Spirulina maxima	13–16	60–71	6–7	The next big superfood (2011)
Spirulina platensis	8–14	46–63	4–9	The next big superfood (2011)
Synechococcus sp.	15	63	11	The next big superfood (2011)
Tetraselmis chuii	25	31–46	12	Brown (1991)

From Dimopoulou, M., Kolonas, A., Stagos, D., Gortzi, O., 2025. A review of the sustainability, chemical composition, bioactive compounds, antioxidant and antidiabetic activity, neuroprotective properties, and health benefits of microalgae. Biomass. 5 (1). https://doi.org/10.3390/biomass5010011.

The cultivation of traditional feed ingredients such as soybeans often involves deforestation and land-use changes, contributing to environmental degradation. Microalgae, on the other hand, can be cultivated in controlled environments without impacting terrestrial ecosystems. The production of soybean meal involves significant land use and deforestation, particularly in regions like the Amazon rainforest. Cultivating microalgae such as *Chlorella* in closed systems eliminates these environmental concerns, providing a more sustainable alternative. Along with that microalgae can mitigate greenhouse gas emissions through carbon sequestration thus reducing the carbon footprint for making fishmeal (Nagarajan et al., 2021).

The long-term viability and economic feasibility of algae-based products must be enhanced in order to develop them as a potential source of food or feed for both people and animals. Both the manufacturing route and the technical process of R&D efforts to attain commercial and environmental viability can be analyzed using the Techno-economic assessment (TEA) and life cycle assessment (LCA) techniques (Quinn and Davis, 2015). A techno-economic assessment was carried out on the full life cycle of a 100-hectare microalgal production plant based on six possible alternative locations. The cost of algal production in Spain has been estimated to be 3.4 euros per kilogram of dry biomass, with an anticipated decrease to 0.5 Euros/kg in 10 years due to higher photosynthetic efficiency and photobioreactors with shorter light paths. Over the next 15 years (or longer), the manufacturing of high-value metabolites (such pigments) might bring in 657 million euros in earnings (Ruiz et al., 2016).

Microalgae are rich in bioactive compounds that can enhance fish health, improving immune response, growth rates, and overall well-being. Traditional feed ingredients may lack these additional health benefits. *Tetraselmis suecica* contains bioactive compounds like carotenoids and polyphenols, which have antioxidant properties, boosting the immune system and stress tolerance in fish. These compounds are not typically found in traditional ingredients like soybean meal.

3. Digestibility and nutrient absorption

It's crucial to comprehend how the nutritional value of digestible feed components is determined. Before being employed in aquafeeds, microalgal products are evaluated for nutritional digestibility. The great diversity is a result of the numerous varieties of microalgal organisms, ranging from prokaryotes to eukaryotes. Additionally, evidence of metabolic profile variability within species has been found. Due to variations in their physical and chemical makeup, different microalgal species have varying levels of nutritional and energy digestibility.

3.1 Factors affecting digestibility

The digestibility of microalgal biomass in fish feed is influenced by various intrinsic and extrinsic factors. Understanding these factors is essential for optimizing the use of microalgae as a feed ingredient. Different species of microalgae have varying cell wall structures and nutrient profiles that can affect digestibility. The composition and structure of the cell wall play a crucial role in determining how easily fish can digest microalgal biomass. Some microalgae have thick and rigid cell walls, while others have more permeable structures (Niccolai et al., 2019). Peptidoglycan layers are found in the cell wall of prokaryotic (cyanobacterial) microalgae, whereas cellulosic layers are found in eukaryotic microalgae. Fish favor microalgae with multilayer cell walls made of peptidoglycan (murein) over those made of cellulose (Teuling et al., 2017). *Chlorella vulgaris* has a tough cell wall composed of cellulose and hemicellulose, which can hinder digestibility. Its rigid cell wall requires mechanical or enzymatic disruption to make the nutrients more accessible. On the other hand, *Spirulinaplatensis* has a more digestible cell wall composed of polysaccharides that are easier for fish to break down and absorb nutrients from.

Nutrient contents including proteins, lipids, and carbohydrates in microalgae can vary significantly between species, affecting their digestibility and nutritional value. *Tetraselmissuecica* is rich in proteins and carbohydrates, with a more digestible cell wall, making it easier for fish to absorb the nutrients. *Nannochloropsis* sp. is known for its high lipid content, particularly omega-3 fatty acids like EPA, which are beneficial for fish health (Valente et al., 2019). However, its cell wall is also robust, requiring pretreatment to improve digestibility.

3.2 Biomass pretreatment

Pretreatment methods can significantly enhance the digestibility of microalgal biomass by breaking down the cell walls and improving nutrient availability.

Techniques such as bead milling, ultrasonication, and high-pressure homogenization can physically break down the cell walls, increasing the accessibility of intracellular nutrients. Bead milling of *Chlorella vulgaris* can effectively disrupt its tough cell walls, enhancing the release of proteins and other nutrients for better absorption by fish (Selvaraj et al., 2025).

The use of specific enzymes, such as cellulases and proteases, can degrade the cell walls and complex molecules, improving the digestibility of microalgae. Enzymatic treatment of *Nannochloropsis* with cellulases and proteases can break down its cell wall components, enhancing the release of lipids and proteins for efficient absorption.

Fermentation processes involving beneficial microbes can break down complex carbohydrates and proteins in microalgal biomass, increasing its digestibility and nutritional value. Fermentation of *Chlorella vulgaris* using lactic acid bacteria can enhance its digestibility by breaking down cell wall components and releasing bioavailable nutrients.

3.3 Impact on fish growth, health, and overall performance

Efficient nutrient absorption minimizes feed waste and reduces the environmental impact of aquaculture. Microalgal biomass provides a nutrient-dense feed source, ensuring that fish receive the essential nutrients required for their overall well-being.

Chlorella vulgaris, with its high protein and carbohydrate content, offers a balanced nutritional profile. When incorporated into fish feed, *Chlorella* enhances the nutrient utilization efficiency, reducing feed conversion ratios (FCR) and minimizing waste. This results in more sustainable aquaculture practices and improved water quality.

The inclusion of microalgae in fish feed can also positively impact the quality of fish flesh, improving its texture, taste, and nutritional value. Fish fed with diets containing *Tetraselmis suecica* exhibit enhanced flesh quality, with increased levels of beneficial fatty acids like EPA and DHA. These omega-3 fatty acids are known to improve the taste and texture of fish fillets, making them more appealing to consumers. Additionally, the presence of bioactive compounds in microalgae can enhance the color and freshness of fish flesh.

Certain microalgae have been shown to positively influence the reproductive performance of fish, enhancing their breeding success and larval quality. *Isochrysis galbana* is rich in essential fatty acids and has been used as a feed supplement in broodstock diets. Fish fed with *Isochrysis* demonstrate improved reproductive performance, with higher egg production, better fertilization rates, and enhanced larval survival and growth.

By addressing the impact of digestibility and nutrient absorption on various aspects of fish growth, health, and performance, the inclusion of microalgal biomass in fish feed can contribute to more sustainable and nutritious aquaculture practices. The enhanced growth rates, improved health, and superior flesh quality achieved through the use of microalgae highlight their potential as a valuable feed ingredient in the aquaculture industry.

4. Environmental and economic benefits

4.1 Reduction in feed waste and eutrophication risk

The integration of microalgal biomass into fish feed formulations can significantly reduce feed waste and mitigate the risk of eutrophication. Traditional feed ingredients, such as fishmeal and soybean meal, often lead to inefficient nutrient utilization, resulting in excess nutrients being released into aquatic environments. This can cause eutrophication, a process where water bodies become enriched with nutrients, leading to harmful algal blooms and degradation of water quality.

Microalgae offer a nutrient-dense alternative to traditional feed ingredients, ensuring more efficient nutrient utilization and reducing feed waste. The balanced nutritional profile of microalgae, including high protein content, essential amino acids, and beneficial fatty acids, contributes to optimal fish growth and health.

Spirulina platensis has a protein content of approximately 60%–70%, with a balanced amino acid profile that supports efficient protein synthesis in fish. Studies have shown that fish fed with diets containing *Spirulina* exhibit improved nutrient utilization, resulting in lower feed conversion ratios (FCR) and reduced feed waste.

The improved digestibility of microalgal biomass ensures that fish can absorb and utilize a higher proportion of the nutrients present in their feed. This reduces the amount of undigested feed and excreted nutrients, minimizing the risk of eutrophication.

Chlorella vulgaris contains high levels of proteins and carbohydrates, which, when properly digested, provide essential nutrients for fish growth. Pretreatment methods such as enzymatic hydrolysis can enhance the digestibility of *Chlorella*, ensuring that fish absorb a greater proportion of the nutrients, reducing feed waste and nutrient excretion.

Reducing feed waste and nutrient excretion has a positive impact on water quality, as it decreases the amount of excess nutrients entering aquatic environments. This helps prevent the onset of eutrophication and maintains the ecological balance of water bodies.

The inclusion of *Nannochloropsis* in fish feed has been shown to improve nutrient utilization efficiency, leading to lower excretion of nitrogen and phosphorus. This reduction in nutrient discharge helps maintain water quality and prevents the proliferation of harmful algal blooms that can result from eutrophication (Chattaraj et al., 2026a).

4.2 Cost-effectiveness of microalgal feed

While the initial costs of microalgal cultivation and processing may be higher, the long-term economic benefits can outweigh these expenses. Microalgal biomass offers a renewable and sustainable feed source that can stabilize feed prices and reduce dependency on finite marine resources.

The production of fishmeal and fish oil is subject to price fluctuations due to overfishing and changing marine conditions. In contrast, microalgae like *Chlorella vulgaris* can be cultivated consistently in controlled environments, providing a stable and potentially cost-effective feed ingredient. Additionally, advancements in cultivation and processing technologies can further reduce the cost of microalgal biomass production.

The projected cost of making fish feed based on microalgae was 0.43 USD/kg, while the cost of traditional feed (commercial) was 0.59 USD/kg. When compared to traditional commercial feed, the formulated feed demonstrated a notable decrease in feed costs of 0.16 USD (26.1%), which represents a substantial profit margin for fish farmers (Yadav et al., 2019).

4.3 Sustainability and environmental impact

Microalgal cultivation is inherently sustainable, as it does not compete with terrestrial agriculture for arable land and can be grown using non-potable water sources. This reduces the environmental impact associated with traditional feed ingredients, such as deforestation, land-use change, and freshwater consumption.

The cultivation of microalgae such as *Nannochloropsis* in photobioreactors or open pond systems can be conducted on non-arable land, utilizing minimal freshwater resources. This contrasts with the production of soybean meal, which often involves deforestation and significant water usage. By integrating microalgal biomass into fish feed, the aquaculture industry can reduce its environmental footprint and promote more sustainable practices.

The reliance on fishmeal and fish oil as primary feed ingredients contributes to the overfishing of small pelagic fish species, putting pressure on marine ecosystems. By substituting these ingredients with microalgal biomass, the demand for wild-caught fish can be alleviated, supporting marine conservation efforts.

Microalgae like *Phaeodactylum tricornutum* are rich in omega-3 fatty acids, making them an excellent alternative to fish oil. By incorporating *Phaeodactylum* into fish feed, the aquaculture industry can reduce its dependence on wild-caught fish, helping to preserve marine biodiversity and prevent overfishing.

Microalgal biomass production can be integrated into a circular economy framework, where waste streams from other industries are utilized as nutrient sources for microalgal cultivation. This promotes resource efficiency and reduces waste generation.

Wastewater from agricultural or industrial processes can be used to cultivate microalgae like *Chlorella vulgaris*. The microalgae can then be harvested and processed into fish feed, creating a closed-loop system that minimizes waste and maximizes resource utilization. This approach not only enhances the sustainability of microalgal production but also provides economic benefits by reducing the costs associated with wastewater treatment and nutrient inputs.

5. Case study: Sustainable aquaculture practices

5.1 Successful examples of partial and complete replacement of traditional feed ingredients

Several case studies highlight the effectiveness of microalgal biomass in reducing feed waste and minimizing eutrophication risk in aquaculture operations.

The economic viability of using microalgal biomass in fish feed is a critical factor for its widespread adoption in the aquaculture industry. While the initial costs of microalgal production can be high, advancements in technology and economies of scale have the potential to make microalgal feed cost-effective in the long run. A study by Ansari et al. (2021) estimated that the production cost of microalgae ranges from \$1.50 to \$3.00 per kg of dry biomass, depending on the cultivation method and scale. As production scales up, the cost per unit of microalgal biomass decreases due to improved efficiency and reduced per-unit expenses. According to a techno-economic analysis by Stănescu et al. (2024), the cost of microalgal biomass can be reduced to \$0.80 to \$1.20 per kg when produced at large scales using optimized cultivation systems. A study by Kumar Sarker and Kaparaju (2024) highlighted that the integration of microalgae into aquaculture feed can lead to a 10%–20% reduction in overall feed costs over a 5-year period due to improved nutrient utilization and reduced feed waste.

Utilizing wastewater for microalgal cultivation can further reduce production costs and enhance sustainability. A case study by Stănescu et al. (2024) demonstrated that integrating microalgal cultivation with wastewater treatment can reduce production costs by up to 30% while providing a sustainable nutrient source for microalgae. A study by Kumar Sarker and Kaparaju (2024) estimated that microalgal cultivation can sequester up to 2.5 kg of CO_2 per kilogram of dry biomass produced, contributing to the reduction of greenhouse gas emissions. Research by Ansari et al. (2021) showed that microalgae can be grown in brackish water or wastewater, reducing freshwater consumption by up to 90% compared to soybean cultivation.

Microalgal biomass production can be integrated into a circular economy framework, utilizing waste streams from other industries as nutrient sources. This promotes resource efficiency and reduces waste generation. A study by Byung Sun Yu et al. (2024) highlighted that using agricultural or industrial wastewater for microalgal cultivation can reduce production costs by up to 30% while providing a sustainable nutrient source.

In a study conducted by Kumar Sarker and Kaparaju (2024), tilapia diets were supplemented with *Nannochloropsis* to replace a portion of fishmeal and fish oil. The inclusion of *Nannochloropsis* resulted in improved growth rates, enhanced immune response, and better feed conversion ratios (FCR). The fish exhibited similar or superior health parameters compared to those fed with traditional fishmeal and fish oil-based diets.

A study by Ansari et al. (2021) evaluated the partial replacement of soybean meal with *Chlorella* in rainbow trout diets. The inclusion of *Chlorella* led to improved protein digestibility and nutrient absorption. The fish showed enhanced growth performance and better muscle quality, with increased levels of essential fatty acids like EPA and DHA.

A study by Stănescu et al. (2024) investigated the complete replacement of traditional feed ingredients with *Spirulina platensis* in European sea bass diets. The complete replacement with *Spirulina* resulted in comparable growth rates, improved immune function, and reduced disease incidence. The fish exhibited high levels of antioxidants and bioactive compounds, contributing to better health and stress tolerance.

5.2 Impact on fish muscle composition, disease resistance, stress tolerance, texture, taste and digestibility

Consumers' acceptance regarding farmed fish depends on the firmness of the fillet, and soft fillets are discouraged in the food market. It has been demonstrated that microalgae decrease fillet gaping. When compared to a control diet, a 5% *Schizochytrium* sp. diet enhanced fillet quality in Atlantic salmon by decreasing gaping (Kousoulaki et al., 2016). In a different study, Striped Jack (*Pseudocaranxdentex*) had better texture and flavor that fed a fish feed containing 5% *Spirulina* (Watanabe et al., 1990). Nile Tilapia (*Oreochromis niloticus*) exhibited increased hardness, muscle quality, and fibrousness when fed fish feed with 2% *Spirulina* (Mustafa, 1995). Salmon fillets made from *Schizochytrium limacinum* had the same flavor and smell as fillets made from regular fish oil (Katerina et al., 2020).

Numerous vitamins, particularly type E, can be found in high concentrations in microalgae (Madeira et al., 2017). Microalgae are rich in calcium, potassium, iron, copper, sodium, sulfur, zinc, phosphorus, and magnesium (Dineshbabu et al., 2019). Incorporation of mineral-rich microalgae into salmon diets improved the flavor and texture of the fish (Nagappan et al., 2021; Guedes et al., 2015).

The nutritionally important PUFAs may be found in high concentrations in a variety of microalgae. These include *Phaeodactylum*, *Nannochloropsis*, *Isochrysis*, *Nitzschia*, and *Diacronema*, which produce eicosapentaenoic acid (EPA); Porphyridium, which produce arachidonic acid; *Desmodesmus* sp., which produce alpha-linolenic acid; and *Schizochytrium* and *Crypthecodinium*, which produce docosahexaenoic acid (DHA) (Lu et al., 2021; Nagappan and Kumar Verma, 2018). Salmon are fed thraustochytrid *Schizochytrium* sp. only for PUFA in the commercial sector (Ren et al., 2010). Additionally, thraustochytrids were given to finfish larvae and utilized to enrich zooplankton (Barclay and Zeller, 1996).

Colorful seafoods attracts consumers, and microalgal biomasses are known to affect fish pigmentation. The aquaculture business uses both extracted astaxanthin and whole cell *Haematococcus pluvialis* as feed additives, with a range of 1.5% (Sun et al., 2017). High lutein microalgae, such as *Chlorella* sp. and *Scenedesmus* sp. are employed as feed additives primarily for their pigments. Yellow catfish (*Pelteobagrus fulvidraco*) exhibit a vivid yellow coloring when it fed 4% defatted *Spirulina* and 0.4% *Spirulina* lipid extract (Liu et al., 2021). Also, gilthead seabream showed yellow coloration when fed a diet that contains 2.5% *Phaeodactylum tricornutum*, a diatom with a high fucoxanthin content (Ribeiro et al., 2017; Dawood et al., 2018).

Microalgae contain non-starch polysaccharides, such as cellulose, pectin, and algenan, which are often indigestible (Gong et al., 2018). However, because of presence of mucopeptides in the cell wall of *Spirulina* sp., fish may easily eat it (Becker, 2007; Bleakley and Hayes, 2017).

Chlorella sp., *Desmodesmus* sp., *Nannochloropsis* sp., and *Tetraselmis* sp. are examples of microalgal species that have thick cellulose cell-wall and protein digestibility that ranges from 80%–89%, 54%–67%, 69%–81%, and 70%–73%, respectively, in aquatic species. In contrast, non-cellulosic species such as *Spirulina* sp. and *Isochrysis* sp. exhibited comparable greater protein digestibility of 86% (Niccolai et al., 2019).

In comparison to the control diet, the fish weight gain and feed conversion ratio improved by 13% and 11%, respectively, for the cold pasteurization and bead milling diets. The protein digestibility of diets comprising physically processed *Chlorella vulgaris* and *Tetraselmis* sp. and enzymatically processed *Nannochloropsis oceanica* was higher than that of the untreated control (Batista et al., 2020).

The nutritional advantages of a microalga are not same in all treated fishes because of their differences in eating habits and digestive physiologies. Since shrimps do not have an acidic stomach, it results in poor digestion of complex forms of algal nutrients. Consequently, a comprehensive examination of microalgae digestibility is required (Tibbetts et al., 2017).

Furthermore, it has been demonstrated that European sea bass larvae enhance the release of digestive enzymes when algae (*I. galbana*) are present in the rearing tanks (Cahu et al., 1998).

To shield microalgal cells from the damaging effects of heavy metals, microalgae increase antioxidant activity by boosting the activities of peroxidase, ascorbate peroxidase, glutathione reductase, and superoxide dismutase (SOD). Phenolic compounds and carotenoids found in microalgae are known to exhibit radical scavenging and antioxidant capabilities (Safafar et al., 2015; Sørensen et al., 2023).

Natural antioxidants found in microalgae have a strong scavenging activity that may have important anti-aging and disease-fighting effects by avoiding oxidative cell damage, such as DNA and protein oxidation (Karawita et al., 2007).

Modification of the host immune system has been associated with *Arthrospira*, which is an abundant source of phenolic acids, carotenoids, phycocyanin and tocopherols. Antioxidants found in microalgal biomass have the ability to improve fish and crustacean species' physiological states and immune systems (Watanuki et al., 2006).

Immunoglobulin M (IgM), immunoglobulin D (IgD), interleukin-22 (IL-22), and chemokine levels have been reported to rise when gibel carp are fed a low-level addition of Chlorella sp. (0.4%–1.2% per kg) (Zhang et al., 2014).

Similar improvements in stress resistance and survival rate were observed in guppies (*Poecilia reticulata*) that fed a diet containing the microalga *Parietochloris incisa*. These improvements were associated with elevated lysozyme levels.

Numerous species, such as zebrafish (*Danio rerio*), rainbow trout (*O. mykiss*), swordtail (*Xiphophorus hellerii*), gold fish (*Carassius auratus*), basa fish (*Pangasius bocourti*), yellow tail cichlid (*Pseudotropheus acei*) and Nile tilapia (*Oreochromis niloticus*) (Table 9.2), have been shown to benefit from the inclusion of microalgae for improved seed production and fish reproduction. Fatty acids, amino acids, minerals, carotenoids, vitamins E and C, and other important nutrients are the main components for reproduction (e.g., fertility, fertilization, incubation success, enhancing egg quality, and larval development). High LC PUFA concentrations have also been shown to affect steroidogenesis, metabolism, and maturation, preserving reproductive function while enhancing egg quality (Mohamed et al., 2024).

6. Challenges and future prospects

6.1 Current limitations and challenges in the use of microalgal biomass; challenges regarding microalgal toxins

Microalgae are difficult to dry. Incorrect drying alters their nutritional and physical qualities, which in turn reduces their use as feed. Some microalgae have thick cell

Table 9.2 Effects of microalgae biomass in aquafeeds.

Feeder species	Microalgal feed species	Effect of microalgal feed on feeder	References
Puntius gelius	*Arthrospira* sp.	Fish growth rates have increased significantly.	Hajiahmadian (2012)
Oreochromis niloticus	*Arthrospira platensis*	Enhanced fish health through tissue protection and antioxidant effects.	Ibrahem et al. (2013)
Oreochromis niloticus	*Arthrospira* sp.	Growth and feed consumption were unaffected, in contrast to the corn-gluten meal control.	Hussein et al. (2013)
Macrobrachium rosenbergii	*Chlorella vulgaris*	*M. rosenbergii* were resistant to *Aeromonas hydrophila* infection, grew more quickly, and had a stronger immune response.	Maliwat et al. (2017)
Gadus morhua	*Nannochloropsis* sp.	Increased feed intake and fish growth was seen	Walker and Berlinsky (2011)
Salmo salar *Cyprinus carpio*	*Nannochloropsis* sp.	In terms of development and feed intake, microalgal food performed better than fish food, suggesting that it might be substituted for fish meal.	Kiron et al. (2012)
Sparus aurata	*Nannochloropsis gaditana*	Increased defensive activity.	Cerezuela et al. (2012)
Litopenaeus vannamei	*Schizochytrium* sp.	Activity of digestive enzyme, content of fatty acid and shrimp survival were not unaffected; but their growth rate was significantly higher than the control group.	White and Ryan (2015)
Oreochromis niloticus	*Schizochytrium* sp.	The survival rate did not alter substantially.	Sarker et al. (2016)

From Ahmad, A., W. Hassan, S., Banat, F., 2022. An overview of microalgae biomass as a sustainable aquaculture feed ingredient: food security and circular economy. Bioengineered, 13 (4), 9521–9547. https://doi.org/10.1080/21655979.2022.2061148.

wall and extracellular polysaccharides that interfere with nutritional absorption causing poor digestibility and salt accumulation regarding use as fish feed. Several microalgae have a high carbohydrate and low protein content because of their tough cell walls that inhibit the digestion of proteins and fats (Skrede et al., 2011).

It is not advised to use microalgae biomass in aquafeeds because of the high levels of toxins and trace elements present, according to a few studies. It is frequently suggested that the protein left over after fat is removed to create biofuels be used in animal feed. It's likely that the microalgae used to make biofuels are unfit for human consumption and that the need for inexpensive fuel production may lead to the residue of harmful proteins. It is better to use selected algal biomass for higher-value goods like aquafeed first, and then use any leftover chemicals to make biofuel (Ahmad et al., 2022; Li et al., 2015). Successful largescale growth of commercially relevant microalgae species is necessary to provide massive volumes of microalgal biomass for the aquafeed industry (Hannon et al., 2010).

The economic and environmental impacts of commercial-scale microalgal culture customized for feed applications, are crucial. There haven't been many case studies published on the subject yet. The majority of aquaculture farms produce live feed from small-scale microalgal cultivation systems. Scaling up the cultivation process is crucial for reducing environmental consequences including carbon footprints from energy consumption, according to life cycle analysis (Taelman et al., 2015).

In many nations, there are no formal government regulations concerning the usage of microalgae as a feed element. In terms of taking action against climate change and establishing renewable energy facilities, the European Union (EU) has always been at the forefront of these initiatives. Microalgae as a feed ingredient includes "spill-over research" from the application of biofuel, processing of microalgal biomass, fostering collaboration between academia and industry, diversifying microalgal technology for mitigating climate change, providing subsidies to companies that promote carbon capture for microalgal cultivation, and collaborating with other nations for efficient microalgal production. Other nations need to organize and create state regulations for the growth of microalgae as a feed element, particularly those that are at the cutting edge of algal technology (Nagarajan et al., 2021; Enzing et al., 2014).

6.2 Future research directions and technological advancements

The cell walls of microalgae may differ in structure and compositions. This highlights the necessity of looking for commercial strains that are easily handled in order to disrupt cells. Tests have shown that only very small number of microalgal species are digested. Investigating how stress and environmental factors affect the chemical constitution of microalgae is also crucial. For instance, lack of nitrogen may result in accumulation of carbohydrate and PUFA, which fish may eat. Different species of microalgae have different metabolic responses to challenges like nitrogen shortage. Nitrogen deficit causes some strains to have higher carbohydrate levels and reduced protein levels (Nagappan et al., 2020; Karuppaiyan et al., 2022).

An extensive range of microalgal species are necessary for aquaculture. It's feasible that microalgal species with superior growth or nutritional qualities are more effective. It is important to look at the nutritional value of different microalgae. This implies that in order for digestive enzymes to access the algal components, the cell wall must be broken down (Becker, 2004; Berge et al., 2013).

Before going on the market, microalgal material must be examined for harmful compounds in order to enhance fish's health. For composite feed, the needs of industrial processing during production must be taken into account in addition to nutritional value and digestibility. It is widely recognized that one of the most important factors influencing physical quality in extruded fish feed is its composition (Sørensen et al., 2012).

Microalgal screening for digestibility has only been carried out on a small number of fishes, such as African catfish, tilapia, and salmonids. Since the aquaculture sector is growing quickly, more and new fish species need to have their microalgal digestibility examined. Determining anti-nutritional elements such as digestive enzyme inhibitors (such as the terpene caulerpenyene) and variables that reduce the bioavailability of nutrients in microalgae should also be included.

The majority of microalgal digestibility investigations utilize room-temperature extrusion or the cold-pelletization techniques. Microalgae's digestive enzymes may be harmed by the high pressure and high temperature extrusion process, which also breaks down anti-digestive components. Therefore, it would be intriguing to investigate how the extrusion process simultaneously affects the components that induce and hinder digestion in microalgae.

Addressing the execution process, several studies do not adequately detail the pellet processing conditions. For example, the type of pellet produced and the temperature and pressure at which the extrusion operation is conducted are not specified in several studies. These factors have a major impact on the microalgae's energy and nutritional digestibility. Thus, such data should be included in future research. Numerous studies have assessed the digestibility of the fish diet rather than the digestion of individual microalgae. Thus, additional information on microalgal digestibility by direct and indirect evaluation is needed. The nutrient and energy digestibility of microalgae in fish, however, appears promising based on the limited literature (Nagappan Annamalai et al., 2021). A recent study looked at the physical impacts of microalgae on extruded fish feed, but more research is needed (Gong et al., 2018).

Energy-intensive microalgal biomass harvesting drives into high production costs. Techniques for microalgal biorefineries may lower the cost of producing microalgal feed components. High-value biochemical components, like lipids and carotenoids, are extracted from biomass using this method and subsequently utilized in fish and animal feed. Fish feed might be generated from the remaining biomass. Fish digestion of the leftover biomass from biorefinery processes must therefore be investigated (Ahmad et al., 2022).

With the current rate of growth, the algal sector is predicted to contribute to the development of a blue bioeconomy supported by its cost-effective, land-free cultivation techniques. In order to transfer laboratory technology into sustainable algal products, the future has solely depended on algae biorefining and ongoing research and development initiatives by nationally and internationally supported research programs (Ullmann and Grimm, 2021).

Due to the numerous and diverse advantages of algal metabolites, the market for algal products is rising decisively on a global scale. The microalgal market is currently fragmented. As the industry grows, it experiences recurring changes in terms of alliances, newer technology, the introduction novel commodities, etc. These changes are mostly related to the pharmaceutical and neutraceutical industries, biofuels, and health and dietary products. The market for microalgal products is anticipated to grow to a maximum of 18 billion dollars by 2028 (Mancini, 2022) whereas another study shows annual increment of the same at the rate of 12.5% up to 2033 (Fig. 9.2). *Spirulina* sp. is currently the most widely employed species in the aquaculture business and contributes significantly to the global market due to its enormous health advantages to humans.

According to a Business Insider article (2014), being regarded as "superfood" algae are predicted to play a significant role in the global market in the future (which includes the health, pharmaceutical, cosmetic, and neutraceutical industries). They will also be the catalyst for the biggest shift towards the "circular-bioeconomy" (Uma et al., 2023).

6.3 Potential for large-scale implementation

Many efforts have been made in recent years to bring down the cost of microalgae biomass in order to produce a cheaper and affordable feed ingredient. Moreover, potential aqua feed market segmentation has been emerging as motivation to extend its geography in recent years (Fig. 9.3). In addition, dried algal powder was added to fish food formulation at a 15% inclusion level, and flue gas was used as a low-cost carbon source for *Chlorella* cultivation. According to the cost estimation, the price of algae was 0.54 USD/kg, which is less than the price of fishmeal, which

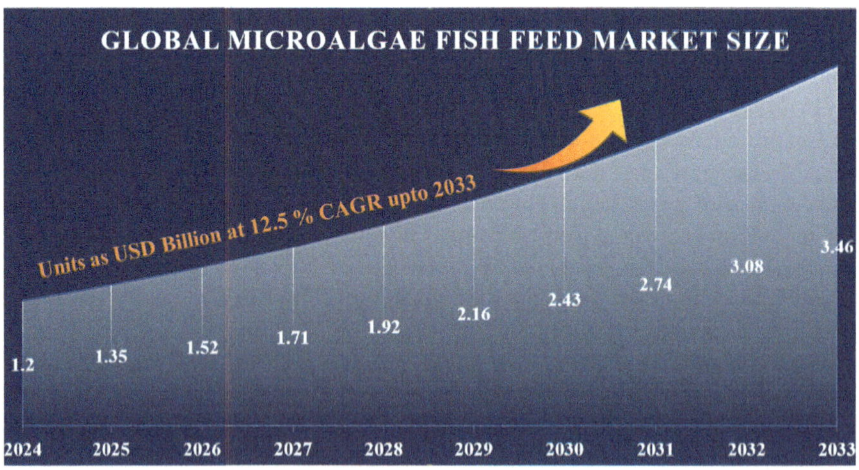

FIGURE 9.2 Projected global microalgae fish feed market size.

Expected growth of global microalgae fish feed market.

- From Microalgae Fish Feed Market Size. 2025. In Growth, Market Share & Forecast 2033. Verified Market Reports. 2025.

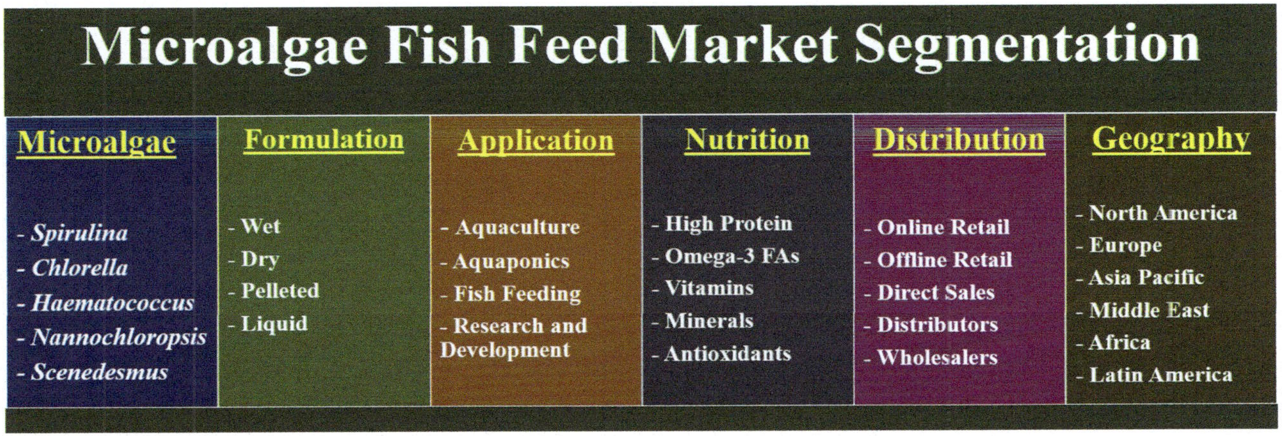

FIG. 9.3 Microalgae fish feed market segmentation.

Geographical market distribution of microalgae-based fish feed.

- From Microalgae Fish Feed Market Size. 2025. In Growth, Market Share & Forecast 2033. Verified Market Reports.

was 0.84 USD/kg. Compared to traditional feed, the overall cost was significantly lower by 0.16 USD (26.1%) (Yadav et al., 2020). Furthermore, other cutting-edge technologies have also been investigated as potential routes to obtain cost-effective microalgal biomass, such as the generation of microalgae on biofilm and the replacement of artificial medium with wastewater (Lu et al., 2015). Cultivating algae in a desert with enough solar energy is also considered an innovative technology to lower the cost of algae production with the installation of photovoltaic technology (Rasheed et al., 2022).

It should be highlighted, however, that there is a significant gap between industry application and scholarly study. Some revolutionary technologies created in laboratories have not yet been marketed in practice because of the difficulties of real-world scenarios. In countries with poor industries, using flue-gas technology is difficult to cultivate microalgae for feeding fish. Additionally, the generation of algae from wastewater has very severe factors for microenvironment regulation, nutrient profile balancing, and microbial control. Accordingly, looking at microalgae to be a more affordable source of protein than conventional fishmeal is not entirely fair. More practicable technology should be made available in the future to encourage the use of microalgae for fish feeding (Lu et al., 2023a).

It has been shown that the marine algae *Staurosira* and *Desmodesmus* can produce biomass using seawater in a hybrid cultivation method that also uses photobioreactors and open ponds. With no residual discharge, the CO_2 utilization was measured at 78% efficiency. The most effective method for producing protein from microalgae was found to be the heterotrophic growth of *Chlorella vulgaris* using hydrolyzed food waste as a carbon source (Smetana et al., 2017; Nagarajan et al., 2021).

The studies by Chattaraj et al. (2024) demonstrated that BSG and chlorella meal can effectively replace the traditional protein components in the diet for *C. reba* without having an adverse effect on growth or nutrient utilization. This is the first study to evaluate and establish the use of BSG meals and algae together as sources of protein in fish feed (Chattaraj et al., 2024).

7. Conclusion

With their high nutritional profile and environment friendly nature microalgae have emerged as a successful alternative to traditional aquafeeds in terms of both sustainability and economic benefits. Thus, the valorization of microalgal biomass for use in fish feed offers great potential for using it in sustainable aquaculture industries.

There has been significant improvement in pretreatment and processing methods of algal biomass that in return enhances digestibility to assure efficient nutrient utilization and reduced feed waste. Micro algae derived high quality proteins, essential amino acids, good fats including omega-3 fatty acids and other bioactive compounds provide increased growth rate, good health and overall performance of the fishes.

The use of microalgae in fish feed reduces dependency on finite marine resources, lowers the risk of eutrophication, and promotes sustainable aquaculture practices. Integrated cultivation systems and circular economy models further enhance resource efficiency and cost-effectiveness.

With all these benefits, the potential of microalgae to revolutionize the aquafeed industries can be visualized through numerous case Studies that demonstrate the practical Applications of microalgae in fish feed formulation.

Ongoing research and technological advancements are poised to overcome current challenges related to production costs, scalability, and nutrient variability.

The effective valorization of microalgal biomass represents a paradigm shift in the aquaculture industry, offering a sustainable and nutritious alternative to conventional feed ingredients. By leveraging the unique properties of microalgae, the aquaculture sector can reduce its environmental footprint, promote marine conservation, and ensure the long-term viability of fish farming operations. Continued innovation, research, and collaboration will be crucial in realizing the full potential of microalgal biomass and driving the industry toward a more sustainable future.

References

Ahmad, A., Hassan, S.W., Banat, F., 2022. An overview of microalgae biomass as a sustainable aquaculture feed ingredient: food security and circular economy. Bioengineered 13 (4), 9521–9547. https://doi.org/10.1080/21655979.2022.2061148.

Al-Muhtaseb, A.H., Jamil, F., Sarwer, A., Al-Maawali, S., 2022. Valorization of Microalgal Biomass for Food Valorization of Microalgal Biomass and Wastewater Treatment. Elsevier., pp. 81–112. https://doi.org/10.1016/B978-0-323-91869-5.00016-8

Ali, S.S., Al-Tohamy, R., Al-Zahrani, M., Schagerl, M., Kornaros, M., Sun, J., 2025. Advancements and challenges in microalgal protein production: a sustainable alternative to conventional protein sources. Microb. Cell Fact. 24 (1). https://doi.org/10.1186/s12934-025-02685-1.

Ansari, F.A., Guldhe, A., Gupta, S.K., Rawat, I., Bux, F., 2021. Improving the feasibility of aquaculture feed by using microalgae. Environ. Sci. Pollut. Control Ser. 28 (32), 43234–43257. https://doi.org/10.1007/s11356-021-14989-x.

Barclay, W., Zeller, S., 1996. Nutritional enhancement of n-3 and n-6 fatty acids in rotifers and Artemia nauplii by feeding spray-dried *Schizochytrium* sp. J. World Aquacult. Soc. 27 (3), 314–322. https://doi.org/10.1111/j.1749-7345.1996.tb00614.x.

Batista, S., Pintado, M., Marques, A., Abreu, H., Silva, J.L., Jessen, F., Tulli, F., Valente, L.M. P., 2020. Use of technological processing of seaweed and microalgae as strategy to improve their apparent digestibility coefficients in European seabass (*Dicentrarchus labrax*) juveniles. J. Appl. Phycol. 32 (5), 3429–3446. https://doi.org/10.1007/s10811-020-02185-2.

Becker, W., 2004. Microalgae in Human and Animal Nutrition Handbook of Microalgal Culture. Wiley-Blackwell.

Becker, E.W., 2007. Micro-algae as a source of protein. Biotechnol. Adv. 25 (2), 207–210. https://doi.org/10.1016/j.biotechadv.2006.11.002.

Berge, G.M., Hatlen, B., Odom, J.M., Ruyter, B., 2013. Physical treatment of high EPA Yarrowia lipolytica biomass increases the availability of n-3 highly unsaturated fatty acids when fed to Atlantic salmon. Aquac. Nutr. 19 (1), 110–121. https://doi.org/10.1111/anu.12092.

Bernaerts, T.M.M., Gheysen, L., Kyomugasho, C., Jamsazzadeh Kermani, Z., Vandionant, S., Foubert, I., Hendrickx, M.E., Van Loey, A.M., 2018. Comparison of microalgal biomasses as functional food ingredients: focus on the composition of cell wall related polysaccharides. Algal Res. 32, 150–161. https://doi.org/10.1016/j.algal.2018.03.017.

Bleakley, S., Hayes, M., 2017. Algal proteins: extraction, application, and challenges concerning production. Foods 6 (5), 33. https://doi.org/10.3390/foods6050033.

Brown, M.R., 1991. The amino-acid and sugar composition of 16 species of microalgae used in mariculture. J. Exp. Mar. Biol. Ecol. 145 (1), 79–99. https://doi.org/10.1016/0022-0981(91)90007-J.

Cerezuela, R., Fumanal, M., Tapia-Paniagua, S.T., Meseguer, J., Moriñigo, M.A., Esteban, M. A., 2012. Changes in cellular and humoral immune parameters of gilthead seabream (*Sparus aurata* L.) fed diets supplemented with vitamin C and n-3 HUFA. Fish Shellfish Immunol. 33 (2), 348–353. https://doi.org/10.1016/j.fsi.2012.05.019.

Chattaraj, S., Behera, B.K., Das Mohapatra, P.K., 2024. Effective valorization of Chlorella biomass and brewers spent grain substituting fish meal and soybean meal in the diet of herbivorous fish Cirrhinus reba for higher growth, digestibility and sustainable cultivation. J. Appl. Phycol. 36 (1), 217–232. https://doi.org/10.1007/s10811-023-03106-9.

Chattaraj, S., Das Mohapatra, P.K., 2024. Immunostimulatory and antagonistic potential of the methanolic extract of Oedogonium intermedium SCB in Cirrhinus reba challenged with Aeromonas hydrophila. Aquac. Int. 32 (3), 2523–2548.

Chattaraj, S., Ganguly, A., Mitra, D., et al., 2026a. Sustainable valorization of algal biomass in fish nutrition and dietetics: a critical review and scientometric study. Waste Biomass Valor. https://doi.org/10.1007/s12649-026-03619-7.

Chattaraj, S., Sahoo, S.S., Behera, S., Thatoi, H., 2026b. Sustainable aquafeed development using Spirulina: effects on growth performance of Catla catla. Algal Res. 104447.

Dawood, M.A.O., Koshio, S., Esteban, M.Á., 2018. Beneficial roles of feed additives as immunostimulants in aquaculture: a review. Rev. Aquacult. 10 (4), 950–974. https://doi.org/10.1111/raq.12209.

Dimopoulou, M., Kolonas, A., Stagos, D., Gortzi, O., 2025. A review of the sustainability, chemical composition, bioactive compounds, antioxidant and antidiabetic activity, neuroprotective properties, and health benefits of microalgae. Biomass 5 (1), 11. https://doi.org/10.3390/biomass5010011.

Dineshbabu, G., Goswami, G., Kumar, R., Sinha, A., Das, D., 2019. Microalgae–nutritious, sustainable aqua- and animal-feed source. J. Funct. Foods. 62.

Enzing, C.P.M., Barbosa, M., Sijtsma, L., Vigani, M., Parisi, C., Cerezo, R., 2014. Microalgae-Based Products for the Food and Feed Sector: An Outlook for Europe. Publications Office of the European Union.

Ganguly, A., Chattaraj, S., Ganguly, M., Chattaraj, M., Banerjee, A., Mandal, A., Ali Khan, M., Sen, S.K., Das Mohapatra, P.K., 2025. Effect of three probiotic Bacillus strains supplemented feeds on growth, carcass composition and blood parameters of Clarias magur (Hamilton). J. Appl. Aquac. 37 (1), 19–44.

Gong, Y., Guterres, H.A.D.S., Huntley, M., Sørensen, M., Kiron, V., 2018. Digestibility of the defatted microalgae Nannochloropsis sp. and Desmodesmus sp. when fed to Atlantic salmon, *Salmo salar*. Aquac. Nutr. 24 (1), 56–64. https://doi.org/10.1111/anu.12533.

Guedes, A.C., Sousa-Pinto, I., Malcata, F.X., 2015. Application of microalgae protein to aquafeed handbook of marine microalgae: biotechnology advances. Elsevier Inc., pp. 93–125. https://doi.org/10.1016/B978-0-12-800776-1.00008-X

Hajiahmadian, M.S., 2012. Use of probiotics in aquaculture: a review. Sci. J. Fisher. Marine Sci. 1 (1), 1–7.

Hannon, M., Gimpel, J., Tran, M., Rasala, B., Mayfield, S., 2010. Biofuels from algae: challenges and potential. Biofuels 1 (5), 763–784. https://doi.org/10.4155/bfs.10.44.

Hussein, E.E.S., Dabrowski, K., El-Saidy, D.M.S.D., Lee, B.J., 2013. Enhancing the growth of Nile tilapia larvae/juveniles by replacing plant (gluten) protein with algae protein. Aquac. Res. 44 (6), 937–949. https://doi.org/10.1111/j.1365-2109.2012.03100.x.

Ibrahem, M., Mohamed, M.F., Ibrahim, M., 2013. The role of Spirulina platensis (*Arthrospira platensis*) in growth and immunity of Nile tilapia (*Oreochromis niloticus*) and its resistance to bacterial infection. J. Agric. Sci. 5 (6). https://doi.org/10.5539/jas.v5n6p109.

Karawita, R., Senevirathne, M., Athukorala, Y., Affan, A., Lee, Y.-J., Kim, S.-K., Lee, J.-B., Jeon, Y.-J., 2007. Protective effect of enzymatic extracts from microalgae against DNA damage induced by H2O2. Mar. Biotechnol. 9 (4), 479–490. https://doi.org/10.1007/s10126-007-9007-3.

Karuppaiyan, S., Sivakumar, P., Elumalai, S., 2022. Enhancing biomass and biochemical compositions of *Scenedesmus quadricauda* through nitrogen limitation and its potential for biofuel production. Bioresour. Technol. Rep. 18, 101037. https://doi.org/10.1016/j.biteb.2022.101037.

Katerina, K., Marit Berge, G., Mørkøre, T., Krasnov, A., Grete, B., Trine, Y., Mats, C., John, S., Bente, R., 2020. Microalgal *Schizochytrium limacinum* biomass improves growth and filet quality when used long-term as a replacement for fish oil, in modern salmon diets. Front. Mar. Sci. 7. https://doi.org/10.3389/fmars.2020.00057.

Kiron, V., Phromkunthong, W., Huntley, M., Archibald, I., De Scheemaker, G., 2012. Marine microalgae from biorefinery as a potential feed protein source for Atlantic salmon, common carp and whiteleg shrimp. Aquac. Nutr. 18 (5), 521–531. https://doi.org/10.1111/j.1365-2095.2011.00923.x.

Kousoulaki, K., Mørkøre, T., Nengas, I., Berge, R.K., Sweetman, J., 2016. Microalgae and organic minerals enhance lipid retention efficiency and fillet quality in Atlantic salmon (*Salmo salar* L.). Aquaculture 451, 47–57. https://doi.org/10.1016/j.aquaculture.2015.08.027.

Krimpen, P., Bikker, I.M., der Meer, van, Peet-Schwering, Vereijken, J.M., 2013. Cultivation, processing and nutritional aspects for pigs and poultry of European protein sources as alternatives for imported soybean products. Lelystad 150.

Kumar Sarker, N., Kaparaju, P., 2024. Microalgal Bioeconomy: a green economy approach towards achieving sustainable development goals. Sustainability 16 (24), 11218. https://doi.org/10.3390/su162411218.

Cahu, C.L., Zambonino Infante, J.L., Péres, A., Quazuguel, P., Le Gall, M.M., 1998. Algal addition in sea bass (*Dicentrarchus labrax*) larvae rearing: effect on digestive enzymes. Aquaculture 161 (1–4), 479–489. https://doi.org/10.1016/s0044-8486(97)00295-0.

Li, J., Liu, Y., Cheng, J.J., Mos, M., Daroch, M., 2015. Biological potential of microalgae in China for biorefinery-based production of biofuels and high value compounds. New Biotechnol. 32 (6), 588–596. https://doi.org/10.1016/j.nbt.2015.02.001.

Liu, C., Li, Y., Chen, Z., Yuan, L., Liu, H., Han, D., Jin, J., Yang, Y., Hu, Q., Zhu, X., Xie, S., 2021. Effects of dietary whole and defatted Arthrospira platensis (*Cyanobacterium*) on growth, body composition and pigmentation of the yellow catfish *Pelteobagrus fulvidraco*. J. Appl. Phycol. 33 (4), 2251–2259. https://doi.org/10.1007/s10811-021-02445-9.

Lu, Q., Zhou, W., Min, M., Ma, X., Chandra, C., Doan, Y.T.T., Ma, Y., Zheng, H., Cheng, S., Griffith, R., Chen, P., Chen, C., Urriola, P.E., Shurson, G.C., Gislerød, H.R., Ruan, R.,

2015. Growing *Chlorella* sp. on meat processing wastewater for nutrient removal and biomass production. Bioresour. Technol. 198, 189–197. https://doi.org/10.1016/j.biortech.2015.08.133.

Lu, Q., Li, H., Xiao, Y., Liu, H., 2021. A state-of-the-art review on the synthetic mechanisms, production technologies, and practical application of polyunsaturated fatty acids from microalgae. Algal Res. 55, 102281. https://doi.org/10.1016/j.algal.2021.102281.

Lu, Q., Lu, Y., Yang, L., 2023. Challenging problems of applying microalgae for aquaculture environment protection andnutrition supplementation: a long road traveled and still a far way to go. Front. Bioeng. Biotechnol. 11. https://doi.org/10.3389/fbioe.2023.1151440.

Madeira, M.S., Cardoso, C., Lopes, P.A., Coelho, D., Afonso, C., Bandarra, N.M., Prates, J.A. M., 2017. Microalgae as feed ingredients for livestock production and meat quality: a review. Livest. Sci. 205, 111–121. https://doi.org/10.1016/j.livsci.2017.09.020.

Majluf, P., Matthews, K., Pauly, D., Skerritt, D.J., Palomares, M.L.D., 2024. A review of the global use of fishmeal and fish oil and the Fish In: Fish Out metric. Sci. Adv. 10 (42). https://doi.org/10.1126/sciadv.adn5650.

Maliwat, G.C., Velasquez, S., Robil, J.L., Chan, M., Traifalgar, R.F., Tayamen, M., Ragaza, J. A., 2017. Growth and immune response of giant freshwater prawn *Macrobrachium rosenbergii* (De Man) postlarvae fed diets containing *Chlorella vulgaris* (Beijerinck). Aquac. Res. 48 (4), 1666–1676. https://doi.org/10.1111/are.13004.

Mancini, R., 2022. The Blue Economy in the Mediterranean Region and Opportunities for the Algae Industry. IAI Papers.

Mohamed, M.S., Mohamed, A., El-Hais, A.M., Gomaa, A., Mansour, A.T., 2024. Impact of dietary lipids on the reproductive performance and egg quality of aquaculture species: a comprehensive review 34, 101890. https://doi.org/10.1016/j.aqrep.2024.101890.

Mustafa, M.G., 1995. A review: dietary benefits of algae as an additive in fish feed. Isr. J. Aquac. 47, 155–162.

Nagappan, S., Kumar Verma, S., 2018. Co-production of biodiesel and alpha-linolenic acid (omega-3 fatty acid) from microalgae, *Desmodesmus* sp. MCC34. Energy Sources, Part A Recovery, Util. Environ. Eff. 40 (24), 2933–2940. https://doi.org/10.1080/15567036.2018.1514434.

Nagappan Annamalai, S., Das, P., Thaher, M.I.A., Abdul Quadir, M., Khan, S., Mahata, C., Jabri, H.Al, 2021. Nutrients and energy digestibility of microalgal biomass for fish feed applications. Sustainability 13 (23), 13211. https://doi.org/10.3390/su132313211.

Nagappan, S., Devendran, S., Tsai, P.C., Jayaraman, H., Alagarsamy, V., Pugazhendhi, A., Ponnusamy, V.K., 2020. Metabolomics integrated with transcriptomics and proteomics: evaluation of systems reaction to nitrogen deficiency stress in microalgae. Process Biochem. 91, 1–14. https://doi.org/10.1016/j.procbio.2019.11.027.

Nagappan, S., Das, P., AbdulQuadir, M., Thaher, M., Khan, S., Mahata, C., Al-Jabri, H., Vatland, A.K., Kumar, G., 2021. Potential of microalgae as a sustainable feed ingredient for aquaculture. J. Biotechnol. 341, 1–20. https://doi.org/10.1016/j.jbiotec.2021.09.003.

Nagarajan, D., Varjani, S., Lee, D.-J., Chang, J.-S., 2021. Sustainable aquaculture and animal feed from microalgae – nutritive value and techno-functional components. Renew. Sustain. Energy Rev. 150, 111549. https://doi.org/10.1016/j.rser.2021.111549.

Niccolai, A., Zittelli, G.C., Rodolfi, L., Biondi, N., Tredici, M.R., 2019. Microalgae of interest as food source: biochemical composition and digestibility. Algal Res. 42, 101617. https://doi.org/10.1016/j.algal.2019.101617.

Quinn, J.C., Davis, R., 2015. The potentials and challenges of algae based biofuels: a review of the techno-economic, life cycle, and resource assessment modeling. Bioresour. Technol. 184, 444–452. https://doi.org/10.1016/j.biortech.2014.10.075.

Rasheed, R., Thaher, M., Younes, N., Bounnit, T., Schipper, K., Nasrallah, G.K., Jabri, H.Al, Gifuni, I., Goncalves, O., Pruvost, J., 2022. Solar cultivation of microalgae in a desert environment for the development of techno-functional feed ingredients for aquaculture in Qatar. Sci. Total Environ. 835, 155538. https://doi.org/10.1016/j.scitotenv.2022.155538.

Ren, L.J., Ji, X.J., Huang, H., Qu, L., Feng, Y., Tong, Q.Q., Ouyang, P.K., 2010. Development of a stepwise aeration control strategy for efficient docosahexaenoic acid production by Schizochytrium sp. Appl. Microbiol. Biotechnol. 87 (5), 1649–1656. https://doi.org/10.1007/s00253-010-2639-7.

Ribeiro, A.R., Gonçalves, A., Barbeiro, M., Bandarra, N., Nunes, M.L., Carvalho, M.L., Silva, J., Navalho, J., Dinis, M.T., Silva, T., Dias, J., 2017. Phaeodactylum tricornutum in finishing diets for gilthead seabream: effects on skin pigmentation, sensory properties and nutritional value. J. Appl. Phycol. 29 (4), 1945–1956. https://doi.org/10.1007/s10811-017-1125-3.

Ruben, M.O., Akinsanola, A.B., Okon, M.E., Shitu, T., Jagunna, I.I., 2025. Emerging challenges in aquaculture: current perspectives and human health implications. Veter. World 18 (1), 15–28. https://doi.org/10.14202/vetworld.2025.15-28.

Ruiz, J., Olivieri, G., De Vree, J., Bosma, R., Willems, P., Reith, J.H., Eppink, M.H.M., Kleinegris, D.M.M., Wijffels, R.H., Barbosa, M.J., 2016. Towards industrial products from microalgae. Energy Environ. Sci. 9 (10), 3036–3043. https://doi.org/10.1039/c6ee01493c.

Safafar, H., Van Wagenen, J., Møller, P., Jacobsen, C., 2015. Carotenoids, phenolic compounds and tocopherols contribute to the antioxidative properties of some microalgae species grown on industrial wastewater. Mar. Drugs 13 (12), 7339–7356. https://doi.org/10.3390/md13127069.

Sarker, P.K., Gamble, M.M., Kelson, S., Kapuscinski, A.R., 2016. Nile tilapia (*Oreochromis niloticus*) show high digestibility of lipid and fatty acids from marine *Schizochytrium* sp. and of protein and essential amino acids from freshwater *Spirulina* sp. feed ingredients. Aquac. Nutr. 22 (1), 109–119. https://doi.org/10.1111/anu.12230.

Sarker Pallab, K., 2023. Microorganisms in fish feeds, technological innovations, and key strategies for sustainable aquaculture. Microorganisms 11 (2), 439. https://doi.org/10.3390/microorganisms11020439.

Selvaraj, D., Sanjiv, R., Shakira Banu, S., Arivazhagan, M., 2025. Recent advances in microalgae harvesting and dewatering: a techno-economic evaluation perspective. Bioresour. Technol. Rep. 31, 102201. https://doi.org/10.1016/j.biteb.2025.102201.

Skrede, A., Mydland, L., Ahlstrøm, Ø., Reitan, K., Gislerød, H., Øverland, M., 2011. Evaluation of microalgae as sources of digestible nutrients for monogastric animals. J. Anim. Feed Sci. 20 (1), 131–142. https://doi.org/10.22358/jafs/66164/2011.

Smetana, S., Sandmann, M., Rohn, S., Pleissner, D., Heinz, V., 2017. Autotrophic and heterotrophic microalgae and cyanobacteria cultivation for food and feed: life cycle assessment. Bioresour. Technol. 245 (Pt A), 162–170. https://doi.org/10.1016/j.biortech.2017.08.113.

Sørensen, I., Rose, J.K.C., Doyle, J.J., Domozych, D.S., Willats, W.G.T., 2012. The charophycean green algae as model systems to study plant cell walls and other evolutionary adaptations that gave rise to land plants. Plant Signal. Behav. 7 (1), 1–3. https://doi.org/10.4161/psb.7.1.18574.

Sørensen, M., Kousoulaki, K., Hammerø, R., Kokkali, M., Kleinegris, D., Marti-Quijal, F.J., Barba, F.J., Palihawadana, A.M., Egeland, E.S., Johnsen, C.A., Romarheim, O.H., Bisa, S., Kiron, V., 2023. Mechanical processing of *Phaeodactylum tricornutum* and *Tetraselmis chui* biomass affects phenolic and antioxidant compound availability, nutrient digestibility and deposition of carotenoids in Atlantic salmon. Aquaculture 569, 739395. https://doi.org/10.1016/j.aquaculture.2023.739395.

Sun, H., Liu, B., Lu, X., Cheng, K.W., Chen, F., 2017. Staged cultivation enhances biomass accumulation in the green growth phase of *Haematococcus pluvialis*. Bioresour. Technol. 233, 326–331. https://doi.org/10.1016/j.biortech.2017.03.011.

Sun Yu, B., Pyo, S., Lee, J., Han, K., 2024. Microalgae: a multifaceted catalyst for sustainable solutions in renewable energy, food security, and environmental management. Microb. Cell Fact. 23 (1). https://doi.org/10.1186/s12934-024-02588-7.

Taelman, S.E., De Meester, S., Van Dijk, W., Da Silva, V., Dewulf, J., 2015. Environmental sustainability analysis of a protein-rich livestock feed ingredient in The Netherlands: microalgae production versus soybean import. Resour. Conserv. Recycl. 101, 61–72. https://doi.org/10.1016/j.resconrec.2015.05.013.

Teuling, E., Schrama, J.W., Gruppen, H., Wierenga, P.A., 2017. Effect of cell wall characteristics on algae nutrient digestibility in Nile tilapia (*Oreochromis niloticus*) and African catfish (*Clarus gariepinus*). Aquaculture 479, 490–500. https://doi.org/10.1016/j.aquaculture.2017.06.025.

Tibbetts, S.M., Yasumaru, F., Lemos, D., 2017. In vitro prediction of digestible protein content of marine microalgae (*Nannochloropsis granulata*) meals for Pacific white shrimp (*Litopenaeus vannamei*) and rainbow trout (*Oncorhynchus mykiss*). Algal Res. 21, 76–80. https://doi.org/10.1016/j.algal.2016.11.010.

The next big superfood could be green and slimy, 2011. Business Insider.

Tibbetts, S.M., Milley, J.E., Lall, S.P., 2015. Chemical composition and nutritional properties of freshwater and marine microalgal biomass cultured in photobioreactors. J. Appl. Phycol. 27 (3), 1109–1119. https://doi.org/10.1007/s10811-014-0428-x.

Ullmann, J., Grimm, D., 2021. Algae and their potential for a future bioeconomy, landless food production, and the socio-economic impact of an algae industry. Org. Agric. 11 (2), 261–267. https://doi.org/10.1007/s13165-020-00337-9.

Uma, V.S., Usmani, Z., Sharma, M., Diwan, D., Sharma, M., Guo, M., Tuohy, M.G., Makatsoris, C., Zhao, X., Thakur, V.K., Gupta, V.K., 2023. Valorisation of algal biomass to value-added metabolites: emerging trends and opportunities. Phytochem. Rev. 22 (4), 1015–1040. https://doi.org/10.1007/s11101-022-09805-4.

Valente, L.M.P., Custódio, M., Batista, S., Fernandes, H., Kiron, V., 2019. Defatted microalgae (*Nannochloropsis* sp.) from biorefinery as a potential feed protein source to replace fishmeal in European sea bass diets. Fish Physiol. Biochem. 45 (3), 1067–1081. https://doi.org/10.1007/s10695-019-00621-w.

Walker, A.B., Berlinsky, D.L., 2011. Effects of partial replacement of fish meal protein by microalgae on growth, feed intake, and body composition of Atlantic cod. N. Am. J. Aquac. 73 (1), 76–83. https://doi.org/10.1080/15222055.2010.549030.

Watanabe, T., Liao, W., Takeuchi, Y., Yamamoto, H., Vassallo-Agius, R., 1990. Effect of dietary Spirulina supplement on growth performance and flesh lipids of cultured striped jack. J. Tokyo Univ. Fish 77, 231–239.

Watanuki, H., Ota, K., Tassakka, A.C.M.A.R., Kato, T., Sakai, M., 2006. Immunostimulant effects of dietary *Spirulina platensis* on carp, *Cyprinus carpio*. Aquaculture 258 (1–4), 157–163. https://doi.org/10.1016/j.aquaculture.2006.05.003.

White, R.L., Ryan, R.A., 2015. Long-term cultivation of algae in open-raceway ponds: lessons from the field. Ind. Biotechnol. 11 (4), 213–220. https://doi.org/10.1089/ind.2015.0006.

Yadav, G., Meena, D.K., Sahoo, A.K., Das, B.K., Sen, R., 2020. Effective valorization of microalgal biomass for the production of nutritional fish-feed supplements. J. Clean. Prod. 243. https://doi.org/10.1016/j.jclepro.2019.118697.

Zhang, Q., Qiu, M., Xu, W., Gao, Z., Shao, R., Qi, Z., 2014. Effects of dietary administration of chlorella on the immune status of gibel carp, *Carassius auratus* gibelio. Ital. J. Anim. Sci. 13 (3), 653–656. https://doi.org/10.4081/ijas.2014.3168.

CHAPTER

Harnessing microalgae for next-generation aquafeeds: Nutritional, environmental, and economic impacts

10

Sanjeeb Kumar Mandal[1], Gadde Padma Priyasha[1], Srikanth Muthyala[1], Disha Penchala[1], Kiran Yellappa Vajanthri[1], B. Sumithra[1] and G. Sri Thatchayani[2]

[1]*Department of Biotechnology, Chaitanya Bharathi Institute of Technology, Hyderabad, Telangana, India;* [2]*HCL Technologies, ELCOT IT Park, Madurai, Tamil Nadu, India*

1. Introduction

Sustainable development of fish feed is a vital part of the process of progressing the aquaculture of the future. As the demand for fish is getting higher all over the world, aquaculture is becoming a major source of nutritious and economical resources for the growing populations. Nevertheless, in most cases, the traditional fish feeds are highly dependent on fishmeal and fish oil that are taken from wild-caught fish. This leads to various environmental concerns like depletion of ecosystem and overfishing. In response to these issues, sustainable development of fish feed focuses on the implementation of alternatives that are renewable and eco-friendly raw materials, which not only provide the nutritional needs of the given species but also lower the environmental impact of aquaculture production (Table 10.1 and Fig. 10.1) (Chattaraj et al., 2024).

2. Importance of sustainable and nutrient-rich fish feed development

Developing sustainable fish feed is a crucial part of expanding global aquaculture and managing it responsibly. The continued reliance of the industry on fishmeal and fish oil obtained from wild-caught fisheries causes a great deal of stress to marine ecosystems and contributes to the depletion of resources (Julie, 2024). The use of microalgal biomass and various other alternative field ingredients is not only nutritionally valuable but also a way to gradually phase out the dependency on these finite resources (Chattaraj et al., 2026a,b). Sustainable feed formulations utilize raw materials that are locally available and Agro industrial by products, which, in

Fish Feed Formulation with Agroindustrial Wastes and Byproducts. https://doi.org/10.1016/B978-0-443-41728-3.00007-1

Table 10.1 Optimization techniques for microalgal cultivation, processing, and application in aquafeeds.

Technique	Usefulness/Benefits	References
Open Pond Systems	Low construction and operational costs; suitable for large-scale cultivation.	Barros et al. (2019)
Photobioreactors (PBRs)	Controlled environment, reduced contamination, better regulation of temperature, pH, and light; consistent and high-quality biomass.	Singh et al. (2021)
Co-cultivation (Microalgae + Fungi/ Bacteria)	Enhances biomass yield, lipid production, and natural flocculation; simplifies harvesting (e.g., *Chlorella* + *Aspergillus oryzae*pelletization).	Barros et al. (2019)
Cell Disruption—Bead Milling, High-Pressure Homogenization, Pulsed Electric Field, Enzymatic Hydrolysis	Breaks tough cell walls; improves nutrient digestibility and bioavailability in aquafeeds.	Paterson et al. (2023)
Physical Methods—Sonication, Microwave-Assisted Extraction	Efficient nutrient release without degrading sensitive compounds.	Singh et al. (2021)
Ionic Liquids (ILs) + Ultrasound/Microwave	Low-energy extraction of lipids and pigments; supports biorefinery approach.	Barros et al. (2019)
Real-Time Monitoring—Sensors (pH, DO, Temp, Light, Nutrients)	Enables immediate adjustments to optimize photosynthesis, growth rate, and yield.	Chaudhary et al. (2024)
Real-Time Quality Control—Imaging, Flow Cytometry, Spectroscopy	Early contamination detection; biochemical profiling of pigments, proteins, and chlorophyll for consistent biomass quality.	Chaudhary et al. (2024)
AI & Machine Learning for Process Optimization	Predictive modeling of growth, anomaly detection, decision support for cultivation adjustments.	Chaudhary et al. (2024)

Table 10.1 Optimization techniques for microalgal cultivation, processing, and application in aquafeeds.—*cont'd*

Technique	Usefulness/Benefits	References
Feed Formulation with Microalgae	Replaces fishmeal/oil; provides proteins (EAAs), lipids (PUFAs), pigments, vitamins, minerals; eco-friendly diets.	Paterson et al. (2023)
Pre-treatments for Digestibility	Bead milling and enzymatic hydrolysis improve nutrient release, reduce anti-nutritional factors.	Paterson et al. (2023)
Eco-Design in Feed	Circularity by using effluent-grown microalgae, minimizing waste, sustainable feed sourcing.	Paterson et al. (2023)
Resource & Energy Efficiency—Bio flocculation, Co-cultivation with Filamentous Fungi, Immobilization	Reduces harvesting costs and energy demand; simplifies processing.	Barros et al. (2019)
Drying Methods (Solar, Freeze, Spray Drying)	Trade-offs in cost, nutrient retention, and scalability; innovations reduce carbon footprint.	Singh et al. (2021)
Water & Nutrient Recycling	Minimizes input losses, prevents contamination, enhances sustainability.	Chaudhary et al. (2024)
Biorefinery & Process Integration	Extracts multiple products (proteins, lipids, pigments, antioxidants) from one batch, maximizing revenue.	Singh et al. (2021)

addition to fully meeting the nutritional requirements of fish being farmed, make ecosystems healthier has they provide nourishment to the pond microbiome and natural food webs (Wiegertjes et al., 2023). This holistic strategy strengthens the aquaculture systems, making the more stable and resilient, there by enhancing disease resistance and growth. Besides this, the economic side of the feed ingredients is very important as the feed is the most significant cost of aquaculture operations. Hence the price and availability of sustainable alternatives are what determine the success of aquaculture ventures globally. Besides that, sustainable feed development is in line with the food system objectives and thus, by limiting environmental impacts, resource use efficiency gets improved, and livelihoods get enhanced through new industry opportunities (WorldFish, 2022; Chattaraj and Das Mohapatra, 2024).

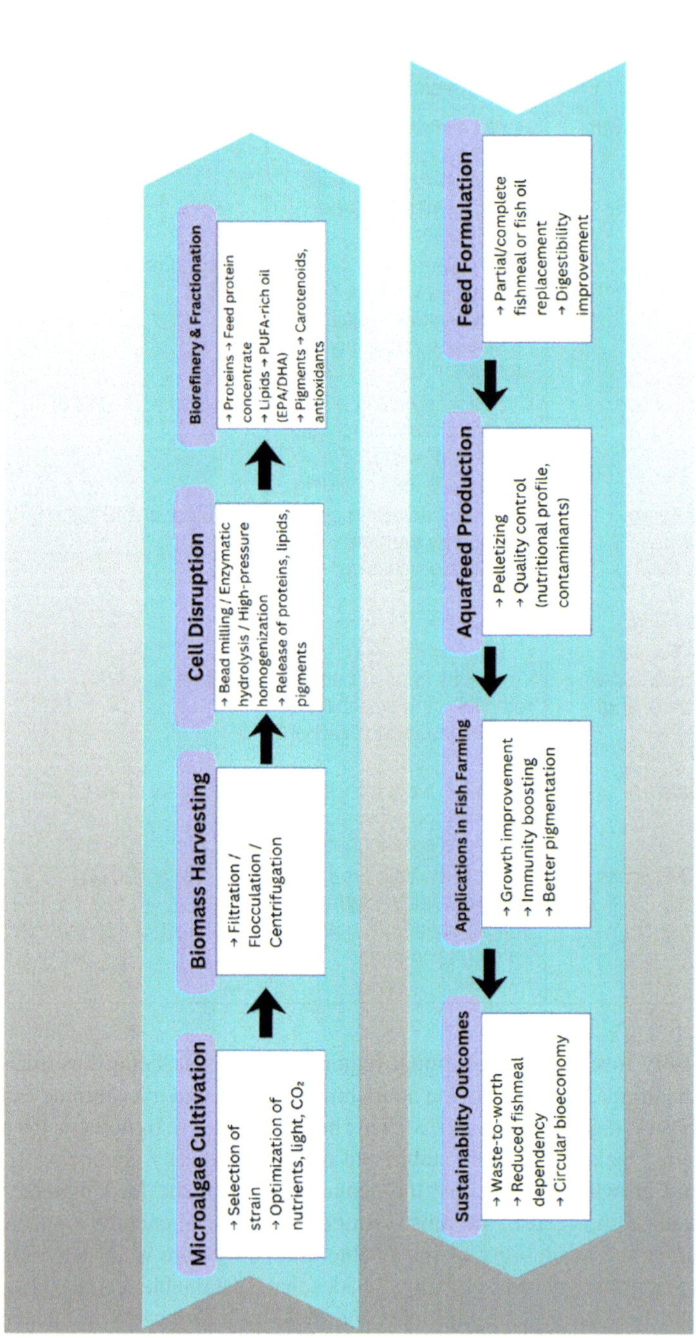

FIGURE 10.1
Overview of microalgal valorization for aquafeeds.

3. Objectives and scope of the chapter

This chapter seeks to explore in-depth the potential of microalgal biomass to serve as a nutritionally rich and environmentally friendly source of fish feed. It features a comprehensive literature review of microalgal consumption in aquatic food production, with a particular focus on the species most frequently used, their growth factors, and the methods of their valorization to gain a deeper insight into the subject matter. The chapter will focus primarily on providing a detailed description of how microalgal nutritional compositions contribute to fish becoming healthy, high-quality food products. Microalgae promise to supply proteins, polyunsaturated fatty acids, vitamins, pigments, and bioactive compounds that serve as the building blocks of the growth, health status, and quality of fish products (Nagappan et al., 2021a, 2021b).

In this chapter we will discussing about the environmental and economic aspects of sustainability with a strong focus on the integration of agro-industrial residues has nutrient sources for microalgal growth to promote circular bioeconomic practices while simultaneously reducing the cost of production. Issues such as scaling up of microalgal culture, difficulties in harvesting process, and the resistance of the potential consumers to the new products will be discussed to give a complete picture of situation. In addition to that this chapter also discuss the next-generation ideology and ongoing research, which is closely end to address the existing shortcomings and realizing the potential use of microalgal biomass in worldwide aquafeed markets. This synopsis will be of assistance to the scientists, industry stakeholders and decision makers in lining up their actions toward sustainable aquaculture development.

4. Understanding microalgal valorization for aquafeeds

4.1 Definition and core principles

Microalgal valorization refers to the mechanisms of effectively turning microalgae biomass into different products for a variety of industrial usages (Wang et al., 2025). It is an easy methodthatincludes harvesting the biomass, which is done along with cultivation, processing and biorefinery techniques to extract compounds such as proteins, lipids, pigments, and vitamins. They facilitate nutrition and health of the fish (Singh et al., 2021; Yadav et al., 2019).

The fundamental concept of microalgal valorization relies on the environmentally friendly model (Melikoglu, M., 2025). In this method, the waste and nutrient streams (like agro-industrial wastewater) are recycled into microalgal biomass. Then the biomass is turned into high-value products. It minimizes the waste and promotes sustainability (Patel et al., 2021). This "waste-to-worth" strategy not only improves the resources but also reduces the pollution of the environment and supports the production of ecological fish feeds (Chaudhary et al., 2024).

In that case, one can say microalgae are the source of the essential nutrients for fish feed as they are photosynthetic microorganisms that can make biomass by using sunlight, water, and carbon dioxide (Nagappan et al., 2021a, 2021b). Moreover, microalgal valorization is also about the deployment of facilities to harvest and extract that are scalable in order to reduce energy and economic costs which the microalgae may become the practicable substitutes for fishmeal, and fish oil in aquafeeds (Warapat et al., 2022). The effective valorization processes help digestibility and nutrient bioavailability, which in turn are very necessary for feeding conversion ratios and enhancing fish health.

4.2 Historical evolution of microalgal use in aquaculture

The use of microalgae in aquafeeds dates back to several decades. At that time, microalgae were first used as live feed for larval stages of shellfish and fish (Milstein et al., 2014). The early research was mainly concentrated on the nutritional benefits of species like *Isochrysis galbana* and Chaetoceros in the growth of bivalve molluscs (Tredici, 2004). Eventually, the focus shifted to the production of dried or processed microalgal biomass for direct feeding or as a supplement to traditional fishmeal. Valorization implements the "waste-to-worth" concept by integrating microalgal production with waste management. The wastewater from agriculture and aquaculture can be used for this process as well. It not only increases the environmental impact of nutrient pollution but also produces biomass that can be a viable substitute for the conventional feed ingredients such as fishmeal and fish oil (Oswald et al., 2022; Chaudhary et al., 2024). Microalgae are fast growers, are biochemically versatile, and have a high photosynthetic efficiency, hence making them ideal for this model. Microalgae proteins can be equivalent to fishmeal in amino acid profiles, whereas their PUFAs, mainly EPA and DHA, are highly valuable for fish health and meat quality (Nagappan et al., 2021a, 2021b; Morais et al., 2020). Developments in microalgal biorefinery technologies include the use of genetically modified organisms for high production strains and the incorporation of waste nutrient streams, further making these more viable and sustainable sources for aquafeed. The other two factors that keep the research and industry adoption going include acceptance of regulations and global demand for seafood products that are eco-friendly.

5. Smart and sustainable approaches in microalgal feed systems

The increase in microalgal feed systems represents one of the important factors that determine the sustainability of aquaculture. Considering:

1. Cultivation and processing
2. Using real-time monitoring and quality control
3. Application of green feed formulation practices
4. Enhanced resource and energy efficiency

The microalgal biomass can be converted into premium fish feed (Chen et al., 2021). These significant areas are discussed in detail in this section.

5.1 Optimization strategies for cultivation and processing

The optimization of microalgal cultivation improves the nutrient quality, biomass yield and decrease the production costs (Figueroa-Torres et al., 2021). Traditional systems such as open ponds are still widely used due to their low construction and operating costs. Open pond systems have to deal with problems such as:

1. changing environmental conditions
2. contamination risks
3. low cell concentrations

and difficulties in controlling factors like:

1. temperature
2. light
3. intensity

These issues may lead to lower growth rates and biomass quality. These problems have been solved by using photobioreactors (PBRs). They provide regulated conditions that result in increased productivity, contamination reduction, and provision for light, temperature, pH regulation. Even though PBRs have higher capital and running costs, their potential for the production of high-quality, stable biomass is a great advantage for aquafeed applications (Barros et al., 2019; Singh et al., 2021). Another way to optimize the production process is to use co-cultivation strategies. If the microalgae are grown along with the fungi or bacteria, these interactions can not only increase biomass yield, lipid production but also make the harvesting and processing steps easier due to natural flocculation. Co-cultivating of *Chlorella* with fungi like *Aspergillus oryzae* has been demonstrated to not only increase biomass yield but also facilitate downstream processing by pelletization is one of the examples. Processing methods are also very important for the feed value. Different pre-treatment methods, such as bead milling, high-pressure homogenization, pulsed electric fields, and enzymatic hydrolysis, help to break the microalgal cells that are of a strong nature. This releases nutrients that were intracellular and also improves digestibility for fish. Also, there are physical methods such as sonication and microwave-assisted extraction, which can be used effectively to release nutrients, thus improving bioavailability without the need to scare sensitive compounds. Innovative extraction methods with the use of iconic liquids together with the ultrasound or the microwave irradiation allow for obtaining lipids and pigments in an effective way and at the same time to reduce the energy consumption. These modern processing technologies give way to the production of value-added bioproducts which are produced alongside feed biomass, thus facilitating the implementation of biorefinery concept.

5.2 Real-time monitoring and quality control mechanism

For the commercial aquafeed applications, maintaining the quality of microalgal biomass is a very important aspect. This is only possible through real-time monitoring of microalgal growth, nutritional value, identification and control of factors that impede the growth. For effective control, sensors that monitor and control the growth conditions like pH, temperature, dissolved oxygen, light, and nutrients should be triggered to respond automatically and within the appropriate limits.The automated systems can be used to maintain, optimize and control of biomass and photosynthesis process. Modern imaging techniques and flow cytometry methods can be used as quality control measures to assess and monitor for the contamination and compromised cultures. The spectroscopic techniques like fluorescence, near-infrared spectroscopy not only assist in the rapid evaluation of chlorophyll, carotenoids, and proteins, but also help in production and quality control (Chaudhary et al., 2024). The predictive modeling and decision-making processes are streamlined through the collection and integration of data by software utilizing the artificial intelligence and machine learning. These technologies improve operational efficiency and product uniformity by predicting growth patterns, identifying anomalies, and suggesting operational changes. Adhering to quality assurance protocols and specifically testing for microbial and contaminant safety ensures harvested biomass is of the appropriate regulatory and nutritional standards for feed production.

5.3 Feed formulation, digestibility, and eco-design

The digestibility, nutrition and environmental effects while formulating aquafeed with microalgae must be considered. Microalgae enhance the quality of end products and fish health by providing compounds like protein, essential amino acids, lipids, vitamins, minerals and pigments. Some biomass pre-treatments that help break down cell walls improve digestibility. Studies indicate that bead milling of *Tetraselmis* and *Chlorella* microalgae significantly improves the digestibility of the proteins (Chattaraj et al., 2024). With the enzymatic hydrolysis of biomass, the nutrient availability increases while the anti-nutritional factors decrease (Paterson et al., 2023). Microalgae allow aquafeeds to be formulated with less fishmeal and fish oil, easing the pressure on marine resources. The formulations of aquafeed are made by considering key features like, optimal cost, nutritional value and flavor as inclusion rates are empirical to fish species and their developmental stages (Rust et al., 2011). The use of sustainably sourced feedstuffs, reduction of waste, and incorporation of industry byproducts are supported by the eco-design in feed formulation to enhance the circularity. Microalgae cultivation on nutrient-rich effluents can help close nutrient loops, feed waste recycling, and decrease environmental impacts associated with aquaculture feed production.

5.4 Techniques for resource and energy efficiency

Strategically using the resources and energy available becomes essential for the microalgae biomass production to be economically viable and sustainable.

Energy-consuming activities, particularly harvesting and drying, need innovative strategies. Innovative strategies for the energy consuming activities, mainly like harvesting and drying, are needed. 30% of the total production costs is constituted by harvesting the biomass (Kabeyi and Olanrewaju, 2025). Certain practices like bioflocculation, co-cultivation with filamentous fungi that assist in pellet formation and substrate immobilization reduce energy costs and streamline downstream processing. Methods of drying-such as solar drying, freeze drying, and spray drying-are a trade-off between cost, nutrient retention, and scalability. The development of low-energy drying strategies has been based on continuous processes in order to reduce the biomass preparation carbon footprint. Real-time nutrient monitoring prevents overuse and contamination, while recycling of water and nutrients within the growing system minimizes input losses. In an integrated approach of wastewater treatment and production of microalgae, residential nutrients serve both biomass and bioremediation purposes. The process integration biorefinery strategy extracts multiple high-value products, including proteins, lipids, pigments, and antioxidants from single batches of microalgae. This would fully valorize the biomass components and minimize revenue streams.

6. Waste stream integration and circular nutrient management

Sustainable aquafeed production from microalgal biomass relies heavily on the integration of waste streams and circular nutrient management. The use of agro-industrial waste streams for microalgal cultivation, circular models in fish feed supply chains, and lifecycle assessment with material selection to ensure economic and environmental sustainability are all covered in this section.

6.1 Circular models in fish feed supply chains

The goal of circular economy models is to convert linear supply chains into regenerative systems with minimal waste and efficient resource reuse. This entails streamlining the procurement, processing, distribution, and reuse of waste in fish feed supply chains in order to establish closed-loop systems that support sustainability objectives. For instance, by-products of fish processing like fish skin, bones, and offal are increasingly valued into fish meal, oil, collagen, and other feed ingredients, which lessens the impact on the environment and creates economic value for producers and communities. (Seafish, 2025). Smallholder farmers and cooperatives are involved in locally adapted circular supply chains, which combine supply and demand to provide economies of scale and better access to high-quality feed inputs (WorldFish, 2022). Micro-franchises and cooperative-led feed manufacturing are examples of models that lower transaction costs, improve feed quality control, and bolster robust rural economies (Obwanga and Lewo, 2017; Ngugi et al., 2017). Through real-time transparency, traceability, and optimized logistics, digital technologies play a growing role in circular supply chains, lowering spoilage and enabling resource sharing among stakeholders (Seafish, 2025).

7. Integration of agro-industrial waste streams for microalgal cultivation

Coupling microalgal cultivation with agro-industrial waste streams provides a synergistic opportunity for nutrient recycling and cost reduction. Nutrient-rich effluents from agriculture, aquaculture, food processing, and wastewater treatment facilities convert environmental liabilities into feedstock inputs by supplying nitrogen, phosphorus, and micronutrients required for microalgal growth (Chaudhary et al., 2024; Blue Biotechnology, 2024). One example is the use of fish pond effluent for the growth of Chlorella and Nannochloropsis to improve water quality and simultaneously produce biomass. Different types of agro-industrial wastes as an organic carbon source for mixotrophic microalgal growth have been noted to enhance productivity and simultaneously decrease operational inputs, and they include dairy wastewater, food processing residues, and glycerol from biodiesel production.These circular integrated systems bring on board the circular bioeconomy principles that minimizes waste pollution, locks up resource losses in production of microalgal feed ingredients, and closes nutrient loops. Feed variability management, however, is still a limiting factor, as will also be the risks from pollution, regulatory effects of wastes, and delivery of safe, uniform feeds; standardized procedures will have to be followed and state-of-the-art control measures installed (Tham et al., 2023; Osundeko et al., 2023).

7.1 Lifecycle assessment and material selection considerations

A significant aspect of LCA is to quantify the inputs determining the environmental footprint of microalgal biomass production and fish feed formulation. This will further support the justification of critical, evidence-based decisions relating to materials, selection, and process design. While LAs consider the accountability of fossil resources, greenhouse gases, water, and land use, the comparisons of microalgae and traditional fish feed ingredients, such as fish and soy meal, are very informative (Morais et al., 2020; Singh et al., 2021). The integration of microalgal cultivation with waste stream valorization will reduce the carbon footprints, the dependency on overfished stocks and the rate of biodiversity loss (Chaudhary et al., 2024). To ensure sustainability, otimization of the energy needed for harvesting and drying of microalgae needs to be done due to its high-cost concerns. By balancing feed formulation with sustainable inclusion rates, and in combination with other sustainable feed ingredients, trimming feed costs will be easier to achieve.

8. Case studies and industrial implementations

The usage of microalgae biomass as a sustainable resource in aquafeeds production has been adopted in various regions of aquaculture (Ahmad et al., 2022). The extensive use of microalgae and their subsequent valorization technologies and techniques is documented with recognition of aquaculture practices and the scientific literature proving their success. Several examples and microalgae success technologies valorization practices aquaculture documented success practices, science literature review.

8.1 Sector-wise examples (freshwater, marine, ornamental fish)

8.1.1 Freshwater aquaculture

Globally, the most important freshwater aquaculture species are Nile tilapia (*Oreochromis niloticus*), and common carp (*Cyprinus carpio*), and *Clarias gariepinus* (catfish) and these have been the focus of microalgae feed research. A milestone study by Sarker et al. (2020) madc an important lcap by formulating a diet that did not include fishmeal and fish oil for Nile tilapia, by mixing the defatted biomass of *Nannochloropsis oculata* and DHA-rich *Schizochytrium* sp. The feed not only achieved the performance targets of tilapia on a conventional diet but also exceeded the feed conversion rates and the nutritional quality of the fillet compared to the conventional diet. The fish-free formulation delivered nearly double the DHA content in fillets, enhancing the omega-3 fatty acid profile for improved human consumption benefits. Economic analysis indicated the fish-free diet was cost-competitive, positioning microalgal biomass as a commercially viable option for intensively farmed freshwater species. Other research work on carp has shown that a partial replacement of fishmeal by *Chlorella* and *Spirulina* biomass improved growth performance, immune parameters and resistance to disease challenges. Dietary microalgal carotenoids like astaxanthin in the catfish feed are reported to enhance pigmentation, reproductive performance, and antioxidant status (Nurhayati et al., 2022). This observation suggests that microalgae have a wide potential for application as nutritional enhancers in the major freshwater cultured species.

8.1.2 Marine aquaculture

Marine organisms, such as Atlantic salmon (*Salmo salar*), seabass (*Dicentrarchus labrax*) and shrimp (*Litopenaeus vannamei*) have derived significant benefit from the inclusion of microalgal feed ingredients. The use of *Schizochytrium* sp. as a source of plant-based oil in replacement of fish oil, being high in omega-3 fatty acids, with salmon feeds for which continued growth and acceptable product quality, and health status have been achieved even when the total fish oil replacement was fully replaced (Tocher et al., 2021). Application of *Isochrysis galbana* and *Tetraselmis* species in nursery diets contributes to higher survival rate and faster metamorphosis in marine finfish and crustaceans indicating the importance of microalgal lipids and pigments during early life (Ma et al., 2024). Shrimp aquaculture has been greatly enhanced by the immunostimulatory effects of microalgal polysaccharides and microelements in reducing disease challenge on an intensive culture system with a view to sustain productivity (Ramesh and Hemalatha, 2023). Carotenoid-rich Haematococcus pluvial is biomass enhances shell pigmentation and the antioxidant defense system in commercial laying hens.

8.1.3 Ornamental fish

The ornamental fish industry is a major beneficiary of coloration and microalgal carotenoid can play an important role for the same. Natural coloration of astaxanthin and β-carotene from microalgae, help in giving attractive colors to the mollies preferred by hobbyists and collectors (Camacho et al., 2019). Moreover, microalgae supplementation increases the health and stress resistance of species such as goldfish (*Carassius auratus*) and guppies (*Poecilia reticulata*) by stimulating antioxidant enzyme activities and immune status, essential for successful captive breeding (Li et al., 2022).

8.2 Technologies and valorization methodologies

In order to efficiently utilize the microalgal biomass, developing processes for optimized cultivation, harvesting, processing and integration technologies are necessary. These technologies assist in processing raw microalgal biomass into nutritionally complete fractions and ingredients with high market value for commercial feeds.

8.2.1 Cultivation systems

Photoautotrophic growth in open raceway ponds still predominates because of cost-effectiveness toward mass cultivation. However, photobioreactors (PBRs) have become popular for higher biomass productivity, practice preventing contamination effect and desirable cultivation environment at the cost of expensive capital investment. Hybrid systems combining the benefits from both approaches are being investigated (Barros et al., 2019). Latest developments to promote optimal growth include cultivation mixotrophically, by combining photosynthesis with assimilation of organic carbon sources, to increase both biomass productivities and concentrations of the target bio-compounds. Addition of nutrient rich wastewaters (e.g., from aquaculture or agriculture) to medium provides a unique solution for recycling nutrients as well as generation of biomass, in line with the circular economy concept (Chaudhary et al., 2024).

8.2.2 Harvesting and processing

Microalgal cell harvesting from culture medium is an expensive and energy-intensive process. Bio-flocculation, fungal-algal co-cultivation that produces pelletized biomass, and immobilization on substrates are examples of innovative harvesting techniques that lower energy intensity and speed up downstream processes (Nagappan et al., 2021a, 2021b). Fish digestibility improves when mechanical and enzymatic pretreatments are added because they increase cell disruption and improve the extractability of proteins and lipids. Before using leftover protein-rich material for aquafeeds, emerging biorefinery techniques maximize biomass utility by extracting lipids, carotenoids, and bioactive peptides for the nutraceutical, pharmaceutical, and cosmetics industries (Singh et al., 2021). These integrated protocols support the concept of zero-waste and increase the economic viability.

8.2.3 Formulation and quality assurance

Nutritional specificities of the target species, dietary balance, and inter-feed constituent relationships must all be precisely calibrated before integrating the microalgal feed components into aquafeed formulation. Aquafeeds typically use microalgae for full and partial replacement of fish oils and fishmeal because they present alternatives for meats, oils, and other dietary essentials such as vitamins, minerals, omega-3 fatty acids EPA and DHA, amino acids, and pigments. Process that enhances nutrient bioavailability in addition to improving feed acceptability, stability, and nutritional value also include drying, milling, and pelletizing of microalgal feeds. Biochemical assays, chromatography, and spectroscopy are used for quality control with a view to assessing the safety and value of feeds against international standards and their ethnicity (Tham et al., 2023).

8.3 Success stories and impact assessment

Industry-led initiatives and research collaborations have made the commercial use of microalgal biomass in aquafeeds viable, assuring benefits related to fish health, sustainability, and economic performance. Among the main achievements is the microalgae-blended feed for Nile tilapia from Sarker et al. (2020). It totally replaced fishmeal and fish oil with defatted *Nannochloropsis* biomass and *Schizochytrium* sp. whole cells. Besides improving growth performance, this feed enhanced the omega-3 content of the fillet and exhibited favorable economic conversion ratios, indicating market-ready formulations. Further, the use of defatted biomass from microalgal oil extraction through co-product valorization contributes to waste management at an industrial level. At the industrial level, companies like Alltech and Solazyme have microalgal ingredient solutions for aquafeeds that are of quality and consistency over the long run and are scalable. Their products lead to the improved immune system, better growth, and environmental sustainability in farmed fish (Alltech, 2022). Small- and medium-sized aquaculture enterprises in Asia and Africa have, to a different extent, incorporated microalgal biomass in their local feed, thus enhancing local feed self-sufficiency and reducing the costs of imported fishmeal (Khan et al., 2021). Socioeconomic evaluations narrate the positive effect of such practices on rural livelihoods, gender equity, and the resilience of fish farming systems. Environmental impact assessments have been, without exception, in favor of the microalgal feed ingredient usage, which would alleviate the overfished source of industrial fishmeal, lead to a reduction in aquaculture associated greenhouse gas emissions, and support biodiversity conservation (Morais et al., 2020). Cost evaluations are aware of the situation in which a high initial investment and production costs remain as a hurdle, but by the integration of circular biorefinery practices and by the optimization of production systems, it is expected that the costs will soon go down to the level of conventional ingredients (Nagappan et al., 2021a, 2021b).

9. Challenges in large-scale adoption

Microalgal biomass production for aquafeeds has numerous obstacles as it moves from small-scale research, pilot projects to full-scale commercial volume and ranges across organizational, technological, and financial domains. The industry cannot be altered without first removing the barriers, even if microalgae offer a nearly perfect option as a sustainable and nutrient-rich feed element. The current research studies serve as a source for this section, which, after a deep dive, lays out the core issues that interact in three broad categories:

- technological and operational barriers
- economic and investment constraints
- organizational readiness and workforce development

9.1 Technological and operational barriers

On a large scale, the technical problems that have been the main concern of laboratory research for a long time are the main factors that restrict the application of microalgal cultivation in open environments and have impacts on productivity, cost-efficiency, and product quality. One cannot make a perfect choice between open raceway ponds, closed photobioreactors (PBRs) or biofilm systems for production due to variables such as cost, control, and area expansion. For example, open ponds are a low-cost solution, but their drawbacks include low biomass concentrations (often less than 0.5 g/L), susceptibility to pollution, seasonal fluctuations and inefficient use of CO_2 and nutrients (Novoveská et al., 2023). On the other hand, PBRs have much higher capital and operating costs, although they allow higher cell densities (up to 30+ g/L), exact environmental control and less pollution (Barros et al., 2019; Skifa et al., 2024). Microalgal biomass harvesting is still a major obstacle. Because cultures are diluted, processing them requires a lot of water, which makes centrifugation, filtration, or flocculation resource-intensive and expensive (Nagappan et al., 2021a, 2021b). Fungal-algal co-cultures and bio-flocculation are two novel harvesting methods that show promise for lowering energy input, but they are still being optimized for consistency and scale (Warapat et al., 2022). Other main challenges include contamination control in large-scale outdoor systems. Viruses, grazers, and invasive microbes can rapidly destabilize cultures, causing loss of biomass and erratic production. While closed systems minimize risks, they add complexity and capital costs that are prohibitive to most growers, according to Lu et al. (2023). The scale and complexity of the related monitoring and adjustment of cultivation parameters such as light, temperature, pH, and CO_2 concentration warrant automation investments, often unaffordable for small and medium enterprises, according to Novoveská et al. (2023).

In particular, the downstream processing of microalgae biomass for drying, cell disruption, and extraction of bioactive substances is also one of the limiting factors for scaling up the production process due to the high energy requirements and inefficiencies of the process. Keeping the nutritional content and biochemicals intact during processing is of paramount importance for retaining the quality of aqua feed, but this, in turn, makes operational decisions more complex (Tham et al., 2023). Additionally, the digestibility factor in fish feed is hampered by the microalgae's hard cell wall. The use of pre-treatment technologies like enzymatic hydrolysis is becoming more widespread. These technologies raise the method's processing complexity and production costs (Gong et al., 2020). Furthermore, fish digestion and feed conversion efficiency may be impacted by the presence of anti-nutritional factors (ANFs), such as protease inhibitors and cell wall fibers in particular species. This could result in strain and species selection as well as processing changes (Lu et al., 2023).

9.2 Economic and investment constraints

The reason that microalgal biomass is not often used for marine feed is due to the economic aspect of success. Because of the resource-intensive harvesting and drying steps, processing demands, area expansion and the production costs are still much higher than those of conventional products like fishmeal and soybean meal (Occhipinti et al., 2024). The initial capital expenditures (CAPEX) for closed photobioreactors and related facilities usually are in the range of millions of dollars, thus, the gate-keeping effect raises such financial obstacles for potential newcomers. The operational expenditures (OPEX) kept coming from the energy consumption, labor, nutrients, and waste management (Novoveská et al., 2023). Economies of scale help reduce costs, but only facilities producing thousands of tons per year approach competitive pricing (Nagappan et al., 2021a, 2021b). Additionally, because the microalgal aquafeed industry is still in its infancy, there is little market confidence, which limits private investment. Investor skepticism has spread to bioproducts due to the failure of previous algal biofuel projects, necessitating strong proof of profitability and scalability in order to draw funding (Occhipinti et al., 2024). Furthermore, feed market price volatility and increased perceived risk of microalgal ingredients continue to dampen adoption rates. Current aquafeed manufacturers also face practical operational hurdles transitioning to alternative protein sources without sacrificing product quality or disrupting supply.

Despite these challenges, integrated biorefinery approaches that extract a range of products from a single cultivation batch (proteins, lipids, pigments, bioactives, among others) will improve economic sustainability. In addition, mixotrophic or heterotrophic modes of cultivation and valorization of wastes for nutrients have the potential for reducing the overall cost of production by lowering carbon and nutrient inputs (Patel et al., 2021). Each will require policy incentives, public-private partnerships, and targeted funding of research to lower financial barriers and drive innovation.

9.3 Organizational readiness and workforce development

Besides technology and finance, an effective scale-up of microalgal biomass production for aquafeeds calls for qualified labor, organizational infrastructure, and enabling conditions. Trained psychologists, bioengineers, and aquafeed nutritionists with focused expertise in microalgae culture and processing as well as feed formulations are currently scarce (Novoveská et al., 2023). Several companies lack the interdisciplinary teams of biologists, engineers, chemists, and data scientists needed to drive operational efficiencies and problem-solving. By developing specific courses and degree programs in algal biotechnology, educational institutions have started to bridge the gap and nurture the next generation of experts. At the industrial level, there is still a need for continuous professional development and knowledge transfer. Organizational readiness encompasses adaptive management skills, willingness to innovate, and the integration of automated monitoring and control systems. Smaller firms often have gaps in productivity and product consistency due to the inability to afford investment in digital infrastructure, which can ease operations through reducing labor costs (Tham et al., 2023). Cultural and legal obstacles also hinder its adoption. Misconceptions about the safety and quality of food and feed made from algae always seem to influence public opinion. Uncertainties restrict market release because microalgal compounds have different countries that have regulatory laws governing them (Ahmad et al., 2022). Well-defined regulatory channels, industry-academia partnerships, and cooperative networks can all enhance the prospects for establishing trust, expanding, and promoting the skill development in the workforce essential to sustain a sustainable microalgal aquafeed industry.

10. Role of training, policy, and industry initiatives

Its sustainable adoption and growth depend on interrelated pillars: industry-led programmes driving innovation and standardization, international policies and regulatory frameworks that support it, and the development and education of skills. Each of these components provides a different part of the overall framework necessary for overcoming organizational, financial, and technological challenges to allow robust industry growth consistent with socioeconomic and environmental goals.

10.1 Education and skill development

Accompanying the industry growth will be the need for education and training to provide a suitably skilled workforce to lead and expand the microalgae-based aquafeed industry. Production and processing of microalgal biomass, unlike conventional aquaculture inputs, require specialized knowledge in areas of phycology, biotechnology, engineering, and nutrition. In the absence of appropriate training programs, scaled-up applications of microalgae also risk biomass quality inconsistency, operational inefficiency, and non-compliance with safety and regulatory requirements. These training courses are at various levels, showing the importance of the same. For instance, the ICAR- Central Marine Fisheries Research Institute (ICAR-CMFRI, 2024) provides practical training courses regarding the method of culture of marine microalgae for hatchery technicians and professionals in aquaculture.

These contain minimum topics to be covered, such as harvesting, water quality management, integration of live feeds, isolation of strains, and monoculture and batch culture techniques. These demanding courses ensure that better knowledge is transferred in practical business settings.International efforts have centered on the issue of training, as well. The Algae Technology Educational Consortium, ATEC, has a modular online training program, the Algae Cultivation Extension Short Course called ACES, which covers the essentials of growing macro- and microalgae, scaling methodology, biochemical composition, and environmental impacts (ATEC, 2024). These tools democratize access to knowledge globally and help build the talent pipeline the industry needs to grow.The EIT Food-developed European INSPIRE Restorative Aquaculture program and The Seaweed Academy bring together active stakeholder engagement, coaching for entrepreneurship, knowledge about algal biotechnology, and production systems in aquaculture. This suggests a multidisciplinary approach which should give new employees the technical skills and business acumen to lead sustainable aquaculture projects and create innovations. Formal education should be augmented by practical skills workshops and on-the-job training. Short courses covering various aspects of the technology of microalgal culture, harvesting, and processing are held regularly at universities, research centers, and industry consortia. Such applied learning is critical to technicians and managers for grasping the particular demands of scale-up, quality control, and product formulation.

Organizational capacity building should emphasize continuing education and adaptability relevant to industry evolution. Collaborative platforms, knowledge exchange networks, and industry-academic partnerships allow ongoing skill development in this respect and help ensure the capability of the workforce to keep up with innovation.

10.2 International policy initiatives and regulatory frameworks

The commercialization and wide application of microalgal biomass in aquafeeds depend on clear regulations and favorable policy environments. Considering that microalgae are categorized as novel feed ingredients in most countries, they usually undergo strict safety testing, approval processes, and quality standards that may differ significantly between countries. International organizations, such as the Food and Agriculture Organization of the United Nations, Codex Alimentarius Commission, and European Food Safety Authority, make scientific guidelines and risk assessment procedures available that could support a harmonization effort for standards pertaining to microalgal feed ingredients. Such schemes pave the way toward food safety, label transparency, and consumer confidence, while ensuring access to international markets. National legislation usually sets contaminant levels, nutritional adequacy, toxin-free, and microbiological safety. There are case-by-case authorization processes that are protecting the consumers, while this slows down the entry of products into the market (Ahmad et al., 2022). Some countries have created pre-approved microalgal species or specific positive lists considered safe for animal feed to streamline approvals through the FDA, USDA, and EFSA. Policy initiatives support hubs of innovation, pilot demonstration projects, and research funding. Public-private partnerships create an ecosystem of innovation where enterprise, government, and academia come together to reduce stress, simplify procedures, and provide access to new markets.At the global level, several integrated multi-level strategies support the use of microalgae aquafeed in circular bio economics. The European Union's Green Deal focuses on the development of bio-based feed and nutrient recycling, with direct incentives regarding waste valorization and nutrient recovery (European Commission, 2024). Similarly, national blue economy strategies embed aquaculture sustainability goals in industrial symbiosis that fosters microalgal cultivation for multi-sectoral nutrient reuse. Trade facilitation agreements, which increasingly recognize sustainable feed certifications, facilitate the export-import flows of ingredients derived from microalgae (Algae Biomass Organization, 2024). There are still discrepancies over definitions, safety assessments, and standards for labeling that hinder global commercialization; these are being addressed in ongoing efforts to harmonize regulations through international forums.

10.3 Standards and incentives for long-term adoption

Industry-wide initiatives, financial incentives, and established standards hasten the pace at which microalgae move from experimental uses to commercial utilization. Certification and sustainability standards put in place a reliable framework for manufacturers, processors, and consumers alike.Among the sustainability certification requirements, such as Aquaculture Stewardship Council (ASC) and GlobalGAP, emphasis is being placed on traceability and ecological impact with responsible sourcing. By following these recommendations, feed producers will be able to include more sustainable microalgal ingredients in their products and attract environmentally conscious customers.Financial incentives such as grants, subsidies, tax breaks, and low-interest loans stimulate investment in feed innovation and microalgal cultivation technology. For instance, many countries in Europe provide feed innovation grants to biorefineries and farms operating on a pilot scale (European Commission, 2024). Recognizing the environmental co-benefits, such as carbon sequestration and nutrient pollution reduction, from microalgal production systems, carbon credits and ecosystem service payments can add other sources of financial returns.Industry consortia and innovation clusters spur adoption through the sharing of best practices, collaborative investments, and jointly developed technology roadmaps. Various platforms, such as the Algae Biomass Organization and the Blue Economy Innovation Networks, have therefore started to bring stakeholders together and advance policy support, regulatory clarity, and market expansion.Besides, standardization of analytical methods, biomass specifications, and quality benchmarks also give support to industry trust and interoperability of supply chain actors. According to ISO, 2023, the protection for feed users and consumers' quality standards are provided based on nutrient specification, contaminant limits, and microbiological safety.Education on the benefits of sustainability and continued communication on the value of microalgal feed create the potential for transitions to circular aquaculture at the ecosystem-wide scale.

11. Current research patterns and prospects

With the rapidly advancing technology, increasing use of data-driven tools, and new research influencing its course toward sustainability and scalability, the microalgae-based aquafeed sector is at a turning point in its history.

11.1 New developments and genetic engineering

For their natural adaptability, nutritional value, and rapid growth, microalgae have seen quite a wide number of researches on metabolic and genetic engineering to fully realize their potential as aquafeed ingredients. Because of developments in genetic modification that are concerned with increased biomass productivity, increased production of essential fatty acids mainly EPA and DHA, improved lipid content, and reduced antinutritional factors, microalgal biomass is being more and more prepared for fish nutrition (Singh et al., 2021; Tham et al., 2023). Precise changes enabled by technologies such as CRISPR-Cas9 genome editing have enhanced carotenoid synthesis, lipid biosynthesis pathways, and growth rates of microalgae strains like *Chlorella* and *Nannochloropsis* in various environmental conditions (Morais et al., 2022). Metabolic engineering techniques, which use systems biology and multi-omics to guide the design of the strain, complement genetic engineering by redirecting carbon flux toward metabolites of interest. Synthetic biology enables the accelerated development of strains that are customized, with the ability to make microalgae dynamically adapt to process conditions such as light intensity or nutrient stress by generating genetic circuits that uniquely control biosynthetic pathways. When integrated with transcriptomic and proteomic tools, these could lead to rapid screening of high-performance strains for aquafeed applications. In this, cultivation technology developments combine such elite strains with mixotrophic growth modes and optimized photobioreactor designs to enhance biomass yield and nutrient density (Barros et al., 2019). Microalgae are coupled with agro-industrial effluents for the development of reproducible nutrient environments that are favorable for the engineered strain apart from the usage of valorization of waste-to-worth (Chaudhary et al., 2024).

11.2 Making decisions based on data in feed formulation

There is not any important information that can be lost, which is not present in more contemporary Feed formulation, quantitative analysis along with experimental data, fish modeling, growth prediction, and nutritional requirements have gained significant importance (Joshi et al., 2021). Feed manufacturing parameters and algal inclusion and overall cost can be modeled to optimize fish health and growth

(Nagappan et al., 2021a, 2021b). Utilizing AI and Machine Learning in tools to analyze data to optimize cost and growth. Predictive algorithm using fish metabolism data and other feed ingredients can estimate digestibility, bioavailability, and overall feed palatability. Consequently, this approach enables the nutrition-ist to formulate feed for the individual species for specific environment and life stage (Tham et al., 2023). Sensor data in aquaculture systems can be used to develop feed systems models, minimizing feed waste and environmental pollution.Also, integrated supply chain analytics work out the raw material procurement and inventory management for microalgal biomass by harmonizing availability, quality variation, and production costs of Patel et al. (2021). Using digital twins to simulate cultivation and processing workflows can help identify bottlenecks and suggest process modifications to accelerate research-to-production pipelines.Moreover, the incorporation of multi-omics data-genomics, metabolomics, and proteomics-into feed design has provided new understandings of the impact of microalgal bioactives on fish physiology and immune response and allows targeted nutrient enhancements. Data-driven breeding programs for aquaculture species complement feed formulation through the selection of strains that perform best on microalgal diets.

11.3 Research deficits and prospects for progress

Despite the significant advances made so far, several knowledge gaps are limiting the full potential of microalgal biomass in aquafeeds. First, there is still an economic problem with large-scale production of microalgal biomass. Further research is required regarding systems of cultivation that can balance cost and productivity, in particular with regards to changing climatic and geographical conditions (Novoveská et al., 2023). Advancement of low-energy harvesting, drying, and preservation methods appropriate for aquafeed can reduce operational costs efficiently. Apart from that, the strain variability, cultivation conditions, and processing further add to the variability in microalgal biomass composition, hence making feed formulation and regulatory approval more complex. Safety parameters and nutritional profiles of biomass need to be described and certified using standardized processes (Tham et al., 2023). Third, full comprehension of the long-term impacts of supplementation of microalgal feed on the health, reproduction, and product quality of fishes requires more controlled studies across various species. This will involve determining whether there is accumulation of residual pollutants or toxins from such waste-based farming systems.The fourth would be the need for research studies regarding maximized multi-product valorization through integrated biorefineries from an engineering, economic, and sustainability point of view. Production of aquafeed ingredients with other compatible products such as pharmaceuticals, biofuels, or cosmetics involves many opportunities and challenges which must be critically appraised (Singh et al., 2021). Lastly, full life cycle assessments based on environmental, social, and economic considerations will guide sustainable industry growth in the context of changing regulations aimed at better aligning with global climate objectives.

12. Conclusion

The utilization of microalgal biomass is both an environmentally ethical and nutritionally dense approach to implement sustainable aquaculture and circular bioeconomy practices. The incorporation of microalgae into aquafeeds for numerous species and production systems continues to improve growth, health, and profitability, and aquaculture stakeholders' acceptance is growing. However, for true industry wide adoption, challenges pertaining to scalability, cost structure, digestibility, bio-contamination, and policies around microalgal biomass need to be overcome. To address these challenges, investments in microalgal biomass assimilation for aquaculture, the internet of things for agricultural practices, and genetic editing for microalgae will be required. To enable women, youth, and smallholder farmers to partake in this microalgae biomass value chain, training, certification, and legislation will need to be balanced, equitable and inclusive. The increasing consumer demand for high value, sustainably produced seafood creates market and innovation opportunities to absorb the products of these complete microalgae value chain. The bioplastics, pigments, bio-actives and microalgae biomass feed additives is creating a more prominent bioeconomy to microalgae. The collaboration, innovation and transparency around microalgal biomass is valorization is to be commended.

References

Ahmad, A., Hassan, W., Banat, F., 2022. An overview of microalgae biomass as a sustainable aquaculture feed ingredient: food security and circular economy. Bioengineered 13 (4), 9521–9547.

Algae Biomass Organization, 2024. Industry Ecosystem and Innovation Networks for Algae Production.

Alltech, 2022. AquaFeed Innovation Center: enhancing fish growth and health through microalgae-enriched feeds. Alltech Aquac. Rep. 50–65.

Barros, A.I.R.N.A., et al., 2019. Cultivation of microalgae for aquaculture feed: production, harvesting, and downstream processing. Algal Res. 40, 101498.

Camacho, F., et al., 2019. Industrial applications of microalgae in the ornamental fish sector: pigmentation and growth enhancement. J. Appl. Phycol. 31 (6), 3679–3690.

Chattaraj, S., Behera, B.K., Das Mohapatra, P.K., 2024. Effective valorization of Chlorella biomass and Brewers spent grain substituting fish meal and soybean meal in the diet of herbivorous fish Cirrhinus reba for higher growth, digestibility and sustainable cultivation. J. Appl. Phycol. 36 (1), 217–232.

Chattaraj, S., Das Mohapatra, P.K., 2024. Immunostimulatory and antagonistic potential of the methanolic extract of Oedogonium intermedium SCB in Cirrhinus reba challenged with Aeromonas hydrophila. Aquac. Int. 32 (3), 2523–2548.

Chattaraj, S., Ganguly, A., Mitra, D., et al., 2026a. Sustainable valorization of algal biomass in fish nutrition and dietetics: a critical review and scientometric study. Waste Biomass Valor. https://doi.org/10.1007/s12649-026-03619-7.

Chattaraj, S., Sahoo, S.S., Behera, S., Thatoi, H., 2026b. Sustainable aquafeed development using Spirulina: effects on growth performance of Catla catla. Algal Res. 104447.

Chaudhary, R., et al., 2024. Marine microalgae for bioremediation and waste-to-worth valorization. Environ. Biotechnol. 18 (2), 96–110.
Chen, F., Leng, Y., Lu, Q., Zhou, W., 2021. The application of microalgae biomass and bioproducts as aquafeed for aquaculture. Algal Res. 60, 102541.
European Commission, 2024. The European Green Deal and Circular Bioeconomy Strategy.
Figueroa-Torres, G.M., Pittman, J.K., Theodoropoulos, C., 2021. Optimisation of microalgal cultivation via nutrient-enhanced strategies: the biorefinery paradigm. Biotechnol. Biofuels 14 (1), 64.
Gong, Y., et al., 2020. Improving digestibility and nutrient bioavailability of microalgae for aquaculture: advances and challenges. Aquac. Rep. 18, 100517.
ICAR-CMFRI, 2024. Training Program on Marine Microalgae Culture Techniques.
Joshi, P.S., Praveen, B.M., Aithal, P.S., 2021. Introduction to the fish nutrition, feed formulation, and feeding conversion. Biosci. Discov. 12 (4), 208–216.
Julie, M.P., 2024. Sustainable fish feed development using by-products from wild fisheries. Int. J. Res. Publ. Rev.
Kabeyi, M.J.B., Olanrewaju, O.A., May 2025. Current and future trends in energy harvesting technology and applications. In: 2025 13th International Conference on Smart Grid (icSmartGrid). IEEE, pp. 679–687.
Khan, A.J., et al., 2021. Integration of microalgal biomass into small-scale fish farming in Bangladesh. Aquac. Exten. Ser. 3–17.
Lu, Q., et al., 2023. Challenging problems of applying microalgae for aquaculture feed: a review. Front. Bioeng. Biotechnol. 10, 1151440.
Ma, M., et al., 2024. Microalgae as feed sources and feed additives for sustainable aquaculture: an overview. Rev. Aquacult. 16 (1), 30–45.
Melikoglu, M., 2025. Waste valorization strategies with inputs for microalgae biorefineries: a global review. Clean. Water(Washingt.), 100125.
Milstein, A., et al., 2014. Marine microalgae: culture, harvesting, and utilization in aquaculture. Aquac. Sci. 23 (4), 177–192.
Morais, M., et al., 2020. Microalgae as a sustainable ingredient in aquafeeds. Nutr. Aquac. 12 (1), 15–29.
Nagappan, S., et al., 2021a. Potential of microalgae as a sustainable feed ingredient for aquaculture. Aquaculture 542, 123–138.
Nagappan, S., et al., 2021b. Potential of microalgae as a sustainable feed ingredient for aquaculture. Aquac. Rep. 19 (3), 57–72.
Novoveská, L., et al., 2023. Overview and challenges of large-scale cultivation of microalgae and cyanobacteria. Front. Plant Sci. 14, 1235798.
Obwanga, G., Lewo, E., 2017. Micro-franchise supply chains for aquaculture inputs in Kenya. Aquacult. Econ. Manag. 25 (2), 123–134.
Occhipinti, P.S., et al., 2024. Current challenges of microalgae applications: exploiting their potential for sustainable development. J. Sci. Food Agric. 104 (4), 1294–1306.
Osundeko, O., et al., 2023. Advances in microalgal research for valorization of industrial effluents and biomass. Bioenergy J. 15 (5), 489–510.
Patel, A., et al., 2021. Microalgae integrated biorefineries: approaches for sustainable production. Renew. Energy Rev. 140, 110749.
Rust, M.B., Barrows, F.T., Hardy, R.W., Lazur, A., Naughten, K., Silverstein, J., 2011. The Future of Aquafeeds.

Sarker, P.K., Kapuscinski, A.R., McKuin, B., Fitzgerald, D.S., Nash, H.M., Greenwood, C., 2020. Microalgae-blend tilapia feed eliminates fishmeal and fish oil and improves growth and omega-3 content. Sci. Rep. 10 (1), 19328.

Seafish, 2025. Driving a circular economy in seafood supply chains. In: Seafish Responsible Sourcing Report, pp. 1–35.

Singh, J., et al., 2021. Microalgal biorefinery concept for production of biofuels and bioproducts. Bioresour. Technol. 319, 124190.

Skifa, R., et al., 2024. Photobioreactors versus open ponds: a comparative analysis. Algal Res. 66, 103780.

Tham, P.E., et al., 2023. Insights of microalgae-based aquaculture feed: a review. Algal Res. 68, 102927.

Tredici, M.R., 2004. Mass production of microalgae: photobioreactors. Aquac. Eng. 32 (3), 345–356.

Wang, Z., Tian, J., Wang, Y., Tang, Y., Cui, S., 2025. Structure-guided valorization of microalgal biomass for renewable and versatile applications. Bioresour. Technol., 133123

Warapat, J., et al., 2022. Techniques for microalgal harvesting and processing in aquaculture. Adv. Bioeng. 39, 232–245.

Wiegertjes, G., et al., 2023. Nutritional pond feeds: feeding fish and ecosystems. In: World-Fish Center Report, pp. 1–45.

WorldFish, 2022. Developing Sustainable Fish Feed for Healthy People and Planet.

Yadav, S.K., et al., 2019. Circular bioeconomy through microalgae: a review. Clean. Prod. 230, 1040–1055.

CHAPTER

11 Antinutritional factors and mycotoxin contamination in nonconventional feed ingredients: Mitigation strategies

Joydev Maity and Sunita Pradhan

Department of Fishery Sciences, Vidyasagar University, Midnapore, West Bengal, India

The world is apprehensive about the availability, environmental sustainability, and cost escalation of traditional fish feed ingredients and is searching for alternatives. Utilizing plant-based nonconventional ingredients can be a successful replacement for traditional ingredients. However, utilization of plant-derived nonconventional sources of fish feed ingredients has been limited due to the presence of natural plant toxins, also referred to as antinutritional factors (ANFs). Along with ANFs, the use of crops in feed can also lead to a high risk of fungal contamination, which in turn increases the incidence of mycotoxicosis in fish. In the following chapter, we will discuss the most important antinutritional factors and mycotoxins found in plant-based ingredients and finished feed, their impact on the health of fish and humans, and how to mitigate the impact.

1. Introduction

Aquaculture is the fastest-growing food production sector in the world, producing more than 50% of the world's fish consumed. The growth is fueled by growing demand for aquatic commodities, depletion of wild fish resources through overfishing, and the desire for sustainable food resources (FAO, 2022). Aquaculture is a key factor in food security and rural livelihood in India. India is one of the world's prominent producers of aquaculture products, and freshwater aquaculture, most notably carp farming, is the most prevalent. Aquaculture development in India is facilitated by a favorable climate, an abundance of water resources, and government policies encouraging sustainable aquaculture.

The increasing size of aquaculture operations has led to increased reliance on formulated feeds to meet the nutritional requirements of cultured species. Traditionally, these feeds have been designed based on traditional ingredients like fishmeal and soybean meal (Chattaraj et al., 2026a). The sustainability of these ingredients

Fish Feed Formulation with Agroindustrial Wastes and Byproducts. https://doi.org/10.1016/B978-0-443-41728-3.00016-2

is, however, compromised by various factors like overfishing, competition with other livestock sectors, and fluctuating market prices.

To address these issues, there is a growing emphasis on the application of nonconventional feed ingredients that are based on locally available plant and animal by-products. These options hold potential benefits, including economic viability, reduced environmental impacts, and promotion of circular bioeconomy principles (Chattaraj et al., 2026b).

2. Nonconventional fish feed ingredients

Fish feed ingredients that are not commonly utilized in the production of feed compared to traditional ingredients are referred to as nonconventional fish feed ingredients. They have high-quality nutrients and are likely to be less expensive than the traditional feed ingredients (Chattaraj et al., 2024). Despite their potential, nonconventional ingredients have not been utilized in fish feed manufacturing because of the following reasons:

- They are not well known or understood.
- No productive research into the mode of production with a focus on the commodification of them.
- They are not available as such.
- They are toxic or poisonous.
- Nonconventional fish feed ingredients can be either plant or animal source origin.

2.1 Nonconventional plant-derived feed ingredients

Plant sources of fish foods comprise leaf protein or leaf meal, aquatic macrophytes, cultivable pulses, like Mucuna beans, yam beans, bread beans, winged beans, or any legume ornamental with the potential to produce pods containing seeds.

- **Leaf diets:** Leaves have varied protein contents, which can yield an infinite and cheap source of nutrients for fish. Leaves with high nutritional value content are leaves of cassava (*Manihot esculenta*), pawpaw (*Carica papaya*), pineapple (*Ananas comosus*), Groundnut (*Arachis. hypogea*), soyabean (*Glycine max*), plantain (*Musa paradisica*), *Moringa oleifera*, *alfalfa*, and *Leucaena leucocephala*. Antinutritional factors (ANFs) such as protease inhibitors, mimosines, tannins, and saponins are present in the leaves, which affect nutrient intake.
- **Oilseed meals and cakes:** These are residues of oil extraction processes, such as sunflower meal, rapeseed meal, cottonseed cake, etc. They are high in protein content but may contain ANFs such as gossypol and phytic acid that affect mineral bioavailability.
- **Nonconventional pulses:** Mucuna beans, broad beans, sword beans, winged beans, yam beans etc., are some of the nonconventional pulses. They are

moderate in protein content and rich in carbohydrate, hence easily gelatinizable. Hydrogen cyanide and trypsin inhibitors are the toxic compounds present in them.

- **Agro-industrial by-products:** Wheat bran and rice bran are inexpensive and accessible. They are good in fiber and energy content, but may be rich in ANFs and mycotoxins if poorly stored and processed.
- **Water plants:** These consist of various kinds of rooted, submerged, free floating aquatic vegetation. Duckweed (*Lemna minor*), water hyacinth *(Eichhornia crassipes), Ceratophyllum, nymphaea, Azolla pinnata* etc., are high-yielding, high-protein plants. However, being high in fiber content and containing ANFs such as oxalates, these need to be processed suitably before they can be added to feeds.

2.2 Nonconventional animal-based feed ingredients

These are products derived from any living organism; they could either be aquatic or land-based. Some examples are crab waste, crab meal, blood meal, insect meal, earthworm meal, and animal manures like cow dung, pig and poultry droppings, etc.:

- **Crab meal, crab waste:** The shell fraction of the crab, once the meat-eating portion has been removed, is dried, ground, and the resulting product is called crab meal. Crab shells are rich in chitin, a nitrogenous compound whose structure is close to cellulose. Crab meal is a significant source of carotenoid pigments.
- **Fish protein concentrate (FPC), fish hydrolyzed:** Can be made from any fish or fish waste. Prepared from fish in two methods, by removing the oil and without removing the oil and water from the fish; followed by screening or settling out the bones and drying. Nutrient value of FPC varies with the species used, but is always richer in protein, around 85%–95%, and poorer in ash content than the conventional fish meal used.
- **Fish ensiling:** Fish silage production is a good option when compared to fishmeal, particularly in places where small quantities of fishery waste or bycatch are generated. From the fishery waste, simple and cheap ways of fishery waste preservation can be prepared, so that it may be fed later. This can be particularly useful in the case of the availability of small quantities of fishery waste.
- **Poultry by-product meal:** Poultry processing produces a considerable amount of waste, such as offal, blood, and bird heads. These wastes are processed to create poultry by-product meal. Protein content is good with a balanced amino acid pattern.
- **Animal wastes:** Animal feces, especially cow dung, piggery and poultry droppings, can be applied as direct feed for fish; early morning feces, free from contamination by urine, are dropped on a designated section of the pond as fish feed. Oven-dried or sundried feces were also utilized by some fish culturists to mix with other ingredients when producing fish feed.

- **Earthworm meal:** Processed earthworms are either oven dried, smoked over a kiln, or pulverized using a grinder. The most well-known application of earthworms today is as ornamental fish feed and fish bait.
- **Bloodmeal:** Blood is easily obtained in slaughterhouses on a daily basis. Prepared by boiling the blood and then drying in an oven or smoking over or Sundry. Bloodmeal is rich in protein content and a supplementary source of lysine, but poor in methionine.
- **Insect diets:** Black soldier fly larvae (*Hermetia illucens*) have high lipids and protein content. They are efficient converters of organic residues into biomass but may be chitin-contaminated, which affects digestibility (Camperio et al., 2026).
- **Silkworm pupae meal:** A by-product of the silk trade, silkworm pupae are high in fat and protein. Seasonal availability and possible allergenic chemicals are their limitations.

While these nontraditional ingredients offer potential substitutes, their use in aquafeeds must consider the presence of ANFs and the potential for mycotoxin contamination. Being aware of these issues and implementing appropriate mitigation strategies is required to maintain fish health and aquaculture product safety.

2.3 Antinutritional factors

Antinutritional factors (ANFs) are naturally occurring chemical compounds, mainly in plants (and some animal tissues), harmful to the digestion, absorption, and utilization of nutrients. These chemicals, such as tannins, phytates, protease inhibitors, lectins, saponins etc., are defense chemicals of plants against herbivores and disease-causing microbes (Chattaraj et al., 2025). ANFs are undesirable in aquaculture and animal feed because of their potential to decrease the quality of nutrients in diets, disrupt enzyme functions, bind nutrients, damage intestinal tissue, and, in the long term, affect the health of animals (Tacon and Metian, 2015). Data on the chemical nature, origin, and physiological action of ANFs is important to design animal feed products with safety and efficacy.

Some potential unconventional plant sources of fish nutrients have been found to contain considerable amounts of toxic compounds (Table 11.1).

2.4 Why are ANFs detrimental to fish health?

In aquatic animals, especially omnivorous and carnivorous fish, elevated ANF levels may result in:

- Lower feed consumption owing to unpalatability.
- Compromised nutrient intake, leading to nutrient deficiencies.
- Metabolic disturbances, influencing energy and protein metabolism.
- Tissue damage, particularly in the gastrointestinal tract.
- Compromised immune mechanisms, which leave fish vulnerable to infections and disease.

Table 11.1 ANFs in non-conventional feed ingredients and their effects on fish health.

Origin of Ingredients	ANFs	Ingredients	Effects on fish health
Plant origin	Protease inhibitor	Most legumes, soybean etc.	Inhibit proteolytic enzyme. Pancreatic hypertrophy. Increase the requirements of sulfur containing amino acids.
	Lectins	Most legumes	Agglutination of vertebrate RBC. Reduce the absorption of nutrient in the gut
	Mimosine	Ipil-Ipil	Disturbing of thyroxin secretion.
	β-glycine, β-conglycine	Soyabean meal	Gastro intestinal hyposensitivity reaction and digestive disturbance
	Saponins	Soyabean meal, groundnut, lupin, sunflower, alfa-alfa	Interference with the absorption of lipid, cholesterol, bile acid and vit. A and B. Lower the surface tension and hemolyzed red blood cells
	Estrogens	Crucifers plants, rape seed and mustard seed Wheat, rice, chickpea, alfa-alfa, soyabean, groundnut, lupin	Produce elevated zinc in the liver and bones. Increase the deposition of calcium, phosphorous and manganese in the bone
	Cyanogens	Cassava, linseed, chickpea, and pigeon pea	Reduce oxygen carrying capacity and inactivate cytochrome oxidase
	Gossypol	Cotton seed	Liver and kidney abnormality. Reduction in heamocrit, hemoglobin and plasma protein levels
	Tannin	Mustard oil cake, sorghum, sal seed meal, mango seed kernel	Bind with enzymes or binds with feed components like protein or minerals. Reduce the absorption of vit B12
	Antivitamin factors	Raw soyabean, most of the raw legumes	Destroy vitamin absorption

Continued

Table 11.1 ANFs in non-conventional feed ingredients and their effects on fish health.—*cont'd*

Origin of Ingredients	ANFs	Ingredients	Effects on fish health
	Antienzymes	Potato Wheat, oats, and rye Sunflower seed	Inhibit enzyme activity
	Phytate	Leaf meals	Act as demineralize phytate bound with minerals P, Ca, Zn, Mg, Fe, Cu, and make them unavailable
	Oxalates	Leaves of plant ingredients	Bind with calcium and make insoluble
	Alkaloids	Found in plant	Interfere with nerve functioning
	Oligosaccharide and nonstarch polysaccharide	Grains, legume, and cereals	Binding to bile acid. Obstructing action against the digestive enzyme
	Oxy-carotenoid	Corn gluten meal	Give yellow color to the white muscle
Animal origin	Thaiminase	Raw fish more specifically freshwater fishes	Hydrolyze thiamine
	Histamine	Fish meal due to improper storage	Food poisoning
	Gizzerosine	Fishmeal	Gizzard erosion
	Peroxidized lipid	Raw material	Free radical, toxic compound
	Avidine	Raw egg white	Bind with biotin and inactivate
	Prion(pathogen)	Meat and Bonemeal	Bovine spongiform encephalopathy (BSE)

- Damage to organs, especially the liver and kidneys, with chronic exposure.

Severity of effect depends on the type of fish, quantity and quality of ANFs, and for how long they are exposed (Tacon and Metian, 2015). Although herbivorous fish can tolerate plant foods, excessive ANFs are harmful (Francis et al., 2001).

3. Some typical antinutrients and how to decrease them

Antinutrients are of four classes: 1. compounds that affect the digestion and utilization of protein, such as protease inhibitors, tannins, and lectins; 2. compounds that affect utilization of minerals such as phytates, gossypol pigments, oxalates, and glucosinolates; 3. antivitamins; 4. other compounds such as mycotoxins, mimosine, cyanogens, nitrate, alkaloids, photosensitizing compounds, phytoestrogens, and saponins (Francis et al., 2001).

3.1 Protease inhibitors

Protease inhibitors (PIs) are bioactive proteins present in high concentrations in seeds of most legumes and cereals, especially in soybeans. They lower the digestibility of dietary proteins by inhibiting digestive enzymes (Norton, 1991; Giri and Kachole, 2004).

3.1.1 Mode of action

- Enzyme inhibition: Kunitz trypsin inhibitors encapsulate one molecule of trypsin or chymotrypsin per inhibitor, whereas Bowman–Birk inhibitors are able to inhibit two molecules of either or one of each in combination (Norton, 1991).
- Complex formation: PIs interact with food proteins to create nondigestible complexes, limiting amino acid availability (Friedman et al., 2003).

3.1.2 Sources

- Soybean meal: 2–6 mg TI/g (average 4 mg/g) (Synder and Kwon, 1987).
- Other legumes/cereals: Lupin, Rapeseed, Jatropha seed meals contain different levels of PI (Makkar and Becker, 1999).

3.1.3 Mitigation measures

- Heat treatment: Autoclaving (15–30 min) unfolds most PIs but needs to be under strict control to prevent loss of amino acids (Tacon and Metian, 2015).
- Diet adaptation: Most fish (e.g., carp, seabream, tilapia) adapt by increasing protease secretion if TI < 5 mg/g (Wee and Shu, 1989; Krogdahl et al., 1994).

3.2 Phytate

Phytate (inositol hexakisphosphate) is the major phosphorus storage compound in plant seeds and constitutes 0.5%–5% dry weight (Loewus, 2002). Phytate inhibits mineral and protein absorption in nonruminants to a large degree (Walter et al., 2002).

3.2.1 Mechanism of action

- **Mineral chelation:** Chelates di- and trivalent ions (Zn^{2+}, $Fe^{2+/3+}$, Ca^{2+}, Mg^{2+}, Mn^{2+}, Cu^{2+}), precipitating as insoluble complexes (Mueller, 2001).
- **Protein binding:** Forms phytate–protein complexes that resist enzymatic hydrolysis, impairing amino acid uptake (Richardson et al., 1985).

3.2.2 Sources

- Soybean meal: 10–15 g/kg phytate.
- Rapeseed meal: 50–75 g/kg.
- Safflower meal: ~24 g/kg (Spinelli et al., 1983; Hossain and Jauncey, 1993).

3.2.3 Mitigation strategies

- Phytase supplementation: Enhances Phytate availability from 9.7%–48.4% to 46.2%–75.6% (Riche and Brown, 1996).
- Processing: Autoclaving, fermentation, and milling decrease phytate by as much as 74% (Duffus and Duffus, 1991; Hossain and Jauncey, 1990).

3.3 Glucosinolates

Glucosinolates are thiocarbamoyl moieties of sulfur-glucosides, occurring almost exclusively in the Brassicaceae (e.g., rapeseed, mustard) and in plant defense molecules (Duncan, 1991).

3.3.1 Mechanism of action

- Hydrolysis products: Thioglucosidase, when tissues are injured, breaks down glucosinolates to yield isothiocyanates and nitriles, which can produce thyroid disruption, liver hemorrhage, and nephrotoxicity (Duncan, 1991).

3.3.2 Sources

- Rapeseed meal: Rich in glucosinolates; canola varieties (<3 mmol/kg) are low-glucosinolate crops (Duncan, 1991).
- Mustard oil cake: Also, the glucosinolate content is high.

3.3.3 Mitigation strategies

- Thermal processing: Pressure-cooking wet reduces glucosinolates from 40 to 26 mmol/kg (Burel et al., 2000).
- Aqueous extraction: Successful in Moringa kernels (Makkar and Becker, 1997).
- Coextrusion (e.g., Colzapro): Reduces glucosinolate effect on thyroid structure (Gomes et al., 1993).

3.4 Saponins

Saponins are amphiphilic triterpenoid or steroid glycosides found in a wide variety of plant materials (e.g., legumes 18–41 mg/kg; soybean flour ~67 mg/kg) (Fenwick et al., 1991). Saponins have been employed as fish poisons for centuries owing to their surfactant-like toxicity to gill epithelium (De et al., 1987).

3.4.1 Mechanism of action

- Membrane disruption: Their surfactant properties disrupt gill respiratory epithelium, resulting in compromised gas exchange and death (De et al., 1987).
- Gut interaction: Saponins also create saponin–protein complexes that lower protein digestibility and may inactivate digestive enzymes such as trypsin and chymotrypsin (Shimoyamada et al., 1998; Potter et al., 1993).
- Deterrent effects: Irritant character and bitter taste reduce intake of feed (Liener, 2003).

3.4.2 Sources

- Legume seeds: 18–41 mg/kg.
- Defatted roasted soybean flour: ~67 mg/kg (Fenwick et al., 1991).
- Tea seed cake: Comprises 7%–8% saponins and is extremely toxic at 100 ppm in water (De et al., 1987).
- Lupin and alfalfa: Varying amounts (0.3%–1.5%), associated with lowered growth in tilapia and trout (De la Higuera et al., 1988; Olvera-Novoa et al., 1990; Yousif et al., 1994).

3.4.3 Mitigation strategies

- Aqueous extraction: Most saponins are water-soluble; simple washing removes them effectively.
- Formulation: Maintenance of dietary saponin below 1 g/kg averts depression in growth among most cultured fishes.
- Tannin complexation: Antinutrient interaction counteracts toxicity (Freeland et al., 1985)

3.5 Lectins

Lectins are nonenzymatic glycoproteins capable of binding specific carbohydrate moieties on cell surfaces, derived from the Latin word "legere" ("to select") (Fereidoon, 2014). Also termed hemagglutinins when sugar specificity is unknown, they agglutinate erythrocytes without altering carbohydrate structures and are prevalent in legumes, cereals, tubers, and oil seeds (Mueller, 2001; Felix and Mello, 2000).

3.5.1 Mechanism of action

- Gut binding: Lectins bind to enterocyte glycoproteins, disrupting mucosal integrity and nutrient uptake, and may lead to epithelial lesions (Grant, 1991).
- Immune reactions: Function as antigens eliciting IgG/IgM production, resulting in food allergy and systemic immune response (Felix and Mello, 2000).
- Hemagglutination: Some lectins bind to erythrocytes, leading to clumping and anemia, and half of dietary lectins are pan-hemagglutinins (Fereidoon, 2014).

3.5.2 Sources

- Soybean agglutinin (SBA): ~60 HAU/mg protein; also binds strongly to salmonid intestinal brush border, linked to distal enteritis in Atlantic salmon (Hendriks et al., 1990; Van der Ingh et al., 1991).
- Jatropha seed meal: Native lectin activity ~102 HAU, inactivated by moist heat (Aregheore et al., 1998).
- Cottonseed and legumes: Variable levels of lectins with the ability to bind 0.01% free gossypol and inhibit nutrient transport (Fereidoon, 2014).

3.5.3 Mitigation measures

- Heat/aqueous treatment: Autoclaving or boiling (100°C, 10 min) is capable of inactivating lectins (Grant, 1991; Aregheore et al., 1998).

- Formulation of diet: Balancing with other food and proper processing prevents mucosal injury due to lectin.

3.6 Tannins

Tannins are polyphenolic secondary metabolites that are usually classified as hydrolyzable or condensed tannins, present in most plant feeds. Tannins disrupt digestion by binding proteins, enzymes, and minerals, and can inhibit vitamin B12 absorption (Liener, 1989)

3.6.1 Mechanism of action

- Enzyme binding: Tannins react with digestive enzymes (e.g., trypsin, chymotrypsin, amylase, lipase) and react with dietary proteins, lowering protein digestibility and increasing fecal nitrogen (Felix and Mello, 2000; Kyriazakis and Whittenmore, 2006).
- Toxic metabolites: Hydrolyzable tannins are metabolized into lower-molecular-weight compounds taken up into circulation, which may cause chronic organ toxicity (kidney, liver).

3.6.2 Sources

- Quebracho wood: Powdered condensed tannin at 2% was acceptable to carp; hydrolyzable tannic acid at the same level caused refusal to feed after 28 days (Becker and Makkar, 1999).
- Copra meal: ~2.4% condensed tannins causing growth depression in tilapia and *Labeo rohita* fingerlings at 20%–25% inclusion (Jackson et al., 1982; Mukhopadhyay and Ray, 1999b).
- Pea and rapeseed meals: Possess moderate tannin content that is acceptable to most species.
- Broad bean (*Vicia faba*): Condensed tannins of high concentration decreased in vitro protein digestibility more in carp gut conditions than in trout (Grabner and Hofer, 1985).

3.6.3 Mitigation measures

- Physical processing: Dehulling eliminates tannin-containing seed coats; autoclaving or alkali treatment eliminates tannins (Griffiths, 1991).
- Biological treatment: Lactic acid bacteria fermentation lowered sesame meal tannin levels by half (Mukhopadhyay and Ray, 1999a).
- Chemical additives: Tannin-binding compounds and oxidizing agents such as polyethylene glycol reduce tannin toxicity (Makkar et al., 1995; Makkar and Becker, 1996).

3.7 Oxalates

Oxalates are oxalic acid salts that bind stably with minerals like calcium, magnesium, sodium, and potassium in plants (Liener, 2005). Even though sodium and potassium oxalates are soluble, calcium oxalate is relatively insoluble and precipitates in the form of pointed crystals in the urinary system at high concentration (Nachbar et al., 2000). Sequestered in plant tissues under normal conditions, processing or digestion can cause free oxalate to be released, bind dietary nutrients, and irritate the gut, and also cause deficiency of nutrients in case of its excess consumption (Noonan and Savage, 1999).

3.7.1 Mechanism of action

- Mineral chelation: Oxalate chelates divalent cations (Ca^{2+}, Mg^{2+}, and Fe^{2+}), precipitating as insoluble complexes that lower mineral availability to nonruminant animals such as fish; Zn^{2+} is less affected.
- Gut irritation: Free oxalate may irritate the gastrointestinal mucosa, disrupting nutrient absorption.

3.7.2 Sources

- Leafy and tuberous plants: Oxalis and Rumex species contain acid oxalate ($HC_2O_4^-$) in acidic sap (~pH 2), primarily as potassium oxalate.
- Goosefoot family: These plants contain neutral sap (~pH 6) containing soluble sodium oxalate and insoluble calcium/magnesium oxalates.
- Ingredient feeds: Certain vegetables and by-products may contain high levels of oxalates that will disrupt mineral uptake (Liebman and Al-Wahsh, 2011).

3.7.3 Mitigation measures

- Nutritional supplementation: Bioavailable calcium salts can reverse oxalate-induced sequestration of calcium.
- Processing: Boiling, soaking, or fermenting plant material can leach soluble oxalates.
- Ruminant analogy: Unlike ruminants that break down soluble oxalates through ruminal microflora, monogastric fish do not have this route, highlighting the need for extrinsic mitigation.

3.8 Alkaloids

Alkaloids are heterocyclic ring secondary metabolites with basic and nitrogen properties, derived from amino acid precursors (Petterson et al., 1991; Felix and Mello, 2000). They are widespread in some grain legumes, e.g., Lupinus albus, with high digestible protein (30%–50%). Quinolizidine alkaloids, on the other hand, are disruptive to nervous system function, lowering feed palatability and intake in fish like rainbow trout (De la Higuera et al., 1988; Bangoula et al., 1993).

3.8.1 Mechanism of action

- Neurotoxicity: Quinolizidine and tropane alkaloids disrupt synaptic transmission, which can lead to increased heart rate, paralysis, and death when administered in high doses (Fernando et al., 2012).

- Membrane disruption: Some alkaloids, for example, glycoalkaloids, inhibit cholinesterase and destabilize gastrointestinal cell membranes, leading to impairment of nutrient absorption and epithelial cell damage.

3.8.2 Sources

- Lupin seed meal is variable in content of quinolizidine alkaloids; for improved safety, low-alkaloid varieties (<20 mg/kg) have been developed (De la Higuera et al., 1988; Burel et al., 1998, 2000).
- Other plant sources: The majority of vascular plants (15%–20%) synthesize alkaloids, typically accumulating in the form of plant acid salts (oxalic, malic) and glycosylated derivatives (Felix and Mello, 2000).

3.8.3 Mitigation strategies

- Breeding: Low-alkaloid varieties like sweet lupins can be cultivated successfully to decrease dietary alkaloid consumption.
- Processing: Soaking, water extraction, and extrusion (e.g., 145°C compared to 120°C) enhance alkaloid extraction and improve digestibility and growth rate in trout (Bangoula et al., 1993).

3.9 Cyanogens

Cyanogenic glycosides are glycosides of the plant that upon enzymic hydrolysis by β-glucosidase, when plant cells are burst, release hydrogen cyanide (HCN) (Davies, 1991; Ufodike and Matty, 1983).

3.9.1 Mode of action

- Respiratory inhibition: HCN inhibits cytochrome oxidase by blocking cellular respiration, leading to anoxia and cardiac arrest.
- Antithyroid effect: Thiocyanate, a byproduct of detoxification, can interfere with thyroid function and is also teratogenic (Mishra et al., 2024).

3.9.2 Sources

- Roots and leaves of cassava: have up to 71 ppm total cyanide in dried meal (Ng and Wee, 1989)
- Linseed (flaxseed): Cyanogenic and causes stunted growth in trout (Hossain and Jauncey, 1989).
- Pulses: Some grain legumes also have cyanogenic glucosides.

3.9.3 Mitigation measures

- Processing: Wilting, soaking, and fermentation decrease cyanide content in leaves and roots.
- Dietary adaptation: There is species-specific tolerance; certain fish can withstand low concentrations of cyanide without growth loss (Ng and Wee, 1989).

3.10 Mimosine

Mimosine is a nonproteinogenic amino acid, which forms at about 3%–5% of the leaf protein in *Leucaena leucocephala* and is known as an unusual amino acid with proven antinutritional activity (Liener, 1980; D'Mello, 1991b).

3.10.1 Mode of action

- Reproductive disruption: Mimosine causes reproductive disruption and shows teratogenic activity in animal models.
- Protein digestibility: Poorly digested by tilapia, which may inhibit absorption of mimosine but still impact growth and fry yield, with males being more tolerant than females (Santiago et al., 1988).

3.10.2 Sources

- Leucaena leaf meal: Primary feed ingredient containing mimosine; inclusion of over 40% of the diet significantly reduces fry yield in tilapia.

3.10.3 Mitigation strategies

- Processing: Water soaking of Leucaena leaf meal for long periods lowers mimosine levels.
- Selection and breeding: Leucaena varieties with low mimosine content could enhance feed safety and utilization.

3.11 Nonsteroidal estrogenic agents

Nonsteroidal estrogenic compounds, isoflavones and coumestans, are ubiquitous in seeds such as soybean, safflower, linseed, and cottonseed. These plant constituents, in their native form, possess estrogenic activity either by estrogen receptor binding or by conversion to active metabolites (Farnsworth et al., 1975; Liener, 1980).

3.11.1 Mode of action

- Receptor interaction: Isoflavones such as genistein, daidzein, and coumestrol act as mimics of natural estrogens, stimulating vitellogenin synthesis and vitellogenesis in fish (Pelissero et al., 1991b; Kaushik et al., 1995).
- Physiological effects: Increased plasma vitellogenin and hepatosomatic hypertrophy were observed in fish like Siberian sturgeon and trout that were fed phytoestrogen-rich diets, exhibiting steroid-like effects (Pelissero et al., 1991a; Mambrini et al., 1999).

3.11.2 Sources

- Soybean meal: Contains approximately 0.25% total isoflavones (genistein, daidzein, biochanin A) (Pelissero et al., 1991a).
- Other seed meals, such as linseed, safflower, and cottonseed, contain varying levels of estrogenic glycosides.

3.11.3 Mitigation strategies

- Heat treatment: Most nonsteroidal estrogens are heat-labile and are substantially inactivated by normal feed processing temperatures.
- Choice of ingredients: The use of low-phytoestrogen varieties or counterbalancing with nonestrogenic protein sources can reduce endocrine disruption (Pelissero and Sumpter, 1992).

3.12 Antigenic compounds

Protein fractions of legume seeds and cereals like glycinin and β-conglycinin of soybeans can act as antigenic agents (allergens) and induce targeted and nontargeted immune responses and intestinal mucosal lesions (D'Mello, 1991a; Rumsey et al., 1993).

3.12.1 Mode of action

- Immune stimulation: Antigenic proteins engage immune cell receptors, triggering IgE- and IgG-dependent immunity that may induce enteritis-like alterations in the distal intestine (Krogdahl et al., 2000).
- Mucosal damage: Allergen-induced inflammation disrupts villi architecture and impairs nutrient absorption and digesta transit.

3.12.2 Sources

- Soybean meal: Immunologically active glycinin (G) and β-conglycinin (bC) are present, which are not destroyed even by normal toasting and defatting processes (Rumsey et al., 1993).
- Cereal proteins: Other cereal-based foods could contain less-characterized antigenic proteins.

3.12.3 Mitigation measures

- Processing innovations: Novel hydrolysis, fermentation, or protease procedures can decrease antigenicity below standard toasting and defatting.
- Ingredient choice: Utilization of soy protein concentrates with minimal residual antigenic activity, as indicated by lack of agglutinating activity in trout sera (Kaushik et al., 1995).

3.13 Antivitamin factors

Antivitamins are thermolabile compounds in most of the plant-containing feedstuffs such as soybean meal, alfalfa leaf meal, and oilseed meals. They are activity or absorption inhibitors of the necessary vitamins, and they could threaten the use of nutrients (Liener, 1980).

3.13.1 Mechanism of action

- Enzyme inhibition: Some antivitamins block vitamin-requiring enzymes by attaching to the active vitamin form.
- Competitive binding: Others bind to vitamins to form complexes (e.g., thiamine-binding factors) that inhibit their absorption or cell uptake.

3.13.2 Sources

- Soybean meal: Exhibits antivitamin properties that adversely affect biotin, thiamine, and riboflavin.
- Alfalfa leaf meal: High in thiaminase enzymes that break down thiamine.
- Oilseed meals: May contain lipophilic antivitamins that affect fat-soluble vitamin A and E stability.

3.13.3 Mitigation strategies

- Heat treatment: Enzyme inactivation by routine toasting and autoclaving inactivates most antivitamins, and hence heat-treated seed meals are relatively antivitamin-free.
- Supplementation: Certain vitamins may reverse residual antivitamin impacts through fortification.

3.14 Gossypol

Gossypol is a dimeric polyphenolic pigment of high concentration in Gossypium plant pigment glands. Gossypol in meal is present almost entirely as a free, biologically active compound, but may be partially bound to protein and therefore be inactivated during processing (D'Mello and Macdonald, 1998).

3.14.1 Mode of action

- Toxicity: Free gossypol can cause organ damage, hepatic and renal pathology, cardiac failure, and death (Herman, 1970; Berardi and Goldblatt, 1980).
- Reproductive effects: Gossypol causes abnormalities in sperm, decreases sperm yields, and disrupts sex ratios in fish and mammals; it interacts with hormonal pathways and reproductive tissues (Randel et al., 1992; Salaro et al., 2000).
- Protein binding: Binds to create indigestible gossypol–protein complexes, reducing bioavailability of essential amino acids such as lysine and methionine, influencing growth and fat metabolism.

3.14.2 Sources

- Glanded cottonseed meal: Contains ~1.3% free gossypol.
- Glandless varieties: Low content of gossypol (~0.01%), safer for high inclusion levels.
- Byproducts: Cottonseed meal and extracts in full-fat form can provide gossypol levels adequate to inhibit growth in rainbow trout, carp, and tilapia (Roehm et al., 1967; Robinson et al., 1984).

3.14.3 Mitigation strategies

- Dietary iron supplementation: Ferrous gossypolate complex formation detoxifies gossypol in the gastrointestinal tract, increasing safety for gossypol-sensitive organisms.

- Ingredient selection: Use of prepressed, solvent-extracted "glandless" cottonseed meals with supplemental limiting amino acids promotes vigorous growth even when fed at high inclusion rates (Reigh, 1999; Jackson et al., 1982).
- Processing: Food processing decreases free gossypol content by heat treatment, solvent removal, and protein binding.

3.15 Oxycarotenoids

Oxycarotenoids such as zeaxanthin and lutein are pigmented xanthophylls found in feedstuffs such as corn gluten meal. Their deposition in fish muscle has the potential to cause yellow coloration of white-fleshed fish, reducing market acceptability in certain markets (Skonberg et al., 1998).

3.15.1 Mechanism of action

- Tissue deposition: Diet-derived oxycarotenoids are deposited into muscle tissue via intestinal absorption, changing fillet color.

3.15.2 Sources

- Corn gluten meal: Contains variable levels of zeaxanthin and lutein that may deposit yellow pigments on fish flesh.

3.15.3 Mitigation strategy

- Bleaching: Addition of soy flour to corn gluten meal will successfully bleach the carotenoids prior to feed manufacture (Park et al., 1997).

4. Mycotoxins

Mycotoxins are poisonous secondary metabolites generated by some fungi, particularly species in Aspergillus, Fusarium, and Penicillium. Mycotoxins have the tendency to typically infect farm produce like grains, nuts, and oilseeds, particularly when the climate is warm and wet. The mycotoxins carry enormous health risks for both human and fish consumption, affecting food quality and aquaculture output.

4.1 Aflatoxins and sterigmatocystin

Aflatoxins were the initial mycotoxins to be identified following an outbreak of acute aflatoxicosis ("Turkey X disease") in the 1960s, which resulted in the death of approximately 100,000 turkeys (Blount, 1961; Richard, 2008). Aflatoxins are now the most researched and best-characterized mycotoxins. The most significant aflatoxins in crops are B1, B2, G1, and G2, which are mainly synthesized by *Aspergillus flavus*, with *A. parasiticus* and *Aspergillus nomius* also being able to synthesize them. Various filamentous fungi, such as *Penicillium*, *Rhizopus*, *Mucor*, and *Streptomyces*, are also aflatoxin producers (Agag, 2004). Sterigmatocystin (STC), a biosynthetic precursor of aflatoxin B1, is also a highly toxic compound with genotoxic and carcinogenic activities, albeit less toxic than aflatoxins (EFSA, 2013).

4.1.1 Source

Aflatoxins, especially aflatoxin B1, are mainly produced by Aspergillus flavus but can also be produced by A. parasiticus and *Aspergillus nomius*. Sterigmatocystin, which is a precursor to aflatoxin B1, is mainly synthesized by *Aspergillus versicolor* and *Aspergillus nidulans*, as well as other Aspergillus species (Rank et al., 2011). Aflatoxins have the ability to infest a broad range of feedstuffs such as maize, corn, wheat, cottonseed, and nuts (CAST, 2003). Groundnut meal, maize, and cottonseed meal are the primary sources of nutrients in animal feeds, and contamination is possible both pre and postharvest, particularly under stressful conditions like heat, drought, or infestation by insects (FAO, 2001; Viegas et al., 2018).

4.1.2 Mode of action

Aflatoxin B1 has its toxic effects mainly through its metabolism to a reactive epoxide that reacts with DNA, resulting in mutations and hepatocellular carcinoma (Wild and Gong, 2010). It also induces immunosuppression, anemia, and liver dysfunction. Sterigmatocystin acts in the same way, especially attacking liver and kidney tissues, causing DNA damage, chromosomal aberrations, and oxidative stress (EFSA, 2013). Both toxins bioaccumulate in fish muscle and organs, with implications for food safety and human exposure following the intake of contaminated fish.

Exposure to aflatoxins B1 causes stunted growth, anemia, impaired blood coagulation, liver damage, reduced immune resistance, and enhanced susceptibility to infection, causing enhanced death in animals and humans (Marijani et al., 2019; Tacon, 1992). Sterigmatocystin exposure also leads to liver and kidney damage, albeit less toxic compared to aflatoxins (Viegas et al., 2018; Scudamore et al., 1996).

4.1.3 Effect on fish

Clinical symptoms of aflatoxicosis in fish consist of pale gills, decreased blood clotting, stunted growth rates, and failure to gain weight (Agag, 2004). Extreme exposures can produce reduced survival rate, darkened or yellowish bodies, as well as erratic behavior, which are observed in Nile tilapia and juvenile sturgeon (Marijani et al., 2019; Farabi et al., 2006; El-Sayed and Khalil, 2009). Susceptibility varies by species and age; rainbow trout (*Oncorhynchus mykiss*), for instance, are most sensitive, with long-term exposure to even low doses (0.0004 mg/kg) predisposing to liver cancer (Russo and Yanong, 2013). The lethal dose (LD50) of aflatoxin B1 is 0.81 mg/kg body weight in rainbow trout (Bauer et al., 1969), although other species such as red drum and sea bass are also very sensitive (El-Sayed and Khalil, 2009; Zychowski et al., 2013a, 2013b). Yet, channel catfish and Nile tilapia are less sensitive, with LD50 values ranging from 1.0 to 1.3 mg/kg and 11.5 mg/kg body weight, respectively (Jantrarotai et al., 1990; Abdelhamid et al., 2002). Sterigmatocystin exposure causes effects similar to those of aflatoxins and results in DNA damage, liver lesions, as well as greater mortality in Nile tilapia (Abdel-Wahhab et al., 2005; Mahrous et al., 2006).

4.1.4 Effect on humans

Human aflatoxin exposure is predominantly by ingestion of contaminated food, with dietary exposure much greater in developing nations. Exposure can be as high as greater than 100 ng/kg of bodyweight per day in sub-Saharan Africa and lower in developed nations at less than 1.0 ng/kg/day (WHO, 2018). Aflatoxin B1 is a highly carcinogenic toxin and has been implicated in hepatocellular carcinoma, contributing approximately 28% to worldwide cases, mainly in countries such as Africa, where the proportion is even greater (Liu and Wu, 2010). Short-term aflatoxin exposure has the potential to induce liver damage, jaundice, upper abdominal pain, vomiting, diarrhea, and ascites, and acute aflatoxicosis cases have been reported in Kenya (2004) and Tanzania (2016), resulting in scores of deaths (Azziz-Baumgartner et al., 2006; Kamala et al., 2018). Sterigmatocystin, a suspected human carcinogen (group 2B), is also hepatotoxic but lacks adequate data on its effect on human health, and thus more studies are needed (EFSA, 2013).

4.2 Fumonisins

Fumonisin B1 is the most toxic fumonisin. It is produced by several Fusarium species, among which the common producer is Fusarium verticillioides, and also by *Fusarium proleferatum* and *Fusarium nygamai. Alternaria alternata* is another fumonisin B1 producer. Fumonisin contamination is most common in maize, and its by-products, with this toxin found in 80%–100% of corn samples in Mozambique, Burkina Faso, China, and Malaysia (Reddy and Salleh, 2011; Warth et al., 2012; Wang et al., 2000).

4.2.1 Source

Regulations for the upper limits of fumonisins in cereals have been given only for cocontamination with fumonisin B1 and B2. The upper limit permissible in unprocessed maize (excluding unprocessed maize that is to be processed by wet milling) is 4000 μg/kg. In feedstuff, fumonisins must not exceed 60 mg/kg in maize and maize products, whereas in complete feed for fish they must not exceed 10 mg/kg (CEC, 2006a, 2006b).

4.2.2 Mode of action

Fumonisins, especially fumonisin B1, interfere with sphingolipid metabolism by inhibiting ceramide synthase. This interference causes an increase in the ratio of free sphinganine to free sphingosine, which can be responsible for apoptosis, cell proliferation, and potentially carcinogenesis (Yildirim et al., 2007).

4.2.3 Effects on fish

Susceptibility appears to be species-specific, as with aflatoxins. Channel catfish seem moderately susceptible. Exposure to fumonisin B1 resulted in liver histopathologic alteration, growth inhibition, and low survival on testing doses greater than 320 mg/kg. Low survival was caused by Cytophaga infection (Lumlertdacha et al., 1995), which may be indicative of immune suppression. Yildirim et al. (2007) noted that doses of 20.0 mg/kg or higher slowed growth rate and elevated the free sphinganine to sphingosine ratio (Yildirim et al., 2007), indicating compromised lipid metabolism. Brown et al. (1994), on the other hand, did not have any effects except decreased weight gain in catfish that were fed 313.0 mg/kg for 5 weeks (Brown et al., 1994), rendering conclusions inconclusive.

Nile tilapia exhibited effects comparable to those reported by Yildirim et al. (2007), such as decreased growth rate and sphingolipid metabolism disruption (Tuan et al., 2003), which reflects moderate sensitivity.

Common carp (*Cyprinus carpio*) are very sensitive to fumonisin injury. Administration of 10.0 and 100.0 mg/kg of fumonisin B1 for 42 days resulted in damage to the liver, pancreas, kidney, heart, brain, and blood vessels (Petrinec et al., 2004). Histopathological alterations were noted in the brains and vasculature of juvenile carp (Kovačić et al., 2009).

On the whole, fumonisins cause organ damage, immune suppression, weight loss, and metabolic changes associated with cancer and death. At present, there are no reports of fumonisin accumulation in fish muscle, and this implies minimal food safety risk to human consumers.

4.2.4 Effects on humans

Fumonisins are a risk to humans mainly by chronic exposure. Fumonisin B1 is a Group 2B carcinogen (IARC classification, 2020), meaning it is a suspected human carcinogen. It is associated with a higher incidence of esophageal and hepatic cancer in China (Wang et al., 2000; Sun et al., 2007; Yoshizawa et al., 1994) and Africa (Rheeder et al., 1992), where contamination is widespread. Fumonisin exposure during pregnancy is also linked with increased risk of neural tube defects in children (Missmer et al., 2006).

4.3 Zearalenone

Zearalenone is largely synthesized by *Fusarium* spp., most notably *F. graminearum*, though also *Fusarium culmorum*, *Fusarium equiseti*, and *Fusarium crookwellense* (Bennett and Klich, 2003). These fungi mostly infect preharvest in corn and other crops (CAST, 2003). Zearalenone is a potent estrogenic compound, i.e., a mycoestrogen which impacts the reproductive capacity of a wide range of animals (Zhang et al., 2018).

4.3.1 Source

European Commission has fixed the maximum level of zearalenone in foodstuffs as 350 μg/kg in unprocessed maize except for unprocessed maize to be processed by wet milling and 100 μg/kg in other unprocessed cereals (CEC, 2006a, 2006b). In feeds, maize byproducts should not exceed 3 mg/kg of deoxynivalenol, but for other cereals and cereal products, the limit is 2.0 mg/kg (CEC, 2006a, 2006b). Complete feedstuffs have no limit fixed for the rest of the animals except mammals.

4.3.2 Mode of action

Zearalenone has a similar structure to natural estrogen and may bind with estrogen receptors on animal and human cells. Binding causes hormone imbalances that can interfere with reproduction functions and may result in pathological alterations, such as hyperplastic and neoplastic changes (Tomaszewski et al., 1998).

4.3.3 Effects on fish

Zearalenone had built up in high concentrations in rainbow trout ovaries (up to 7.1 μg/kg) (Wozny et al., 2013), but the reproductive effects of this build-up are still not well understood. Short-term zearalenone exposure to zebrafish (*Danio rerio*) lowered reproductive capacity by reducing spawning frequency (Schwartz et al., 2010) and relative fecundity between generations (Schwartz et al., 2013). Besides its reproductive impact, zearalenone (2 mg/kg) seems to accelerate the growth and improve the feed efficiency of rainbow trout, although its immunomodulatory impacts might have affected fish health negatively (Wozny et al., 2019).

Also, trace levels of this toxin were found in the liver and intestines (<2 μg/kg) of rainbow trout (Wozny et al., 2013). No accumulation, however, was observed in muscle tissue, suggesting a low risk to human consumers via consumption of fish.

4.3.4 Effects on humans

Because of its structural analogy to estrogen, zearalenone is able to bind to estrogen receptors in human cells, causing hormonal disturbances and pathologies of the reproductive system. It has been found in hyperplastic and neoplastic endometrial tissues (Tomaszewski et al., 1998), which implies a possible role in carcinogenesis. Nevertheless, the IARC presently classifies zearalenone as not classifiable as to its carcinogenicity to humans (Group 3) (IARC classification, 2020), meaning that there is not enough evidence to make a final conclusion. More studies are needed to determine its health impacts.

How does fish feed or feed ingredients get contaminated?

Mycotoxin contamination of aquafeeds may result from.

- Preharvest phases (fungal infection of crops)
- Improper drying postharvest
- Storage under high humidity and temperature

Nontraditional feed ingredients, especially agro-industrial by-products, are very prone to fungal colonization and resulting mycotoxin production if not managed properly (Zychowski et al., 2013a, 2013b).

Precautionary measures during feed preparation to prevent mycotoxin contamination:

- Ingredient selection: Utilize high-quality, visibly mold-free raw materials.
- Moisture control: Ensure moisture content levels below critical values (~10–12%) during feed formulation.

Use of mycotoxin binders: Add adsorbents such as bentonite, zeolites, or yeast cell walls to feed formulations (Huwig et al., 2001).

Precautionary measures during feed storage to prevent mycotoxin contamination:

- Proper drying: Dry ingredients and feeds properly prior to storage.
- Controlled storage conditions: Store feeds in cool, dry, well-ventilated conditions.
- Regular monitoring: Monitor feed intermittently for evidence of mold growth and mycotoxin content.

Use of preservatives: Use organic acids such as propionic acid to prevent fungal development when storing (Jard et al., 2011).

References

Abdelhamid, A.M., Magouz, F.I., Salem, M.F.E., Mohamed, A.A., Mohsen, M.K., 2002. Effect of graded levels of aflatoxin B1 on growth performance and biochemical, chromosomal and histological behaviour of Nile tilapia Oreochromis niloticus. In: Proceedings of the 1st Conf. Animal & Fish Prod., Mansoura, Egypt, 24–25 September 2002, pp. 231–250.

Abdel-Wahhab, M.A., Hasan, A.M., Aly, S.E., Mahrous, K.F., 2005. Adsorption of sterigmatocystin by montmorillonite and inhibition of its genotoxicity in the Nile tilapia fish (*Oreachromis nilaticus*). Mutat. Res. Genet. Toxicol. Environ. Mutagen. 582, 20–27.

Agag, B., 2004. Mycotoxins in foods and feeds 1-aflatoxins. Assiut. Univ. Bull. Environ. Res. 7, 173–206.

Aregheore, E.M., Makkar, H.P.S., Becker, K., 1998. Assessment of lectin activity in a toxic and a non-toxic variety of Jatropha curcas using latex agglutination and haemagglutination methods and inactivation of lectin by heat treatments. J. Sci. Food Agric. 77, 349–352.

Azziz-Baumgartner, E., Lindblade, K., Gieseker, K., Rogers, H.S., Kieszak, S., Njapau, H., Schleicher, R., McCoy, L.F., Misore, A., DeCock, K., et al., 2006. Case-control study of an acute aflatoxicosis outbreak, Kenya, 2004. Environ. Health Perspect. 114, 1779–1783.

Bangoula, D., Parent, J.P., Vellas, F., 1993. Nutritive value of white lupin Lupinus albus ar Lutop. fed to rainbow trout *Oncorhynchus mykiss*. Effects of extrusion cooking. Reprod. Nutr. Dev. 33, 325–334.

Bauer, D.H., Lee, D.J., Sinnhuber, R.O., 1969. Acute toxicity of aflatoxins B1 and G1 in the rainbow trout (*Salmo gairdneri*). Toxicol. Appl. Pharmacol. 15, 415–419.

Becker, K., Makkar, H.P.S., 1999. Effects of dietary tannic acid and quebracho tannin on growth performance and metabolic rates of common carp (*Cyprinus carpio* L.). Aquaculture 175, 327–335.

Bennett, J., Klich, M., 2003. Mycotoxins. Clin. Microbiol. Rev. 16, 497–516.

Berardi, L.C., Goldblatt, L.A., 1980. Gossypol. In: Huisman, J., Van der Poel, A.F.B., Liener, I.E. (Eds.), Recent advances of research in antinutritional factors in legume seeds. Pudoc, Wageningen, pp. 184–237.

Blount, W.P., 1961. Turkey "X" disease. Turkeys 9, 52–55.

Brown, D.W., Mccoy, C.P., Rottinghaus, G.E., 1994. Experimental feeding of Fusarium moniliforme culture material containing fumonisin B1 to channel catfish, *Ictalurus punctatus*. J. Vet. Diagn. Investig. 6, 123–124.

Burel, C., Boujard, T., Corraze, G., Kaushik, S.J., Boeuf, G., Mol, K.A., Geyten, S.V.D., Kühn, E.R., 1998. *Incorporation of high levels of extruded lupin in diets for rainbow trout *Oncorhynchus mykiss*: nutritional value and effects on thyroid status. Aquaculture 163, 325–345.

Burel, C., Boujard, T., Tulli, F., Kaushik, S.J., 2000. Digestibility of extruded peas, extruded lupin, and rapeseed meal in rainbow trout Oncorhynchus mykiss. and turbot Psetta maxima. Aquaculture 188, 285–298.

Camperio, J., Carroza-Meza, C.H., Suarez, J., Benetti, D., 2026. Global research trends of black soldier fly larvae (Hermetia illucens) meal in aquaculture from a scientometric perspective (2007–2025). Aquac. Nutri. 2026 (1), 5560332.

Chattaraj, S., Behera, B.K., Das Mohapatra, P.K., 2024. Effective valorization of Chlorella biomass and Brewers spent grain substituting fish meal and soybean meal in the diet of herbivorous fish Cirrhinus reba for higher growth, digestibility and sustainable cultivation. J. Appl. Phycol. 36 (1), 217–232.

Chattaraj, S., Chattaraj, M., Mitra, D., Ganguly, A., Thatoi, H., Das Mohapatra, P.K., 2025. 16S amplicon sequencing of the gastrointestinal microbiota of Cirrhinus reba and isolation of an autochthonous probiotic using culture based approaches. Syst. Microbiol. Biomanuf. 5 (1), 156–170.

Chattaraj, S., Ganguly, A., Mitra, D., et al., 2026a. Sustainable valorization of algal biomass in fish nutrition and dietetics: a critical review and scientometric study. Waste Biomass Valor. https://doi.org/10.1007/s12649-026-03619-7.

Chattaraj, S., Sahoo, S.S., Behera, S., Thatoi, H., 2026b. Sustainable aquafeed development using Spirulina: effects on growth performance of Catla catla. Algal Res. 104447.

Commission of the European Communities, 2006a. Commission recommendation of 17 August 2006 on the presence of deoxynivalenol, zearalenone, ochratoxin A, T-2 and HT-2 and fumonisins in products intended for animal feeding. O_. J. Eur. Commun. L229, 7.

Commission of the European Communities, 2006b. Commission regulation (EC) No 1881/2006 of 19 December 2006 setting maximum levels for certain contaminants in foodstuffs. O_. J. Eur. Commun. L364, 5.

Council for Agricultural Science and Technology, 2003. Mycotoxins: Risks in Plant, Animal, and Human Systems. Council for Agricultural Science and Technology, Ames, IA, USA.

D'Mello, F.J.P., 1991a. Antigenic proteins. In: D'Mello, F.J.P., Duffus, C.M., Duffus, J.H. (Eds.), Toxic substances in crop plants. The Royal Society of Chemistry, Thomas Graham House, Science Park, Cambridge CB4 4WF, Cambridge, pp. 108–125.

D'Mello, F.J.P., 1991b. Toxic amino acids. In: D'Mello, F.J.P., Duffus, C.M., Duffus, J.H. (Eds.), Toxic Substances in Crop Plants. The Royal Society of Chemistry, Thomas Graham House, Science Park, Cambridge CB4 4WF, Cambridge, pp. 22–48.

D'Mello, J.P.F., Macdonald, A.M.C., 1998. Fungal toxins as disease elicitors. In: Environmental Toxicology: Current Developments. Journal Rose Edition. Gordon and Breach Science Publishers, The Netherlands, pp. 253–289.

Davies, R.H., 1991. Cyanogens. In: D'Mello, F.P.J., Duffus, C.M., Duffus, J.H. (Eds.), Toxic substances in crop plants. The Royal Society of Chemistry, Thomas Graham House, Science Park, Cambridge CB4 4WF, Cambridge, pp. 202–225.

De la Higuera, M., Garcia-Gallego, M., Cardenete, G., Suarez, M.D., Moyano, F.J., 1988. Evaluation of lupin seed meal as an alternative protein source in feeding of rainbow trout *Salmo gairdneri*. Aquaculture 71, 37–50.

De, D.K., Nath, D., Sen, P.R., 1987. Preliminary studies on tea seed-cake as a fish toxicant. Indian J. Anim. Sci. 57 (7), 781–783.

Duffus, C.M., Duffus, J.H., 1991. In: D'Mello, F.J.P., Duffus, C.M., Duffus, J.H. (Eds.), Toxic substances in crop plants. The Royal Society of Chemistry, Thomas Graham House, Science Park, Cambridge CB4 4WF, Cambridge, pp. 1–21.

Duncan, A.J., 1991. Glucosinolates. In: D'Mello, F.J.P., Duffus, J.P., Duffus, C.M. (Eds.), Toxic substances in crop plants. The Royal Society of Chemistry, Thomas Graham House, Science Park, Cambridge CB4 4WF, Cambridge, pp. 126–147.

EFSA Panel on Contaminants in the Food Chain, 2013. Scientific Opinion on the risk for public and animal health related to the presence of sterigmatocystin in food and feed. EFSA J 11, 3254.

El-Sayed, Y.S., Khalil, R.H., 2009. Toxicity, biochemical effects and residue of aflatoxin B1 in marine water-reared sea bass (*Dicentrarchus labrax* L.). Food Chem. Toxicol. 47, 1606–1609.

FAO, 2022. The state of world fisheries and aquaculture 2022: Towards blue transformation. Rome.

Farabi, S.M.V., Yousefian, M., Hajimoradloo, A., 2006. Aflatoxicosis in juvenile *Huso huso* fed a contaminated diet. J. Appl. Ichthyol. 22, 234–237.

Farnsworth, N.R., Bingel, A.S., Cordell, G.A., Crane, F.A., Fong, H.H.S., 1975. Potential value of plants as sources of new antifertility agents I. J. Pharm. Sci. 64, 535–598.

Felix, J.P., Mello, D., 2000. Farm Animal Metabolism and Nutrition. CABI, United Kingdom.

Fenwick, G.R., Price, K.R., Tsukamoto, C., Okubo, K., 1991. Saponins. In: D'Mello, F.J.P., Duffus, C.M., Duffus, J.H. (Eds.), Toxic substances in crop plants. The Royal Society of Chemistry, Thomas Graham House, Science Park, Cambridge CB4 4WF, Cambridge, pp. 285–327.

Fereidoon, S., 2014. Beneficial health effects and drawbacks of antinutrients and phytochemicals in foods. Appl. Microbiol. Biotechnol. 97, 45–55.

Fernando, R., Pinto, M.D.P., Pathmeswaran, A., 2012. Goitrogenic food and prevalence of Goitre in Sri Lanka. J. Food Sci. 41, 1076–1081.

Francis, G., Makkar, H., Becker, K., 2001. Antinutritional factors present in plant-derived alternate fish feed ingredients and their effects in fish. Aquaculture 199, 197–227. https://doi.org/10.1016/S0044-8486(01)00526-9.

Freeland, W.J., Calcott, P.H., Anderson, L.R., 1985. Tannins and saponin: interaction in herbivore diet. Biochem. Syst. Ecol. 13 (2), 189–193.

Friedman, M., Henika, P.R., Mackey, B.E., 2003. Effectof feeding solanidine, solasodine and tamatidine to nonpregnant and pregnant mice. Food Chem. Toxicol. 41, 61–71.

Giri, A.P., Kachole, M.S., 2004. Amylase inhibitors of pigeon pea (*Cajanus cajan*) seeds. Phytochemistry 47, 197–202.

Gomes, E.F., Corraze, G., Kaushik, S., 1993. Effects of dietary incorporation of a co-extruded plant protein rapeseed and peas. on growth, nutrient utilisation and muscle fatty acid composition of rainbow trout (*Oncorhynchus mykiss*). Aquaculture 113, 339–353.

Grabner, M., Hofer, R., 1985. The digestibility of the proteins of broad bean Vicia faba. and soybean Glycine max. under in vitro conditions stimulating the alimentary tracts of rainbow trout *Salmo gairdneri*. and carp *Cyprinus carpio*. Aquaculture 48, 111–122.

Grant, G., 1991. Lectins. In: D'Mello, F.J.P., Duffus, C.M., Duffus, J.H. (Eds.), Toxic substances in crop plants. The Royal Society of Chemistry, Thomas Graham House, Science Park, Cambridge, pp. 49–67.

Griffiths, D.W., 1991. Condensed tannins. In: D'Mello, F.J.P., Duffus, C.M., Duffus, J.H. (Eds.), Toxic substances in crop plants. The Royal Society of Chemistry, Thomas Graham House, Science Park, Cambridge, pp. 180–201.

Hendriks, H.G.C.J.M., Van der Ingh, T.S.G.A.M., Krogdahl, A., Olli, J., Koninkx, J.F.J.G., 1990. Binding of soybean agglutinin to small intestinal brush border membranes and bush border membrane enzyme activities in Atlantic salmon *Salmo salar*. Aquaculture 91, 163–170.

Herman, R.L., 1970. Effects of gossypol on rainbow trout *Salmo gairdneri* Richardson. J. Fish. Biol. 2, 293–303.

Hossain, M.A., Jauncey, K., 1989. Nutritional evaluation of some Bagladeshi oilseed meals as partial substitutes for fishmeal in the diet of common carp, *Cyprinus carpio* L. Aquacult. Fish. Manage. 20, 255–268.

Hossain, M.A., Jauncey, K., 1990. Detoxification of linseed and sesame meal and evaluation of their nutritive vaölue in the diet of carp *Cyprinus carpio* L. Asian Fish Sci. 3, 169–183.

Hossain, M.A., Jauncey, K., 1993. The effect of varying dietary phytic acid, calcium and magnesium levels on the nutrition of common carp, *Cyprinus carpio*. In: Kaushik, S.J., Luquent, P. (Eds.), Fish Nutrition in Practice. Proceedings of International Conference, Biarritz, France, June 24–27, 1991, pp. 705–715.

Huwig, A., Freimund, S., Käppeli, O., Dutler, H., 2001. Mycotoxin detoxication of animal feed by different adsorbents. Toxicol. Lett. 122 (2), 179–188.

International Agency for Research on Cancer List of Classifications.

Jackson, A.J., Capper, B.S., Matty, A.J., 1982. Evaluation of some plant proteins in complete diets for the tilapia *Sarotherodon mossambicus*. Aquaculture 27, 97–109.

Jantrarotai, W., Lovell, R.T., Grizzle, J.M., 1990. Acute toxicity of aflatoxin B1 to channel catfish. J. Aquat. Anim. Health 2, 237–247.

Jard, G., Liboz, T., Mathieu, F., Guyonvarc'h, A., Lebrihi, A., 2011. Review of mycotoxin reduction methods in foods. Food Control 22 (11), 1448–1457.

Joint FAO, WHO Expert Committee on Food Additives, 2001. Evaluation of Certain Mycotoxins in Food: Fifty-Sixth Report of the Joint FAO/WHO Expert Committee on Food Additives. World Health Organization, Geneva, Switzerland.

Kamala, A., Shirima, C., Jani, B., Bakari, M., Sillo, H., Rusibamayila, N., De Saeger, S., Kimanya, M., Gong, Y.Y., Simba, A., et al., 2018. Outbreak of an acute aflatoxicosis in Tanzania during 2016. World Mycotoxin J. 11, 311–320.

Kaushik, S.J., Cravedi, J.P., Lalles, J.P., Sumpter, J., Fauconneau, B., Laroche, M., 1995. Partial or total replacement of fishmeal by soybean protein on growth, protein utilization, potential estrogenic or antigenic effects, cholesterolemia and flesh quality in rainbow trout, *Oncorhynchus mykiss*. Aquaculture 133, 257–274.

Kovăcíc, S., Pepeljnjak, S., Petrinec, Z., Klarić, M.Š., 2009. Fumonisin B1 neurotoxicity in young carp (*Cyprinus carpio* L.). Arh. Hig. Rada Toksikol. 60, 419–426.

Krogdahl, A., Bakke-Mckellep, A.M., Roed, K.H., Baeverfjord, G., 2000. Feeding Atlantic salmon *Salmo salar* L. soybean products: effects on disease resistance furunculosis., and lysozyme and IgM levels in the intestinal mucosa. Aquac. Nutr. 6, 77–84.

Krogdahl, A., Lea, T.B., Olli, J.J., 1994. Soybean proteinase inhibitors affect intestinal trypsin activities and amino acid digestibilities in rainbow trout *Oncorhynchus mykiss*. Comp. Biochem. Physiol. 107A, 215–219.

Kyriazakis, I., Whittenmore, C.T., 2006. Whittenmore's science and practice of pig production. Wiley-Blackwell, Oxford.

Liebman, M., Al-Wahsh, I.A., 2011. Probiotics and other key determinants of dietary oxalate absorption. Adv. Nutr. 2, 254–260.

Liener, I.E., 2005. Implications of antinutritional components in soybean foods. Food Sci. (N. Y.) 34, 31.

Liener, I.E., 1980. Toxic Constituents of Plant Foodstuffs. Academic Press, New York 10003, NY, pp. 1 502.

Liener, I.E., 1989. Antinutritional factors in legume seeds: state of the art. In: Huisman, J., Van der Poel, A.F.B., Liener, I.E. (Eds.), Recent Advances of Research in Antinutritional Factors in Legume Seeds. Pudoc, Wageningen, pp. 6–14.

Liener, I.E., 2003. Phytohemagglutinins: their nutritional significance. J. Agric. Food Chem. 22, 17.

Liu, Y., Wu, F., 2010. Global burden of aflatoxin-induced hepatocellular carcinoma: a risk assessment. Environ. Health Perspect. 118, 818–824.

Loewus, F.A., 2002. Biosynthesis of phytate in food grains and seeds. In: Food Phytates. CRC Press, Boca Raton, FL, pp. 53–61.

Lumlertdacha, S., Lovell, R.T., Shelby, R.A., Lenz, S.D., Kemppainen, B.W., 1995. Growth, hematology, and histopathology of channel catfish, *Ictalurus punctatus*, fed toxins from Fusarium moniliforme. Aquaculture 130, 201–218.

Mahrous, K.F., Khalil, W.K.B., Mahmoud, M.A., 2006. Assessment of toxicity and clastogenicity of sterigmatocystin in Egyptian Nile tilapia. Afr. J. Biotechnol. 5, 1180–1189.

Makkar, H.P.S., Becker, K., 1996. Effect of pH, temperature, and time on inactivation of tannins and possible implications in detannification studies. J. Agric. Food Chem. 44, 1291–1295.

Makkar, H.P.S., Becker, K., 1997. Nutrients and antiquality factors in different morphological parts of the Moringa oleifera tree. J. Agric. Sci. Cambridge 128, 311–322.

Makkar, H.P.S., Becker, K., 1999. Nutritional studies on rats and fish carp, *Cyprinus carpio*. fed diets containing unheated and heated Jatropha curcas meal of a non-toxic provenance. Plant Foods Hum. Nutr. 53, 183–192.

Makkar, H.P.S., Blümmel, M., Becker, K., 1995. Formation of complexes between polyvinyl pyrrolidone and polyethylene glycol with tannins and their implications in gas production and true digestibility in in vitro techniques. Br. J. Nutr. 73, 897–913.

Mambrini, M., Roem, A.J., Cravedi, J.P., Lalles, J.P., Kaushik, S.J., 1999. Effects of replacing fishmeal with soy protein concentrate and of DL-methionine supplementation in high-energy, extruded diets on the growth and nutrient utilization of rainbow trout, *Oncorhynchus mykiss*. J. Anim. Sci. 77, 2990–2999.

Marijani, E., Kigadye, E., Okoth, S., 2019. Occurrence of fungi and mycotoxins in fish feeds and their impact on fish health. Int. J. Microbiol.

Mishra, S., Devi, S., Nottanalan, H., Kantal, D., 2024. Unraveling the complexities of antinutritional factors: implications for fish nutrition and feeding practices.

Missmer, S.A., Suarez, L., Felkner, M., Wang, E., Merrill, A.H., Rothman, K.J., Hendricks, K. A., 2006. Exposure to fumonisins and the occurrence of neutral tube defects along the Texas-Mexico border. Environ. Health Perspect. 114, 237–241.

Mueller, I., 2001. Analysis of hydrolysable tannins. Anim. Feed Sci. Technol. 91, 3–20.

Mukhopadhyay, N., Ray, A.K., 1999a. Effect of fermentation on the nutritive value of sesame seed meal in the diets for rohu, Labeo rohita Hamilton., fingerlings. Aquacult. Nutr. 5, 229–236.

Mukhopadhyay, N., Ray, A.K., 1999b. Utilisation of copra meal in the formulation of compound diets for rohu, Labeo rohita, fingerlings. J. Appl. Ichthyol. 15, 127–131.

Nachbar, M.S., Oppenheim, J.D., Thomas, J.O., 2000. Lectins in the US diet: isolation and characterization of alectin from the tomato (Lycopersicon). J. Biol. Chem. 255, 2056.

Ng, W.K., Wee, K.L., 1989. The nutritive value of cassava leaf meal in pelleted feed for Nile tilapia. Aquaculture 83, 45–58.

Noonan, S.C., Savage, G.P., 1999. Oxalic acid and itseffects on humans. Asia Pac. J. Clin. Nutr. 8, 64–74.

Norton, G., 1991. Proteinase inhibitors. In: D'Mello, F.J.P., Duffus, C.M., Duffus, J.H. (Eds.), Toxic substances in crop plants. The Royal Society of Chemistry, Thomas Graham House, Science Park, Cambridge CB4 4WF, Cambridge, pp. 68–106.

Olvera-Novoa, M.A., Campos, S.G., Sabido, M.G., Palacios, C.A.M., 1990. The use of alfalfa protein concentrates as a protein source in diets for tilapia *Oreochromis mossambicus*. Aquaculture 90, 291–302.

Park, H., Flores, R.A., Johnson, L.A., 1997. Preparation of fish feed ingredients: reduction of carotenoids in corn gluten meal. J. Agric. Food Chem. 45, 2088–2092.

Pelissero, C., Bennetau, B., Babin, P., Le Menn, F., Dunogues, J., 1991b. The estrogenic activity of certain phytoestrogens in the Siberian sturgeon *Acipenser baeri*. J. Steroid Biochem. Mol. Biol. 38 (3), 293–299.

Pelissero, C., Le Menn, F., Kaushik, S., 1991a. Estrogenic effect of dietary soybean meal on vitellogenesis in cultured Siberian sturgeon *Acipenser baeri*. Gen. Comp. Endocrinol. 83, 447–457.

Pelissero, C., Sumpter, J.P., 1992. Steroids and "steroid-like" substances in fish diets. Aquaculture 107, 283–301.

Petrinec, Z., Pepeljnjak, S., Kovacic, S., Krznaric, A., 2004. Fumonisin B1 causes multiple lesions in common carp (*Cyprinus carpio*). Dtsch Tierarztl Wochenschr 111, 358–363.

Petterson, D.S., Harris, D.J., Allen, D.G., 1991. Alkaloids. In: D'Mello, F.J.P., Duffus, C.M., Duffus, J.H. (Eds.), Toxic Substances in Crop Plants. The Royal Society of Chemistry, Thomas Graham House, Science Park, Cambridge CB4 4WF, Cambridge, pp. 148–179.

Potter, S.M., Jimenez-Flores, R., Pollack, J., Lone, T.A., Berber-jimenez, M.D., 1993. Protein saponin interaction and its influence on blood lipids. J. Agric. Food Chem. 41, 1287–1291.

Randel, R.D., Chase, C.C., Wyse, S.J., 1992. Effects of Gossypol and cottonseed products on reproduction of mammals. J. Anim. Sci. 70, 1628–1638.

Rank, C., Nielsen, K.F., Larsen, T.O., Varga, J., Samson, R.A., Frisvad, J.C., 2011. Distribution of sterigmatocystin in filamentous fungi. Fungal Biol. 115, 406–420.

Reddy, K., Salleh, B., 2011. Co-occurrence of moulds and mycotoxins in corn grains used for animal feeds in Malaysa. J. Anim. Vet. Adv. 10, 668–673.

Reigh, R.C., 1999. Production characteristics of pond-raised channel catfish *Ictalurus punctatus* fed diets with and without animal protein for three growing seasons. J. World Aquacult. Soc. 30, 154–160.

Rheeder, J.P., Marasas, W.F.O., Tiel, P.G., Sydenham, E.W., Shepard, G.S., van Schalkwyk, D., 1992. Fusarium moniliforme and fumonisins in corn in relation to human esophageal cancer in Transkei. Postharvest Pathol. Mycotoxins 82, 353–357.

Richard, J.L., 2008. Discovery of aflatoxins and significant historical features. Toxin Rev. 27, 171–201.

Richardson, N.L., Higgs, D.A., Beames, R.M., McBride, J.R., 1985. Influence of dietary calcium, phosphorous, zinc and sodium phytate level on cataract incidence, growth and histopathology in juvenile chinook salmon *Oncorhynchus tshawytscha*. J. Nutr. 115, 553–567.

Riche, M., Brown, P.B., 1996. Availability of phosphorus from feedstuffs fed to rainbow trout, *Oncorhynchus mykiss*. Aquaculture 142, 269–282.

Robinson, E.H., Rawles, S.D., Oldenburg, P.W., Stickney, R., 1984. Effects of feeding glandless or glanded cottonseed products and gossypol to Tilapia aurea. Aquaculture 38, 145–154.

Roehm, J.N., Lee, D.J., Sinnhuber, R.O., 1967. Accumulation and elimination of dietary gossypol in the organs of rainbow trout. J. Nutr. 92, 425–428.

Rumsey, G.L., Hughes, S.G., Winfree, R.A., 1993. Chemical and nutritional evaluation of soy protein preparations as primary nitrogen sources of rainbow trout *Oncorhynchus mykiss*. Anim. Feed Sci. Technol. 40, 135–151.

Russo, J.R., Yanong, R.P.E., 2013. Molds in fish feeds and aflatoxicosis. UFIFAS Ext. FA95, 1–3.

Salaro, A.L., Toledo, M.R., Guimaraes, J.D., Luz, K.R., Souto, E.F., Miranda, N.T., Ribeito, J. I., 2000. Effect of cottonseed meal on the reproductive physiology of male of Nile tilapia. In: Fitzsimmons, K., Filho, J.C. (Eds.), Tilapia Aquaculture in the 21st Century. Proceedings of the Fifth International Symposium on Tilapia Aquaculture, September 3–7, 2000, Rio de Janero-RJ, Brazil, pp. 24–29.

Santiago, C.B., Aldaba, M.B., Laron, M.A., Reyes, O.S., 1988. Reproductive performance and growth of Nile tilapia Oreochromis niloticus. broodstock fed diets containing Leucaena leucocephala leaf meal. Aquaculture 70, 53–61.

Schwartz, P., Bucheli, T.D., Wettstein, F.E., Burkhardt-Holm, P., 2013. Life-cycle exposure to the estrogenic mycotoxin zearalenone A_ects zebrafish (*Danio rerio*) development and reproduction. Environ. Toxicol. 28, 276–289.

Schwartz, P., Thorpe, K.L., Bucheli, T.D., Wettstein, F.E., Burkhardt-Holm, P., 2010. Short-term exposure to the environmentally relevant estrogenic mycotoxin zearalenone impairs reproduction in fish. Sci. Total Environ. 409, 326–333.

Scudamore, K.A., Hetmanski, M.T., Clarke, P.A., Barnes, K.A., Startin, J.R., 1996. Analytical methods for the determination of sterigmatocystin in cheese, bread and corn products using HPLC with atmospheric pressure ionization mass spectrometric detection. Food Addit. Contam. 13, 343–358.

Shimoyamada, M., Ikedo, S., Ootsubo, R., Watanabe, K., 1998. Effects of soybean saponins on chymotryptic hydrolyses of soybean proteins. J. Agric. Food Chem. 46 (12), 4793–4797.

Skonberg, D.I., Hardy, R.W., Barrows, F.T., Dong, F.M., 1998. Color and flavour analysis of fillets from farm-raised rainbow trout *Oncorhynchus mykiss*. fed low-phosphorus feeds containing corn or wheat gluten. Aquaculture 166, 269–277.

Spinelli, J., Houle, C.R., Wekell, J.C., 1983. The effect of phytates on the growth of rainbow trout Salmo gairdneri. fed purified diets containing varying quantities of calcium and magnesium. Aquaculture 30, 71–83.

Sun, G., Wang, S., Hu, X., Su, J., Huang, T., Yu, J., Tang, L., Gao, W., Wang, J.S., 2007. Fumonisin B1 contamination of home-grown corn in high-risk areas for esophageal and liver cancer in China. Food Addit. Contam. 24, 181–185.

Synder, H.E., Kwon, T.W., 1987. Soybean Utilization. Van Nostrand Reinhold, New York.

Tacon, A.G.J., 1992. Nutritional Fish Pathology Morphological Signs of Nutrient Deficiency and Toxicity in Farmed Fish, 85. FAO, Rome, Italy.

Tacon, A.G.J., Metian, M., 2015. Feed matters: satisfying the feed demand of aquaculture. Rev. Fish. Sci. Aquac. 23, 1–10.

Tomaszewski, J., Miturski, R., Semczuk, A., Kotarski, J., Jakowicki, J., 1998. Tissue zearalenone concentration in normal, hyperplastic and neoplastic human endometrium. Ginekol. Pol. 69, 363–366.

Tuan, N.A., Manning, B.B., Lovell, R.T., Rottinghaus, G.E., 2003. Responses of Nile tilapia (*Oreochromis niloticus*) fed diets containing different concentrations of moniliformin or fumonisin B1. Aquaculture 217, 515–528.

Ufodike, E.B.C., Matty, A.J., 1983. Growth responses and nutrient digestibility in mirror carp *Cyprinus carpio*. fed different levels of cassava and rice. Aquaculture 31, 41–50.

van der Ingh, T.S.G.A.M., Krogdahl, A., Olli, J.J., Hendriks, H.G.C.J.M., Koninkx, J.F.J.G., 1991. Effects of soybean-containing diets on the proximal and distal intestine in Atlantic salmon *Salmo salar*.: a morphological study. Aquaculture 94, 297–305.

Viegas, C., Nurme, J., Piecková, E., Viegas, S., 2018. Sterigmatocystin in foodstuffs and feed: aspects to consider. Mycology.

Walter, H.L.,L., Fanny, C., Charles, Christian, 2002. Minerals and phytic acid interaction: is it a real problem for human nutrition. Int. J. Food Sci. Tech. 37, 727–739.

Wang, H., Wei, H., Ma, J., Luo, X., 2000. The fumonisin B1 content in corn from North China, a high-risk area of esophageal cancer. J. Environ. Pathol. Toxicol. Oncol. 19, 139–141.

Warth, B., Parich, A., Atehnkeng, J., Bandyopadhyay, R., Schuhmacher, R., Sulyok, M., Krska, R., 2012. Quantitation of mycotoxins in food and feed from Burkina Faso and Mozambique using a modern LC-MS/MS multitoxin method. J. Agric. Food Chem. 60, 9353–9363.

Wee, K.L., Shu, S.-W., 1989. The nutritive value of boiled full-fat soybean in pelleted feed for Nile tilapia. Aquaculture 81, 303–314.

WHO department of Food Safety and zoonoses Aflatoxins. Food Saf. Dig. 6, 2018, 20–23.

Wild, C.P., Gong, Y.Y., January 2010. Mycotoxins and human disease: a largely ignored global health issue. Carcinogenesis 31 (1), 71–82. https://doi.org/10.1093/carcin/bgp264.

Wozny, M., Obremski, K., Jakimiuk, E., Gusiatin, M., Brzuzan, P., 2013. Zearalenone contamination in rainbow trout farms in north-eastern Poland. Aquaculture 416–417, 209–211.

Wozny, M., Obremski, K., Hliwa, P., Gomułka, P., Rózynski, R., Wojtacha, P., Florczyk, M., Segner, H., Brzuzan, P., 2019. Feed contamination with zearalenone promotes growth but a_ects the immune system of rainbow trout. Fish Shellfish Immunol. 84, 680–694.

Yildirim, M., Manning, B.B., Lovell, R.T., Grizzle, J.M., Rottinghaus, G.E., 2007. Toxicity of moniliformin and fumonisin B1 fed singly and in combination in diets for young channel catfish *Ictalurus punctatus*. J. World Aquac. Soc. 31, 599–608.

Yoshizawa, T., Yamashita, A., Luo, Y., 1994. Fumonisin occurrence in corn from high- and low-risk areas for human esophageal cancer in China. Appl. Environ. Microbiol. 60, 1626–1629.

Yousif, O.M., Alhadhrami, G.A., Pessarakli, M., 1994. Evaluation of dehydrated alfalfa and salt bush Atriplex. leaves in diets for tilapia *Oreochromis aureus* L. Aquaculture 126, 341–347.

Zhang, G.-L., Feng, Y.-L., Song, J.-L., Zhou, X.-S., 2018. Zearalenone: a mycotoxin with Different toxic effect in domestic and laboratory animals' Granulosa cells. Front. Genet. 9, 667.

Zychowski, K.E., et al., 2013a. Investigation of the effect of mycotoxins on aquatic animals: a review. Toxins 5 (10), 1836–1851.

Zychowski, K.E., Rodrigues Ho_mann, A., Ly, H.J., Pohlenz, C., Buentello, A., Romoser, A., Gatlin, D.M., Phillips, T.D., 2013b. The e_ect of aflatoxin-B1 on red drum (Sciaenops ocellatus) and assessment of dietary supplementation of NovaSil for the prevention of aflatoxicosis. Toxins 5, 1555–1573.

CHAPTER

12

Evaluation of feed additives (like de-oiled rice bran, mustard oil cakes, brewers spent grains) and supplements (brewers spent yeast) to enhance nutritional value and palatability of feed

Hanane Boutaj[1,2], Salma Latique[3], Sakho Mahamadou[1], Youssef Ait Rahou[2], Meddich Abdelilah[2], Cherkaoui El Modafar[2], Noureddine El Aouad[1], Divya Gunsola[4], Devvret Verma[5], Debasis Mitra[4] and Marika Pellegrini[6]

[1]*Laboratory of Life and Health Sciences, Faculty of Medecine and Pharmacy of Tangier, Abdelmalek Essaadi University, Tetouan, Morocco;* [2]*Centre d'Agrobiotechnologie et de Bioingénierie, Unité de Recherche Labellisée CNRST (Centre AgroBiotech-URL-CNRST-05), Université Cadi Ayyad, Marrakesh, Morocco;* [3]*Laboratory of Drugs Sciences, Faculty of Medicine, Pharmacy and Dentistry, Sidi Mohamed Ben Abdellah University, Fes, Morocco;* [4]*Department of Microbiology, Graphic Era (Deemed to be University), Dehradun, Uttarakhand, India;* [5]*Department of Biotechnology, Graphic Era (Deemed to be University), Dehradun, Uttarakhand, India;* [6]*Department of Life, Health and Environmental Sciences, University of L'Aquila, L'Aquila, Italy*

1. Introduction

1.1 The global challenge of feed security and sustainability

The global livestock sector, the pillar of the world's food security and economic development, is facing the challenges of history, meeting the growing need for animal protein during the environmental and economic constraints of the 21st century. With the number of humans projected to nearly reach the level of 10 billion by the year 2050, the production of animal-derived products is projected to increase by 60%–70% by the year 2050 (The future of food and agriculture – Alternative pathways to 2050, 2018). This acceleration exerts immense pressure on traditional feed ingredients such as soybean meal and maize, the production of which is often accompanied by deforestation, high water usage, and high greenhouse gas emissions

Fish Feed Formulation with Agroindustrial Wastes and Byproducts. https://doi.org/10.1016/B978-0-443-41728-3.00020-4

(Steinfeld et al., 2006). Furthermore, the price volatility and geopolitical risk of the main producing countries threaten the economic sustainability of animal production systems in the world (Henchion et al., 2021). This range of driving forces emphasizes the urgency of seeking and integrating alternative, sustainable, and economic ingredients into the supply chains of feedstocks to ensure their long-term sustainability and reduce their ecological footprint.

1.2 The circular bio-economy: Turning waste into resources

A paradigm change to a circular bio-economy is a revolutionary response to these demands. This paradigm is centered on the valorization of agro-industrial by-products, turning streams of waste into valuable resources with minimal environmental effects (Jurgilevich et al., 2016). Billions of tons of residual biomass are generated every year by the agricultural and food sectors, the greater part of which remains unharnessed and ends up as litter, generating pollution and nutrient loss (Mirabella et al., 2014). Byproducts, such as de-oiled rice bran (DORB), mustard oil cake (MOC), brewer's spent grains (BSG), and brewer's spent yeast (BSY) offer exemplary examples of these latent resources. These materials offer bright futures as alternatives to traditional feed ingredients that are rich in fiber, proteins, vitamins, and bioactive molecules. The strategic exploitation of these by-products in the production of animal feeds can significantly enhance sustainability by ending the conflict between the production of feeds and foods, reducing the carbon footprint of the production of the feed matrix, and injecting economic value into streams of waste (Ivanović et al., 2023; Xiong et al., 2021). This vision is perfectly aligned with the United Nations' Sustainable Development Goals (SDGs), particularly the 12th goal (Responsible Consumption and Production) and Goal 13 (Climate Action). This chapter critically reviews the role of some agro-industrial by-products, DORB, MOC, BSG, and BSY, in enhancing the nutritional quality and functional properties of animal feeds. In addition to the substitution level, our focus was on the role of these by-products as feed supplements and additives that improve palatability, digestive health, and overall farm animal performance. The key objectives of this chapter are to provide a detailed scientific account of the nutritive profile, positive bioactive compounds, and inherent limitations (e.g., antinutritional factors) of DORB, MOC, BSG, and BSY. Describe the mechanisms of action by which these ingredients enhance the palatability of feeds, absorbability of nutrients, and the health of animals. Describe practical techniques for their successful integration into combinations of feed for various livestock species, including inclusion levels and necessary treatments. We discuss standardized methods of evaluation and future directions of study to further enhance their usage in sustainable livestock production systems. By combining extant studies and emphasizing their applicability, this chapter aims to offer a foundation for researchers, nutritionists, and industry stakeholders who remain committed to the development of sustainable animal agriculture through the precepts of the circular economy.

2. The science of feed additives and supplements

2.1 Definitions and regulatory landscape

Within the context of animal nutrition, the terms "feed additive" and "feed supplement" feed supplement' have distinct yet occasionally overlapping meanings. Feed additives are broadly defined as substances intentionally added to feed to perform specific technological or zootechnical functions, such as enhancing feed stability, altering organoleptic properties, or favorably affecting animal performance and health (Regulation (EC) No 1831/2003 of the European Parliament and of the Council of September 22, 2003 on additives for use in animal nutrition," 2003). They are often used in microquantities. In contrast, a feed supplement typically refers to a concentrated source of nutrients (e.g., protein, vitamins, and minerals) used to correct deficiencies in a diet or to provide a specific nutritional boost and is included in larger quantities (e.g., percentage of the diet) (NRC, 2016).

The use of these materials is governed by a stringent and complex regulatory landscape designed to ensure animal, human consumer (through residues in animal products), and environmental safety. Key regulatory frameworks include the following.

The European Union Regulation (EC) No 1831/2003 on additives for use in animal nutrition categorizes additives into functional groups (e.g., technological, sensory, nutritional, and zootechnical).

The U.S. Food and Drug Administration (FDA) Center for Veterinary Medicine regulates animal feed and feed ingredients under the Federal Food, Drug, and Cosmetic Act.

A critical regulatory hurdle for many by-product ingredients is Generally Recognized as Safe (GRAS) status or authorization for a specific zootechnical purpose, which requires comprehensive scientific dossiers to demonstrate efficacy and safety (Das et al., 2023).

2.2 Mechanisms of action: Beyond basic nutrition

2.2.1 Nutritional enhancement (macro and micronutrients)

The mode of action of most of the supplements was simple supplementation of dietary levels of the key nutrients. By-products, such as BSY and MOC, serve as very good protein corrective agents for the imbalance of the amino acid composition of cereal-based diets (Akin et al., 2025). DORB provides energy, lipids, and indispensable fatty acids, along with a unique profile of bioactive molecules, such as γ-oryzanol and tocotrienols, which possess antioxidant activity (Gul et al., 2015). BSG, which contains high amounts of fiber and β-glucans, principally contributes to the energy pool of ruminants and might have an impact on the digestion of monogastrics. This nutrient supplementation directly addresses the requirements of the animals by virtue of maintenance, development, and production and serves as the prime role of these ingredients.

2.2.2 Palatability modulators (flavor, aroma, texture)

Feed consumption is a critical determining factor for animal performance and is greatly influenced by palatability. Ingredients can have the capability of impacting palatability via different mechanisms. Some by-products contain soluble sugars, peptides, or volatiles that enhance the smell and taste of feed. For instance, BSY contains high nucleotide and free amino acid contents, such as glutamic acid, which have been established as umami taste promoters that can enhance the attractiveness of feed and voluntary poultry and pig feed consumption (Jung and Batal, 2012). Ingredients such as DORB and BSG can affect pellet quality, pellet hardness, and mouthfeel of the feed, which may either encourage or deter consumption (Dozier et al., 2010). Thus, physical form optimization is crucial for maximizing consumption.

2.2.3 Functional properties: Gut health and immunomodulation

Beyond nutrition, many byproducts exert functional effects that enhance animal health and reduce the need for therapeutic antibiotics. The mechanisms are diverse; components such as mannan oligosaccharides (MOS) in BSY and arabinoxylans in BSG can selectively stimulate the growth of beneficial gut bacteria (e.g., *Lactobacillus* and *Bifidobacterium*), improve gut microbial balance, and inhibit the colonization of pathogens (Giannenas et al., 2020). β-Glucans from BSG and BSY can bind to specific receptors on immune cells (e.g., macrophages), priming the innate immune system, and enhancing the animal's resistance to diseases (Lowry et al., 2005). Short-chain fatty acids (SCFAs) produced from the fermentation of fiber in BSG and DORB serve as energy sources for colonocytes and can improve villus height and crypt depth, leading to better nutrient absorption and a stronger intestinal barrier (Jha and Berrocoso, 2016).

2.3 Economic and environmental impact assessment

The incorporation of agro-industrial byproducts into feed chains presents compelling economic and environmental advantages. Byproducts are typically cheaper than conventional feedstuffs. Their use in least-cost formulation software can significantly reduce feed expenditure, which represents 60%–70% of the total production costs in intensive systems (Shurson, 2020). This improved the economic resilience of livestock operations. Furthermore, creating a market for these "waste" products generates additional revenue streams for primary industries (e.g., rice milling, oil extraction, and brewing). From an LCA perspective, valorizing byproducts dramatically reduces the environmental footprint of both the source and livestock industries. It diverts waste from landfills (avoid methane emissions); reduces the land, water, and fertilizer burden associated with cultivating dedicated feed crops; and lowers the overall carbon footprint per unit of animal product produced (Salomone et al., 2017). This directly contributes to the sustainability metrics of a circular bioeconomy.

3. Critical analysis of selected agro-industrial by-products

The four by-products, DORB, MOC, BSG, and BSY, exhibit a wide range of nutritional compositions, as detailed in Table 12.1. This variance in protein, fiber, and key amino acid contents dictates their specific applications and limitations in animal feed formulation, which are explored in detail in the following subsections.

Table 12.1 Composition ranges & key amino acids of DORB/MOC/BSG/BSY.

	Protein %	Crude fiber%	Ether extract %	Ash %	Amino acids	References
DROB	15–18	15–19	15–20	15–21	Lysine (2.7%–4.4%) Methionine (1.6%–2.3%) Threonine (3.1%–4.3%)	Feedipedia (2025); Huang et al. (2018); Ranjan et al. (2022); Sharma et al. (2018)
MOC	33–39	7	8.1–10.7	8–10	Lysine (4%) Methionine (1.5%) Threonine (3.7%)	Feedipedia (2025); Jahan et al. (2013); Sarker et al. (2015); Sharma (2013); Singh et al. (2013)
BSG	27–33	0.1–4.4	1.9–9.9	6.5–11.9	Lysine (2.1%–3.7%) Methionine (1.0%–1.9%) Threonine (2.6%–3.5%)	Feedipedia (2025); Lynch et al. (2016); Mussatto et al. (2006)
BSY	39–57	0–4	0–8.2	5.1–9.3	Lysine (4.6%–7.6%) Methionine (1.3%–2.2%) Threonine (3.7%–5.6%)	Feedipedia (2025); Marson et al. (2020); Podpora et al. (2016)

3.1 De-oiled rice bran (DORB)

3.1.1 Origin, production, and global availability

DORB, which is generated after the extraction of oil from raw rice bran, is a major by-product of the rice milling industry. Raw rice bran is a mixture of the pericarp, aleurone layer, germ, and subaleurone layer removed during the polishing of brown rice to produce white rice (Sharif et al., 2014). Global production of rice exceeds 740 million tons annually, resulting in the generation of approximately 63–76 million tons of rice bran (Nidhishree et al., 2024). Although a significant portion is stabilized and used for oil extraction, the resultant DORB represents a vast and underutilized resource. Its availability is directly correlated with rice production, making it a staple by-product in major rice-producing regions, such as East Asia, Southeast Asia, and the Indian subcontinent. The economic viability of DORB as a feed ingredient is enhanced by its stable shelf life after oil extraction and defatting, which reduces the lipase activity responsible for the rapid rancidity in raw bran (Tan et al., 2023).

3.1.2 Comprehensive nutritional profile

DORB is valuable owing to its well-balanced nutrient profile, which makes it a valuable addition to animal feed. As Table 12.1 shows, DORB contains a moderate level of crude protein (15%–18%) and a high fiber fraction (15%–19% crude fiber), making it a medium-grade protein source and fiber source, respectively. Nutrient quality is also defined by residual lipid-associated antioxidants that remain after the extraction of the lipid component and a well-balanced profile of the amino acids it contains. It is low in lysine (2.7%–4.4%) but contains high methionine (1.6%–2.3%) and threonine (3.1%–4.3%) contents, and could thus serve as a quality protein complement when it is compounded with high-lysine protein foods such as soybean meal (Huang et al., 2018; Ranjan et al., 2022; Feedipedia, 2025; Sharma et al., 2018). One of the strengths of the nutrient profile is the availability of γ-oryzanol, a very active antioxidant and bioactive compound that is unique to rice bran (Yılmaz Tuncel, 2023; Ranjan et al., 2023).

The amino acid profile is thought to be poor in lysine; otherwise, it is well complemented by the remaining indispensable amino acids, providing a good complement protein source when paired with lysine-rich foods (Meza et al., 2024). The residual oil fraction is particularly rich in unsaturated fatty acids (oleic and linoleic acids) as well as the primary reservoir of highly active lipid-soluble antioxidants, particularly γ-oryzanol and tocoperols, which are responsible for its functional properties beyond plain nutrition (Huang et al., 2024).

3.1.3 Anti-nutritional factors and mitigation strategies

A major hurdle in utilizing agro-industrial by-products is the presence of ANFs, which vary by ingredients and require targeted mitigation strategies (Table 12.2). For DORB, the primary constraints in monogastric diets are phytic acid (myo-inositol hexaphosphate) and high fiber content (notably hemicellulose and lignin). Phytic acid chelates essential minerals, such as phosphorus, calcium, zinc, and iron, reducing their bioavailability and potentially leading to impaired growth and

Table 12.2 ANFs and mitigation strategies by ingredient.

Ingredients	ANFS factor	Mitigation strategies	References
DROB	Pytates, trypsin inhibitors, impact protein digestion.	The application of microbial fermentation processes (e.g., *Rhizopus oryzae*) significantly reduces phytates and other protein inhibitors; stabilization and reduction of ANF through heat treatment; exogenous enzymes (phytases).	Irakli et al. (2020); Kaur et al. (2015); Ranjan et al. (2019)
MOC	Glucosinolates break down into isothiocyanates and oxazolidinedione (toxic to the thyroid); tannins and phytates	Prolonged immersion (8 hours) hydrolysis affects 76% of glucosinolates (traditional practice in India); heat treatment to inactivate myrosinase; fermentation by bacteria and yeast; elimination through germination.	Okunade et al. (2015); Tyagi (2002); Yan et al. (2025)
BSG	High percentage of fiber (lignin and cellulose) and phytates, reducing digestibility	Preservation and partial degradation of fibers through silage with lactic acid bacteria; increased protein bioavailability through solid fermentation (Trichoderma, bacteria, and Aspergillus fungi); fibrolytic enzymes and physical extrusion to reduce ANF and increase fiber solubility.	Marcus and Fox (2021); Puligundla and Mok (2021); Stahn et al. (2023)
BSY	Absence of major classic ANFs. High levels of purines (nucleic acids).	Breaking down the cell wall and releasing nutrients through autolysis and drying. No significant effects of ANF have been reported in the literature.	Fundo et al. (2023)

mineral deficiencies (Vashishth et al., 2017). In addition, the presence of trypsin inhibitors may limit protein digestion.

Several effective mitigation strategies have been developed:

- Exogenous phytase supplementation in feed is the most common approach for effectively hydrolyzing phytic acid and releasing bound minerals (Joudaki et al., 2023; Vashishth et al., 2017).
- Heat stabilization (e.g., autoclaving and extrusion) inactivates lipase, disrupts fiber structures, and reduces certain ANFs, thereby enhancing nutrient digestibility.
- Solid-state fermentation (SSF) with fungi such as Aspergillus spp. or bacteria significantly reduces phytic acid and crude fiber while simultaneously increasing protein content and improving the amino acid profile through microbial synthesis (Shi et al., 2015).
- Microbial fermentation with *Rhizopus oryzae* has been shown to markedly decreases phytates and enhances protein content (Kaur et al., 2015; Ranjan et al., 2019; Irakli et al., 2020).

Together, these processing techniques mitigate ANFs in DORB, enabling safer and more effective use in animal feed. In summary, effective mitigation strategies for DORB, as outlined in, include SSF to break down phytates and fibers, heat treatment (extrusion) to stabilize the product and limit endogenous enzymes, and the use of exogenous enzymes such as xylanase in feed to improve digestibility post-processing (Rahman et al., 2015; Yang et al., 2016).

3.1.4 Key research findings in ruminant and monogastric diets

DORB is an excellent source of energy and protein for ruminants. Fiber is effectively degraded by the rumen microbiota, and the protein contributes to the rumen undegradable protein (UDP) fraction, supplying amino acids post-ruminally. DORB can be effectively incorporated into ruminant diets at substantial levels. For example, in crossbred dairy cows, the replacement of maize grain with DORB (up to 30% of the concentrate mixture) did not adversely affect dry matter intake, nutrient digestibility, milk yield, or milk composition, demonstrating its value as an economical protein and energy source (Singh et al., 2000). Its favorable price often makes it a cornerstone of least-cost-formulated total mixed rations (TMR). In monogastrics, inclusion levels are limited owing to the fiber and phytic acid content.In broiler diets, inclusion of rice bran (notably de-oiled or whole) is typically capped at approximately 10%–15%, because increasing levels (e.g., 20%–25%) tend to depress growth performance and feed efficiency, likely due to diluted energy density and reduced mineral availability. For instance, a controlled study found that diets containing 25% rice bran led to significantly poorer body weight gain and feed conversion than diets containing 10% or lower levels (Alewi et al., 2022). However, with proper enzyme supplementation (phytase and xylanase), these levels can be maintained effectively, and bioactive compounds may confer health benefits (Selle et al., 2009). In laying hens, DORB can be used at similar levels, without affecting egg production or quality. Growingfinishing pigs can tolerate the inclusion of up to 20% defatted rice bran (DFRB) in their diet without compromising growth performance. Feed efficiency (gain: feed) may improve linearly with increasing DFRB levels. Some digestibility metrics (GE, DM, and NDF) also respond positively, particularly at moderate inclusion rates (Huang et al., 2021). Overall, the recommended inclusion levels varied among species (Table 4). Poultry diets generally cap DORB at 10%–15% unless supplemented with enzymes, whereas pigs can handle up to 20% without performance loss. This underscores the greater ability of swine to handle the fiber content of DORB than poultry (Huang et al., 2021).

3.2 Mustard oil cake (MOC)/rapeseed meal

3.2.1 Origin and processing variations

Mustard oil cake (MOC) and rapeseed meal (RSM) are protein-rich by-products derived from the extraction of oil from seeds of the Brassicaceae family, primarily *Brassica juncea* (mustard) and *Brassica napus* (rapeseed) (Usman et al., 2023; Xiong et al., 2024). While the terms are often used interchangeably, MOC typically refers to cake from indigenous mustard varieties, whereas RSM is associated with modern, cultivated rapeseed and canola (a cultivar low in erucic acid and GSLs) (Lin et al., 2013). Global production is substantial and linked to the vegetable oil industry, with China, Canada, India, and the European Union being major producers (USDA, 2025). The nutritional quality of the final meal was profoundly influenced by the oil extraction process. Traditional mechanical expelling produces cakes with higher residual oil content (10%–15%), whereas solvent extraction yields a leaner meal (<3% oil) with a higher concentration of protein and ANFs per unit mass (Li et al., 2017; Östbring et al., 2019). Heat treatment during processing is a critical variable, as it can denature proteins and mitigate some ANFs; however, it must be carefully controlled to avoid impairing protein digestibility through Maillard reactions (Oliveira et al., 2020).

3.2.2 Nutritional profile: A high-protein source

MOC/RSM is primarily valued as a high-protein ingredient (33%–39%), making it a prime candidate for replacing conventional protein sources (Table 12.1). Its amino acid profile is a key defining feature: it is notably rich in sulfur-containing amino acids, such as methionine (approximately 1.5%), often making it superior to soybean meal in this regard. However, the data also revealed its primary limitation: a lower lysine content (approximately 4%) and potentially reduced bioavailability due to processing, which must be accounted for through careful diet formulation and supplementation (Jahan et al., 2013; Sharma, 2013; Singh et al., 2013; Sarker et al., 2015; Feedipedia, 2025). Nevertheless, it remains a cost-effective and nutritionally valuable alternative to conventional protein sources in balanced diets (Tang et al., 2024).

3.2.3 Navigating anti-nutritional factors: Glucosinolates, erucic acid, and tannins

The utilization of MOC/RSM is constrained by several potent ANFs that can adversely affect animal health and performance, if not properly managed.

- **GSLs:** These were the most significant ANFs. Upon enzymatic hydrolysis by myrosinase (released upon seed crushing), GSLs break down into bioactive metabolites, such as isothiocyanates, oxazolidinethiones, and nitriles. These compounds are goitrogenic and inhibit iodine uptake by the thyroid gland, leading to goiter and reduced growth rates. They also impart a characteristic pungent taste that can reduce palatability (Alexander et al., 2008).
- **Erucic Acid:** A long-chain fatty acid present in the residual oil of traditional MOC and high-erucic acid RSM. High dietary levels have been associated with myocardial lipidosis and heart lesions in monogastric animals (Roine et al., 1960; Abdellatif and Vles, 2004; Kramer et al., 1983). The development of low-erucic acid canola varieties (*Canola*) has largely mitigated this issue in modern RSM (Downey and Craig, 1964; Stefansson, 1983; Stefansson et al., 1961).
- **Tannins (Condensed):** In seed hulls, tannins can bind dietary proteins and digestive enzymes, forming indigestible complexes that reduce protein digestibility and nitrogen retention (Butler, 1989; Mueller-Harvey, 2006; Khajali and Slominski, 2012).
- **Phytic Acid:** As many plant meals, phytic acid chelates minerals, such as phosphorus, zinc, and calcium, reducing their bioavailability (Gupta et al., 2015).

The presence of ANFs, particularly GSLs, as highlighted in Tables 13.2 and is the primary determinant of the safe inclusion level of MOC. The breakdown products of GSLs are toxic to the thyroid glands. Therefore, mitigation measures are essential. Traditional practices, such as prolonged immersion (hydrolyzing up to 76% of GSLs), are still used. Industrially, heat treatment to inactivate the myrosinase enzyme is commonly used to prevent GSL hydrolysis. However, the most advanced and effective strategy is biological processing through solid-state fermentation using bacteria or yeast, which not only degrades GSLs but also improves their overall nutritional profile (Tyagi, 2002; Okunade et al., 2015; Yan et al., 2025).

3.2.4 Detoxification methods: Physical, chemical, and biological

To enhance the safety and utility of MOC/RSM, various detoxification strategies have been employed to reduce the burden on ANFs, particularly GSLs.

Physical Methods: Heat Treatment (autoclaving, toasting) is a common and effective method for denaturing the myrosinase enzyme and preventing the hydrolysis of GSLs into their toxic breakdown products (Fenwick et al., 1986). However, excessive heat can damage protein quality (Oliveira et al., 2020). Water Washing can leach water-soluble GSLs, but it is less practical on an industrial scale because of its high water usage and effluent production (KLOSS et al., 1994; Liu et al., 1994).

Chemical Methods: Treatment with alkalis (e.g., NaOH, $Ca(OH)_2$) or ferroussulfate ($FeSO_4$) can effectively complex with and neutralize GSL breakdown products. For instance, ferrous sulfate reacts with isothiocyanates to form inert complexes (Srivastava et al., 2020).

Biological Methods, such as solid-state fermentation (SSF), are considered one of the most efficient and eco-friendly approaches. SSF using specific fungi (e.g., *Aspergillus niger* and *R. oryzae*) or bacteria can utilize the meal as a substrate (Bhargav et al., 2008). These microorganisms produce enzymes that degrade GSLs, tannins and phytic acids. Furthermore, SSF can enrich the meal with microbial protein, improve its amino acid profile, and enhance overall digestibility, thereby adding value beyond detoxification (Shi et al., 2015; Costa-Silva et al., 2022). The choice of detoxification method depends on the cost, efficiency, scalability, and intended use of the final product.

The detoxification methods for MOC are summarized in Table 12.3, ranging from physical (prolonged soaking in water and heat treatment) to biological (SSF). While traditional soaking effectively hydrolyzes GSLs, modern approaches favor controlled heat treatment to inactivate myrosinase or advanced SSF using specific fungi or bacteria to not only degrade GSLs but also enrich the meal, adding value beyond detoxification (Stahn et al., 2023).

3.2.5 Application in livestock feeds and performance outcomes

The successful incorporation of MOC and RSM into livestock diets is contingent upon understanding species-specific tolerance levels, which are largely determined by the residual concentration of GSLs and the efficacy of detoxification methods.

Table 12.3 Summary of processing methods to enhance value and mitigate limitations.

Ingredients	Method	Main effects	References
DROB/ BSG	Fermentation (bacteria or fungi)	Phytates, tannins (ANFs), and fibers are broken down; free amino acids and proteins increase. Allows DROB and BSG to be recycled into useful, highly nutritious ingredients.	Ranjan et al. (2019)
MOC	Prolonged soaking in water	The enzymatic reaction causes hydrolysis of glucosinolates in the MOC. Increased palatability	Stahn et al. (2023)
DROB/ BSG	Heat treatment (extrusion)	Limitation of endogenous enzymatic reaction. Reduction of ANF. Removal of some of the fibers. Stabilization of the product by blocking rancidity.	Rahman et al. (2015)
DROB	Exogenous enzymes (Xylanase)	Degradation of plant cell walls improving the digestibility of BSG and DROB. Hydrolysis of phytates	Yang et al. (2016)
BSY	Autolysis + drying	The production of lactic acid partially breaks down the fibers, thereby releasing nutrients.	Fundo et al. (2023)

Modern double-low (low erucic acid, low GSL) canola meal has a significantly broader application range than the traditional MOC (Bonnardeaux, 2007).

Ruminants: Ruminants, particularly dairy cattle, were the most tolerant livestock category for MOC/RSM inclusion. Rumen microbiota can degrade and adapt to GSLs, thereby mitigating their goitrogenic effects. Canola meal is a highly valued protein source because of its excellent rumen-undegradable protein (RUP) value and amino acid profile (Swanepoel, 2016). Meta-analyses have confirmed that soybean meal can be replaced at inclusion levels of up to 11% of dietary dry matter (DM) without adverse effects on intake, milk yield, or milk composition (Paula et al., 2018; Martineau et al., 2013). In some studies, milk protein yield increased, likely because of a better methionine supply (Evans et al., 2016). Traditional MOC with higher GSL content must be restricted (<10% DM) to avoid off-flavors in milk and subclinical thyroid effects (Papas et al., 1979). Beef cattle and sheep also utilize MOC/RSM effectively, with canola meal comprising up to 20% of the concentrate mix, without compromising growth or feed efficiency (Nair et al., 2015).

Poultry: Poultry are more sensitive to GSLs than ruminants, and require stricter inclusion limits. Double-low canola meal is safe at inclusion levels up to 15% in broiler diets without impairing growth, feed conversion, or carcass quality, provided diets are enzyme-supplemented and lysine-balanced (Payvastegan et al., 2017). Higher inclusions (>20%) reduce palatability, growth depression, and thyroid hypertrophy due to GSL intake (Alexander et al., 2008). In layer hens, canola meal can be used at 10%–15% with no adverse effects on egg production or quality and no transfer of GSLs into eggs (Kopacz et al., 2021). However, in some cases, very high inclusions may impart a negative flavor to the eggs. Non-detoxified MOC are not recommended for broiler feed.

Swine: Swine tolerance falls between that of ruminants and poultry. Canola meal can be included in 5%–12% of growing-finishing pig diets to maintain growth rates and carcass quality (Hansen et al., 2020). In sow diets, inclusion of up to 30% fiber is possible, with added benefits of satiety during gestation due to high fiber content (Velayudhan and Nyachoti, 2017).

Aquaculture: The application of MOC/RSM in aquafeeds is driven by the need to reduce reliance on fishmeal. Detoxified meals can be included at levels of 15%–30% in the diet without significantly compromising growth performance, provided the feed is meticulously balanced (Karolina and Justyna, 2025). Higher levels can depress growth and cause histopathological changes in the liver and thyroid (Karolina and Justyna, 2025). Solid-state fermented MOC shows promise for higher inclusion levels by improving palatability and nutrient digestibility (Shi et al., 2015).

MOC and RSM represent valuable protein resources for animal nutrition when detoxified or replaced with double-low varieties and adhering to species-specific inclusion guidelines. When implemented correctly, they can reduce feed costs without compromising animal health or productivity, thereby contributing significantly to the economic and environmental sustainability of livestock production. Table 4 summarizes the tolerance levels across species, highlighting the stricter limits in poultry compared to ruminants and swine (Rao et al., 2005; Cheva-Isarakul et al., 2001).

3.3 Brewers' spent grains (BSG)

3.3.1 Brewery by-product: Composition and stability challenges

Brewers' spent grains (BSG) are the major solid by-product of the beer brewing process, constituting approximately 85% of the total by-products generated (Mussatto et al., 2006). These are obtained after the mashing and lautering stages, during which fermentable sugars are extracted from malted barley and other grains. Globally, the brewing industry produces approximately 39 million tons of BSG annually, presenting a significant disposal challenge and an equally significant opportunity for valorization (Zhang, 2024).

The chemical composition of BSG is intrinsically linked to the raw materials and brewing technology used, it is consistently characterized by its high moisture content (70%–85%), which is its primary stability challenge. This high moisture, combined with a rich nutrient profile, makes BSG highly perishable and susceptible to rapid microbial spoilage (e.g., by lactic acid bacteria, yeasts, and molds) and necessitates immediate stabilization post-production to avoid significant dry matter and nutrient losses (Zhang, 2024). Once dried, the composition of the BSG is consistently valuable but fiber dominant. Table 13.1 shows a protein content of 27%–33%, which is significant but is overshadowed by its high fiber content (0.1%–4.4% crude fiber, though Neutral Detergent Fiber would be much higher). This high fiber content directly influences the energy density and application. Its amino acid profile is reasonable, with lysine levels (2.1%–3.7%) comparable to those of many cereals, but its value is often locked behind a fibrous matrix, necessitating processing for monogastric species (Feedipedia, 2025; Mussatto et al., 2006; Lynch et al., 2016).

3.3.2 High fiber content: A double-edged sword (role of β-glucans and arabinoxylans)

The fiber fraction is the most defining feature of BSG, acting as both its greatest nutritional asset and its main limitation.

Structural Role and Limitation: The cell walls of barley kernel, largely intact in BSG, are composed of complex polysaccharides, primarily arabinoxylans (AX) and β-glucans, along with cellulose and lignin (Zhang, 2024). This fibrous matrix creates a physical barrier that impedes access to digestive enzymes in monogastric animals, reducing the availability of intracellular nutrients, such as proteins and starch. This high fiber content directly lowers the metabolizable energy value of BSG in poultry and swine (Song et al., 2025).

Functional Benefits: Conversely, this same fiber fraction is highly beneficial for ruminants, serving as an effective source of physically effective fibers that stimulate rumination, salivation, and rumen health. Furthermore, both β-glucans and AX are classified as dietary fibers with a prebiotic potential. They can resist digestion in the small intestine and can be fermented by beneficial bacteria (e.g., *Bifidobacterium* and *Lactobacillus*) in the hindgut of monogastrics and the rumen, producing short-chain fatty acids (SCFAs) such as acetate, propionate, and butyrate. These SCFAs serve as energy sources for the host and have been shown to improve gut health, morphology, and integrity (Jha et al., 2019; Song et al., 2025).

Anti-nutritional Aspects and Mitigation: While classic ANF, the high content of lignin, cellulose, and phytates in BSG still reduces nutrient digestibility, particularly in monogastrics (Table 13.2). Several mitigation strategies have been applied to overcome this problem. Ensiling lactic acid bacteria offers cost-effective preservation and partial fiber degradation. For more significant improvements, solid-state fermentation with fungi such as *Aspergillus* or *Trichoderma* enhances protein bioavailability. In addition, fibrolytic enzymes (e.g., xylanases) or physical processes, such as extrusion in feed mills, can reduce fiber complexity, increase solubility, and improve nutrient utilization in monogastric animals (Marcus and Fox, 2021; Puligundla and Mok, 2021; Stahn et al., 2023).

3.3.3 Utilization in ruminant versus monogastric nutrition

The utilization of BSG is fundamentally dictated by the digestive physiology of animals. BSG is an excellent feed ingredient in dairy and beef cattle. In ruminants, rumen microorganisms efficiently ferment fibrous components, allowing high inclusion levels in Total Mixed Rations (TMR), often up to 30% of dietary dry matter, without negatively affecting dry matter intake or milk production. Its high protein bypass value can also be an advantage, as it supplies amino acids directly to the small intestine (Chelkapally et al., 2025).

In poultry, BSG serves primarily as a source of bulk and functional fiber rather than energy because of its high fiber and low energy content. It can be included at levels of 10%–40% in layer diets, where its fiber content helps control energy density and reduces feather pecking without adversely affecting egg production or quality (Gebremedhn et al., 2019). However, inclusion levels are typically lower, at 10%–20%, because higher levels depress growth performance and feed efficiency due to nutrient dilution. Enzyme supplementation (xylanases and β-glucanases) is crucial for breaking down the fibrous matrix and improving nutrient availability at these inclusion levels (Denstadli et al., 2010).

In pigs, the inclusion levels depend on the production stage. Gestating sows can tolerate up to 50% BSG in their diets, and its fiber content promotes satiety and benefits gut health. In contrast, growing-finishing pigs are usually restricted to <10% to avoid reduced growth rates (Mukasafari et al., 2018).

As summarized in Table 4, ruminants can effectively utilize high levels of BSG, while monogastrics are more limited owing to fiber-related constraints. In pigs, high inclusion levels (e.g., 30%) can lead to faster satiety and reduced weight gain, even though the animals remain healthy (Stahn et al., 2023). In poultry, dried BSG has been shown to reduce growth performance in broilers, underscoring the importance of processing or maintaining low inclusion rates in species with high metabolic demand (Pires (Filho et al., 2021).

3.3.4 Valorization techniques: Drying, ensiling, and fractionation

To overcome the stability challenge and enhance its nutritional application, especially in monogastrics, BSG must be processed. The primary valorization techniques categorized in Table 13.3 aim to preserve the material, break down its fibrous matrix, and fractionate it into higher-value streams. These include drying, ensiling with lactic acid bacteria for preservation, and SSF to increase protein bioavailability and break down the fibers (Ranjan et al., 2019).

Drying: Thermal drying (e.g., rotary drum, belt dryers) is the most common method for reducing the moisture content to <10%, creating a stable, storable product. However, it is energy-intensive and expensive, often negating the economic advantages of the raw materials. This can also lead to Maillard reactions, which may reduce protein digestibility (Chetrariu and Dabija, 2020).

Ensiling: A cost-effective alternative to drying. BSG can be ensiled alone or in combination with other feed. Natural lactic acid bacteria ferment soluble sugars, producing lactic acid and lowering pH, which preserves the material anaerobically for several months. Ensiled BSG is ideal for use on or near ruminant farms (Terefe, 2022).

Fractionation: This represents the future of BSG valorisation. Mechanical (e.g., milling and sieving) and biological (enzymatic) processes can separate BSG into value-added streams.

- A **protein-rich fraction** for monogastric and aquafeeds.
- A **fiber-rich fraction** for ruminant feed or as a source of prebiotic oligosaccharides and bioactive compounds in the food industry. This approach allows for the optimal and high-value use of each component, maximizing the economic return from this by-product (Sun et al., 2024; Osei et al., 2025).

3.4 Brewers' spent yeast

3.4.1 Distinction from BSG: A high-value microbial biomass

Brewer spent yeast (BSY) is a by-product of the fermentation and maturation stages of beer production. Unlike Brewers' Spent Grains (BSG), which is a plant-derived fiber-rich material, BSY consists of microbial biomass (*Saccharomyces cerevisiae*) that has fulfilled its primary role in converting sugars to alcohol and carbon dioxide. This fundamental difference in origin renders BSY a concentrated source of microbial nutrients and bioactive compounds, elevating it from a mere filler to a high-value functional feed supplement (Jaeger et al., 2020).

Following fermentation, BSY is typically separated from beer by centrifugation or filtration. It has a high moisture content similar to that of BSG, but is notably richer in protein and other micronutrients. Its composition is more consistent than that of BSG but is influenced by the yeast strain and number of re-pitching cycles.

This microbial nature is key to the unique nutritional and functional properties of animal nutrition.

3.4.2 Nutritional powerhouse: Nucleotides, B-vitamins, and bioavailable amino acids

Table 13.1 clearly distinguishes BSY from the other by-products, showing an exceptionally high crude protein content (39%–57%) that rivals high-quality protein meals. More importantly, its amino acid profile is exceptional for a by-product, which is particularly rich in the critical amino acid lysine (4.6%–7.6%). This high lysine content makes BSY an excellent supplement to balance the amino acid profile of cereal-based diets. Furthermore, it contains substantial levels of methionine (1.3%–2.2%) and threonine (3.7%–5.6%), confirming its status as a well-rounded, high-value nutritional resource that offers far more than protein (Podpora et al., 2016; Marson et al., 2020; Zeko-Pivač et al., 2023; Feedipedia, 2025).

Beyond proteins, BSY is one of the richest natural sources of nucleotides and nucleosides (e.g., inosine monophosphate). These compounds are not considered essential nutrients, but are conditionally essential during periods of rapid growth, stress, or gastrointestinal challenges. Dietary nucleotides can reduce the energy cost of de novo synthesis, supporting enhanced cell proliferation, immune function, and gut repair (Zeko-Pivač et al., 2023; Vieira et al., 2013). BSY is exceptionally rich in B-complex vitamins including thiamine (B1), riboflavin (B2), niacin (B3), and folate (B9), which are crucial cofactors in energy metabolism. It is also a valuable source of trace minerals such as selenium, chromium, and zinc, which are often present in highly bioavailable organic forms complexed within yeast cells (Anonymous, 2024).

Unlike many plant-derived byproducts, BSY contains no major classic antinutritional factors (ANFs), which simplifies its dietary inclusion (Table 13.2). The main consideration is its high purine (nucleic acid) content, which can pose a metabolic concern in certain contexts, but is generally well tolerated. Consequently, processing focuses not on detoxification but on enhancing nutrient availability. As shown in Table 13.3, the key method is autolysis, often combined with drying, which uses the yeast enzymes to break down the robust cell wall and release the internal pool of nucleotides, vitamins, and amino acids, making them accessible for digestion (Fundo et al., 2023).

3.4.3 Functional role as a palatability enhancer and prebiotic (mannan oligosaccharides—MOS)

In addition to basic nutrition, BSY exerts significant functional effects that enhance feed quality and animal health. High levels of free amino acids (especially glutamic acid), small peptides, and nucleotides in BSY act as potent flavor enhancers (umami taste). Numerous studies have demonstrated that the inclusion of even small amounts (1%–3%) of BSY in poultry and aquaculture diets significantly improves feed palatability and voluntary intake, a critical factor for early-growth-stage animals and during periods of stress (Pelusio et al., 2023; Park et al., 2020; Tao et al., 2022; Sampath et al., 2023). The yeast cell wall, which constitutes approximately 10%–30% of the dry weight, is rich in Mannan Oligosaccharides (MOS) and β-glucans. MOS functions as a prebiotic by MOS possesses lectin-like properties that allow it to agglutinate and block the adhesion of pathogenic bacteria (e.g., *Escherichia coli*, *Salmonella* spp.) that express type-1 fimbriae to the gut mucosa, facilitating their excretion and By preventing pathogen colonization, MOS indirectly promotes a beneficial gut microbiota balance (Elend and Kramer, 2015; Bzducha-Wróbel et al., 2013).

3.4.4 Evidence of immunostimulatory effects in aquaculture and poultry

β-Glucans present in the BSY cell wall are powerful immunostimulants. β-(1,3)-(1,6)-Glucans are recognized by specific pattern-recognition receptors (e.g., Dectin-1) on immune cells such as macrophages and neutrophils. This activation enhances phagocytic activity, oxidative burst capacity, and cytokine production, creating a state of heightened alertness that improves the animal's ability to resist pathogens without causing overt inflammation (Stier et al., 2014). In aquaculture, where antibiotics are increasingly restricted, BSY is a valuable functional ingredient. Dietary inclusion of 1%–4% BSY has consistently improved survival rates in fish and shrimp challenged with pathogens like *Vibrio* and *Aeromonas* with benefits linked to enhanced palatability, improved gut health from MOS, and systemic immunostimulation from β-glucans (Andrews et al., 2011; Ayiku et al., 2020). In poultry, BSY or its derivatives have been shown to strengthen vaccination responses, reduce morbidity and mortality, and improve gut morphology by increasing villus height and reducing crypt depth. These changes translate into better nutrient absorption and overall performance, especially under challenging conditions (Mendieta et al., 2018; Soren et al., 2024). BSY's value of BSY lies not only in its protein content but also in its unique portfolio of nucleotides, vitamins, MOS, and β-glucans, which collectively enhance gut health, immune competence, and feed palatability. Although it is typically used at lower inclusion levels (2%–4%) in monogastrics for its functional properties, higher levels are possible in ruminants. For example, Table 4 shows that BSY successfully replaced soybean meal in the diet of lactating ewes, with no negative effects on milk yield and a significant improvement in milk fat quality. This illustrates how BSY functions as a high-value additive in monogastrics, but can serve as a practical rumen-friendly protein source at higher inclusion levels in ruminants (Oancea et al., 2023).

4. Synergistic integration into feed formulations

Integrating agro-industrial byproducts into animal feed is not simply a matter of substitution; it represents a strategic improvement in feed formulation, with processing playing a central role. As shown in Fig. 12.1, techniques such as heating, fermentation, and extrusion bring about physicochemical changes that enhance nutrient availability, improve palatability, deactivate ANFs, reduce microbial contamination, and

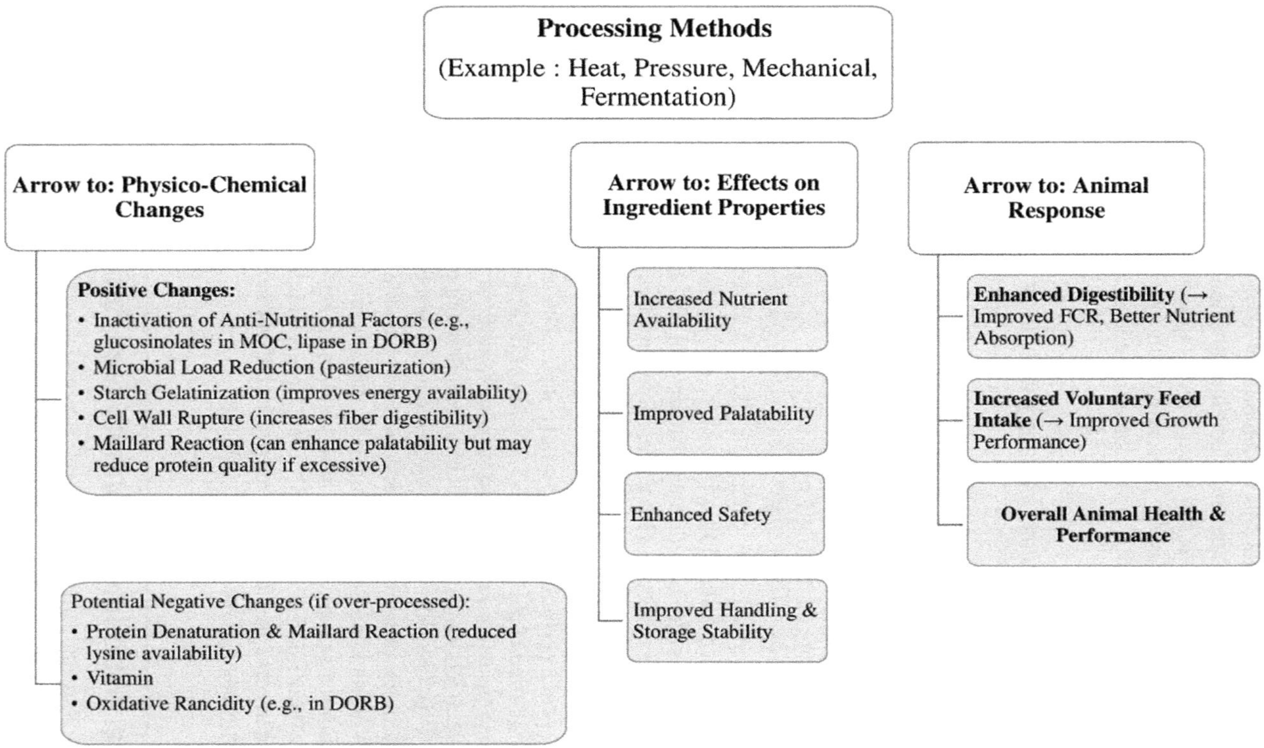

FIGURE 12.1 How processing shifts digestibility & palatability (conceptual pathway).

Processing transforms raw foods, enhancing digestibility and palatability.

break down cell walls. When applied correctly, these processes lead to tangible benefits, such as improved digestibility, higher voluntary feed intake, better feed conversion efficiency, and overall enhanced animal performance. Simultaneously, the framework emphasizes potential risks, such as protein denaturation or rancidity, highlighting the need for careful optimization. This provides the foundation for later discussions on least-cost formulation and practical deployment of processed by-products in feed systems.

4.1 Principles of least-cost formulation with by-products

The least-cost formulation (LCF) is a mathematical optimization approach designed to produce nutritionally adequate feed at the lowest possible cost. The inclusion of agro-industrial byproducts has become integral to modern LCF, extending beyond cost efficiency to embrace the broader goals of the circular economy. However, using these materials introduces additional challenges compared with conventional feed ingredients, requiring more nuanced formulation strategies (Akintan et al., 2024). Accurate matrix values for each byproduct (DORB, MOC, BSG, and BSY) are essential. This includes not only proximate analysis, but also amino acid profiles, fatty acids, vitamin and mineral content, and, critically, levels of ANFs. These data must be updated regularly to account for the natural variability stemming from the source material and processing methods.

The model must incorporate constraints specific to by-products:

- **Maximum Inclusion Levels:** Set based on ANF content (e.g., GSLs in MOC), fiber content (e.g., NDF in BSG, DORB), or fat level.
- **Palatability Limits:** To ensure voluntary feed intake is not compromised.
- **Functional Constraints:** Ensuring that the final blend has appropriate physical properties (pelletability and bulk density) (Alexander et al., 2008).

The true cost of a by-product includes its nutritional value, handling, and processing costs (e.g., drying BSG), and any required mitigation strategies (e.g., enzyme supplementation for MOC/DORB). Formulators must calculate their value in use, not just their purchase price per ton.

4.2 Case studies: Successful diet formulations

4.2.1 Ruminant total mixed rations

Ruminants are ideal candidates for byproduct integration because of their microbial digestive system. A successful case study involved formulating a TMR for mid-lactation Holstein cows, in which a blend of by-products replaced traditional ingredients.

Formulation: 15%–19% DORB (energy/fiber source), 33%–39% solvent-extracted MOC (protein source), and 27%–33% dried BSG (protein/fiber source) were incorporated, replacing a portion of the corn grain and soybean meal.

Outcome: The formulation achieved a 15% reduction in feed cost per kilogram without compromising the milk yield, milk fat, or protein content. The high rumen-degradable protein from BSG and MOC synergizes with the energy from DORB to support efficient microbial protein synthesis in the rumen (Singh et al., 2000; Swanepoel, 2016; Jha et al., 2019; Song et al., 2025).

Consideration: The diet was balanced for metabolizable protein and supplemented with iodine to counter potential goitrogenic effects of MOC (Payvastegan et al., 2017).

4.2.2 Poultry layer and broiler diets

The integration of by-products into poultry diets is more constrained but highly valuable. A previous study demonstrated that 5% fermented MOC (to reduce GSLs) and 3% BSY could be included in a broiler grower diet. BSY provides nucleotides and vitamins that support immune function, while fermented MOC offers a cost-effective protein source (Jung and Batal, 2012; Tang et al., 2024). This combination resulted in an improved feed conversion ratio (FCR) and lower mortality compared to a standard corn-soybean meal diet, offsetting the slightly reduced growth rate from the fiber content (Hansen et al., 2020; Badsha et al., 2025).

Laying Hens: Research showed that 10%–15% DORB (stabilized and with phytase enzyme) can be used in layer diets. The balanced amino acid profile of DORB complemented other ingredients, and its antioxidant content (γ-oryzanol) may have contributed to the improved lipid stability in eggs. Egg production and quality are maintained, significantly reducing the feed cost per dozen eggs (Gul et al., 2015; Kang and Kim, 2016; Chowdhury et al., 2024).

4.2.3 Aquafeed for Finfish and Crustaceans

The high cost of fishmeal drives innovation in aquafeeds. A formulation for Nile tilapia utilized a blend of 15% fermented MOC and 10% DORB to replace 25% of fishmeal protein. The SSF of MOC drastically reduces GSLs and improves protein digestibility (Badsha et al., 2025). DORB provides both energy and essential fatty acids. The formulated diet supported growth rates comparable to high-fishmeal control, while dramatically improving the economic efficiency of production (Badsha et al., 2025). BSY is a premium ingredient in shrimp feed because of its attractive properties (nucleotides and β-glucans) and immunostimulatory effects (Suphantharika et al., 2003). Inclusion levels of 4% BSY have been shown to improve palatability, growth rates, and resistance to pathogens, such as *Vibrio* spp., reducing the need for antibiotics (Hamidoghli et al., 2023).

4.3 Addressing technical challenges

4.3.1 Bulk density, flowability, and storage

The physical properties of the by-products present significant engineering challenges in feed mills. By-products such as BSG and DORB have low bulk density

(300–400 g/L) compared to cereals (600–700 g/L) (Naibaho and Korzeniowska, 2021). This can cause:

- **Transportation inefficiency:** Higher volume for the same weight.
- **Storage bin overflow:** Prematurely filling storage capacity.
- **Metering errors:** Volumetric feeders can under-dose low-density ingredients.
- **Solution:** Pelleting or densification through extrusion can significantly increase the bulk density and improve handling (Hoover et al., 2014).

Many by-products are powdery or have irregular particle sizes, leading to poor flowability, bridging in hoppers, and segregation in the mixed feed. Processing steps, such as pelleting the entire diet or using liquid binders, can enhance flow (Wan et al., 2021). The high residual oil in expeller-pressed cakes (e.g., MOC) and moisture in wet by-products (e.g., fresh BSG) render them susceptible to oxidative and microbial spoilage. Drying to a safe moisture content, use of antioxidants (e.g., ethoxyquin and natural tocopherols from DORB itself), and storage under cool and dry conditions are essential for maintaining nutrient quality and preventing mycotoxin formation (Keller et al., 2012; Nyhan et al., 2023; Błaszczyk et al., 2013).

4.3.2 Processing strategies for nutritional improvement and ANF mitigation

The successful utilization of by-products depends on targeted processing methods that address their limitations while maximizing their nutritional value. Table 13.3 summarizes the main approaches, their effects on nutrient quality and ANF reduction, and their practical applications.

Fermentation is particularly effective for DORB and BSG, as it breaks down phytates and fibers while increasing protein concentration (Rahman et al., 2015; Ranjan et al., 2019). Heat treatments and extrusion help stabilize products and reduce ANFs, whereas exogenous enzymes provide a precise means of boosting digestibility directly within feed formulations. For MOC, traditional soaking remains an effective low-cost method for reducing GSLs (Stahn et al., 2023). In contrast, the processing of BSY focuses on breaking down its robust cell wall to release valuable intracellular nutrients. Therefore, choosing the appropriate processing method is essential for maximizing the nutritional contribution of each by-product (Fundo et al., 2023).

4.4 From waste to resource: Processing pathways of key by-products

The transformation of raw agro-industrial by-products into stable and valuable feed ingredients requires targeted processing steps designed to stabilize the material, reduce ANFs, and enhance nutrient availability. The specific pathway varies according to the inherent properties of each byproduct. Fig. 12.2 provides a comparative overview of the typical process flows for DORB, MOC, BSG, and BSY, highlighting how the methods are uniquely tailored to each material while achieving the common goal of producing safe, stable, and nutritious feed ingredients.

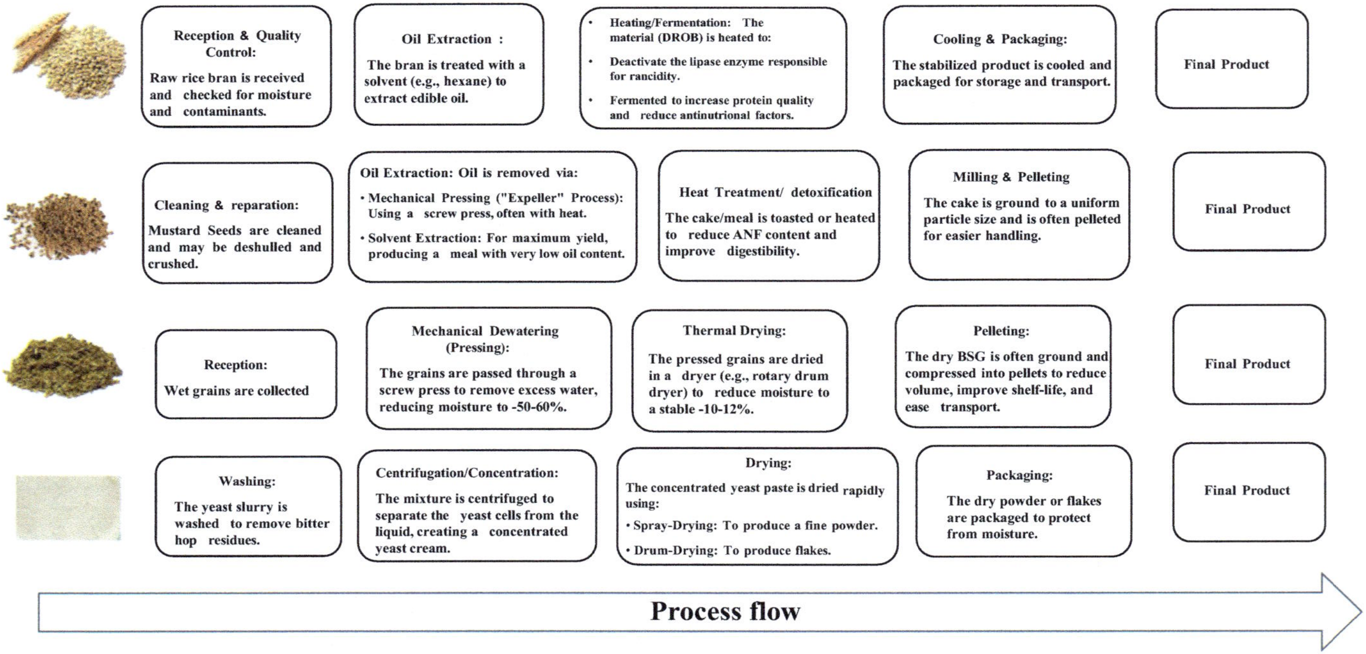

FIGURE 12.2 Process flow—from by-product to feed ingredient (DORB, MOC, BSG, BSY).

From by-product waste to nutrient-rich feed: DORB, MOC, BSG, BSY transformation.

- **DORB:** Oil extraction followed by stabilization through heating or fermentation inactivates lipase and reduces ANFs.
- **MOC:** Cleaning and oil extraction (expeller or solvent) are followed by heat treatment for detoxification before milling.
- **BSG:** Owing to its high perishability, BSG is mechanically dewatered, thermally dried for stability, and often pelleted for improved handling.
- **BSY:** Processing emphasizes purification (washing to remove hop bitters), concentration via centrifugation, and drying into a powder or flakes.

This overview demonstrates that, while each by-product requires a distinct processing approach, all methods converge on the objectives of safety, stability, and enhanced nutritional value.

5. Standardized methods for evaluation

The accurate evaluation and application of agro-industrial by-products in animal feed necessitate robust, standardized methods for evaluating their nutritional quality, palatability, and chemical composition. Reliable data from these assays forms the essential foundation for least-cost formulation software and for making informed decisions on inclusion levels.

5.1 In Vitro digestibility assays

In vitro digestibility techniques are indispensable for rapidly predicting nutrient bioavailability without the expense and ethical considerations of live animal trials. These methods simulate the physiological conditions of the digestive tract of an animal.

The Two-Stage Technique (DaisyII, ANKOM Systems): This widely used standardized protocol simulates ruminant digestion. In the first stage, samples were incubated in buffered rumen fluid to mimic microbial fermentation, while in the second stage, acid-pepsin digestion was applied to replicate abomasal conditions. The undigested fraction provides measures of in vitro dry matter digestibility (IVDMD) and organic matter digestibility (IVOMD), both of which strongly correlate with the in vivo results (Tilley and Terry, 1963; Tassone et al., 2020).

Enzymatic Methods:For monogastrics Enzyme-based assays are suitable for monogastrics. These typically involve sequential incubations with pepsin (stomach simulation) and pancreatin (small-intestine simulation). The resulting digestibility values for protein (IVPD) and dry matter depend heavily on standardized enzyme concentrations, pH, and incubation times, making strict protocol adherence crucial for reproducibility (Bryan and Classen, 2020). Such systems are especially valuable for assessing detoxification techniques such as MOC fermentation.

5.2 Palatability assessment protocols

Palatability, a measure of feed acceptability and voluntary intake, is critical for successful adoption of new ingredients. Standardized behavioral tests were used to quantify the data.

Double-Choice (Two-Choice) Preference Test: This fundamental protocol. Animals were simultaneously offered two feeders: one containing a control diet and the other containing a test diet (e.g., control vs. diet with 10% BSG). The positions of the feeders were regularly switched to avoid place preference. The intake from each feeder was measured over a set period (e.g., 24–72 h). The preference ratio was calculated as follows: (Intake of Test Diet)/(Total Intake from Both Diets). A ratio significantly above 0.5 indicates a preference for the test diet, while a ratio below 0.5 indicates aversion (Cottone et al., 2008; Gaillard and Stratford, 2016; AFB, 2020).

Free-Choice Consumption Test: A simpler approach in which animals were offered only the test diet ad libitum for a period, and their daily intake was measured and compared to their intake of a standard control diet in the previous period. Although less complex, it requires a longer duration and can be confounded by changes in animal physiology over time (Aldrich and Koppel, 2015). These tests are essential for evaluating ingredients such as MOC (which can be bitter) and BSY (which can be highly palatable), providing data to determine the maximum inclusion levels before intake is depressed.

5.3 Analytical techniques for key bioactive compounds

Precise chemical analysis is non-negotiable for quality control and understanding the functional properties of the by-products.

High-Performance Liquid Chromatography (HPLC): A core method for quantifying both antinutritional and bioactive compounds. For example, HPLC with UV/DAD is the standard for measuring GSLs in MOC and phytic acid in DORB and BSG, while HPLC-MS/MS provides high sensitivity for γ-oryzanol, tocopherols, nucleotides, and B-vitamins (Stöggl et al., 2005; Jaeger et al., 2020).

Gas Chromatography-Mass Spectrometry (GC-MS): Analysis of volatile and semi-volatile compounds. It is commonly applied to fatty acid profiling in DORB and MOC oils as well as for detecting off-flavors or strong odors in spoiled or untreated by-products (Schwarzinger et al., 2022; Lietzow, 2021).

Near-Infrared Spectroscopy (NIRS): NIRS is a rapid, nondestructive, and high-throughput technology that is revolutionizing feed analysis. Calibration models were developed using reference data from primary techniques, such as HPLC. Once validated, NIRS can predict multiple constituents simultaneously and within minutes.

- *Proximate Analysis:* Moisture, crude protein, fat, fiber.
- *Key Parameters:* Predict glucosinolate content in MOC, oil content in DORB, and fiber fractions (NDF, ADF) in BSG. Its major advantage is its potential for

at-line or in-line quality control in feed mills, allowing for real-time formulation adjustments based on the actual composition of a delivered batch of by-products (Kumar et al., 2010; Wittkop et al., 2012).

6. Future perspectives and research directions

The valorization of agro-industrial by-products is poised at a critical juncture, driven by the converging pressures of sustainability, food security, and economic efficiency. Future advancements will hinge on the integration of sophisticated technologies and holistic assessment frameworks to overcome existing barriers and to fully unlock the potential of these resources.

6.1 Emerging technologies: Microbial fermentation, enzyme treatment, and biofortification

The future of by-product upgrading lies in moving beyond simple detoxification to value-added transformations.

SSF: Research is shifting toward designing defined microbial co-cultures (e.g., fungi for fiber degradation alongside yeasts for protein enrichment) to simultaneously target multiple ANFs and enhance the nutritional profile of by-products. For instance, SSF of MOC with *A. niger* not only reduces GSLs, but also produces exogenous phytases and xylanases in situ, increasing phosphorus availability and energy value (Santos, 2011; Shi et al., 2015).

Tailored Enzyme Cocktails: The next generation of enzyme treatments will extend beyond generic phytases and xylanases. Precision enzyme technology involves designing custom cocktails based on the specific substrate structure of a byproduct batch (e.g., the specific arabinoxylan profile of BSG from a particular brewery) to maximize nutrient liberation and digestibility (Lynch et al., 2021; Nyhan et al., 2023).

Microbial Biofortification: This cutting-edge approach uses fermentation to actively enrich by-products and not just to remove antinutrients. Examples include the use of yeast strains to accumulate selenium into a more bioavailable organic form within BSY, or engineering microbes to produce essential amino acids (e.g., lysine and methionine) during the fermentation of DORB, directly addressing their inherent amino acid limitations (Kieliszek et al., 2015; Chen et al., 2019; Zhou et al., 2020).

6.2 The role of omics technologies in understanding nutrient utilization

Advances in omics technologies have deepened our understanding of how by-products interact with animal physiology and the gut microbiota.

- *Nutrigenomics* explores how bioactive components influence gene expression in host tissues, such as the liver and intestine, revealing effects on metabolism, immunity, and overall health (Hasan et al., 2019).
- *Microbiomics* leverages 16S rRNA sequencing and metagenomics to analyze how byproducts reshape the gut microbiota, thereby clarifying the prebiotic effects of BSG fiber or BSY-derived mannan-oligosaccharides (Bonifácio-Lopes et al., 2022; Calvete-Torre et al., 2023).
- *Metabolomics* provides a functional snapshot of an animal's physiological response by profiling metabolites in the blood, urine, and digesta, offering biomarkers for intake, digestibility, and early metabolic stress (Shibutami and Takebayashi, 2021).

6.3 Life cycle assessment (LCA) for environmental credentials

To truly validate the sustainability claims of by-product utilization, robust and standardized environmental accounting is required.

Beyond Simple Substitution: Future LCAs must move from attributional LCAs (which provide a static snapshot) to consequential LCAs. These models assess the system-wide consequences of diverting a by-product from one fate (e.g., landfill) to another (animal feed), including indirect effects on land-use change and markets for competing ingredients (Salemdeeb et al., 2017).

Multi-Indicator Assessment: Comprehensive LCAs should evaluate a suite of environmental impact categories beyond global warming potential (carbon footprint), including

- *Water scarcity footprint:* Given the high water footprint of conventional crop production (Pfister et al., 2017).
- *Land use:* Highlighting the advantages of by-products that require no additional land.
- *Eutrophication potential:* Assessing the impact of nutrient runoff from animals fed these diets. This provides a more holistic view of environmental trade-offs (Poore and Nemecek, 2018).

6.4 Market trends and commercialization barriers

Despite their potential, significant barriers impede the widespread commercialization of by-products as standardized feed ingredients.

Market Trends: Key drivers include the rising cost of conventional feed, stringent sustainability regulations, and growing consumer demand for circular economic practices in animal production. This creates a growing market pull for certified and sustainable feed ingredients (Li et al., 2025).

Commercialization Barriers.

- *Variability:* The single greatest challenge is batch-to-batch inconsistency in nutrient and ANF composition, which discourages adoption by feed mills that require formula consistency (Pesti and Choct, 2023).
- *Infrastructure and Logistics:* The lack of decentralized collection, drying, and processing facilities makes handling perishable by-products such as wet BSG economically unviable for many (Kamboh, 2017).
- *Regulatory Hurdles:* The Approving new feed ingredients or novel processing methods (e.g., fermentation with a specific GMO strain) can be slow, costly, and stifling innovation (Augustin et al., 2024).
- *Economic Valuation:* The lack of a universal framework for pricing ecosystem services (e.g., carbon sequestration from waste avoidance) means that the true value of by-products is often underestimated in purely cost-based models (Costanza et al., 2014; Ghisellini et al., 2016).

Therefore, future research must focus not only on biological efficacy but also on developing solutions to these market barriers, such as rapid at-line quality monitoring (e.g., NIRS), business models for decentralized preprocessing hubs, and policy advocacy for streamlined regulatory pathways.

7. Conclusion

The integration of agro-industrial by-products (DORB, MOC, BSG, and BSY) into animal feed formulations reflects a shift toward a more sustainable and economically resilient circular bioeconomy. These materials are far more abundant than substitutes; they are nutrient-rich resources with functional properties that can strengthen feed systems. However, their effective use requires overcoming inherent challenges such as ANFs and variability in composition through tailored processing technologies. The valorization of by-products is driven by the convergence of advanced bioprocessing (e.g., solid-state fermentation and precision enzymes), cutting-edge data analytics (e.g., omics technologies and life cycle assessment), and innovative supply chain strategies that address variability and logistics. By embracing these approaches, the animal production sector can meaningfully reduce its environmental footprint, build greater economic resilience, and contribute to global food security by transforming waste into valuable resources.

Conflicts of Interest

The authors declare no conflicts of interest.

Data Availability Statement

The data supporting the findings of this study are available upon request from the corresponding author.

References

Abdellatif, A.M.M., Vles, R.O., 2004. Pathological effects of dietary rapeseed oil in rats. Ann. Nutr. Metabol. 12 (5), 285–295. https://doi.org/10.1159/000175302.

AFB international, 2020. The Importance of p-value in Palatability Assessments. Available url: https://www.afbinternational.com/wp-content/uploads/2022/09/The-importance-of-p-value-in-palatability-assessments_WP_2020.pdf.

Akin, M., Jukic, M., Lukinac, J., Yilmaz, B., Özogul, F., Rocha, J.M., 2025. Valorization and functionalization of cereal-based industry by-products for nutraceuticals. In: Nutraceutics from Agri-Food by-Products, pp. 173–222.

Akintan, O., Gebremedhin, K.G., Uyeh, D.D., 2024. Animal feed formulation—Connecting technologies to build a resilient and sustainable system. Animals 14 (10), 1497.

Aldrich, G.C., Koppel, K., 2015. Pet food palatability evaluation: a review of standard assay techniques and interpretation of results with a primary focus on limitations. Animals 5 (1), 43–55. https://doi.org/10.3390/ani5010043.

Alewi, M., Edea, C., Demeke, S., Tesfaye, E., 2022. Effect of different levels of rice (Oryza sativa) bran on the growth performance of broiler chicken. Green. J. Agric. Sci. 12, 120–130.

Alexander, J., Auðunsson, G.A., Benford, D., Cockburn, A., Cravedi, J.P., Dogliotti, E., Di Domenico, A., Férnandez-Cruz, M.L., Fürst, P., Fink-Gremmels, J., Van, C., 2008. Glucosinolates as undesirable substances in animal feed scientific panel on contaminants in the food chain. EFSA J. 6 (1), 1–76.

Andrews, S.R., Sahu, N.P., Pal, A.K., Mukherjee, S.C., Kumar, S., 2011. Yeast extract, brewer's yeast and spirulina in diets for Labeo rohita fingerlings affect haemato-immunological responses and survival following Aeromonas hydrophila challenge. Res. Vet. Sci. 91 (1), 103–109. https://doi.org/10.1016/j.rvsc.2010.08.009.

Anonymous, 2024. Brewer's Yeast. Available url: https://www.drugs.com/npp/brewer-s-yeast.html?.

Augustin, M.A., Hartley, C.J., Maloney, G., Tyndall, S., 2024. Innovation in precision fermentation for food ingredients. Crit. Rev. Food Sci. Nutr. 64 (18), 6218–6238. https://doi.org/10.1080/10408398.2023.2166014.

Ayiku, S., Shen, J.-F., Tan, B.-P., Dong, X.-H., Liu, H.-Y., 2020. Effects of dietary yeast culture on shrimp growth, immune response, intestinal health and disease resistance against Vibrio harveyi. Fish Shellfish Immunol. 102, 286–295. https://doi.org/10.1016/j.fsi.2020.04.036.

Badsha, A., Shadin, A.S., Sarmina, Y., Al-Amin, S., 2025. Effect of fermented rice bran and mustard oil cake on growth performance and economic analysis in monosex tilapia, Oreochromis niloticus. Int. J. Fish. Aqua. Stud. 13 (2), 101–105. https://www.fisheriesjournal.com/archives/2025/vol13issue2/PartB/13-2-6-193.pdf?.

Bhargav, S., Panda, B.P., Ali, M., Javed, S., 2008. Solid-state fermentation: an overview. Chem. Biochem. Eng. Q. 22 (1), 49–70.

Błaszczyk, A., Augustyniak, A., Skolimowski, J., 2013. Ethoxyquin: an antioxidant used in animal feed. Int. Journal Food Sci. 2013 (1), 585931.

Bonifácio-Lopes, T., Catarino, M.D., Vilas-Boas, A.A., Ribeiro, T.B., Campos, D.A., Teixeira, J.A., Pintado, M., 2022. Impact of circular brewer's spent grain flour after in vitro gastrointestinal digestion on human gut microbiota. Foods 11 (15). https://doi.org/10.3390/foods11152279.

Bonnardeaux, J., 2007. Uses for Canola Meal, 16. Available from: https://library.dpird.wa.gov.au/pubns/29.

Bryan, D.D.S.L., Classen, H.L., 2020. In vitro methods of assessing protein quality for poultry. Animals 10 (4). https://doi.org/10.3390/ani10040551.

Butler, L.G., 1989. Effects of Condensed Tannin on Animal Nutrition. Springer Science and Business Media LLC, pp. 391–402. https://doi.org/10.1007/978-1-4684-7511-1_24.

Bzducha-Wróbel, A., Kieliszek, M., Błażejak, S., 2013. Chemical composition of the cell wall of probiotic and brewer's yeast in response to cultivation medium with glycerol as a carbon source. Eur. Food Res. Technol. 237 (4), 489–499.

Calvete-Torre, I., Sabater, C., Montilla, A., Moreno, F.J., Riestra, S., Margolles, A., Ruiz, L., 2023. Physicochemical characterization and microbiota modulatory potential of brewer's spent grain and arabinoxylan-derived fractions: a valorization study. LWT 185, 115107. https://doi.org/10.1016/j.lwt.2023.115107.

Chelkapally, S.C., Terrill, T.H., Ogunade, I.M., Estrada-Reyes, Z.M., Pech-Cervantes, A.A., 2025. Meta-analysis of the effects of the dietary inclusion of brewers grains on feed intake, milk yield, and feed efficiency of lactating dairy cows. JDS Commun. 6 (3), 299–303. https://doi.org/10.3168/jdsc.2024-0626.

Chen, Z., Liu, G., Zhang, J., Bao, J., 2019. A preliminary study on l-lysine fermentation from lignocellulose feedstock and techno-economic evaluation. Bioresour. Technol. 271, 196–201. https://doi.org/10.1016/j.biortech.2018.09.098.

Chetrariu, A., Dabija, A., 2020. Brewer's spent grains: possibilities of valorization, a review. Appl. Sci. 10 (16), 5619. https://doi.org/10.3390/app10165619.

Cheva-Isarakul, B., Tangtaweewipat, S., Sangsrijun, P., 2001. The effect of mustard meal in laying hen diets. Asia.-Austra. J. Animal Sci. 14 (11), 1605–1609.

Chowdhury, A., Kumar, A.Y.N., Kumar, R., Maurya, V.K., Mahesh, M.S., Singh, A.K., Yadav, P.K., Ghosh, M., 2024. Optimization of microwave parameters to enhance phytochemicals, antioxidants and metabolite profile of de-oiled rice bran. Res. Square. https://doi.org/10.21203/rs.3.rs-4503656/v1.

Costa-Silva, V., Anunciação, M., Andrade, E., Fernandes, L., Costa, A., Fraga, I., Barros, A., Marques, G., Ferreira, L., Rodrigues, M., 2022. Biovalorization of grape stalks as animal feed by solid state fermentation using white-Rot fungi. Appl. Sci. 12 (13), 6800. https://doi.org/10.3390/app12136800.

Costanza, R., de Groot, R., Sutton, P., van der Ploeg, S., Anderson, S.J., Kubiszewski, I., Farber, S., Turner, R.K., 2014. Changes in the global value of ecosystem services. Glob. Environ. Change 26 (1), 152–158. https://doi.org/10.1016/j.gloenvcha.2014.04.002.

Cottone, P., Sabino, V., Steardo, L., Zorrilla, E.P., 2008. Intermittent access to preferred food reduces the reinforcing efficacy of chow in rats. Am. J. Physiol. Regul. Integr. Comp. Physiol. 295 (4), R1066. https://doi.org/10.1152/ajpregu.90309.2008.

Das, R.S., Dong, G., Tiwari, B.K., Garcia-Vaquero, M., 2023. Food Safety Concerns of Alternative Proteins and Regulatory Guidelines for Their Commercialization in the Human Food Market Future Proteins: Sources, Processing, Applications and the Bioeconomy. Elsevier, Ireland, pp. 469–508. https://doi.org/10.1016/B978-0-323-91739-1.00022-2.

Denstadli, V., Westereng, B., Biniyam, H.G., Ballance, S., Knutsen, S.H., Svihus, B., 2010. Effects of structure and xylanase treatment of brewers' spent grain on performance and nutrient availability in broiler chickens. Br. Poult. Sci. 51 (3), 419–426. https://doi.org/10.1080/00071668.2010.495745.

Downey, R.K., Craig, B.M., 1964. Genetic control of fatty acid biosynthesis in rapeseed (Brassica napus L.). J. Am. Oil Chem. Soc. 41 (7), 475–478. https://doi.org/10.1007/bf02670026.

Dozier, W.A., Behnke, K.C., Gehring, C.K., Branton, S.L., 2010. Effects of feed form on growth performance and processing yields of broiler chickens during a 42-day production period. J. Appl. Poult. Res. 19 (3), 219–226. https://doi.org/10.3382/japr.2010-00156.

Elend, C., Kramer, M., 2015. Prebiotic Animal Feed Product. Available from: https://patentimages.storage.googleapis.com/af/58/97/f842ed6163d05d/US20150366238A1.pdf.

Evans, E., Callum, C., Dyck, B., 2016. Review: new findings regarding the feeding value of canola meal for dairy cows. J. Adv. Dairy Res. 4, 151.

Feedipedia, F., 2025. List of Feeds | Feedipedia. Available from: https://www.feedipedia.org/.

Fenwick, G.R., Spinks, E.A., Wilkinson, A.P., Heaney, R.K., Legoy, M.A., 1986. Effect of processing on the antinutrient content of rapeseed. J. Sci. Food Agric. 37 (8), 735–741. https://doi.org/10.1002/jsfa.2740370805.

Fundo, J.F., Deuchande, T., Rodrigues, D.A., Pimentel, L.L., Vidigal, S.S.M.P., Rodríguez-Alcalá, L.M., Pintado, M.E., Amaro, A.L., 2023. Induced autolysis of engineered yeast residue as a means to simplify downstream processing for valorization—a case study. Fermentation 9 (7). https://doi.org/10.3390/fermentation9070673.

Gaillard, D., Stratford, J.M., 2016. Measurement of behavioral taste responses in mice: two-bottle preference, lickometer, and conditioned taste-aversion tests. Curr. Protocols Mouse Biol. 6 (4), 380–407. https://doi.org/10.1002/cpmo.18.

Gebremedhn, B., Niguse, M., Hagos, B., Tesfamariam, T., Kidane, T., Berhe, A., Gebresilassie, L., Gebreegziabher, L., Gebremariam, T., Gebremeskel, Y., 2019. Effects of dietary brewery spent grain inclusion on egg laying performance and quality parameters of Bovans Brown chickens. Science 21 (2). https://doi.org/10.1590/1806-9061-2018-0765.

Ghisellini, P., Cialani, C., Ulgiati, S., 2016. A review on circular economy: the expected transition to a balanced interplay of environmental and economic systems. J. Clean. Prod. 114, 11–32.

Giannenas, I., Sidiropoulou, E., Bonos, E., Christaki, E., Florou-Paneri, P., 2020. The History of Herbs, Medicinal and Aromatic Plants, and Their Extracts. Elsevier BV, pp. 1–18. https://doi.org/10.1016/b978-0-12-814700-9.00001-7.

Gul, K., Yousuf, B., Singh, A.K., Singh, P., Wani, A.A., 2015. Rice bran: nutritional values and its emerging potential for development of functional food - a review. Bioact. Carbohydr. Dietary Fibre 6 (1), 24–30. https://doi.org/10.1016/j.bcdf.2015.06.002.

Gupta, R.K., Gangoliya, S.S., Singh, N.K., 2015. Reduction of phytic acid and enhancement of bioavailable micronutrients in food grains. J. Food Sci. Technol. 52 (2), 676–684.

Hamidoghli, A., Lee, Y., Hwang, S., Choi, W., Choi, Y.H., Bai, S.C., 2023. Evaluation of yeast hydrolysate in a low-fishmeal diet for Whiteleg shrimp (Litopenaeus vannamei). Animals 13 (11). https://doi.org/10.3390/ani13111877. Available from: http://www.mdpi.com/journal/animals.

Hansen, J.Ø., Øverland, M., Skrede, A., Anderson, D.M., Collins, S.A., 2020. A meta-analysis of the effects of dietary canola/double low rapeseed meal on growth performance of weanling and growing-finishing pigs. Anim. Feed Sci. Technol. 259. https://doi.org/10.1016/j.anifeedsci.2019.114302.

Hasan, M.S., Feugang, J.M., Liao, S.F., 2019. A nutrigenomics approach using RNA sequencing technology to study nutrient–gene interactions in agricultural animals. Curr. Dev. Nutr. 3 (8), nzz082.

Henchion, M., Moloney, A.P., Hyland, J., Zimmermann, J., McCarthy, S., 2021. Review: Trends for meat, milk and egg consumption for the next decades and the role played by livestock systems in the global production of proteins. Animal 15, 100287. https://doi.org/10.1016/j.animal.2021.100287.

Hoover, A.N., Tumuluru, J.S., Teymouri, F., Moore, J., Gresham, G., 2014. Effect of pelleting process variables on physical properties and sugar yields of ammonia fiber expansion pretreated corn stover. Bioresour. Technol. 164, 128–135. https://doi.org/10.1016/j.biortech.2014.02.005.

Huang, B., Huang, C., Lyu, Z., Chen, Y., Li, P., Liu, L., Lai, C., 2018. Available energy and amino acid digestibility of defatted rice bran fed to growing pigs. J. Anim. Sci. 96 (8), 3138–3150. https://doi.org/10.1093/jas/sky191.

Huang, B., Shi, H., Wang, L., Wang, L., Lyu, Z., Hu, Q., Zang, J., Li, D., Lai, C., 2021. Effects of defatted rice bran inclusion level on nutrient digestibility and growth performance of different body weight pigs. Animals 11 (5), 1374. https://doi.org/10.3390/ani11051374.

Huang, W., Liu, B., Shi, D., Cheng, A., Chen, G., Liu, F., Dong, J., Lan, J., Hong, B., Zhang, S., Ren, C., 2024. Research progress on the quality, extraction technology, food application, and physiological function of rice bran oil. Foods 13 (20), 3262. https://doi.org/10.3390/foods13203262.

Irakli, M., Lazaridou, A., Biliaderis, C.G., 2020. Comparative evaluation of the nutritional, antinutritional, functional, and bioactivity attributes of rice bran stabilized by different heat treatments. Foods 10 (1), 57.

Ivanović, N., Ilić, T., Zrnic-Ćirić, M., Todorović, V., Đuričić, I., Dabetić, N., 2023. Agri-food by-products as a source of sustainable ingredients for the production of functional foods and nutraceuticals. Arhiv za farmaciju 73 (3), 190–204. https://doi.org/10.5937/arhfarm73-44481.

Jaeger, A., Arendt, E.K., Zannini, E., Sahin, A.W., 2020. Brewer's spent yeast (BSY), an underutilized brewing by-product. Fermentation 6 (4). https://doi.org/10.3390/fermentation6040123.

Jahan, D.A., Hussain, L., Islam, M.A., Khan, M., 2013. Comparative study of mustard oil cake and soybean meal based artificial diet for Rohu, Labeo rohita (Ham.) Fingerlings. The Agriculturists 11 (1), 61–66. https://doi.org/10.3329/agric.v11i1.15244.

Jha, R., Berrocoso, J.F.D., 2016. Dietary fiber and protein fermentation in the intestine of swine and their interactive effects on gut health and on the environment: a review. Anim. Feed Sci. Technol. 212, 18–26. https://doi.org/10.1016/j.anifeedsci.2015.12.002.

Jha, R., Fouhse, J.M., Tiwari, U.P., Li, L., Willing, B.P., 2019. Dietary fiber and intestinal health of monogastric animals. Front. Vet. Sci. 6. https://doi.org/10.3389/fvets.2019.00048.

Joudaki, H., Aria, N., Moravej, R., Rezaei Yazdi, M., Emami-Karvani, Z., Hamblin, M.R., 2023. Microbial phytases: properties and applications in the food industry. Curr. Microbiol. 80 (12). https://doi.org/10.1007/s00284-023-03471-1.

Jung, B., Batal, A.B., 2012. Effect of dietary nucleotide supplementation on performance and development of the gastrointestinal tract of broilers. Br. Poult. Sci. 53 (1), 98–105. https://doi.org/10.1080/00071668.2012.659654.

Jurgilevich, A., Birge, T., Kentala-Lehtonen, J., Korhonen-Kurki, K., Pietikäinen, J., Saikku, L., Hanna, S., 2016. Transition towards circular economy in the food system. Sustainability 8 (1), 69. https://doi.org/10.3390/su8010069.

Kamboh, A.A., 2017. Review on preservation and utilization of wet brewery spent grain as concentrate replacement feed for lactating dairy cows. J. Anim. Health Prod. 5 (1), 10–13.

Kang, H.K., Kim, C.H., 2016. Effects of dietary supplementation with rice bran oil on the growth performance, blood parameters, and immune response of broiler chickens. J. Anim. Sci. Technol. 58. https://doi.org/10.1186/s40781-016-0092-6.

Karolina, W.A., Justyna, N., 2025. Rapeseed meal as an alternative protein source in fish feed and its impact on growth parameters, digestive tract, and gut microbiota. Animals 15 (9). https://doi.org/10.3390/ani15091264.

Kaur, S., Sharma, S., Singh, B., Dar, B.N., 2015. Effect of extrusion variables (temperature, moisture) on the antinutrient components of cereal brans. J. Food Sci. Technol. 52 (3), 1670–1676. https://doi.org/10.1007/s13197-013-1118-4.

Keller, L.A.M., Pereyra, C.M., Cavaglieri, L.R., Dalcero, A.M., Rosa, C.A.R., 2012. Fungi and mycotoxins from pre-and poststorage brewer' s grain intended for bovine intensive rearing. Int. Sch. Res. Not. 2012 (1), 396590.

Khajali, F., Slominski, B.A., 2012. Factors that affect the nutritive value of canola meal for poultry. Poult. Sci. 91 (10), 2564–2575. https://doi.org/10.3382/ps.2012-02332.

Kieliszek, M., Błażejak, S., Gientka, I., Bzducha-Wróbel, A., 2015. Accumulation and metabolism of selenium by yeast cells. Appl. Microbiol. Biotechnol. 99 (13), 5373–5382. https://doi.org/10.1007/s00253-015-6650-x.

Kloss, P., Jeffery, E., Wallig, M., Tumbleson, M., Parsons, C., Johnson, L., Reuber, M., 1994. Efficacy of feeding glucosinolate-extracted Crambe meal to broiler chicks. Poult. Sci. 73 (10), 1542–1551. https://doi.org/10.3382/ps.0731542.

Kopacz, M., Alicja Drażbo, A., Śmiecińska, K., Ognik, K., 2021. Performance and egg quality of laying hens fed diets containing raw, hydrobarothermally-treated and fermented rapeseed cake. Animals 11 (11), 3083. https://doi.org/10.3390/ani11113083.

Kramer, J.K., Sauer, F.D., Pigden, W.J. (Eds.), 1983. High and Low Erucic Acid Rapeseed Oils: Production, Usage, Chemistry, and Toxicological Evaluation. Academic Press.

Kumar, S., Chauhan, J.S., Kumar, A., 2010. Screening for erucic acid and glucosinolate content in rapeseed-mustard seeds using near infrared reflectance spectroscopy. J. Food Sci. Technol. 47 (6), 690–692.

Li, Z., Li, Y., Lv, Z., Liu, H., Zhao, J., Noblet, J., Wang, F., Lai, C., Li, D., 2017. Net energy of corn, soybean meal and rapeseed meal in growing pigs. J. Anim. Sci. Biotechnol. 8 (1). https://doi.org/10.1186/s40104-017-0169-1.

Li, S., Kallas, Z., Ornelas Herrera, S.I., Mustapa, M.A.C., Behrendt, L., Hajdu, Z., Michels, E., Meers, E., 2025. Market potential for circular food in Europe based on consumer demand and willingness to pay. Npj Sci. Food 9 (1). https://doi.org/10.1038/s41538-025-00475-y.

Lietzow, J., 2021. Biologically active compounds in mustard seeds: a Toxicological perspective. Foods 10 (9), 2089. https://doi.org/10.3390/foods10092089.

Lin, L., Allemekinders, H., Dansby, A., Campbell, L., Durance-Tod, S., Berger, A., Jones, P.J., 2013. Evidence of health benefits of canola oil. Nutr. Rev. 71 (6), 370–385. https://doi.org/10.1111/nure.12033.

Liu, Y.G., Zhou, M.Q., Liu, M.L., 1994. A survey of nutrients and toxic factors in commercial rapeseed meal in China and evaluation of detoxification by water extraction. Anim. Feed Sci. Technol. 45 (3–4), 257–270.

Lowry, V.K., Farnell, M.B., Ferro, P.J., Swaggerty, C.L., Bahl, A., Kogut, M.H., 2005. Purified β-glucan as an abiotic feed additive up-regulates the innate immune response in immature chickens against Salmonella enterica serovar Enteritidis. Int. J. Food Microbiol. 98 (3), 309–318. https://doi.org/10.1016/j.ijfoodmicro.2004.06.008.

Lynch, K.M., Steffen, E.J., Arendt, E.K., 2016. Brewers' spent grain: a review with an emphasis on food and health. J. Inst. Brew. 122 (4), 553–568.

Lynch, K.M., Strain, C.R., Johnson, C., Patangia, D., Stanton, C., Koc, F., Gil-Martinez, J., O'Riordan, P., Sahin, A.W., Ross, R.P., Arendt, E.K., 2021. Extraction and characterisation of arabinoxylan from brewers spent grain and investigation of microbiome modulation potential. Eur. J. Nutr. 60 (8), 4393–4411. https://doi.org/10.1007/s00394-021-02570-8.

Marcus, A., Fox, G., 2021. Fungal biovalorization of a brewing industry byproduct, brewer's spent grain: a review. Foods 10 (9), 2159. https://doi.org/10.3390/foods10092159.

Marson, G.V., de Castro, R.J.S., Belleville, M.P., Hubinger, M.D., 2020. Spent brewer's yeast as a source of high added value molecules: a systematic review on its characteristics, processing and potential applications. World J. Microbiol. Biotechnol. 36 (7). https://doi.org/10.1007/s11274-020-02866-7.

Martineau, R.D.R.O., Ouellet, D.R., Lapierre, H., 2013. Feeding canola meal to dairy cows: a meta-analysis on lactational responses. J. Dairy Sci. 96 (3), 1701–1714.

Mendieta, C.R., Gomez Verduzco, G., Del Río, J.C.G., Cuevas, A.C., Arce, J.M., Ávila, E.G., 2018. Effect of the addition of Saccharomyces Cerevisiae yeast cell walls to diets with mycotoxins on the performance and immune responses of broilers. J. Poul. Sci. 55 (1), 38–46. https://doi.org/10.2141/jpsa.0170019.

Meza, S.L.R., Cañizares, L., Dannenberg, B., Peres, B.B., Rodrigues, L.A., Mardade, C., de Leon, M.A., Gaioso, C.A., Egea, I., de Oliveira, M., 2024. Sustainable rice bran protein: composition, extraction, quality properties and applications. Trends Food Sci. Technol. 145, 104355.

Mirabella, N., Castellani, V., Sala, S., 2014. Current options for the valorization of food manufacturing waste: a review. J. Clean. Prod. 65, 28–41.

Mueller-Harvey, I., 2006. Unravelling the conundrum of tannins in animal nutrition and health. J. Sci. Food Agric. 86 (13), 2010–2037. https://doi.org/10.1002/jsfa.2577.

Mukasafari, M.A., Ambula, M.K., Karege, C., King'ori, A.M., 2018. Effects of substituting sow and weaner meal with brewers' spent grains on the performance of growing pigs in Rwanda. Trop. Anim. Health Prod. 50 (2), 393–398. https://doi.org/10.1007/s11250-017-1446-x.

Mussatto, S.I., Dragone, G., Roberto, I.C., 2006. Brewers' spent grain: generation, characteristics and potential applications. J. Cereal. Sci. 43 (1), 1–14.

Naibaho, J., Korzeniowska, M., 2021. The variability of physico-chemical properties of brewery spent grain from 8 different breweries. Heliyon 7 (3), e06583. https://doi.org/10.1016/j.heliyon.2021.e06583.

Nair, J., Penner, G.B., Yu, P., Lardner, H.A., McAllister, T., Damiran, D., McKinnon, J.J., 2015. Evaluation of canola meal derived from Brassica juncea and Brassica napus seed as an energy source for feedlot steers. Can. J. Anim. Sci. 95 (4), 599–607. https://doi.org/10.4141/CJAS-2015-055.

National Research Council (NRC), 2016. Nutrient Requirements of Beef Cattle: Eighth Revised Edition. The National Academies Press, Washington, DC, p. 463.

Nidhishree, A.S., Menezes, R.A., Venkatachalam, H., Bhat, K.S., 2024. Rice bran as a sustainable source for value added materials: an overview. Discov. Mater. 4 (1). https://doi.org/10.1007/s43939-024-00159-6.

Nyhan, L., Sahin, A.W., Schmitz, H.H., Siegel, J.B., Arendt, E.K., 2023. Brewers' spent grain: an unprecedented opportunity to develop sustainable plant-based nutrition ingredients addressing global malnutrition challenges. J. Agric. Food Chem. 71 (28), 10543–10564. https://doi.org/10.1021/acs.jafc.3c02489.

Oancea, A.-G., Dragomir, C., Untea, A., Saracila, M., Turcu, R., Cismileanu, A., Boldea, I., Radu, G.L., 2023. The effects of brewer's spent yeast (BSY) inclusion in dairy sheep's diets on ruminal fermentation and milk quality parameters. Agriculture 13 (8), 1605. https://doi.org/10.3390/agriculture13081605.

Okunade, O.A., Ghawi, S.K., Methven, L., Niranjan, K., 2015. Thermal and pressure stability of myrosinase enzymes from black mustard (Brassica nigra L. W.D.J. Koch. var. nigra), brown mustard (Brassica juncea L. Czern. var. juncea) and yellow mustard (Sinapsis alba L. subsp. maire) seeds. Food Chem. 187, 485–490. https://doi.org/10.1016/j.foodchem.2015.04.054.

Oliveira, M.S., Wiltafsky-Martin, M.K., Stein, H.H., 2020. Excessive heating of 00-rapeseed meal reduces not only amino acid digestibility but also metabolizable energy when fed to growing pigs. Journal Animal Sci. 98 (7), skaa219.

Osei, E.D., Naik, A.S., Kinsella, G., Delaney, T., Kirwan, S., 2025. Potential of Brewer's spent grain bioactive fractions as functional ingredients for companion and farm animal foods – a review. Appl. Food Res. 5 (2). https://doi.org/10.1016/j.afres.2025.101138.

Östbring, K., Malmqvist, E., Nilsson, K., Rosenlind, I., Rayner, M., 2019. The effects of oil extraction methods on recovery yield and emulsifying properties of proteins from rapeseed meal and press cake. Foods 9 (1), 19.

Papas, A., Ingalls, J.R., Campbell, L.D., 1979. Studies on the effects of rapeseed meal on thyroid status of cattle, glucosinolate and iodine content of milk and other parameters. J. Nutrition 109 (7), 1129–1139.

Park, J.H., Sureshkumar, S., Kim, I.H., 2020. Egg production, egg quality, nutrient digestibility, and excreta microflora of laying hens fed with a diet containing brewer's yeast hydrolysate. J. Appl. Anim. Res. 48 (1), 492–498.

Paula, E.M., Broderick, G.A., Danes, M.A.C., Lobos, N.E., Zanton, G.I., Faciola, A.P., 2018. Effects of replacing soybean meal with canola meal or treated canola meal on ruminal digestion, omasal nutrient flow, and performance in lactating dairy cows. J. Dairy Sci. 101 (1), 328–339. https://doi.org/10.3168/jds.2017-13392.

Payvastegan, S., Farhoomand, P., Daneshyar, M., Ghaffari, M., 2017. Evaluation of different levels of canola meal on performance, organ weights, Hepatic Deiodinase gene expression and thyroid morphology in broiler chickens. J. Poultry Sci. 54 (4), 282–291. https://doi.org/10.2141/jpsa.0160147.

Pelusio, N.F., Parma, L., Volpe, E., Ciulli, S., Errani, F., Natale, S., De Cesare, A., Indio, V., Carcano, P., Mordenti, O., Gatta, P.P., Bonaldo, A., 2023. Yeast-extracted nucleotides and nucleic acids as promising feed additives for European sea bass (Dicentrarchus labrax) juveniles. Front. Mar. Sci. 10. https://doi.org/10.3389/fmars.2023.1145660.

Pesti, G.M., Choct, M., 2023. The future of feed formulation for poultry: toward more sustainable production of meat and eggs. Animal Nutr. 15, 71–87. https://doi.org/10.1016/j.aninu.2023.02.013.

Pfister, S., Boulay, A.M., Berger, M., Hadjikakou, M., Motoshita, M., Hess, T., Ridoutt, B., Weinzettel, J., Scherer, L., Döll, P., Manzardo, A., Núñez, M., Verones, F., Humbert, S., Buxmann, K., Harding, K., Benini, L., Oki, T., Finkbeiner, M., Henderson, A., 2017. Understanding the LCA and ISO water footprint: a response to Hoekstra (2016) "A critique on the water-scarcity weighted water footprint in LCA". Ecol. Indic. 72, 352–359. https://doi.org/10.1016/j.ecolind.2016.07.051.

Pires Filho, I.C., Broch, J., Eyng, C., Silva, I.M., Souza, C., Avila, A.S., Castilha, L.D., Cirilo, E.H., Tesser, G.L.S., Nunes, R.V., 2021. Effects of feeding dried brewers grains to slow-growing broiler chickens. Livest. Sci. 250, 104561. https://doi.org/10.1016/j.livsci.2021.104561.

Podpora, B., Świderski, F., Sadowska, A., Rakowska, R., Wasiak-Zys, G., 2016. Spent brewer's yeast extracts as a new component of functional food. Czech J. Food Sci. 34 (6), 554–563. https://doi.org/10.17221/419/2015-CJFS.

Poore, J., Nemecek, T., 2018. Reducing food's environmental impacts through producers and consumers. Science 360 (6392), 987–992. https://doi.org/10.1126/science.aaq0216.

Puligundla, P., Mok, C., 2021. Recent advances in biotechnological valorization of brewers' spent grain. Food Sci. Biotechnol. 30 (3), 341–353. https://doi.org/10.1007/s10068-021-00900-4.

Rahman, M.A.U., Rehman, A., Chuanqi, X., Long, Z.X., Binghai, C., Linbao, J., Huawei, S., 2015. Extrusion of feed/feed ingredients and its effect on digestibility and performance of poultry: a review. Int. J. Curr. Microbiol. Appl. Sci. 4 (4), 48–61.

Rama Rao, S.V., Raju, M.V.L.N., Panda, A.K., Sailaja Shashibindu, M., 2005. Utilization of low Glucosinalate and conventional mustard oilseed cakes in commercial broiler chicken diets. Asian-Australas. J. Anim. Sci. 18 (8), 1157–1163. https://doi.org/10.5713/ajas.2005.1157.

Ranjan, A., Sahu, N.P., Deo, A.D., Kumar, S., 2019. Solid state fermentation of de-oiled rice bran: effect on in vitro protein digestibility, fatty acid profile and anti-nutritional factors. Food Res. Int. 119, 1–5. https://doi.org/10.1016/j.foodres.2019.01.054.

Ranjan, A., Kumar, S., Sahu, N.P., Jain, K.K., Deo, A.D., 2022. Strategies for maximizing utilization of de-oiled rice bran (DORB) in the fish feed. Aquac. Int. 30 (1), 99–114. https://doi.org/10.1007/s10499-021-00791-6.

Ranjan, D., Verma, P., Singh, M.B., 2023. Ways to improve the utilization of de-oiled rice bran (DORB) in fish feed. Int. J. Pure Appl. Zool. 11 (4), 181.

Regulation (EC), 2003. No 1831/2003 of the European Parliament and of the Council of 22 September 2003 on additives for use in animal nutrition. Off. J. Eur. Union L 268, 29–43.

Roine, P., Uksila, E., Teir, H., Rapola, J., 1960. Histopathological changes in rats and pigs fed rapeseed oil. Z. Ernahr. 1 (2), 118–124. https://doi.org/10.1007/bf02021352.

Salemdeeb, R., zu Ermgassen, E.K.H.J., Kim, M.H., Balmford, A., Al-Tabbaa, A., 2017. Environmental and health impacts of using food waste as animal feed: a comparative analysis of food waste management options. J. Clean. Prod. 140, 871–880. https://doi.org/10.1016/j.jclepro.2016.05.049.

Salomone, R., Saija, G., Mondello, G., Giannetto, A., Fasulo, S., Savastano, D., 2017. Environmental impact of food waste bioconversion by insects: application of Life Cycle Assessment to process using Hermetia illucens. J. Clean. Prod. 140, 890–905. https://doi.org/10.1016/j.jclepro.2016.06.154.

Sampath, V., Sureshkumar, S., Kim, I.H., 2023. The efficacy of yeast supplementation on monogastric animal performance—a short review. Life 13 (10), 2037.

Santos, T., 2011. Optimisation of Phytase Production by Aspergillus Niger Using Solid State Fermentation, p. 127. Available from: https://mural.maynoothuniversity.ie/id/eprint/4078/1/Optimisation_of_Aspergillus_niger_Masters.pdf?.

Sarker, A.K., Saha, D., Begum, H., Zaman, A., Rahman, M.M., 2015. Comparison of cake compositions, pepsin digestibility and amino acids concentration of proteins isolated from black mustard and yellow mustard cakes. AMB Exp. 5 (1). https://doi.org/10.1186/s13568-015-0110-y. Available from: http://www.amb-express.com/.

Schwarzinger, B., Feichtinger, M., Blank-Landeshammer, B., Weghuber, J., Schwarzinger, C., 2022. Quick determination of erucic acid in mustard oils and seeds. J. Anal. Appl. Pyrolysis 164, 105523. https://doi.org/10.1016/j.jaap.2022.105523.

Selle, P.H., Ravindran, V., Partridge, G.G., 2009. Beneficial effects of xylanase and/or phytase inclusions on ileal amino acid digestibility, energy utilisation, mineral retention and growth performance in wheat-based broiler diets. Anim. Feed Sci. Technol. 153 (3–4), 303–313.

Sharif, M.K., Butt, M.S., Anjum, F.M., Khan, S.H., 2014. Rice bran: a novel functional ingredient. Crit. Rev. Food Sci. Nutr. 54 (6), 807–816. https://doi.org/10.1080/10408398.2011.608586.

Sharma, N., 2013. Chemical composition of oilseed cakes and deoiled cakes in Nepal. Online J. Anim. Feed Res. 3 (1), 74–76.

Sharma, R., Srivastava, T., Saxena, D., 2018. Physico-chemical and functional properties of deoiled rice bran and its utilization in the development of extruded product. Pharma Innov. 7 (5, Part B), 109.

Shi, C., He, J., Yu, J., Yu, B., Huang, Z., Mao, X., Zheng, P., Chen, D., 2015. Solid state fermentation of rapeseed cake with Aspergillus Niger for degrading glucosinolates and upgrading nutritional value. J. Anim. Sci. Biotechnol. 6 (1). https://doi.org/10.1186/s40104-015-0015-2.

Shibutami, E., Takebayashi, T., 2021. A scoping review of the application of Metabolomics in nutrition research: the literature survey 2000–2019. Nutrients 13 (11), 3760. https://doi.org/10.3390/nu13113760.

Shurson, G.C., 2020. What a waste"—can we improve sustainability of food animal production systems by recycling food waste streams into animal feed in an era of health, climate, and economic crises? Sustainability 12 (17), 7071.

Singh, A.S., Jain, V.K., Singh, P., Pathak, N.N., 2000. Effect of feeding wheat bran and deoiled rice bran on feed intake and nutrient utilization in crossbred cows. Indian J. Anim. Sci. 70 (12), 1258–1260.

Singh, S., Singh, R.P., Singh, H.K., Kumar, K., 2013. Oil and cake quality characteristics in new varieties/strains of Indian mustard [Brassica juncea (L.) Czern & Coss.]. Plant Arch. 13 (2), 627–632.

Song, M., Tian, K., Ren, C., Liu, Y., Ye, X., Wang, Y., Xie, J., Zhao, F., 2025. Impact of chemical composition on metabolizable energy and its prediction models in brewer's spent grains for broilers at different ages. Poult. Sci. 104 (8), 105323. https://doi.org/10.1016/j.psj.2025.105323.

Soren, S., Mandal, G.P., Mondal, S., Pradhan, S., Mukherjee, J., Banerjee, D., Pakhira, M.C., Amla, Mondal, A., Nsereko, V., Samanta, I., 2024. Efficacy of Saccharomyces cerevisiae fermentation product and probiotic supplementation on growth performance, gut microflora and immunity of broiler chickens. Animals 14 (6). https://doi.org/10.3390/ani14060866.

Srivastava, K., Bhatt, A., Singh, N., Khare, R., Shukla, R., Chaturvedi, D., Kant, R., 2020. Synthesis of isothiocyanates: a review. Chem. Biol. Interface 10 (2), 34–50.

Stahn, T., Storandt, R., Grebenteuch, S., Rohn, S., May, D., Dolsdorf, C., Pleissner, D., 2023. Utilization of brewer's spent grains and agricultural residues in pig feed formation. Sustainability 15 (18), 13774. https://doi.org/10.3390/su151813774.

Stefansson, B.R., 1983. The Development of Improved Rapeseed Cultivars. Elsevier BV, pp. 143–159. https://doi.org/10.1016/b978-0-12-425080-2.50011-6.

Stefansson, B.R., Hougen, F.W., Downey, R.K., 1961. Note on the isolation of rape plants with seed oil free from erucic acid. Can. J. Plant Sci. 41 (1), 218–219.

Steinfeld, H., Gerber, P., Wassenaar, T., Castel, V., Rosales, M., de Haan, C., 2006. Livestock's long shadow: environmental issues and options. Renew. Resour. J. 24 (4), 15–17.

Stier, H., Ebbeskotte, V., Gruenwald, J., 2014. Immune-modulatory effects of dietary yeast Beta-1, 3/1, 6-D-glucan. Nutr. J. 13 (1), 38.

Stöggl, W., Huck, C., Wongyai, S., Scherz, H., Bonn, G., 2005. Simultaneous determination of carotenoids, tocopherols, and γ-oryzanol in crude rice bran oil by liquid chromatography coupled to diode array and mass spectrometric detection employing silica C30 stationary phases. J. Separ. Sci. 28 (14), 1712–1718. https://doi.org/10.1002/jssc.200500176.

Sun, X., Dou, Z., Shurson, G.C., Hu, B., 2024. Bioprocessing to upcycle agro-industrial and food wastes into high-nutritional value animal feed for sustainable food and agriculture systems. Resour. Conserv. Recycl. 201. https://doi.org/10.1016/j.resconrec.2023.107325.

Suphantharika, M., Khunrae, P., Thanardkit, P., Verduyn, C., 2003. Preparation of spent brewer's yeast β-glucans with a potential application as an immunostimulant for black tiger shrimp, Penaeus monodon. Bioresour. Technol. 88 (1), 55–60. https://doi.org/10.1016/S0960-8524(02)00257-2.

Swanepoel, N., 2016. Effects of Rumen Protected Amino Acid Supplementation on Performance of Holstein Cows Fed Rations Containing High Levels of Canola Meal, p. 171.

Tan, B.L., Norhaizan, M.E., Chan, L.C., 2023. Rice bran: from waste to nutritious food ingredients. Nutrients 15 (11), 2503.

Tang, H., Feng, G., Zhao, J., Ouyang, Q., Liu, X., Jiang, X., Deng, M., Xie, Z., Chen, F., Zhou, X., Li, R., Yin, Y., 2024. Determination and prediction of amino acid digestibility in rapeseed cake for growing-finishing pigs. Animals 14 (19), 2764. https://doi.org/10.3390/ani14192764.

Tao, Z., Yuan, H., Liu, M., Liu, Q., Zhang, S., Liu, H., Jiang, Y., Huang, D., Wang, T., 2022. Yeast extract: characteristics, production, applications and future perspectives. J. Microbiol. Biotechnol. 33 (2), 151.

Tassone, S., Fortina, R., Peiretti, P.G., 2020. In vitro techniques using the DaisyII incubator for the assessment of digestibility: a review. Animals 10 (5), 775.

Terefe, G., 2022. Preservation techniques and their effect on nutritional values and microbial population of brewer's spent grain: a review. CABI Agric. Biosci. 3 (1). https://doi.org/10.1186/s43170-022-00120-8.

The Future of Food and Agriculture – Alternative Pathways to 2050, 2018. Available from: https://openknowledge.fao.org/server/api/core/bitstreams/2c6bd7b4-181e-4117-a90d-32a1bda8b27c/content.

Tilley, J.M.A., Terry, R.A., 1963. A two-stage technique for the in vitro digestion of forage crops. Grass Forage Sci. 18 (2), 104–111. https://doi.org/10.1111/j.1365-2494.1963.tb00335.x.

Tyagi, A.K., 2002. Influence of water soaking of mustard cake on glucosinolate hydrolysis. Anim. Feed Sci. Technol. 99 (1–4), 215–219. https://doi.org/10.1016/S0377-8401(02)00142-6.

USDA, Production Rapeseed. Top Producing Countries. https://www.fas.usda.gov/data/production/commodity/2226000.

Usman, I., Ali, I., Arshad, M.U., Saeed, F., Afzaal, M., Sana, S., Islam, F., Siddiqua, A., Naeem, U., Islam, S., 2023. Valorization of mustard and sesame oilseed cakes for food application through eco-innovative technologies. Food Sci. Nutr. 11 (4), 1818–1825. https://doi.org/10.1002/fsn3.3214.

Vashishth, A., Ram, S., Beniwal, V., 2017. Cereal phytases and their importance in improvement of micronutrients bioavailability. 3 Biotech 7 (1), 42.

Velayudhan, D.E., Nyachoti, C.M., 2017. Effect of increasing dietary canola meal inclusion on lactation performance, milk composition, and nutrient digestibility of lactating sows1. J. Anim. Sci. 95 (7), 3129–3135. https://doi.org/10.2527/jas.2016.1191.

Vieira, E., Brandão, T., Ferreira, I.M., 2013. Evaluation of brewer's spent yeast to produce flavor enhancer nucleotides: influence of serial repitching. J. Agric. Food Chem. 61 (37), 8724–8729.

Wan, Y., Ma, R., Khalid, A., Chai, L., Qi, R., Liu, W., Li, J., Li, Y., Zhan, K., 2021. Effect of the pellet and mash feed forms on the productive performance, egg quality, nutrient metabolism, and intestinal morphology of two laying hen breeds. Animals 11 (3), 701. https://doi.org/10.3390/ani11030701.

Wittkop, B., Snowdon, R.J., Friedt, W., 2012. New NIRS calibrations for fiber fractions reveal broad genetic variation in Brassica napus seed quality. J. Agric. Food Chem. 60 (9), 2248–2256.

Xiong, Y., Zhang, P., Warner, R.D., Fang, Z., 2021. Cereal grain-based functional beverages: from cereal grain bioactive phytochemicals to beverage processing technologies, health benefits and product features. Crit. Rev. Food Sci. Nutr. 62 (9), 2404–2431.

Xiong, C., Zou, X., Phan, C.W., Huang, W., Zhu, Y., 2024. Enhancing the potential of rapeseed cake as protein-source food by γ-irradiation. Biosci. Rep. 44 (3). https://doi.org/10.1042/BSR20231807.

Yan, Z., Liu, Z., Zhou, C., Tan, Z., 2025. Anti-nutritional factors of plant protein feeds for ruminants and methods for their elimination. Animals 15 (8), 1107. https://doi.org/10.3390/ani15081107.

Yang, Y., Fan, Y., Cao, Y., Guo, P., Dong, B., Ma, Y., 2016. Effects of exogenous phytase and xylanase, individually or in combination, and pelleting on nutrient digestibility, available energy content of wheat and performance of growing pigs fed wheat-based diets. Asian-Austra. Journal Animal Sci. 30 (1), 57.

Yılmaz Tuncel, N., 2023. Stabilization of rice bran: a review. Foods 12 (9), 1924. https://doi.org/10.3390/foods12091924.

Zeko-Pivač, A., Habschied, K., Kulisic, B., Barkow, I., Tišma, M., 2023. Valorization of spent brewer's yeast for the production of high-value products, materials, and biofuels and environmental application. Fermentation 9 (3), 208. https://doi.org/10.3390/fermentation9030208.

Zhang, J., 2024. Development of Functional Food Ingredients from By-Products of the Brewing Industry. Doctoral dissertation, South East Technological University, p. 302.

Zhou, H.Y., Wu, W.J., Xu, Y.Y., Zhou, B., Niu, K., Liu, Z.Q., Zheng, Y.G., 2020. Calcium carbonate addition improves L-methionine biosynthesis by metabolically engineered Escherichia coli W3110-BL. Front. Bioeng. Biotechnol. 8. https://doi.org/10.3389/fbioe.2020.00300.

CHAPTER

13 Nanocoating of fish feed formulated with agro-industrial wastes and byproducts

Sumaiya Begum Thuraputheen[1], Nooruddin Thajuddin[1], K.P. Kumaraguru Vasagam[2] and Karthikeyan Ramalingam[1]

[1]*School of Life Sciences, B.S Abdur Rahman Crescent Institute of Science and Technology, Chennai, Tamil Nadu, India;* [2]*Department Nutrition, Genetics and Biotechnology Division, Central Institute of Brackishwater Aquaculture (CIBA), Chennai, Tamil Nadu, India*

1. Introduction

Fish and seafood have long been an important part of traditional diets across Asia and the Pacific, providing essential nutrients and supporting human health. Beyond their nutritional value, fisheries and aquaculture also play a crucial role in sustaining livelihoods and local economies, particularly in coastal and rural communities. Over the past few decades, aquaculture production has increased rapidly and now contributes a significant portion of the global fish supply, with Asian countries leading worldwide production. This rapid growth highlights the increasing importance of aquaculture in ensuring food security and economic stability on a global scale (Suzuki, 2021). To meet this nutritional demand, the industry faces significant problems in producing sustainable and economical feed. Conventional fish feeds depend on fish meal and fish oil from wild-caught fish (Chattaraj et al., 2026a). These ingredients provide optimal nutrition for many species, but their high price limits availability, and environmental impact raise questions about long-term viability (Chattaraj et al., 2026b). Feed costs can make up to 75% of total fish farming expenses, highlighting the financial strain of using standard feed components (Ansari et al., 2021). Moreover, the continued dependence on fishmeal and fish oil places significant pressure on wild fish stocks, particularly small pelagic species that are essential components of marine ecosystems and coastal food systems. Because the harvest of these species is unlikely to increase substantially in the future, reliance on fish-derived feed ingredients may ultimately limit the further expansion of aquaculture production. As a result, reducing dependence on conventional feed resources has become a key priority for achieving more sustainable aquaculture systems (Boyd et al., 2022). In larviculture, nutritional enrichment strategies are commonly applied to improve the quality of live feeds. The nutritional composition of rotifers and *Artemia* can be altered through dietary enrichment, which affects their amino acid and

Fish Feed Formulation with Agroindustrial Wastes and Byproducts. https://doi.org/10.1016/B978-0-443-41728-3.00009-5

fatty acid profiles. Adequate levels of free amino acids in live prey are especially important during the early developmental stages of larvae, when their digestive systems are not fully developed, and easily digestible nutrients are required to support proper growth and survival (Aragão et al., 2004).

At the same time, aquaculture and fish processing industries generate large quantities of by-products, including viscera, heads, fins, skins, and trimmings. In many cases, these materials represent more than half of the total fish biomass, suggesting that a substantial portion of valuable nutrients remains underutilised. Improper disposal of these wastes can contribute to environmental pollution while also representing a loss of potentially valuable resources that could otherwise be converted into feed ingredients (Afreen and Ucak, 2020). Converting fish processing by-products into fishmeal or other feed materials can improve resource efficiency, reduce environmental impacts, and lower production costs. However, fishmeal produced from recycled materials often differs in quality from that derived from whole wild-caught fish, frequently containing lower protein levels, higher ash content, and imbalanced amino acid profiles that may negatively affect fish growth performance (Salin et al., 2018).

Similarly, terrestrial animal by-products such as rendered meat meals, fats, and blood meals obtained from livestock and poultry have been explored as alternative protein sources in aquaculture feeds. Although these ingredients contain substantial amounts of protein and lipids, their application is often limited by issues such as nutrient imbalances, reduced palatability, and deficiencies in certain essential nutrients (Salin et al., 2018). Plant-based protein sources, including soybean meal, cottonseed meal, rapeseed meal, and corn gluten meal, have also been investigated as potential substitutes for fishmeal. However, many plant-derived ingredients contain anti-nutritional factors such as phytates, tannins, protease inhibitors, and glucosinolates, which can reduce nutrient digestibility and bioavailability and may negatively affect fish physiology. In addition, plant proteins are often deficient in essential amino acids such as lysine, methionine, and tryptophan, leading to imbalanced nutrition that may impair growth performance, feed utilization, and gut health (Hussain et al., 2024). Given these challenges, the search for alternative and sustainable feed resources has become increasingly important. Agricultural waste, commonly referred to as agro-waste, includes residues generated during the cultivation and processing of agricultural products such as fruits, vegetables, crops, livestock products, and food industry by-products. Examples include crop residues like corn stalks and sugarcane bagasse, as well as fruit and vegetable processing wastes and animal-derived residues (Barati, 2023). Valorizing these materials into functional feed ingredients could improve resource utilization, reduce environmental pollution, and contribute to more sustainable aquaculture production (Chattaraj et al., 2023, 2024a). However, the nutritional composition of such recycled materials can vary considerably, and appropriate processing and formulation strategies are required to ensure consistent feed quality and optimal growth performance in cultured fish (Afreen and Ucak, 2020; Salin et al., 2018). Nano-encapsulated bioactive compounds can protect sensitive nutrients from degradation in the gastrointestinal tract, deliver them more efficiently to target tissues, and reduce the total feed input required to achieve optimal growth. Additionally, these methods can mitigate reliance on wild-caught fish resources, contributing to both environmental sustainability and economic efficiency cross check (Misra et al., 2023). Despite these advances, several gaps remain in translating nanomaterial-based feed solutions into practice. Understanding nano-bio interactions within the fish gastrointestinal tract is limited, and potential long-term effects on fish health, metabolism, and product safety require careful investigation cross check (Khan et al., 2024).

Despite these advances, several gaps remain in turning the nanomaterial-based feed solutions. Understanding of nano bio interactions within the fish gastrointestinal tract is limited, and potential long term effects on fish health, metabolism, and product safety require careful investigation. Scalable production methods and cost-effective incorporation of agro waste-derived nanomaterials into commercial feed formulations also present technological and regulatory challenges. Addressing these knowledge gaps will be critical to realizing the full potential of nanotechnology in sustainable aquaculture feed development. By combining nutritional, environmental, and technological considerations, valorizing of agro waste into enhanced feed provide a transformative approach, by improving aquaculture productivity, feed efficiency, and resource sustainability. Future research should focus on optimizing nanoformulation methods, evaluating species specific responses, ensuring safety and regulatory compliance to enable broad scale adoption in the aquaculture industry.

2. Agro waste as a potential nutrient source

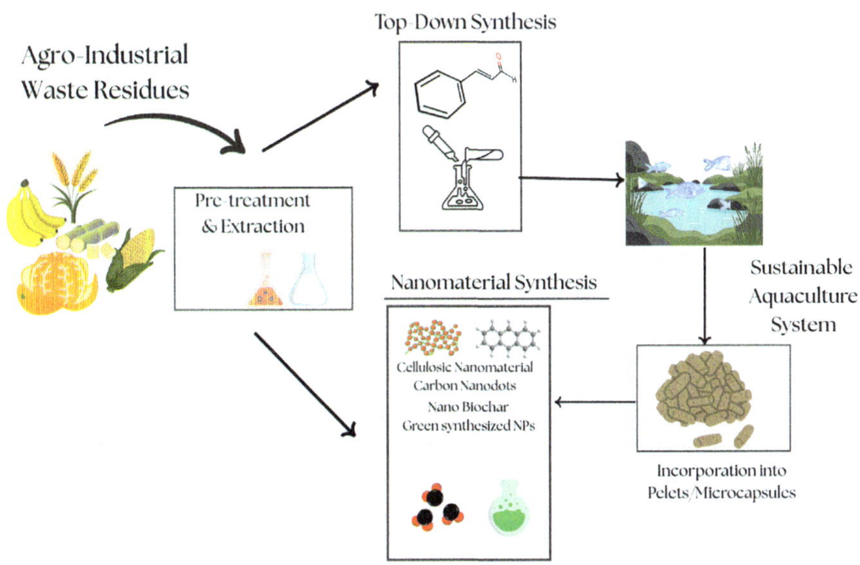

Agro-industrial residues like stems, leaves, husks, and fruit pomace are produced in huge amounts globally. Despite their frequent disposal via burning or landfilling, these materials contain carbohydrates, proteins, dietary fibers, polyphenols, vitamins, minerals, and other bioactive compounds. The nutritional and bioactive composition of such residues offers a promising option for valorization, which can be further used as functional feed ingredients in aquaculture (Bala et al., 2023). Lignocellulosic agricultural residues constitute one of the most abundant renewable biomass resources globally and are primarily composed of cellulose, hemicellulose, and lignin (Chattaraj et al., 2024b). This composition makes them rich in carbon and suitable for conversion into various value-added product (Vélez-mercado et al., 2021). Through biochemical and thermochemical processing methods, these residues can be transformed into biofuels, biochemicals, and other industrial products, thereby reducing reliance on fossil resources and promoting more sustainable production systems (Dutta et al., 2023). Beyond the industrial applications, tapping into agro-waste as a feedstock aligns with circular bioeconomy principles—turning environmental liabilities into high-value resources while cutting down on the pollution caused by uncontrolled waste disposal (Gupta and Kaur, 2025).

Fruit and vegetable peels, such as banana, mango, and orange, are known to contain substantial amounts of antioxidants, dietary fiber, and essential micronutrients (Kumari et al., 2025). Because of this rich composition, these by-products are increasingly being explored as functional ingredients in aquaculture feeds, where they may contribute to improved fish health and growth performance (Ismail et al., 2024). A prime example is pineapple peel powder (PAPP), which is rich in bioactive compounds such as bromelain and phenolic compounds, and has been reported to enhance growth performance, immune responses, and disease resistance in Nile tilapia when included in the diet at a level of 10 g/kg. In addition to these nutritional benefits, the use of pineapple peel powder may also help lower feed costs while reducing the accumulation of agro-industrial waste (Van Doan et al., 2021). Similarly, longan peel powder (LP), another fruit by-product rich in phytochemicals, has been shown to improve growth performance, feed utilisation, immune responses, and antioxidant gene expression in Nile tilapia when incorporated into the diet (Wannavijit et al., 2025). In a related study, the inclusion of 5% banana peel flour (BPF) in the diet of *Labeo rohita* over a 60-day feeding period significantly enhanced growth performance, antioxidant status, and immune responses. The treatment also stimulated cytokine gene expression and upregulated heat shock protein HSP70, ultimately leading to improved resistance against pathogenic infection and higher survival rates (Giri et al., 2016).

Dragon fruit peel (DFP), another agro-industrial by-product, has been evaluated as a functional additive in Nile tilapia diets. Moderate inclusion levels (40–80 g/kg) improved growth performance, feed utilisation, antioxidant capacity, and immune responses, which were associated with the upregulation of genes involved in cytokine production and antioxidant defense (Xuan et al., 2024). Similarly, mango seed powder included at 36–45 g/kg enhanced growth, intestinal health, immune responses, and the expression of genes related to nutrient metabolism and immunity in Nile tilapia (Fontana et al., 2024). Dietary supplementation with apple peel powder, a by-product of food processing that is rich in dietary fiber and bioactive compounds, has been shown to improve growth performance and overall physiological condition in genetically improved farmed Nile tilapia. The inclusion of apple peel powder also influenced lipid metabolism, resulting in lower serum cholesterol and triglyceride levels and improved liver enzyme profiles, which together indicate better metabolic health in the fish (Qiang et al., 2019). In another study, feeds prepared from tropical fruit processing wastes such as pineapple, coconut, and jackfruit residues were found to provide nutritionally adequate diets for Malaysian mahseer while remaining economically viable for large-scale aquaculture production. These fruit-processing by-products contain considerable amounts of carbohydrates, lipids, vitamins, and antioxidant compounds that can support fish growth and contribute to improved health status (Kawata et al., 2021). Waste date fruit meal, a carbohydrate-rich agro-industrial by-product, has been successfully used to replace a substantial portion of conventional feed ingredients in Nile tilapia diets without negatively affecting growth performance. The high carbohydrate content provides a protein-sparing effect, improving protein utilisation efficiency and reducing the dependence on costly protein sources. Although higher inclusion levels slightly increased body lipid deposition, overall feed efficiency remained acceptable (Azaza et al., 2009). Similarly, banana peel flour, an abundant agricultural waste rich in carbohydrates and dietary fiber, significantly improved growth performance and feed conversion efficiency in carp when included at optimal dietary levels. The improved nutrient utilisation suggests that banana peels provide digestible energy and functional compounds beneficial for fish metabolism. Importantly, survival rates were not affected, indicating that this ingredient is safe for use in aquafeeds. Given that banana peels are inexpensive and widely available, their use represents a practical strategy for reducing feed costs while promoting the sustainable recycling of agricultural residues (Melanie et al., 2024). Pumpkin peel powder has similarly demonstrated promising effects in aquaculture nutrition. Its dietary supplementation enhanced growth performance, digestive enzyme activity, immune and antioxidant status, and the expression of genes linked to metabolism and immune regulation in Nile tilapia. These physiological improvements translated into increased resistance and higher survival rates during *Aeromonas veronii* infection (Ibrahim et al., 2025). In another study, diets formulated with mixed vegetable wastes such as cabbage leaves, cauliflower stems, and sponge gourd peels improved growth performance, survival, and feed efficiency in black molly fish relative to conventional diets. The experimental feed also enhanced body coloration due to the presence of natural pigments such as chlorophyll and carotenoids in the plant residues. Fish fed the vegetable-waste-based diet remained disease-free, reflecting an improved overall health status (Patel and Manoj, 2025). Olive pulp has also been evaluated as a functional dietary ingredient in aquafeeds. Fish fed olive pulp-enriched diets exhibited enhanced serum antioxidant capacity along with increased complement activity, peroxidase activity, and cellular immune responses. These changes suggest an overall improvement in metabolic and physiological status (Bahi et al., 2024). Likewise, coffee silverskin, a by-product of coffee roasting rich in dietary fiber and phenolic compounds, has demonstrated beneficial effects when included in Nile tilapia diets. Supplementation at an optimal level of 40 g/kg significantly improved weight gain, specific growth rate, feed conversion efficiency, and survival. In addition, the treatment enhanced several innate immune parameters, including complement activity, lysozyme activity, peroxidase activity, and respiratory burst activity, indicating strengthened nonspecific immune responses (Prakash and Doan, 2022). Collectively, these case studies underscore that processed agro-wastes, when appropriately treated and included at optimal dietary levels, can function as multifunctional feed additives. They provide bulk macronutrients (e.g., carbohydrates and fibers), essential micronutrients, and bioactive compounds (e.g., antioxidants and immunostimulants) that collectively promote growth, improve health, and enhance resilience against stressors and pathogens. Such findings highlight the potential of agro waste valorization not only to reduce feed costs and improve sustainability but also to advance functional feed development with measurable physiological and molecular benefits in cultured fish species.

3. Agro waste derived nanomaterials

Agro wastes serve as abundant and renewable raw materials for the synthesis of nanomaterials, offering a sustainable pathway to valorize residues into high-value applications. These nanostructures, ranging from nanocellulose and carbon-based nanomaterials to metal/metal oxide nanoparticles, nano biochar, and nano silica, exhibit multifunctional properties including nutrient delivery, antimicrobial defense, prebiotic effects, and toxin adsorption, thereby presenting significant potential for aquaculture feed development.

4. Cellulosic nanomaterials

Agricultural residues such as straw, sugarcane bagasse, corn stover, and fruit peels are important sources of lignocellulosic biomass and are widely known as potential raw materials for producing cellulose-based nanomaterials. These residues contain varying proportions of cellulose, hemicellulose, and lignin, which can be processed through chemical or mechanical methods to obtain nanocellulose with useful physicochemical properties (Magagula et al., 2022). Among the different types of nanocellulose, cellulose nanocrystals (CNCs) are commonly synthesized through acid hydrolysis. This process removes the amorphous regions of cellulose while preserving the crystalline domains, yielding highly ordered nanostructures with high crystallinity and large surface area (George and Sabapathi, 2015; Mahmud et al., 2019).

Sugarcane bagasse, in particular, contains substantial amounts of lignocellulosic biomass and is considered a viable feedstock for nanocellulose production.

Converting such agricultural residues into nanocellulose represents an effective value-addition strategy that repurposes low-value biomass into advanced functional materials (Mandal and Chakrabarty, 2011). For example, high-purity cellulose (>90%) extracted from bagasse and corn stover has been converted into granular nanocellulose particles ranging from 200 to 342 nm through enzymatic and oxalic acid hydrolysis. These nanoparticles exhibited lipid-binding capacity and significantly reduced lipase activity and lipid hydrolysis, pointing toward their potential role in regulating lipid metabolism in fish (Chumchoochart et al., 2025).

5. Carbon-based nanomaterials

Carbon quantum dots (CQDs) are nanoscale carbon particles, generally smaller than 20 nm, that can be synthesized from food and agricultural wastes through environmentally friendly and low-cost methods. As zero-dimensional nanomaterials, they feature spherical morphology and possess abundant surface functional groups such as hydroxyl, carboxyl, and amino groups. The presence of these functional groups enhances their water solubility and facilitates interactions with biological systems. Beyond this, CQDs display strong photoluminescence due to quantum confinement effects, which enables their application in sensing, imaging, and monitoring technologies. They also exhibit antioxidant and antimicrobial properties associated with reactive oxygen species scavenging and microbial cell disruption. Their biodegradability, environmental compatibility, and relatively low cytotoxicity make them promising candidates for biological, agricultural, and environmental applications (Aksu and Güzdemir, 2025). Biomass residues such as citrus peels represent valuable carbon sources for the synthesis of CQDs because they contain high levels of carbohydrates, phenolic compounds, and minerals. CQDs produced from these precursors demonstrate strong fluorescence, good stability, and high water solubility, making them suitable for applications in biomedical diagnostics and environmental sensing. Beyond this, nitrogen doping using amino acids can boost the quantum yield, antioxidant capacity, and biological activity of these nanoparticles. Surface functionalization further improves their compatibility with aqueous environments and biological matrices, underscoring the potential of fruit-waste-derived CQDs as multifunctional nanomaterials produced through sustainable approaches (Šafranko et al., 2021). Due to their small size and surface chemistry, carbon quantum dots can interact effectively with biological systems. Their tunable optical properties allow them to be used for imaging and tracking in living organisms, while their redox activity enables interactions with cellular components. Notably, CQDs generally exhibit lower toxicity relative to many metal-based nanomaterials. Their biological effects depend strongly on particle size and surface functional groups, which influence cellular uptake and physiological responses. These characteristics underline the potential of CQDs as bioactive nanomaterials for agricultural and environmental applications (Andleeb et al., 2025). Alongside, fruit-processing residues such as pineapple represent excellent feedstocks for CQD synthesis. Pineapple processing can generate up to 80% solid waste rich in sugars, phenolics, fibers, and bromelain. These residues can be repurposed to produce CQDs and other value-added bio-products such as citric acid and ethanol, thereby supporting circular bioeconomy strategies and improving the sustainable utilisation of agro-industrial waste (Rajamanikandan et al., 2022).

6. Green synthesized metal and metal oxide nanoparticles

Green synthesis of nanoparticles leverages biological systems such as plants, bacteria, fungi, and algae, where biomolecules, including enzymes, amino acids, and proteins, act as reducing and stabilizing agents during nanoparticle formation (Huston et al., 2021). Unlike conventional chemical synthesis methods, green approaches minimize the use of toxic organic solvents and energy-intensive processes, making them more consistent with the principles of sustainable and environmentally responsible nanomaterial production (Huston et al., 2021). Agro-waste extracts are particularly valuable in green synthesis because they contain secondary metabolites such as phenolic acids, flavonoids, tannins, terpenoids, alkaloids, proteins, polysaccharides, and vitamins. These phytochemicals contain functional groups capable of reducing metal ions through redox reactions while simultaneously stabilizing the formed nanoparticles by acting as natural capping agents. Fruit and vegetable residues, in particular, contain high levels of phenolic compounds that enhance both the reduction and stabilization processes. Biomass wastes rich in flavonoids have been reported to chelate and reduce metal ions into nanoparticles, enabling the eco-friendly synthesis of metal and metal oxide nanoparticles such as silver (Ag), copper (Cu/CuO), and zinc oxide (ZnO) (Aswathi et al., 2023; Jovanović et al., 2024; Islam et al., 2025). Similarly, silver nanoparticles synthesized using pineapple peel extracts have shown strong antimicrobial activity at low concentrations against both Gram-positive and Gram-negative pathogens (Baran et al., 2021). Comparable antimicrobial effects have also been reported for silver nanoparticles synthesized from other fruit extracts, pointing to their utility as sustainable alternatives to conventional antibacterial agents (Doan et al., 2025). Overall, these studies demonstrate how agro-waste valorization can contribute to safer and more sustainable nanoparticle production while addressing environmental challenges associated with agricultural waste disposal and supporting eco-friendly innovations in aquaculture systems.

7. Nano biochar

Biochar is a carbon-rich material derived through the thermochemical conversion (pyrolysis) of waste biomass under oxygen-limited conditions. With advances in nanotechnology, bulk biochar can be further refined into nanoscale particles (<100 nm) using techniques such as ball milling, centrifugation, and other size-reduction methods. Compared with conventional biochar, nano-biochar exhibits enhanced specific surface area, adsorption capacity, and mobility in soil systems (Kuppusamy et al., 2016). Beyond these properties, nano-biochar contains abundant surface functional groups and porous structures that enable strong adsorption of nutrients and contaminants. Functionalization of biochar with metal nanoparticles or other modifiers can further enhance its stability and catalytic performance, expanding its applications in environmental remediation and nutrient regulation (Ahmed et al., 2017). For example, peanut shell–derived nano-biochar produced at 450 °C has been reported to form nanoscale particles (~21.86 nm) that improve soil water-holding capacity, increase nutrient availability, and promote plant growth under drought stress conditions (Abd El-Moaty et al., 2024). Agricultural residues such as sugarcane bagasse and other lignocellulosic wastes also serve as sustainable feedstocks for biochar production, demonstrating the potential of agro-waste valorization in developing functional nanomaterials (Bashir et al., 2018). Overall, these findings suggest that waste-derived biochar and its nano-form have potential applications in both sustainable agriculture and aquaculture by enhancing nutrient utilization and contributing to improved environmental management (Sani et al., 2023).

8. Nano silica and ash derived nanoparticles

Agro-waste-derived silica nanoparticles have gained considerable interest as sustainable nanomaterials due to their low production cost, wide availability, and environmental compatibility. Rice husk, a major agricultural by-product, contains approximately 15%–25% silica and has therefore been extensively used as a renewable precursor for the synthesis of SiO_2 nanoparticles (Mahalingam et al., 2023). Various extraction methods, including thermal combustion and sol–gel techniques, have enabled the production of stable, mesoporous, and non-toxic silica nanoparticles from rice husk ash, underscoring the converting lignocellulosic residues into value-added nanomaterials (Mahalingam et al., 2023). Rice husk ash (RHA), a silica-rich agricultural residue, can be processed through alkali extraction followed by sol–gel synthesis to produce silica nanoparticles with average particle sizes of 60–80 nm and a high specific surface area of approximately 78.52 m^2/g. These SiO_2 nanoparticles exhibit porous morphology, chemical stability, and biocompatibility, and have also demonstrated photocatalytic and antibacterial properties (Alhadhrami et al., 2022). Similarly, wood ash generated from tea processing industries has been successfully converted into silica nanoparticles using optimized sol–gel techniques, producing spherical particles with sizes ranging from 20 to 50 nm and crystalline characteristics. This approach highlights the potential of agro-industrial residues as low-cost silica sources while simultaneously addressing waste management challenges (Nivaethaa et al., 2024). Beyond synthesis, silica nanoparticles have shown promising applications in aquaculture nutrition. Dietary supplementation of silica nanoparticles at optimized concentrations (2 mg/kg) significantly improved weight gain, specific growth rate, and muscle composition in *Heteropneustes fossilis* without negatively affecting feed utilization parameters

(Das et al., 2025). Parallel outcomes were reported in *Labeo rohita*, where silica nanoparticle supplementation enhanced feed conversion efficiency, protein digestibility (87.17%), dry matter digestibility, and survival rate, with optimal performance observed at 2 mg/kg inclusion levels. These improvements are largely attributed to the high surface area and adsorption properties of silica nanoparticles, which may enhance nutrient bioavailability and intestinal absorption (Murshed et al., 2023). From a sustainability perspective, the synthesis of nanoparticles from agro-waste aligns with circular economy principles by reducing carbon emissions, lowering energy consumption, and minimizing dependence on synthetic chemical precursors (Flores-Contreras et al., 2024). The valorization of rice husk, wood ash, and other lignocellulosic residues into functional nanomaterials therefore represents a dual-benefit strategy, combining waste management with performance enhancement in aquaculture feed systems.

9. Translational relevance and research gaps

Collectively, agro waste-derived nanomaterials, including nanocellulose, carbon quantum dots, nano biochar, metal/metal oxide nanoparticles, and nano silica, offer multifunctional benefits such as enhanced nutrient bioavailability, toxin adsorption, antimicrobial protection, immune stimulation, and improved feed stability. Their renewable, biodegradable, and biocompatible nature supports circular bioeconomy models by converting low-value waste into high-value aquaculture inputs.

Nevertheless, significant research gaps remain. Critical questions include the long-term impacts of nanomaterial ingestion on fish physiology and product safety, species-specific responses, interactions with gut microbiota, and potential bioaccumulation in edible tissues. Moreover, large-scale, cost-effective production methods and standardised safety assessments are essential before commercial adoption can be realized. Addressing these gaps will be crucial for translating the promise of agro-waste-derived nanomaterials into safe, sustainable, and economically viable solutions for aquaculture feed development.

10. Nanotechnology in fish feed

Nanotechnology has proven to be a transformative tool in aquaculture, particularly in the formulation of functional feeds that enhance nutrient utilisation, growth, immunity, and stress tolerance. The unique physicochemical properties of nanoparticles (NPs), including their extremely high surface-to-volume ratio, enhanced reactivity, stability, and encapsulation ability, make them ideal carriers for nutrients, drugs, and bioactive compounds (Dube, 2024; Riyaz et al., 2025). Unlike conventional feed additives, nano supplements can cross biological barriers more efficiently, interact with cellular components at the molecular level, and provide controlled and targeted release of nutrients, thereby overcoming limitations of traditional feed formulations.

11. Nutritional enhancement

Several studies have demonstrated the effectiveness of nanoparticles in improving nutrient digestibility, absorption, and growth performance in aquaculture species. For example, dietary supplementation of silica nanoparticles in Nile tilapia significantly improved weight gain, specific growth rate, feed efficiency, and apparent protein digestibility, while reducing feed conversion ratio (FCR). These improvements were associated with enhanced intestinal morphology, including increased villi width, surface area, and goblet cell number (Bashar et al., 2021). Nano-sizing of minerals increases their surface area, solubility, and bioavailability, yielding better metabolic efficiency even at lower dietary inclusion levels compared with conventional bulk minerals. Notably nano-selenium supplementation has repeatedly shown improvements in growth performance, feed utilization, antioxidant enzyme activity (GPx, SOD, and CAT), immune responses, and stress tolerance across multiple aquaculture species (Dawood et al., 2021). Similarly, dietary supplementation with chitosan nanoparticles (CNPs) has been shown to significantly enhance growth performance, intestinal villus length and width, antioxidant capacity (GSH and TAC), and innate immune responses such as phagocytic activity and respiratory burst in Nile tilapia. In addition, CNP-treated groups exhibited reduced lipid peroxidation (MDA levels) and increased resistance to *Aeromonas* infection (Hossam-Elden et al., 2024). Dietary zinc nanoparticles have also been reported to improve weight gain, specific growth rate (SGR), feed conversion ratio (FCR), protein efficiency ratio (PER), and hematological parameters in freshwater fish species. Appropriate supplementation levels enhanced antioxidant defense mechanisms and enzymatic activity, although excessive concentrations may pose toxicity risks (Jewel et al., 2024). Moreover, nano-based micronutrient complexes containing Fe, Zn, Cu, and Se have demonstrated improved growth performance, enhanced muscle nutrient composition, and increased essential amino acid levels in tilapia, which are attributed to improved bioavailability and cellular uptake of micronutrients (Khan et al., 2020). Nano-encapsulated vitamin E coated with chitosan has been shown to enhance relative growth rate (RGR), reduce FCR, and stimulate innate immune responses such as lysozyme activity, oxidative burst, and antibacterial activity in Nile tilapia (Farag et al., 2024).

12. Nano selenium: A key antioxidant and immunomodulator

Selenium (Se) is an essential micronutrient that plays a crucial role in selenoprotein synthesis and the regulation of antioxidant enzymes. Selenium nanoparticles (SeNPs) have attracted widespread interest because they exhibit higher bioavailability, greater stability, and lower toxicity relative to conventional inorganic or organic selenium sources. In Nile tilapia, dietary supplementation with SeNPs enhanced growth performance, immune responses, and antioxidant enzyme activity, including CAT, GSH, and GPx, while maintaining stable SOD levels and normal liver histology without signs of toxicity (Thangapandiyan and Monika, 2020). The combination of SeNPs with vitamin E has also shown synergistic effects in improving growth performance and managing oxidative stress, thereby increasing resistance to pathogenic infections (Bazina et al., 2025). Similarly, supplementation of SeNPs together with garlic extract in grass carp diets improved immune parameters and antioxidant activity under stressful high-density culture conditions (Adineh et al., 2021). These findings suggest that nano-selenium functions not only as a micronutrient but also as a multifunctional bioactive compound, particularly under environmental or physiological stress conditions. Beyond growth and immunity, SeNPs have demonstrated protective effects against nanoparticle-induced toxicity. In Nile tilapia, dietary supplementation with SeNPs (1–2.5 mg/kg diet) mitigated aluminum nanoparticle–induced oxidative stress, downregulated pro-inflammatory genes (TNF-β and IL-1β), reduced lipid peroxidation, restored glutathione levels, and alleviated histopathological damage in liver and gill tissues (Abd El Megeed et al., 2025). Further, nano-selenium supplementation has shown benefits during early developmental stages. In European sea bass larvae, nano-selenium–fortified weaning diets significantly enhanced survival rate, growth performance, feed conversion efficiency, glutathione peroxidase (GPx) activity, and gastrointestinal development (El-Dahhar et al., 2024). Positive effects of dietary SeNPs have also been confirmed across several aquaculture species. In *Catla catla*, supplementation with 1.5 mg kg^{-1} SeNP significantly improved growth indices, nutrient digestibility, and mineral absorption (Ahmad et al., 2024). In European sea bass (*Dicentrarchus labrax*), nano-selenium inclusion enhanced survival rate, specific growth rate, feed conversion ratio, antioxidant enzyme activity (GPx), and gastrointestinal development during the larval stage (El-Dahhar et al., 2024). Similarly, in Asian seabass (*Lates calcarifer*), dietary SeNP supplementation improved growth performance, digestive enzyme activity, antioxidant capacity, and immune-related parameters under high salinity conditions (Mohtashemipour et al., 2024). Collectively, these findings highlight nano-selenium as one of the most promising nano-additives for aquafeed applications due to its combined roles in growth promotion, antioxidant defense, and immune regulation.

13. Other nano minerals: Iron and zinc

Nano iron oxide (Fe_2O_3) has demonstrated notable benefits in aquaculture nutrition. Supplementation at levels of 0.2–0.4 mg/kg in Nile tilapia improved growth performance, feed efficiency, intestinal morphology, hematological indices, and immune responses linked to conventional bulk iron. These improvements are attributed to the higher absorption and utilisation efficiency of nano-Fe_2O_3 (Mohammady et al., 2024). Enhanced gut health and stronger innate immune responses were also observed, suggesting that nano iron can function as a dual-purpose additive for both nutritional enhancement and disease resistance (Mohammady et al.,

2024). Nano zinc oxide (ZnO) plays an important role in regulating stress responses in aquaculture species. In African catfish (*Clarias gariepinus*), dietary supplementation with nano-ZnO significantly mitigated heat stress–induced behavioral and physiological disturbances by reducing cortisol levels and restoring normal hematological and biochemical parameters (Mahmoud et al., 2024). Beyond this, green-synthesized ZnO nanoparticles have demonstrated lower toxicity compared to chemically synthesized forms and exerted reduced immunosuppressive effects on skin mucus immune parameters in common carp (*Cyprinus carpio*) (Rashidian et al., 2021). Overall, these findings highlight the potential of nano-minerals such as iron and zinc to enhance fish health and resilience, particularly under intensive aquaculture conditions.

14. Carbon-based nanomaterials in aquafeeds

Carbon nanomaterials, particularly carbon quantum dots (CQDs), are gaining attention in aquaculture due to their antimicrobial and immunomodulatory properties. Spermidine-derived CQDs (CQDSpds) have been reported to enhance growth performance, increase lysozyme expression, and improve resistance to *Edwardsiella tarda* in Japanese eel (*Anguilla japonica*) and giant marbled eel (*Anguilla marmorata*) when included in diets at optimal concentrations of 0.5–1 ppm. Along with promoting growth and immune responses, CQDSpds also exhibited direct antibacterial activity without adversely affecting intestinal morphology, highlighting their safety and potential as functional feed additives in intensive aquaculture systems (Lin et al., 2023). These findings suggest that carbon-based nanomaterials may serve not only as nutritional and immunological enhancers in aquafeeds but also as protective agents against environmental contaminants and pathogenic challenges, which are increasing concerns in modern aquaculture production systems.

15. Green synthesized nanoparticles: Sustainability and environmental benefits

Nanotechnology-based aquafeeds is the shift toward green synthesis approaches, which utilize plant extracts, agro-waste materials, or microbial systems as natural reducing and stabilizing agents for nanoparticle production. These methods are environmentally friendly and reduce the reliance on hazardous chemicals, making nanoparticle synthesis more sustainable.

Notably, ZnO and silica nanoparticles synthesized from *Azolla pinnata* and *Equisetum arvense* have been shown to reduce the bioavailable levels of toxic metals such as Cr, Cu, and Zn in pond sediments by 28%–32%, while simultaneously enriching sediment nutrient content and making it suitable for reuse as fertilizer (Zhu et al., 2025). Similarly, biologically synthesized selenium nanoparticles (SeNPs) exhibited strong antioxidant and antibacterial activity in Nile tilapia, reducing pathogenic bacterial load by 28%–45% and significantly decreasing heavy metal accumulation, with Cd and Hg levels reduced by up to 50%–87% (Saad et al., 2022). In freshwater fish (*Labeo rohita*), green-synthesized ZnO nanoparticles improved growth performance, biochemical parameters, hematological indices, and digestive enzyme activities, demonstrating their potential functional role in aquaculture nutrition (Al-Wakeel et al., 2024). Collectively, these findings highlight several sustainability advantages of green-synthesized nanoparticles, including reduced chemical usage, lower energy requirements, enhanced biocompatibility, and multifunctional activity against pathogens and environmental stressors. Their integration into aquaculture systems, therefore, aligns well with circular bio-economy principles and environmentally responsible production strategies (Saad et al., 2022).

16. Overall impact and future perspectives

Taken together, nanotechnology-based interventions in fish feed, including nano minerals (Se, Zn, Fe), carbon nanomaterials, and green-synthesized nanoparticles, represent a major innovation in aquaculture nutrition. These nanoscale additives enhance nutrient delivery, feed utilisation, growth performance, immune function, and stress resilience. Moreover, their application supports sustainability by lowering feed costs, minimizing resource wastage, reducing environmental toxicity, and decreasing dependence on antibiotics.

However, despite these advances, several critical gaps remain. Optimal inclusion levels, long-term safety, bioaccumulation risks, and ecological impacts of nanoparticle supplementation are not yet fully understood. Standardised trials across diverse aquaculture species, combined with risk assessment frameworks, are urgently needed to establish safe and effective guidelines for nanoparticle use in commercial aquafeeds.

17. Specific examples of fish species and trials

Nile tilapia (*Oreochromis niloticus*): Dietary selenium nanoparticles (Se-NPs) improved growth, feed utilisation, and immune-antioxidant responses without liver toxicity (Al-Wakeel et al., 2024).

Catla (*Catla catla*): Optimal supplementation with Se-NPs (1.5 mg/kg) enhanced nutrient digestibility, mineral absorption, and overall growth performance (Ahmad et al., 2024).

Common carp (*Cyprinus carpio*): Green-synthesized ZnO nanoparticles influenced mucosal immune parameters in the skin mucus of *Cyprinus carpio* and exhibited lower immunosuppressive effects compared with chemically synthesized nanoparticles (Rashidian et al., 2021).

European sea bass (*Dicentrarchus labrax*): Nano-selenium–fortified diets during larval weaning improved survival, growth performance, gastrointestinal development, and antioxidant enzyme activity in *Dicentrarchus labrax* larvae (El-Dahhar et al., 2024).

Grass carp (*Ctenopharyngodon idella*): A Combination of nano-Se with phytogenic additives such as garlic extract, enhanced antioxidant defences, and immune parameters under high stocking density (Adineh et al., 2021).

African catfish (*Clarias gariepinus*): Nano-ZnO supplementation mitigated heat stress in *Clarias gariepinus* by reducing cortisol levels and improving hematological and biochemical stress indicators (Mahmoud et al., 2024).

Asian sea bass (*Lates calcarifer*): Dietary Se-NPs improved growth, digestive enzyme activity, immune responses, and liver health (Mohtashemipour et al., 2024).

18. Ecotoxicology and biosafety considerations of nanomaterials in aquafeeds

Nanotechnology offers significant advantages for aquaculture nutrition, but concerns remain regarding the ecotoxicological impacts, long-term biosafety and potential transfer of nanoparticles (NPs) across trophic levels. Unlike conventional minerals or feed additives, engineered NPs possess physicochemical properties (e.g., high reactivity, nano-size, surface charge) that may lead to unintended biological and environmental interactions (Aksu and Güzdemir, 2025; Mahalingam et al., 2023).

Dose-dependent toxicity represents one of the primary challenges associated with the application of nanoparticles in aquaculture. Although moderate dietary supplementation of nanoparticles such as selenium nanoparticles (Se-NPs) has been shown to enhance growth performance, immune responses, and antioxidant activity in fish, overexposure can lead to adverse physiological effects. At elevated concentrations, nanoparticles may induce oxidative stress, DNA damage, and inflammatory responses in vital organs such as the liver, thereby impairing normal metabolic functions and overall fish health (Van Doan et al., 2021). Similarly, excessive ZnO-NPs accumulation in gill and liver tissues has been linked to impaired respiratory and metabolic functions (Huston et al., 2021). These findings highlight the need for precise dose optimisation and the establishment of safe upper limits for long-term use.

Bioaccumulation and Trophic Transfer: NPs not metabolized within the fish gut can accumulate in tissues (muscle, liver, kidney) and potentially enter the human food chain. For instance, Se-NPs showed accumulation in hepatic tissue when fed at supra-optimal levels, raising food safety concerns (Barati, 2023). Moreover, sediment-bound NPs (e.g., Fe_2O_3, ZnO) may be ingested by benthic organisms, leading to trophic transfer through aquatic ecosystems (Misra et al., 2023). Rice husk (RH), an agricultural byproduct, has been utilized as a bio-adsorbent to remove green-synthesized copper oxide nanoparticles from aquatic systems, resulting in reduced bioaccumulation, oxidative stress, and histological damage in *Oreochromis niloticus*, thereby contributing to enhanced environmental sustainability (Hamed and Badran, 2024).

Nanoparticles may persist in aquatic sediments and alter metal mobility, nutrient dynamics, and ecological interactions within aquatic environments (Zhu et al.,

2025). Nanoparticles released into aquatic environments may interact with microbial communities present in water and sediments, potentially influencing microbial diversity, nutrient cycling processes, and biofilm formation. Such interactions could alter ecological balance and affect key biogeochemical processes within aquaculture ecosystems (Khan et al., 2024). This necessitates life-cycle assessments to evaluate nanoparticle degradation and accumulation under real-world aquaculture conditions.

Synergistic or Antagonistic Interactions: The presence of multiple NPs or co-exposure with pollutants (e.g., pesticides, heavy metals) can amplify toxicological responses. For example, ZnO-NPs were shown to increase copper bioavailability in pond sediments, thereby compounding oxidative stress in fish (Van Doan et al., 2021). Conversely, co-supplementation of Se-NPs with antioxidants (vitamin E, phytogenics) mitigated nanoparticle-induced stress, suggesting that formulation strategies can reduce ecotoxicological risks (Adineh et al., 2021; Bazina et al., 2025). Incorporating advanced analytical approaches such as omics-based technologies, including transcriptomics, metabolomics, and proteomics, together with nanoparticle tracing techniques like inductively coupled plasma–mass spectrometry (ICP-MS) and transmission electron microscopy (TEM), can provide deeper insights into nanoparticle uptake, distribution, and molecular-level interactions within aquatic organisms. Such integrated approaches will enhance the understanding of mechanistic toxicity pathways and facilitate more comprehensive risk assessment frameworks for evaluating the long-term biosafety and sustainable application of nanomaterials in aquaculture systems.

19. Conclusion

Agro wastes represent an underexploited yet highly promising resource for the development of sustainable aquafeeds. Rich in carbohydrates, fibers, proteins, bioactive compounds, and micronutrients, these residues, including fruit and vegetable peels, crop by-products, and food processing discards, can be strategically incorporated into fish diets to improve growth, gut health, antioxidant capacity, and immune resilience while simultaneously reducing dependence on costly conventional feed ingredients. Such valorization directly addresses the dual challenges of aquaculture sustainability: ensuring nutritional adequacy for farmed species and minimizing environmental waste burdens.

Nanotechnology is adding an exciting new dimension to aquaculture. Nanoparticles made from agro wastes or synthesized via green chemistry approaches offer superior bioavailability, stability, and functionality compared with conventional feed additives. For instance, nano minerals such as selenium, zinc, and iron nanoparticles can improve nutrient uptake, enhance antioxidant enzyme activity, and immune responses. Carbon-based nanomaterials and nano silicas improve stress tolerance, gut health, and pellet stability. Beyond nutritional enhancement, nanoparticle-based carriers hold potential for targeted delivery of vaccines, probiotics, or therapeutic agents, thereby reducing the need for antibiotics and aiding disease management in intensive aquaculture systems.

At the same time, integrating nanotechnology into aquafeed needs careful analysis of risk assessment. Optimal dosing regimens, standardised safety evaluations, and long-term ecotoxicological studies are essential to avoid buildup of nanoparticle bioaccumulation and to mitigate the risks to aquatic ecosystems and human consumers. Green synthesis methods using agro-waste-derived phytochemicals, which can reduce energy inputs, minimise hazardous byproducts, and align with circular bioeconomy principles.

Looking forward, combining agro waste valorisation and nanotechnology offers a sustainable aquaculture feed nutrition. Converting low-value residues into high-performance feed components not only lowers production costs but also improves fish health, product quality, and environmental sustainability. Future research should focus on: (i) Understanding the molecular mechanisms underlying nanoparticle–host interactions, (ii) Customising nanomaterial properties to species-specific nutritional and physiological needs, (iii) Scaling up eco-friendly production methods, and (iv) establishing regulatory frameworks to ensure feed safety.

In conclusion, integrating agro waste-derived nutrients and nanomaterials into aquafeeds is a practical, cost-effective, and environmentally responsible strategy that supports global aquaculture growth within the framework of a circular bioeconomy.

References

Abd El Megeed, O.H., Mabrouk, M.M., Gewida, A.G.A., Eissa, M.E.H., Yasin, N.A.E., Gado, W.S., Rashad, M.M., Ali, G.E., 2025. The mitigative effect of selenium nanoparticles against aluminium nanoparticle toxicity in Nile Tilapia (*Oreochromis niloticus*): growth performance, antioxidant status, gene expression, and histopathological changes. Aquacult. Int. 33 (6).

Abd El-Moaty, H.I., El-Dissouky, A., Elhusseiny, A.F., Farag, K.M., Abu-Khudir, R., Alkuwayti, M.A., Al Abdulsalam, N.K., Abdel Rahman, S.M., 2024. Low-cost nano biochar: a sustainable approach for drought stress mitigation in faba bean (*Vicia faba* L.). Front. Plant Sci 15.

Adineh, H., Naderi, M., Nazer, A., Yousefi, M., Ahmadifar, E., 2021. Interactive effects of stocking density and dietary supplementation with nano selenium and garlic extract on growth, feed utilization, digestive enzymes, stress responses, and antioxidant capacity of grass carp, Ctenopharyngodon idella. J. World Aquacult. Soc 52 (4), 789–804. https://doi.org/10.1111/jwas.12747.

Afreen, M., Ucak, I., 2020. Fish processing wastes used as feed ingredient for animal feed and aquaculture feed. J. Surv. Fish. Sci 6 (2), 55–64. https://doi.org/10.18331/sfs2020.6.2.7.

Ahmad, N., Hussain, S.M., Ali, S., Tahir, M.F., Sarker, P.K., Shahid, M., 2024. Nano-selenium supplementation: improving growth, digestibility and mineral absorption in freshwater fish, Catla catla. BMC Vet. Res 20 (1). https://doi.org/10.1186/s12917-024-04291-6.

Ahmed, M.B., Zhou, J.L., Ngo, H.H., Guo, W., Johir, M.A.H., Sornalingam, K., Belhaj, D., Kallel, M., 2017. Nano-Fe0 immobilized onto functionalized biochar gaining excellent stability during sorption and reduction of chloramphenicol via transforming to reusable magnetic composite. Chem. Eng. J. 322, 571–581. https://doi.org/10.1016/j.cej.2017.04.063.

Aksu, M., Güzdemir, Ö., 2025. Food waste-derived carbon quantum dots and their applications in food Technology: a critical review. Food Bioprocess Technol. 18 (8), 6753–6778. https://doi.org/10.1007/s11947-025-03854-1.

Al-Wakeel, A.H., Elbahnaswy, S., Eldessouki, E.A., Risha, E., Zahran, E., 2024. Dietary biogenic selenium nanoparticles improve growth and immune-antioxidant indices without inducing inflammatory responses in Nile tilapia. Sci. Rep. 14 (1). https://doi.org/10.1038/s41598-024-72022-w.

Alhadhrami, A., Mohamed, G.G., Sadek, A.H., Ismail, S.H., Ebnalwaled, A.A., Almalki, A.S. A., 2022. Behavior of silica nanoparticles synthesized from rice husk ash by the sol–gel method as a photocatalytic and antibacterial agent. Materials 15 (22). https://doi.org/10.3390/ma15228211.

Andleeb, N., Zafar, S., Rahim, Z., Iqbal, M.M., Lal, R., Farooq, M.A., 2025. Carbon quantum dots as versatile nanomaterials for improving soil health and plant stress tolerance: a comprehensive review. Planta 262 (2). https://doi.org/10.1007/s00425-025-04758-2.

Ansari, F.A., Guldhe, A., Gupta, S.K., Rawat, I., Bux, F., 2021. Improving the feasibility of aquaculture feed by using microalgae. Environ. Sci. Pollut. Res. 28 (32), 43234–43257.

Aragão, C., Conceição, L.E.C., Dinis, M.T., Fyhn, H.J., 2004. Amino acid pools of rotifers and Artemia under different conditions: nutritional implications for fish larvae. Aquaculture 234 (1–4), 429–445. https://doi.org/10.1016/j.aquaculture.2004.01.025.

Aswathi, V.P., Meera, S., Maria, C.G.A., Nidhin, M., 2023. Green synthesis of nanoparticles from biodegradable waste extracts and their applications: a critical review. Nanotechnol. Environ. Eng. 8 (2), 377–397. https://doi.org/10.1007/s41204-022-00276-8.

Azaza, M.S., Mensi, F., Kammoun, W., Abdelouaheb, A., Brini, B., Kraïem, M., 2009. Nutritional evaluation of waste date fruit as partial substitute for soybean meal in practical diets of juvenile Nile tilapia, *Oreochromis niloticus* L. Aquacult. Nutr 15 (3), 262–272. https://doi.org/10.1111/j.1365-2095.2008.00591.x.

Bahi, A., Bhaskaracharya, R., Esteban, M.A., Guardiola, F.A., 2024. Dietary administration impact of olive pulp on growth performance, metabolic profile, immune status, and antioxidant potential of gilthead seabream (*Sparus aurata* L.). Front. Sustain. Food Syst. 8. https://doi.org/10.3389/fsufs.2024.1395436.

Bala, S., Garg, D., Sridhar, K., Inbaraj, B.S., Singh, R., Kamma, S., Tripathi, M., Sharma, M., 2023. Transformation of agro-waste into value-added bioproducts and bioactive compounds: micro/nano formulations and application in the agri-food-pharma sector. Bioengineering 10 (2). https://doi.org/10.3390/bioengineering10020152.

Baran, A., Keskin, C., Baran, M.F., Huseynova, I., Khalilov, R., Eftekhari, A., Irtegun-Kandemir, S., Kavak, D.E., 2021. Ecofriendly synthesis of silver nanoparticles using *Ananas comosus* fruit peels: anticancer and antimicrobial activities. Bioinorgan. Chem. Appl 2021. https://doi.org/10.1155/2021/2058149.

Barati, S., 2023. Applications of agricultural waste in food industry. J. Biol. Stud. 6 (1), 178–192.

Bashar, A., Hasan, N.A., Haque, M.M., Rohani, M.F., Hossain, M.S., 2021. Effects of dietary silica nanoparticle on growth performance, protein digestibility, hematology, digestive morphology, and muscle composition of Nile Tilapia, *Oreochromis niloticus*. Front. Mar. Sci. (8). https://doi.org/10.3389/fmars.2021.706179.

Bashir, S., Hussain, Q., Akmal, M., Riaz, M., Hu, H., Ijaz, S.S., Iqbal, M., Abro, S., Mehmood, S., Ahmad, M., 2018. Sugarcane bagasse-derived biochar reduces the cadmium and chromium bioavailability to mash bean and enhances the microbial activity in contaminated soil. J. Soils Sediments 18 (3), 874–886. https://doi.org/10.1007/s11368-017-1796-z.

Bazina, W.K., Tawfik, W.A., Abd Elghany, N.A., Saadony, S., Abdul Kari, Z., Alamoudi, M. O., Aly, M.Y.M., Younis, A.E., Mathew, R.T., Eissa, M.E.H., Munir, M.B., Ibrahim, S., Yousef, I., Omar, Y.O.H., Eissa, E.S.H., Abd Elnabi, H.E., 2025. Effects of nano-selenium and/or vitamin E supplementation on growth performance, antioxidant status, histopathology and resistance to Aspergillus flavus in Nile tilapia (*Oreochromis niloticus*). BMC Vet. Res. 21 (1). https://doi.org/10.1186/s12917-025-04471-y.

Boyd, C.E., McNevin, A.A., Davis, R.P., 2022. The contribution of fisheries and aquaculture to the global protein supply: aquaculture and global protein. Food Secur. 14 (3), 805–827.

Chattaraj, S., Ganguly, A., Mitra, D., et al., 2026a. Sustainable valorization of algal biomass in fish nutrition and dietetics: a critical review and scientometric study. Waste Biomass Valor. https://doi.org/10.1007/s12649-026-03619-7.

Chattaraj, S., Ganguly, A., Mitra, D., Mitra, D., Das Mohapatra, P.K., 2023. Study of intestinal bacteria of Cirrhinus reba and characterization of a new probiotic bacteria: an initiative to save the threatened species of Cirrhinus. Appl. Biochem. Microbiol. 59 (6), 833–849.

Chattaraj, S., Mitra, D., Chattaraj, M., Ganguly, A., Thatoi, H., Mohapatra, P.K.D., 2024b. Brewers' spent grain as fish feed ingredient: evaluation of bio-safety and analysis of its impact on gut bacteria of Cirrhinus reba by 16S Metagenomic sequencing. Curr. Res. Microb. Sci. 7, 100286.

Chattaraj, S., Mitra, D., Ganguly, A., Thatoi, H., Mohapatra, P.K.D., 2024a. A critical review on the biotechnological potential of Brewers' waste: challenges and future alternatives. Curr. Res. Microb. Sci. 6, 100228.

Chattaraj, S., Sahoo, S.S., Behera, S., Thatoi, H., 2026b. Sustainable aquafeed development using Spirulina: effects on growth performance of Catla catla. Algal Res. 104447.

Chumchoochart, W., Ketvaraporn, T., Chandet, N., Saenjum, C., Tinoi, J., 2025. Granular nanocellulose derived from sugarcane bagasse and corn stover: production, characterization, and its potential for lipid hydrolysis reduction. Carbohydr. Polym. Technol. Appl. 10, 100851. https://doi.org/10.1016/j.carpta.2025.100851.

Das, P.S., Rohani, Md F., Basim, S.A., Sulivany, Al, Nibir, S.S., Juthi, R.A., Satter, A., Hossain, Md S., Ismael, S.S., 2025. Dietary silica nanoparticle ameliorates the growth performance and muscle composition of stinging catfish, heteropneustes fossilis. Sci. J. Univ. Zakho 13 (1), 33–39. https://doi.org/10.25271/sjuoz.2025.13.1.1394.

Dawood, M.A.O., El Basuini, M.F., Yilmaz, S., Abdel-Latif, H.M.R., Kari, Z.A., Abdul Razab, M.K.A., Ahmed, H.A., Alagawany, M., Gewaily, M.S., 2021. Selenium nanoparticles as a natural antioxidant and metabolic regulator in aquaculture: a review. Antioxidants 10 (9). https://doi.org/10.3390/antiox10091364.

Doan, L., Lam, N.N., Tran, K., Huynh, K.G., 2025. Fruit derived silver nanoparticles synthesis for beginners–a review. Nanocomposites 11 (1), 20–51. https://doi.org/10.1080/20550324.2024.2442270.

Dube, E., 2024. Nanoparticle-enhanced fish feed: benefits and challenges. Fishes 9 (8), 322. https://doi.org/10.3390/fishes9080322.

Dutta, S., Saravanabhupathy, S., Anusha, Rajak, R.C., Banerjee, R., Dikshit, P.K., Padigala, C. T., Das, A.K., Kim, B.S., 2023. Recent developments in lignocellulosic biofuel production with nanotechnological intervention: an emphasis on ethanol. Catalysts 13 (11). https://doi.org/10.3390/catal13111439.

El-Dahhar, A.A., Khaleefa, G.A., El-Zaeem, S.Y., Mourad, M.M., Raheem, M.M.A., 2024. Impact of organic and nanoparticle-selenium fortified larval weaning diet on survival, growth, gastrointestinal, and (GPx) activity of European Sea Bass (*Dicentrarchus labrax*). Aquac. Int. 32 (6), 8501–8517. https://doi.org/10.1007/s10499-024-01576-3.

Farag, E.A.H., Baromh, M.Z., El-kalamwi, N., Sherif, A.H., 2024. Vitamin E nanoparticles enhance performance and immune status of Nile tilapia. BMC Vet. Res. 20 (1). https://doi.org/10.1186/s12917-024-04398-w.

Flores-Contreras, E.A., González-González, R.B., Pablo Pizaña-Aranda, J.J., Parra-Arroyo, L., Rodríguez-Aguayo, A.A., Iñiguez-Moreno, M., González-Meza, G.M., Araújo, R.G., Ramírez-Gamboa, D., Parra-Saldívar, R., Melchor-Martínez, E.M., 2024. Agricultural waste as a sustainable source for nanoparticle synthesis and their antimicrobial properties for food preservation. Front. Nanotechnol. 6. https://doi.org/10.3389/fnano.2024.1346069.

Fontana, C.M., Sumon, M.A.A., Wannavijit, S., Lubis, A.R., Khongdee, N., Linh, N.V., Phimolsiripol, Y., Hoseinifar, S.H., Van Doan, H., 2024. Effects of mango seed (mangifera indica) powder on growth performance, immune response, gut morphology, and gene expression of Nile Tilapia (*Oreochromis niloticus*). Fishes 9 (12).

George, J., Sabapathi, S.N., 2015. Cellulose nanocrystals: synthesis, functional properties, and applications. Nanotechnol. Sci. Appl. 8, 45–54. https://doi.org/10.2147/NSA.S64386.

Giri, S.S., Jun, J.W., Sukumaran, V., Park, S.C., 2016. Dietary administration of banana (*Musa acuminata*) peel flour affects the growth, antioxidant status, cytokine responses, and disease susceptibility of Rohu, *Labeo rohita*. J. Immunol. Res 2016. https://doi.org/10.1155/2016/4086591.

Gupta, N., Kaur, N., 2025. Sustainability and valorisation of agro waste: a review. J. Agroecol. Nat. Res. Manag. 12 (3), 1–12.

Hamed, A., Badran, S.R., 2024. The role of rice husk in *Oreochromis niloticus* safety enhancement by bio-adsorbing copper oxide nanoparticles following its green synthesis: an endeavor to advance environmental sustainability. Sci. Rep. 14 (1). https://doi.org/10.1038/s41598-024-74113-0.

Hossam-Elden, N., Abu-Elala, N.M., AbuBakr, H.O., Luo, Z., Aljuaydi, S.H., Khattab, M., Ali, S.E., Marzouk, M.S., Teiba, I.I., 2024. Dietary chitosan nanoparticles enhance growth, antioxidant defenses, immunity, and aeromonas veronii biovar sobria resistance in Nile tilapia *Oreochromis niloticus*. Fishes 9 (10). https://doi.org/10.3390/fishes9100388.

Hussain, S.M., Bano, A.A., Ali, S., Rizwan, M., Adrees, M., Zahoor, A.F., Sarker, P.K., Hussain, M., Arsalan, M.Z.u.H., Yong, J.W.H., Naeem, A., 2024. Substitution of fishmeal: highlights of potential plant protein sources for aquaculture sustainability. Heliyon 10 (4). https://doi.org/10.1016/j.heliyon.2024.e26573.

Huston, M., DeBella, M., DiBella, M., Gupta, A., 2021. Green synthesis of nanomaterials. Nanomaterials 11 (8), 2130. https://doi.org/10.3390/nano11082130.

Ibrahim, R.E., Elshopakey, G.E., Younis, E.M., Abdelwarith, A.A., Yaseen, A., Khamis, T., Davies, S.J., Abdel Rahman, A.N., 2025. Peel of pumpkin (*Cucurbita pepo*) diets boost growth, digestive-absorptive functions, immune-antioxidant indices, and regulate immunomodulatory genes and mTOR/MAPK-P38/apoptosis signaling pathway in *Oreochromis niloticus*. Aquac. Rep. 40. https://doi.org/10.1016/j.aqrep.2024.102548.

Islam, K.S., Mustak, M.H., Shishir, M.K.H., Karim, M.M., Khan, G.M.A., 2025. Biosynthesis of ZnO nanoparticles using banana peel pectin for antibacterial and photocatalytic applications. S. Afr. J. Chem. Eng 52, 127–140. https://doi.org/10.1016/j.sajce.2025.02.002.

Ismail, S., Dubey, P.K., Mishra, A.A., Ashka, F., 2024. Valorisation of banana peel and mango peel as functional ingredients in baked products: a review. Int. J. Food Sci. Technol. 59 (9), 5938–5950. https://doi.org/10.1111/ijfs.17395.

Jewel, A.S., Haque, A., Akter, N., Akter, S., Satter, A., Sarker, P.K., Marshall, D.J., Paray, B. A., Hossain, M.B., 2024. Effects of dietary supplementation of Zn-nanoparticles on the growth performance and nutritional quality of Asian catfish, *Clarias batrachus*. Front. Sustain. Food Syst. 8. https://doi.org/10.3389/fsufs.2024.1410557.

Jovanović, D., Bognár, S., Despotović, V., Finčur, N., Jakšić, S., Putnik, P., Deák, C., Kozma, G., Kordić, B., Merkulov, D.Š., 2024. Banana peel extract-derived ZnO nanopowder: transforming solar water purification for safer agri-food production. Foods 13 (16), 2643. https://doi.org/10.3390/foods13162643.

Kawata, Y., Yusoff, F.M., Khong, N.M.H., Umi, D.W.A., 2021. The economic feasibility assessment of using tropical fruit wastes in production of fish feed for the Malaysian mahseer, Tor *tambroides* (Bleeker, 1854). J. Mater. Cycles Waste Manag. 23 (3), 1026–1036. https://doi.org/10.1007/s10163-021-01190-y.

Khan, S.K., Dutta, J., Ahmad, I., Rather, M.A., 2024. Nanotechnology in aquaculture: transforming the future of food security. Food Chem. X 24. https://doi.org/10.1016/j.fochx.2024.101974.

Khan, M.Z.H., Hossain, M.M.M., Khan, M., Ali, M.S., Aktar, S., Moniruzzaman, M., Khan, M., 2020. Influence of nanoparticle-based nano-nutrients on the growth performance and physiological parameters in tilapia (*Oreochromis niloticus*). RSC Adv. 10 (50), 29918–29922. https://doi.org/10.1039/d0ra06026g.

Kumari, I., Singh, A.K., Dimri, D.C., Singh, K.P., Nand, V., Shivran, J.S., Gaurav, K., Divya, Gangola, S., Mitra, D. and Chattaraj, S., 2025. Pre-harvest bagging modulates fruit traits, phytochemical profiles and disease incidence in mango (Mangifera indica L.) cv. Mallika. Sci. Rep. 15 (1), 42533.

Kuppusamy, P., Yusoff, M.M., Maniam, G.P., Govindan, N., 2016. Biosynthesis of metallic nanoparticles using plant derivatives and their new avenues in pharmacological applications – an updated report. Saudi Pharm. J. 24 (4), 473–484. https://doi.org/10.1016/j.jsps.2014.11.013.

Lin, Y.T., Pan, Y.F., Han, Y.S., 2023. Effects of adding spermidine carbon quantum dots in feed on growth, intestinal morphology, immunity and disease resistance of Anguilla japonica and Anguilla marmorata. Aquac. Rep. 33, 101847.

Magagula, L.P., Masemola, C.M., Ballim, M.A., Tetana, Z.N., Moloto, N., Linganiso, E.C., 2022. Lignocellulosic biomass waste-derived cellulose nanocrystals and carbon nanomaterials: a review. Int. J. Mol. Sci. 23 (8). https://doi.org/10.3390/ijms23084310.

Mahalingam, V., Sivaraju, M., Shanmugasundaram, P., 2023. Synthesis of silica/hydroxyapatite nanocomposites using rice husk and fish scale for corrosion protection application. Silicon 15 (13), 5603–5613.

Mahmoud, A.S., Sayed, A.E.D.H., Mahmoud, U.T., Mohammed, A.A.A., Darwish, M.H.A., 2024. Impact of zinc oxide nanoparticles on the behavior and stress indicators of African catfish (*Clarias gariepinus*) exposed to heat stress. BMC Vet. Res. 20 (1). https://doi.org/10.1186/s12917-024-04302-6.

Mahmud, M.M., Perveen, A., Jahan, R.A., Matin, M.A., Wong, S.Y., Li, X., Arafat, M.T., 2019. Preparation of different polymorphs of cellulose from different acid hydrolysis medium. Int. J. Biol. Macromol. 130, 969–976. https://doi.org/10.1016/j.ijbiomac.2019.03.027.

Mandal, A., Chakrabarty, D., 2011. Isolation of nanocellulose from waste sugarcane bagasse (SCB) and its characterization. Carbohydr. Polym. 86 (3), 1291–1299. https://doi.org/10.1016/j.carbpol.2011.06.030.

Melanie, K., Nazlia, S., Saputra, J., Febri, S.P., Aprita, I.R., 2024. The effect of banana peel flour (Musa paradisiaca) as additional feed raw materials for common carp (Cyprinus carpio). In: BIO Web of Conferences, 87. https://doi.org/10.1051/bioconf/20248703024.

Misra, V.K., Gupta, S., Singh, C.P., Singh, S., Kumar, A., 2023. Nanoencapsulation of nutrients in fish feed: enhancing bioavailability, absorption, and health benefits for farmed fish. Biol. Forum 15, 632.

Mohammady, E.Y., Elashry, M.A., Ibrahim, M.S., Elarian, M., Salem, S.M.R., El-Haroun, E. R., Hassaan, M.S., 2024. Nano iron versus bulk iron forms as functional feed additives: growth, body indices, hematological assay, plasma metabolites, immune, anti-oxidative ability, and intestinal morphometric measurements of Nile tilapia, *Oreochromis niloticus*. Biol. Trace Elem. Res. 202 (2), 787–799. https://doi.org/10.1007/s12011-023-03708-x.

Mohtashemipour, H., Mohammadian, T., Mozanzadeh, M.T., Mesbah, M., Nejad, A.J., El Basuini, M., 2024. Dietary selenium nanoparticles improved growth and health indices in Asian seabass (*Lates calcarifer*) juveniles reared in high saline water. Aquac. Nutr. 2024 (1). https://doi.org/10.1155/2024/7480824.

Murshed, S., Rahman, M.H., Mahruf, B., Najmunnahar, M.S., Hossain, M.S., 2023. Impact of silica nanoparticles on the digestibility and growth efficiency of rohu (Labeo rohita). J. Agric. Food Environ 04 (04), 24–30. https://doi.org/10.47440/jafe.2023.4405.

Nivaethaa, C., Rajkishore, S.K., Maheswari, M., Sritharan, N., Sathya Moorthy, P., Prasanthrajan, M., Subramanian, P., Ravendran, M., Rajamanickam, U., Rana, P.B., Raveena, R., Premalatha, V., 2024. Synthesis and characterization of silica nanoparticles derived from tea factory generated wood ash. Int. J. Environ. Clim. Change 14 (2), 340–352. https://doi.org/10.9734/ijecc/2024/v14i23949.

Patel, V., Manoj, K., 2025. Impact of sustainable feed from vegetable waste on coloration in black molly (*Poecilia sphenops*). Biotica Res. Today 7 (5), 158–164. https://doi.org/10.54083/brt/7.5.2025/158-164.

Prakash, P., Doan, H.V., 2022. Effect of coffee silverskin enriched diet to enhance the immunological and growth parameters of Nile Tilapia (Oreochromis niloticus). Arch. Razi Inst. 77 (3), 1281.

Qiang, J., Khamis, O.A.M., Jiang, H.J., Cao, Z.M., He, J., Tao, Y.F., Xu, P., Bao, J.W., 2019. Effects of dietary supplementation with apple peel powder on the growth, blood and liver parameters, and transcriptome of genetically improved farmed tilapia (GIFT, *Oreochromis niloticus*). PLoS One 14 (11). https://doi.org/10.1371/journal.pone.0224995.

Rajamanikandan, S., Biruntha, M., Ramalingam, G., 2022. Blue emissive carbon quantum dots (CQDs) from bio-waste peels and its antioxidant activity. J. Clust. Sci. 33 (3), 1045–1053.

Rashidian, G., Lazado, C.C., Mahboub, H.H., Mohammadi-Aloucheh, R., Prokić, M.D., Nada, H.S., Faggio, C., 2021. Chemically and green synthesized ZnO nanoparticles alter key immunological molecules in common carp (*Cyprinus carpio*) skin mucus. Int. J. Mol. Sci. 22 (6). https://doi.org/10.3390/ijms22063270.

Riyaz, R., Iqbal, G., Gargotra, P., Ahmad Ganie, P., 2025. Interventions of nanotechnology-based applications as a novel tool for sustainable aquaculture and fish medicines. Blue Biotechnol 2 (1). https://doi.org/10.1186/s44315-025-00034-w.

Saad, A.M., Sitohy, M.Z., Sultan-Alolama, M.I., El-Tarabily, K.A., El-Saadony, M.T., 2022. Green nanotechnology for controlling bacterial load and heavy metal accumulation in Nile tilapia fish using biological selenium nanoparticles biosynthesized by Bacillus subtilis AS12. Front. Microbiol. 13. https://doi.org/10.3389/fmicb.2022.1015613.

Šafranko, S., Stanković, A., Hajra, S., Kim, H.J., Strelec, I., Dutour-Sikirić, M., Weber, I., Bosnar, M.H., Grbčić, P., Pavelić, S.K., Széchenyi, A., Mishra, Y.K., Jerković, I., Jokić, S., 2021. Preparation of multifunctional n-doped carbon quantum dots from *Citrus clementina* peel: investigating targeted pharmacological activities and the potential application for fe3+ sensing. Pharmaceuticals 14 (9). https://doi.org/10.3390/ph14090857.

Salin, K.R., Arun, V.V., Mohanakumaran, N., Tidwell, J.H., 2018. Sustainable aaquafeed. Sustainable Aquaculture. Springer International Publishing, Cham, pp. 123–151.

Sani, M.N.H., Amin, M., Siddique, A.B., Nasif, S.O., Ghaley, B.B., Ge, L., Wang, F., Yong, J. W.H., 2023. Waste-derived nanobiochar: a new avenue towards sustainable agriculture, environment, and circular bioeconomy. Sci. Total Environ. 905. https://doi.org/10.1016/j.scitotenv.2023.166881.

Suzuki, A., 2021. Rising importance of aquaculture in Asia: current status, issues, and recommendations. *Asian Development Outlook Update Background Paper.* Asian Development Bank, Manila.

Thangapandiyan, S., Monika, S., 2020. Green synthesized zinc oxide nanoparticles as feed additives to improve growth, biochemical, and hematological parameters in freshwater fish Labeo rohita. Biol. Trace Elem. Res. 195 (2), 636–647. https://doi.org/10.1007/s12011-019-01873-6.

Van Doan, H., Hoseinifar, S.H., Harikrishnan, R., Khamlor, T., Punyatong, M., Tapingkae, W., Yousefi, M., Palma, J., El-Haroun, E., 2021. Impacts of pineapple peel powder on growth performance, innate immunity, disease resistance, and relative immune gene expression of Nile tilapia, *Oreochromis niloticus*. Fish Shellfish Immunol 114, 311–319. https://doi.org/10.1016/j.fsi.2021.04.002.

Vélez-mercado, M.I., Talavera-caro, A.G., Escobedo-uribe, K.M., Sánchez-muñoz, S., Luévanos-escareño, M.P., Hernández-terán, F., Alvarado, A., Balagurusamy, N., 2021. Bioconversion of lignocellulosic biomass into value added products under anaerobic conditions: insight into proteomic studies. Int. J. Mol. Sci. 22 (22). https://doi.org/10.3390/ijms222212249.

Wannavijit, S., Outama, P., Le Xuan, C., Fontana, C.M., Paolucci, M., Ahmed Sumon, M.A., El-Haroun, E., Van Doan, H., 2025. Evaluation of longan (*Dimocarpus longan*) peel powder as fruit by-product additive in Nile tilapia (*Oreochromis niloticus*) feed: effects on growth, immunity, and immune-antioxidant gene expressions. Heliyon 11 (1). https://doi.org/10.1016/j.heliyon.2024.e41609.

Xuan, C.L., Nguyen, V.L., Wannavijit, S., Outama, P., Khongdee, N., Sutthi, N., Nguyen, V.V., Hoseinifar, S.H., Srisapoome, P., Van Doan, H., 2024. Modulation of growth, immunity, and immune-antioxidant gene expression in Nile tilapia, *Oreochromis niloticus*, culture under biofloc system by dragon fruit, Hylocereus undatus, peel powder. J. World Aquacult. Soc. 55 (5). https://doi.org/10.1111/jwas.13088.

Zhu, Y., Li, W., Ahmed, W., Mahmood, M., Ali, H.M., Rizwan, M., Bundschuh, J., Akmal, M., Mehmood, S., 2025. Study on application of green-synthesized ZnO and Si nanoparticles in enhancing aquaculture sediment quality. Chem. Biol. Technol. Agric. 12 (1). https://doi.org/10.1186/s40538-024-00716-4.

CHAPTER

14

Case studies Species-specific and rearing method-specific utilisation of agroindustrial wastes and by-products as fish feed

Bijay Kali Mahapatra

ICAR-Central Institute of Fisheries Education, Kolkata, West Bengal, India

1. Introduction

Aquaculture plays a pivotal role in ensuring global food security, with its expansion largely attributed to innovations in feed technology and intensive farming systems. Fish feed, constituting over 50%–60% of the total production cost, is a critical input in aquaculture (Tacon and Metian, 2015). Conventionally, aquafeeds have relied on high-protein ingredients such as fishmeal and soybean meal, valued for their digestibility and amino acid profile. These ingredients, particularly fishmeal, have long been regarded as the "gold standard" in formulated feeds, especially for carnivorous species like salmon and sea bass (Hardy, 2010; Chattaraj et al., 2024a).

However, the reliance on fishmeal and soybean meal has led to ecological and economic challenges (Chattaraj et al., 2025, 2026b). The sustainability of fishmeal is under scrutiny due to overexploitation of wild fish stocks, while soybean cultivation is associated with deforestation, intensive water usage, and carbon emissions (Naylor et al., 2009; FAO, 2020). Moreover, price volatility and supply constraints have made these ingredients less viable, particularly for small-scale and resource-poor farmers in developing nations (El-Sayed, 2020; Chattaraj et al., 2026a; Chattaraj and Das Mohapatra, 2024).

In response to these constraints, there has been a growing interest in exploring agroindustrial by-products (AIBPs) as cost-effective and sustainable alternatives to conventional feed ingredients. AIBPs include materials such as rice bran, wheat bran, mustard oil cake, groundnut cake, fruit pomace, cassava peel, brewery waste, distillery sludge, poultry offal meal, and fish processing residues. These by-products are often locally available, rich in macro- and micronutrients, and can be processed to enhance digestibility and nutrient availability (Amirkolaie, 2011; Sardar et al., 2008). When properly treated and balanced in formulations, AIBPs have shown

Fish Feed Formulation with Agroindustrial Wastes and Byproducts. https://doi.org/10.1016/B978-0-443-41728-3.00011-3

potential to support comparable growth performance and feed efficiency in several farmed fish species (Tiamiyu et al., 2022).

The transition toward AIBPs is also aligned with the broader concept of circular economy and waste-to-wealth strategies in aquaculture. By integrating such by-products into feed formulations, aquaculture systems can reduce environmental burdens, improve resource use efficiency, and promote sustainable farming practices (FAO, 2022).

This chapter aims to document and analyse case studies on species-specific and rearing method-specific utilisation of agroindustrial by-products as alternative fish feed ingredients. Drawing from international and Indian experiences, it provides:

- A classification and nutritional profile of major agroindustrial wastes used in aquafeed;
- Case studies of species such as *Salmo salar* (salmon), *Oreochromis niloticus* (tilapia), *Catla catla* (catla), and *Pangasianodon hypophthalmus* (pangasius);
- System-specific applications in rearing methods including sea cage farming, recirculatory aquaculture systems (RAS), biofloc systems, and integrated farming;
- A critical discussion of the benefits and challenges of AIBP-based feeds;
- Policy and research recommendations to enhance adoption and innovation in sustainable feed practices.

2. Agroindustrial wastes and by-products in aquafeed

2.1 Definition and classification of agroindustrial wastes

Agroindustrial wastes and by-products (AIBPs) are residual materials generated during the processing of agricultural produce and animal products for food, fiber, or bioenergy. These wastes, often underutilized or discarded, hold significant nutritional potential for aquafeed formulation when appropriately treated and processed. Unlike municipal or hazardous waste, AIBPs are biodegradable and rich in organic nutrients, making them suitable for partial or complete replacement of conventional feed ingredients (Amirkolaie, 2011; Chattaraj et al., 2024b).

AIBPs can be broadly classified based on their source of origin:

- **Cereal by-products:** Generated from rice, wheat, maize, and other grains (e.g., rice bran, wheat bran).
- **Oilseed residues:** Resulting from oil extraction industries (e.g., mustard oil cake, soybean cake, groundnut cake).
- **Sugar and starch industry wastes:** Including molasses, bagasse, distillery spent wash, and starch residues.
- **Fruit and vegetable processing waste:** Such as fruit pomace, peels, and pulp.
- **Animal by-products:** Including poultry offal, blood meal, bone meal, feather meal.
- **Fishery waste:** Comprising fishmeal residue, fish frames, viscera, and skin from fish processing units.

- **Root and tuber processing waste:** Such as cassava peels and sweet potato residue.

These categories highlight the diversity and regional specificity of AIBPs. For example, mustard oil cake is commonly used in India and Bangladesh, whereas cassava residues are more prevalent in Southeast Asia and Africa.

2.2 Examples of common agroindustrial by-products used in aquafeed

Table 15.1 represents the agroindustrial by-products that can be used as fish feed ingredients.

2.3 Nutritional composition and variability

The nutritional quality of AIBPs varies significantly depending on the source material, processing method, moisture content, presence of anti-nutritional factors (ANFs), and seasonal variation (Tacon and Metian, 2008). Key aspects include:

- **Crude protein:** Ranges from 10% (e.g., rice bran) to over 60% (e.g., fish waste meal or poultry offal meal). Protein digestibility is influenced by the presence of fiber and ANFs like tannins or phytic acid.
- **Lipid content:** Oilseed cakes and fishery waste are good sources of essential fatty acids, though their composition may be unbalanced compared to marine fish oil.
- **Fiber and ash:** High fiber (e.g., in wheat bran or cassava peels) can reduce digestibility. High ash content in bone meal or spent wash can affect feed palatability.
- **Energy value:** Many AIBPs are energy-dense (e.g., molasses, cassava starch residue), making them suitable carbohydrate sources, especially in omnivorous species.
- **Minerals and vitamins:** Oilseed cakes and bran are sources of calcium, phosphorus, iron, and B-vitamins, though supplementation is often necessary for balanced diets.

Processing methods such as fermentation, enzymatic treatment, or heat application can improve nutrient bioavailability, detoxify ANFs, and increase shelf-life (Hardy, 2010; NRC, 2011).

3. Species-specific case studies

Agroindustrial by-products (AIBPs) offer a promising solution for species-specific aquaculture by meeting the nutritional needs of different fish while reducing feed costs and environmental impacts. The following case studies illustrate the practical application of AIBPs in the diets of selected species across diverse aquaculture systems.

3.1 Salmon (Salmo salar)—Sea cage farming (Norway, Chile)

Salmon farming in Norway and Chile is among the most industrialized aquaculture practices globally, traditionally relying heavily on high-quality fishmeal and fish oil to meet the nutritional demands of this carnivorous species. However, the limited availability and rising cost of marine resources have driven the industry to explore alternative protein and lipid sources:

- **Use of rendered animal fats, poultry by-product meal (PBM), and canola meal:** Studies have shown that PBM can replace up to 30%–40% of fishmeal in salmon diets without compromising growth or feed efficiency (Hatlen et al., 2016). Rendered animal fats (e.g., beef tallow, poultry fat) are used as energy sources, while canola meal provides digestible plant protein and essential amino acids like methionine and lysine.
- **Improved lipid efficiency and reduction in fishmeal use:** Substituting fish oil with animal fats or vegetable oils reduces dependency on marine oils. While total lipid digestibility remains high, tailored dietary fatty acid balancing is essential to ensure optimal health and metabolic function (Klinger and Naylor, 2012).
- **Impact on omega-3 retention and fillet quality:** Replacing marine oils often reduces eicosapentaenoic acid (EPA) and docosahexaenoic acid (DHA) in salmon fillets. However, strategic finishing diets (i.e., switching back to marine oils before harvest) can restore fillet quality (Sprague et al., 2016).

3.2 Tilapia (Oreochromis niloticus)—Biofloc system (India, Brazil)

Tilapia, a hardy and omnivorous species, thrives in biofloc technology (BFT) systems, which enable nutrient recycling and microbial protein utilisation. This system is particularly suitable for smallholder farmers in India and Brazil:

- **Use of brewery waste, fermented fruit peels, cassava peel meal, and sugar industry by-products:** These carbohydrate-rich AIBPs are used to manipulate the C:N ratio in biofloc systems, promoting the growth of heterotrophic bacteria which serve as an in situ protein source (Avnimelech, 2009). Cassava peel meal, after detoxification, provides energy, while fermented fruit peels improve palatability and microbial activity.
- **Enhancement of heterotrophic microbial protein:** Biofloc biomass can contribute 20%–30% of the daily protein requirement of tilapia, effectively lowering external feed input (Hargreaves, 2013).
- **Water quality and nutrient recycling:** AIBPs used in biofloc systems support nitrogen assimilation and reduce ammonia levels. This minimizes water exchange requirements and enhances system sustainability.

3.3 Catla (Catla catla)—Pond and RAS farming (India)

Catla, an Indian Major Carp (IMC), is herbivorous with high dietary carbohydrate tolerance. It is widely cultured in ponds and increasingly in Recirculatory Aquaculture Systems (RAS) for improved biosecurity and water efficiency:

- **Utilisation of mustard oil cake, de-oiled rice bran, and groundnut cake:** These AIBPs are the mainstay of semi-intensive carp aquaculture in India. Mustard oil cake provides protein and residual oil; de-oiled rice bran supplies energy and fiber; groundnut cake is a balanced protein supplement (Jena et al., 2003).
- **Growth performance, digestibility, and FCR improvements:** Balanced use of these by-products yields FCR values between 1.5 and 2.0, depending on feed formulation and culture conditions. Their digestibility can be enhanced through fermentation or enzyme supplementation.
- **Synergy with integrated farming:** Catla culture is commonly integrated with duck or poultry farming, where litter and droppings enrich pond fertilisation, support natural productivity, and reduce external feed needs (Pillay and Kutty, 2005).

3.4 Pangasius (Pangasianodon hypophthalmus)—Intensive pond farming (Vietnam, Bangladesh)

Pangasius is a fast-growing, omnivorous catfish species with high tolerance to low water quality and crowding, making it ideal for intensive farming:

- **Integration of slaughterhouse waste, bone meal, cassava starch:** These ingredients are incorporated in mass-produced floating or sinking pellets. Bone meal serves as a phosphorus source, while slaughterhouse waste contributes high crude protein (50%–60%) and lipids. Cassava starch acts as a binder and carbohydrate energy source (Ali et al., 2020).
- **Cost-effective mass-scale production:** In Vietnam's Mekong Delta, use of AIBPs has helped reduce feed cost by up to 30% without compromising yield, with harvests exceeding 200 t/ha/year (Phan et al., 2019).
- **Water pollution management challenges and mitigation:** While high stocking and nutrient input raise BOD and eutrophication risks, feed quality improvements and sediment management are employed to reduce organic load.

4. Rearing method-specific case studies

Different aquaculture rearing systems demand specific feed strategies to match their water quality dynamics, nutrient retention capabilities, and economic objectives. The incorporation of agroindustrial by-products (AIBPs) into such systems not only reduces feed costs but also supports ecological sustainability by aligning feed inputs with system-specific operational goals.

4.1 Sea cage farming

Sea cage farming is predominantly used for high-value carnivorous species like *Salmo salar* (Atlantic salmon) and *Seriola quinqueradiata* (yellowtail). It involves open-water cages placed in coastal or offshore environments, where nutrient loss from uneaten feed or feces can directly affect the surrounding ecosystem:

- **Use of stabilized animal fats and blood meal for high energy density:** Sea cage diets demand high energy to support fast growth in offshore conditions. Stabilized animal fats (e.g., poultry fat, lard) and blood meal provide concentrated energy and protein, enhancing feed caloric density without increasing pellet bulk (Hatlen et al., 2016; Sørensen, 2012). These by-products have been shown to partially replace fish oil and fishmeal in salmonid diets without compromising growth.
- **Challenges in pellet stability and nutrient leaching:** Sea cage systems require feed pellets to maintain structural integrity in water until consumed. AIBPs such as blood meal and fats can reduce pellet stability if not properly processed. Pellet binders and extrusion techniques are essential to improve water stability and reduce nutrient leaching, which otherwise can lead to benthic enrichment and water pollution (Boonyaratpalin et al., 1998).

4.2 Recirculatory aquaculture system (RAS)

RAS is a closed-loop aquaculture system where water is continuously filtered and reused, enabling high-density fish production with minimal water usage. Feed formulation in RAS must align with waste minimization and water quality control:

- **Importance of low-waste diets:** Nutrient-rich effluents can quickly accumulate in RAS if feeds are not highly digestible and waste is not managed effectively. Diets must be precisely formulated with minimal leaching and high nutrient digestibility (Martins et al., 2010).
- **Application of fruit pulp, cottonseed meal, and okara (soybean residue):** These AIBPs are suitable for RAS feed formulations due to their low cost and good nutritional potential. Fermented fruit pulp provides fermentable carbohydrates, cottonseed meal offers high protein (with proper detoxification), and okara contributes protein and fiber (El-Haroun et al., 2009).
- **Enhanced water stability and reduced nitrogenous waste:** Pelletized feeds incorporating these by-products, when appropriately processed, show improved water stability and reduced ammonia excretion. This ensures stable operation of biofilters and reduced stress on fish.

4.3 Biofloc system

Biofloc Technology (BFT) is a high-density system relying on the manipulation of microbial communities to convert waste nitrogen into microbial protein. It is widely adopted for tilapia and shrimp farming:

- **C:N ratio manipulation using molasses and carbohydrate-rich wastes:** Carbohydrate-rich agro-wastes such as molasses, jaggery, and cassava peel meal are used to elevate the carbon-to-nitrogen (C:N) ratio in the water, stimulating the proliferation of heterotrophic bacteria which assimilate ammonia and form flocs (Avnimelech, 2009).
- **Synergy of biofloc with fermented agro-residues:** Incorporating fermented fruit peels and brewery waste enhances microbial activity and palatability, contributing to nutrient recycling and internal protein generation (Emerenciano et al., 2013).
- **Reduced dependency on formulated feeds:** Biofloc serves as a natural protein source, partially offsetting the need for formulated feed. Studies show that up to 30% of protein requirements can be met through biofloc in tilapia and shrimp culture (Crab et al., 2012).

4.4 Integrated farming (poultry-fish, pig-fish, agroforestry-fish)

Integrated farming systems aim to optimize resource use by combining aquaculture with other agricultural practices. These systems are especially common in South and Southeast Asia, offering a natural platform for the recycling of organic wastes:

- **On-farm recycling of poultry litter, vegetable peels, and cow dung:** Poultry droppings and pig manure are traditionally used to fertilise ponds, stimulating planktonic growth. Vegetable residues and cow dung contribute additional nutrients and organic matter. These materials, when used judiciously, reduce the need for supplementary feed and fertilizer (Edwards et al., 1988).
- **Nutrient flow and ecosystem productivity enhancement:** Integrated systems promote nutrient cycling, enhance natural food availability (e.g., zooplankton, phytoplankton), and increase system resilience. For instance, duck-fish systems contribute nitrogen through droppings while controlling pests and aerating the pond.
- **Case example from West Bengal and Northeast India:** In West Bengal, integrated poultry-fish farming has successfully reduced feed input by 40% while increasing fish yield by 25%–30%. In the Northeast, tribal communities use pig-fish systems where pig waste is channeled directly into ponds, supporting plankton production and tilapia growth (Jena et al., 2003; ICAR, 2021).

5. Benefits and challenges

The utilisation of agroindustrial wastes and by-products (AIBPs) in aquafeed has emerged as a practical approach to enhancing the sustainability and economics of aquaculture. While the benefits are numerous and increasingly well-documented, the adoption of AIBPs is not without its set of challenges. This section critically discusses both sides.

5.1 Benefits

5.1.1 Reduction in feed cost

Feed constitutes the largest operational cost in aquaculture, often accounting for 50%–70% of total expenses (Tacon and Metian, 2015). By incorporating locally available AIBPs—such as rice bran, oilseed cakes, and poultry offal—farmers can significantly reduce dependency on costly fishmeal and soybean meal. Studies in India and Southeast Asia have demonstrated feed cost reductions of up to 30%–40% without compromising fish growth or health (Ali et al., 2020).

5.1.2 Sustainable waste management

AIBPs offer a productive route for recycling large volumes of organic waste generated from agriculture and food industries. Their use in aquaculture not only diverts waste from landfills and waterways but also reduces greenhouse gas emissions associated with waste decomposition and burning (Amirkolaie, 2011). It aligns well with environmental regulations promoting resource recovery and sustainable intensification.

5.1.3 Enhanced local employment and rural empowerment

The decentralized collection, processing, and use of AIBPs in aquafeed can generate employment opportunities in rural and peri-urban areas. Small-scale feed mills, fermentation units, and integrated farming systems facilitate the involvement of local communities, enhancing rural livelihoods and gender inclusivity (FAO, 2022).

5.1.4 Promotion of circular economy in aquaculture

By integrating waste materials into productive aquafeed cycles, a circular economy model is reinforced. This approach ensures efficient resource utilisation, minimizes environmental footprint, and reduces aquaculture's reliance on finite marine resources. It also opens avenues for "green labeling" and eco-certification of aquaculture products (Crab et al., 2012).

5.2 Challenges

5.2.1 Nutritional inconsistency and presence of anti-nutritional factors (ANFs)

AIBPs vary greatly in nutrient composition depending on their source, season, and processing method. Many contain anti-nutritional compounds such as tannins, phytic acid, trypsin inhibitors, and cyanogenic glycosides (notably in cassava peels and cottonseed meal), which can impair nutrient absorption and fish health (Hardy, 2010). Unless adequately treated—through heat processing, fermentation, or enzyme supplementation—these compounds can limit the inclusion rate of AIBPs.

5.2.2 Lack of standardization and processing infrastructure

In many developing regions, the lack of standardization in collection, drying, grinding, and detoxification of AIBPs poses a major constraint. Absence of proper storage facilities leads to microbial contamination, mycotoxin buildup, and nutrient degradation, affecting feed safety and efficacy (NRC, 2011). Investment in small-scale feed processing infrastructure remains critical for quality assurance.

5.2.3 Consumer perception and quality assurance issues

The use of animal by-products or fermented wastes in aquafeed raises concerns among consumers regarding food safety, product taste, and nutritional quality. In export-oriented aquaculture systems (e.g., shrimp, salmon), there are stringent limits on the use of certain feed components due to import regulations and certification norms (El-Sayed, 2020). Overcoming this perception barrier requires transparent labeling, adherence to international feed standards, and education campaigns highlighting sustainability benefits.

6. Conclusion and recommendations

The use of agroindustrial wastes and by-products (AIBPs) as alternative feed ingredients in aquaculture represents a promising strategy for addressing some of the most pressing challenges in the sector—rising feed costs, environmental degradation, and dependency on finite resources like fishmeal and soybean. Through both species-specific and system-specific case studies, this chapter has demonstrated the viability of incorporating AIBPs such as oilseed cakes, fruit and vegetable residues, poultry by-products, and brewery waste in diets for fish such as *Salmo salar*, *Oreochromis niloticus*, *Catla catla*, and *Pangasianodon hypophthalmus* across diverse farming systems.

6.1 Future prospects of agro-waste-based feeds

The future of agro-waste-based feeds lies in the growing need for sustainable intensification of aquaculture. With advancements in feed formulation, microbial fermentation, enzyme supplementation, and pellet processing technologies, many of the limitations associated with nutrient imbalance and anti-nutritional factors can be mitigated. Moreover, the concept of functional feeds—where agro-waste ingredients are enhanced with probiotics, prebiotics, or immune boosters—offers further avenues for value addition (Tiamiyu et al., 2022).

As the circular economy gains global traction, aquaculture stands out as a suitable platform for bioresource recycling. AIBP-based feeds can play a pivotal role in low-input, eco-friendly farming models—particularly in integrated and biofloc systems—thereby promoting both environmental and economic sustainability.

6.2 Need for policy incentives and R&D support

Despite their potential, the widespread adoption of AIBP-based feeds is constrained by the lack of standardised processing infrastructure, quality assurance protocols, and farmer awareness. Therefore, a coordinated effort from governments, research institutions, and industry stakeholders is required:

- **Policy support** should include subsidies for decentralized feed mills, incentives for feed innovation start-ups, and recognition of sustainable feed practices through eco-labeling or carbon credits.
- **Research and development (R&D)** must focus on.
- Nutritional profiling of region-specific agro-wastes.
- Processing techniques for detoxification and preservation.
- Performance trials across species and farming systems.
- Development of regulatory standards and safety guidelines.

Public-private partnerships (PPPs) and community-based feed enterprises can bridge the gap between research innovations and field-level adoption, especially in rural and tribal farming communities.

6.3 Integration with climate-resilient aquaculture

Climate change poses significant risks to aquaculture in the form of temperature fluctuations, disease outbreaks, and water resource scarcity. Agro-waste-based feeds contribute to climate resilience in the following ways:

- **Reduced carbon footprint** by replacing resource-intensive marine ingredients.
- **Enhanced nutrient cycling** within integrated farming systems.
- **Support for low-water-use systems** such as RAS and biofloc, where waste conversion efficiency is crucial.
- **Increased adaptability** through decentralized, locally tailored feed strategies.

Integrating AIBPs into holistic climate-resilient aquaculture models will help build robust food systems that are economically viable, ecologically sound, and socially inclusive.

7. Concluding remarks

For aquaculture to remain sustainable and scalable, it must evolve beyond conventional paradigms of feed production. Agroindustrial by-products, when managed scientifically, offer an invaluable resource in this transformation. With the right blend of innovation, regulation, and participation, AIBP-based feeds can significantly contribute to the future of sustainable aquaculture (Table 14.1).

Table 14.1 Common agroindustrial by-products used in aquafeed.

By-product	Source industry	Primary nutrients	Usage in aquafeed
Rice bran	Rice milling	Crude protein (12%–16%), lipids, B-vitamins	Used for herbivorous fish like rohu, catla
Wheat bran	Wheat flour processing	Crude protein (12%–14%), fiber	Used as filler or energy source
Mustard oil cake	Oil extraction	Crude protein (30%–36%), residual oil	High in protein; used for Indian carps

Table 14.1 Common agroindustrial by-products used in aquafeed.—*cont'd*

By-product	Source industry	Primary nutrients	Usage in aquafeed
Groundnut cake	Oil extraction	Crude protein (40%–50%)	Common in catla and tilapia feed
Soybean meal	Oil extraction	Crude protein (44%–48%), lysine	Popular plant protein substitute for fishmeal
Molasses	Sugar industry	Carbohydrates (45%–55%), trace minerals	Used to manipulate C:N ratio in biofloc
Distillery spent wash	Alcohol production	Soluble proteins, minerals	Use limited due to COD/BOD concerns
Fruit peels/pomace	Fruit juice and pulp industry	Carbohydrates, fiber, vitamins	Used after fermentation or drying
Cassava residue/peels	Starch industry	Carbohydrates, fiber	Requires detoxification (cyanide removal)
Poultry offal meal	Poultry slaughterhouses	Crude protein (50%–60%), fats, minerals	Used for carnivorous species like pangasius
Fish processing waste	Fish filleting industry	Crude protein (60%–70%), omega-3 fats	Used as partial replacement of fishmeal

References

Ali, H., Haque, M.M., Wahab, M.A., 2020. Utilization of animal by-products in aquafeed: a sustainable approach,". Aquac. Rep. 17, 100345.

Amirkolaie, A.K., 2011. Reduction in the environmental impact of waste discharged by fish farms through feed and feeding. Rev. Aquacult. 3 (1), 19–26.

Avnimelech, Y., 2009. Biofloc technology: a practical guide book. The World Aquaculture Society.

Boonyaratpalin, M., Suraneiranat, P., Tunpibal, T., 1998. Replacement of fishmeal with poultry by-product meal in diets for Asian sea bass (*Lates calcarifer*). Aquac. Res. 29 (10), 849–855.

Chattaraj, S., Das Mohapatra, P.K., 2024. Immunostimulatory and antagonistic potential of the methanolic extract of Oedogonium intermedium SCB in Cirrhinus reba challenged with Aeromonas hydrophila. Aquac. Int. 32 (3), 2523–2548.

Chattaraj, S., Behera, B.K., Das Mohapatra, P.K., 2024a. Effective valorization of Chlorella biomass and Brewers spent grain substituting fish meal and soybean meal in the diet of herbivorous fish Cirrhinus reba for higher growth, digestibility and sustainable cultivation. J. Appl. Phycol. 36 (1), 217–232.

Chattaraj, S., Mitra, D., Chattaraj, M., Ganguly, A., Thatoi, H., Mohapatra, P.K.D., 2024b. Brewers' spent grain as fish feed ingredient: evaluation of bio-safety and analysis of its impact on gut bacteria of Cirrhinus reba by 16S Metagenomic sequencing. Curr. Res. Microb. Sci. 7, 100286.

Chattaraj, S., Chattaraj, M., Mitra, D., Ganguly, A., Thatoi, H., Das Mohapatra, P.K., 2025. 16S amplicon sequencing of the gastrointestinal microbiota of Cirrhinus reba and isolation of an autochthonous probiotic using culture based approaches. Syst. Microbiol. Biomanuf. 5 (1), 156–170.

Chattaraj, S., Ganguly, A., Mitra, D., et al., 2026a. Sustainable valorization of algal biomass in fish nutrition and dietetics: a critical review and scientometric study. Waste Biomass Valor. https://doi.org/10.1007/s12649-026-03619-7.

Chattaraj, S., Sahoo, S.S., Behera, S., Thatoi, H., 2026b. Sustainable aquafeed development using Spirulina: effects on growth performance of Catla catla. Algal Res. 104447.

Crab, R., Defoirdt, T., Bossier, P., Verstraete, W., 2012. Biofloc technology in aquaculture: beneficial effects and future challenges. Aquaculture 356–357, 351–356.

Edwards, P., Pullin, R.S.V., Gartner, J.A., 1988. Research and education for the development of integrated crop-livestock-fish farming systems in the tropics. In: ICLARM Studies and Reviews, 16. ICLARM.

El-Haroun, E.R., Azevedo, P.A., Bureau, D.P., 2009. Comparative study on the effects of different plant protein sources with or without enzyme supplementation on growth and nutrient utilization of Nile tilapia. Aquac. Res. 40 (11), 1261–1270.

El-Sayed, A.-F.M., 2020. Tilapia culture, 2nd ed. Academic Press.

Emerenciano, M., Gaxiola, G., Cuzon, G., 2013. Biofloc technology (BFT): a review for aquaculture application and animal food industry. In: Pandey, A. (Ed.), Biomass now—cultivation and utilization. InTechOpen, pp. 301–328.

FAO, 2020. Sustainable Aquaculture: The Role of Feed and Feeding Practices. Food and Agriculture Organization of the United Nations.

FAO, 2022. The State of World Fisheries and Aquaculture 2022: Towards Blue Transformation. Food and Agriculture Organization of the United Nations.

Hargreaves, J.A., 2013. In: Biofloc Production Systems For Aquaculture, 4503. Southern Regional Aquaculture Center Publication, pp. 1–11.

Hardy, R.W., 2010. Utilization of plant proteins in fish diets: effects of global demand and supplies of fishmeal. Aquac. Res. 41 (5), 770–776.

Hatlen, B., Olli, J.J., Tvenning, L., 2016. Growth performance and nutrient digestibility in Atlantic salmon fed diets containing rendered animal fats. Aquac. Nutr. 22 (1), 63–75.

ICAR, 2021. Success Stories in Integrated Fish Farming. Indian Council of Agricultural Research.

Jena, J.K., Aravinda Kumar, S., Ayyappan, S., Muduli, H.K., 2003. Comparative evaluation of production performance in carp polyculture system with different species combinations. J. Appl. Aquacult. 13 (3–4), 73–87.

Klinger, D., Naylor, R., 2012. Searching for solutions in aquaculture: charting a sustainable course. Annu. Rev. Environ. Resour. 37, 247–276.

Martins, C.I.M., Eding, E.H., Verdegem, M.C.J., Heinsbroek, L.T.N., Schneider, O., Blancheton, J.P., Roque d'Orbcastel, E., Verreth, J.A.J., 2010. New developments in recirculating aquaculture systems in Europe: a perspective on environmental sustainability. Aquac. Eng. 43 (3), 83–93.

Naylor, R.L., Hardy, R.W., Bureau, D.P., Chiu, A., Elliott, M., Farrell, A.P., Forster, I., Gatlin, D.M., Goldburg, R.J., Hua, K., Nichols, P.D., 2009. Feeding aquaculture in an era of finite resources. Proc. Natl. Acad. Sci. U S A. 106 (36), 15103–15110.

NRC (National Research Council), 2011. Nutrient Requirements of Fish and Shrimp. The National Academies Press.

Phan, L.T., Bui, T.M., Nguyen, Q.H., 2019. Pangasius farming in Vietnam: financial and environmental sustainability. Aquac. Rep. 13, 100179.

Pillay, T.V.R., Kutty, M.N., 2005. Aquaculture: Principles and Practices, 2nd ed. Blackwell Publishing.

Sardar, P., Abidi, S.F., Jafri, A.K., 2008. Replacement of fishmeal with soybean meal in diets of Indian major carp *Labeo rohita*. Asian Fish Sci. 21 (2), 195–203.

Sørensen, M., 2012. Alternative protein sources as partial fish meal replacements in diets for Atlantic salmon. Anim. Feed Sci. Technol. 173 (2–4), 144–155.

Sprague, M., Dick, J.R., Tocher, D.R., 2016. Impact of sustainable feeds on omega-3 long-chain fatty acid levels in farmed Atlantic salmon, 2006–2015. Sci. Rep. 6, 21892.

Tacon, A.G.J., Metian, M., 2008. Global overview on the use of fishmeal and fish oil in industrially compounded aquafeeds: trends and future prospects. Aquaculture 285 (1–4), 146–158.

Tacon, A.G.J., Metian, M., 2015. Feed matters: satisfying the feed demand of aquaculture. Rev. Fish. Sci. Aquacult. 23 (1), 1–10.

Tiamiyu, L.O., El-Haroun, E.R., Davies, S.J., 2022. Sustainable aquafeeds: a review of ingredients and nutritional strategies. Aquac. Rep. 26, 101330.

CHAPTER

15 Effect of agro-waste incorporated fish feed on the growth of carnivorous fish

Pratikhya Mohanty[1], Partha Bandyopadhyay[2], Snehasish Mishra[1], Saswati Mohanty[1], Puneet K. Singh[1] and Saroj K. Swain[3]

[1]*School of Biotechnology, Campus 11, KiiT Deemed University, Bhubaneswar, Odisha, India;* [2]*Finray Biotech Private Ltd, Ahmedabad, Gujarat, India;* [3]*Central Institute of Freshwater Aquaculture (ICAR-CIFA), Bhubaneswar, Odisha, India*

1. Introduction

Ecological needs of carnivorous fish are very peculiar. On the one hand, there are numerous species that are invasive in nature, thereby being potentially threatening to the biodiversity in particular, and the ecosystem in general. There are also the concerns of deliberate (for high productivity) or indeliberate (through, say, flood incursion) introduction of exotic species, on the other hand. The cases of African catfish (*Clarias gariepinus*) and Tilapia (*Oreochromis mossambicus* and *Oreochromis niloticus*) in Indian inland waters are glaring instances that are a matter of concern equally to fishery professionals, limnologists, and environmental scientists (Sarkar et al., 2022). The introduction of these exotic species in Indian waters allegedly has widespread and significant ecological consequences, such as competition with native species and irreversibly modifying the ecosystem. In an aquaculture ecosystem, the presence of invasive exotic carnivorous fish worries and concerns the ecologists and limnologists. Carnivorous fish is aquacultured due to its high market demand.

Further, carnivorous fish defy the typical food pyramid in the aquaculture ecosystem. In an aquatic food-chain or food-web, herbivorous fish largely feed on phytoplankton such as algae, diatoms, and cyanobacteria as primary producers, and zooplankton such as krill, meroplankton, protozoa, and holoplankton as primary consumers or secondary producers. In contrast, however, carnivorous fish, which require high-protein diet, are often dependent on other non-conventional sources for their food provision (Ganguly et al., 2025). These include insects, crustaceans, shrimps, worms, and other small fish. When cultured in confined encroachments, focusing on quality protein to support natural predation, the carnivorous fish is frequently fed with commercial Hikari pellets or sticks (nutritionally rich scientifically formulated floating pellets especially to feed top-feeding carnivorous fish),

Fish Feed Formulation with Agroindustrial Wastes and Byproducts. https://doi.org/10.1016/B978-0-443-41728-3.00015-0

frozen or live bloodworms, brine shrimp, earthworms as food, or even occasionally supplemented with beef heart or hard-boiled egg yolk. In light of this, there is always a need for extended research and development of sustainable carnivores-focused aquafeeds. This is due to the high dependence of culture fishery on high-protein feeds keeping the general ecology of carnivorous fish under consideration. The feed for carnivores needs to be high in protein content, and therefore the feed formulation often involves the use of proteins from animal origin (primarily, the fish oil and fishmeal) (Chattaraj et al., 2024a, 2026a).

In the Indian context, there is a huge quantity of agri-waste generated every year, which could be considered a good alternate resource in culture fishery (Barik et al., 2000) and feed formulation (Bandyopadhyay et al., 2005), both for the purpose of balanced nutrient source as well as adding bulk and water stability to the feed. With this background, this chapter is an attempt to understand the recent cutting-edge technological advancements in similar lines. It has compiled the data in forms of tables and figures for easy comprehension and providing technically-sound suggestions and recommendations to further reinforce the consorted efforts. Formulated feeds that are species-specific and region-specific face numerous market constraints that include a (10–60 times) higher price in comparison to conventional aquaculture feeds, wide variations in price and packaging (being often available only in half kg pack). In contrast, conventional aquaculture feeds are usually available in bulk (in 22 kg packs). These constraints discourage aqua-feed manufacturers to produce species-specific commercial feeds. Enough research and development and comparisons on the performance of the feeds have not been carried out primarily for this reason and hence, reliable information on species-specific formulated feeds is low or rather non-existent. Plant protein ingredients (such as *Spirulina*, soybean and alfalfa meals) and fiber are incorporated into the diets of herbivorous fish such as goldfish and koi, as well as certain catfishes and cichlid species (Bandyopadhyay et al., 2005; Chapman, 2000; Chattaraj et al., 2026b). Nevertheless, many such feeds seem to lack popularity due to the cost factor and the inability to satisfy the nutritional needs of the aquacultured fish species. Also, these feeds apparently do not demonstrate consistent growth upon regular use (Bandyopadhyay et al., 2005).

2. Agri-wastes as potential feed ingredients

Aquaculture and environment are always intricately linked especially in India, and hence a healthy and sustainable aquaculture necessitates a healthy environment (Mishra, 1998a, 1998b; Mishra and Rath, 1999). Agri-wastes are produced in large amounts worldwide, from production to processing, as a result of extensive farming and food-processing operations (Singh et al., 2022a). Global assessments indicate that nearly one-third of the agricultural output is lost or wasted along the supply chain, amounting to about 1.3 billion tonnes of food each year that never reaches the consumption stage (De Boni et al., 2022; Singh et al., 2022a). If not properly managed, these discarded materials could represent a significant economic loss while adding to ecological imbalance and environmental pollution (Mishra, 1998b). The demand for processed foods has increased in recent times even the global population is growing rapidly, so also the amount of agri-waste is expected to further rise. Therefore, proper waste management is crucial to support sustainable aquaculture (Mishra, 2008), minimize the loss of the valuable resources, and ensure a working circular bioeconomy (Singh et al., 2022b). Agri-waste is generated at the various stages, from agri-farming to food processing. These include cultivation of the crops, crop harvesting, food processing, food manufacturing, and the overall logistics involved at each stage. The major contributors to the residues from farming include cereal straw, oilseed residue, stumps, and roots in the farming field, and the spent/discarded tubers, fruits, and vegetables. Similarly, the most common residual wastes generated during the processing stage include husk and bran of the milled rice, wheat bran during flour production, corn cobs and husks from maize processing, cassava peels and pulp from root crop processing, and so on. Fruit peels and pomace produced during juice and canning operations also form a sizable portion of the waste (Drosou et al., 2023). Additionally, brewery operations produce brewer's spent grain while oil extraction activities generate byproducts such as oil-cakes, palm kernel cake, and so on (Chattaraj et al., 2024b, 2025). All of these have the potential to be used in animal/fish feed and numerous other value-added products. Bandyopadhyay et al. (2005) successfully incorporated agri-residues in fish feed.

Based on their physical form and place of origin, these agri-wastes are classified and categorized. Crop residues are plant materials such as straw, stalks, and leaves remain after harvesting the cereals or cash crops. The left-over materials after primary processing, such as bran, husk, cobs, seed cakes, and pomace, are referred as the processing byproducts. Peels, skins, pulp, and leftovers from fruit and vegetable processing facilities make up the processed food wastes. Depending on the source crop and processing technique, these wastes differ greatly in terms of composition, moisture content, and nutrient profile. The nutritional contents of agri-wastes vary considerably, which affects their appropriateness and limitations as feed materials. Although significant amounts of nutrients such as protein, fiber, carbohydrates, and bioactive substances could be found in many agro-industrial byproducts, the nutrient content levels vary depending on the kind of waste and origin of the crop. For instance, cereal bran (like rice and wheat bran) have higher protein and fiber contents as compared to the pure lignocellulosic wastes such as husks and stalks. The nutritional characteristics of such agri-wastes are detailed in Table 15.1.

Processing and formulation techniques can transform the agri-residues to useful ingredients in sustainable animal and aquaculture diets (Bala et al., 2023), although their nutrient composition in the raw forms could vary greatly. Bran products offer dietary fiber, carbohydrate, mild protein, and some important fatty acids, and are beneficial feed components when complemented and supplemented with other nutritional ingredients. However, the high-fiber content may lead to low digestibility particularly in monogastric animals such as the carnivorous fish and requires pretreatment or (maybe in conjunction with) enzyme (biocatalysts) supplementation. Brewer's spent grain (BSG), a brewing waste of appreciable quantity, similarly

Table 15.1 The proximate of some of the commonly generated agri-wastes.

Residue	% moisture	% crude protein	% crude fiber	% organic matter digestibility	Source literatures
Rice bran	9.3–11.4	11.4–17.4	10.4–20.0	62.0	Shah et al. (2025)
Coconut cake	10.0	18.0	12.0	78.0	Ajila et al. (2012)
Rice hull	6.7–9.7	1.5–2.8	14.3–41.4	37.0	Swidiq et al. (2012)
Brewer grain	9.8–10.8	24.0–27.4	15.9–17.1	60.0	Ajila et al. (2012)
Bagasse	3.9–4.7	2.9–6.9	10.3–39.3	49.0	Jan et al. (2017)
Palm-press fiber	13.8	4.0	36.4	30.8	Mutwedu et al. (2022)
Cocoa-pod husk	89.6	6.0	31.5	45.0	Mirzaei-Aghsaghali and Maheri-Sis (2008)
Wheat bran	9.8	15.5	42.8	–	Shah et al. (2025)
Pineapple waste	6.8	4.9	20.8	76.0	Ajila et al. (2012)

could potentially be used as a feed ingredient due to its high protein and fiber contents after processing it in a technically sound manner (Chattaraj et al., 2024c). BSG is usually composed of malt husks and pericarp remnants that contribute to its considerably high crude fiber and protein contents and its direct use often also necessitates considering moisture management and storage stability issues (Henkin et al., 2025).

Similarly, peels, pomace, and other such fruit and vegetable wastes could have high fiber, carbohydrate, vitamins, and phytonutrients levels but less protein. Although such wastes could provide energy and micronutrients, it was essential to combine them with protein-rich ingredients or enrich them through fermentation to optimize their nutritional profiles and bioavailability before incorporating into the diet feed for formulation. As agri-wastes differ in their nutrient contents, no single waste type on its own could function as a complete and balanced feed ingredient. Rather, these are best utilized in blends or processed (for instance, subjecting these to microbial fermentation) to improve the availability of amino acids, decrease the indigestible fiber fractions, and to increase the overall protein content (Yafetto et al., 2023). Such treatments could possibly enhance the overall quality and digestibility of the formulated feed by bringing down the anti-nutritional elements present in

Table 15.2 A list of the antinutritional factors often encountered in various common agri-wastes.

Agricultural byproduct	Frequently encountered anti-nutritional factor
Castor seed meal	Ricin, hemagglutinin
Guar meal	Protease inhibitors
Peanut meal	Aflatoxin, goitrogen, protease inhibitors, saponins
Soybean meal	Haemagglutinins, goitrin, protease inhibitors, saponins
Cotton seed cake	Gossypol
Rape seed and mustard wastes	Thioglucoside, goitrin, isothiocyanate
Linseed meal	Cyanogens, anti B-6

many phyto-residues. Table 15.2 lists the various anti-nutritional factors (ANFs) that are usually encountered in various agri-wastes.

3. Challenges in the direct inclusion of agri-waste in formulated feeds

Low cost and environmental advantages of the agri-wastes make them suitable substitutes as conventional feed ingredients. Direct incorporation of these materials into fish feed presents considerable obstacles, despite the fact that they offer potential opportunities to make use of the agricultural byproducts and lower the feed cost. Agri-wastes frequently have low digestibility, high ANFs, and are deficient in vital elements needed for fish growth (Obosi and Enyidi, 2025). Fish health, feed efficiency, assimilation efficiency (feed conversion ratio) and overall production performance may all be compromised as a result of these. Thus, it is crucial to comprehend the bottlenecks related to their direct use before adding agri-wastes into the formulated fish feeds. Commercial feeds often claim a minimum crude protein content of 46% although laboratory analyses showed a lower percentage content (about 29.3%). Commercial feeds seemed unpopular and less accepted owing to high retail price due to high labor cost in processing, formulation, packaging, and marketing over and above the cost of the ingredients, and not catering to all the nutritional needs of the target species. Incorporation of local rural potential ingredients in feeds would reflect its importance among the rural farming community that is actively engaged in fish breeding, culture and marketing. The water-stable feeds could be formulated either as sinking type (so that it does not interfere with the feeding habit of fish and does not deteriorate water quality) or floating type. Bandyopadhyay et al. (2005) formulated four sinking-type diets using local agri-produce, two with crude protein content lower than the commercial feeds and two higher, mixed in suitable proportions. The view holds well also to popularize the low-cost formulated feeds among the aquarists.

3.1 Anti-nutritional factors (ANFs) and their effects

Fish digestion and metabolism are impeded by the natural anti-nutritional components phyto-derived products that are found in numerous agri-wastes. Substances such as phytates, tannins, saponins, oxalates, and trypsin inhibitors decrease nutrient availability by binding to proteins and minerals or by blocking vital digestion enzymes. For instance, tannins decrease protein digestibility (and protein efficiency ratio [PER]) by forming complexes with dietary proteins and enzymes, whereas phytates bind with minerals such as calcium, zinc, and phosphorus and render them inaccessible for absorption (Ram et al., 2020). While trypsin inhibitors directly prevented protein breakdown, saponins irritated the gut lining and decreased the feed intake. Further, considerable amounts of agricultural wastes have significant lignin and crude fiber contents that serve as physical barriers thereby limiting nutrient release and absorption (Gupta et al., 2022). ANFs collectively hinder nutrient absorption, reduce digestion efficiency, and might even result in intestinal injury, oxidative stress or compromised immunological responses when the raw or unprocessed agri-waste materials are consumed by fish. Thus, the use of raw, unprocessed or untreated agri-wastes might seriously harm the growth and wellbeing of aquacultured fish. The overall effect including the mortality (Mishra et al., 2000) and pathogenesis (Mahanty et al., 2013) issues may also need to be addressed in the process.

3.2 Low digestibility and assimilation due to poor nutrient complementation

Intrinsic low digestion of agri-wastes poses a significant obstacle in directly incorporating into fish feed. The majority of fish species, particularly the carnivores, lack certain digestive enzymes needed to break down complex plant fibers such as lignin, cellulose, and hemicellulose (Moraes and de Almeida, 2020). Such fish find it challenging to access and use the nutrients in agri-wastes as these are enclosed within the robust cell wall constructs. Additionally, the nonstarch polysaccharides (NSPs) found in many agricultural residues cause the gut to become more viscous and impede nutrient absorption (Wang et al., 2024). Due to poor digestion and undesirable amino acid compositions, the actual usable protein decreases even when crude protein levels seem sufficient. The water quality of aquaculture systems deteriorates as a result of the decreased feed conversion efficiency and increased generation of fecal waste (Boyd, 2017; Mishra, 1998b). Fish with high fiber and low nutrient release must expend more energy for metabolism, which lowers the amount of energy available for development and physiological processes, thereby hampering growth and the overall wellbeing. Thus, agri-wastes have minimal nutritional value when used directly without processing or supplementation.

3.3 Deficiencies of essential nutrients

Significant nutrient shortages that are vital for fish growth and performance are also present in agri-wastes. The majority of agricultural waste materials include low-quality proteins that lack essential amino acids such as tryptophan, lysine, and methionine, whereas fishmeal offers high-quality protein with balanced amino acids (Kari et al., 2023). These deficits impair the overall growth in aquacultured fish species, prevent muscle development and impact protein synthesis. Additionally, many phyto-based wastes lack important fatty acids (especially EPA and DHA) which are critical for cellular integrity, immunological function, and reproductive health. Phytates further limit mineral absorption by binding to calcium, phosphorus, and trace minerals.

Bandyopadhyay et al. (2005) carried out a 60-d feeding trial on goldfish (*Carassius auratus*) fed diets containing various levels (23.34%, 26.21%, 29.30%, 32.24%,

Table 15.3 A gist of a representative cost-benefit analysis considering the various factors.

Aspect	Benefit	Cost/challenge
Economic	**Lower raw material cost:** Agri-wastes are abundant and cheaper than the conventional soybean and fish meals whose prices are highly volatile	**Processing costs:** Requires technology investments to upgrade nutritional value, reduce anti-nutritional factors (ANFs), and form a stable pellet (e.g., in solid state fermentation, pelletization)
Environmental	**Waste management solution:** Repurposing the waste addresses the disposal issues and minimizes environmental pollution (e.g., from burning or burying)	**Potential contamination:** Poorly managed waste could introduce heavy metals or other contaminants into feeds and then the fish. It poses human health risks, if not monitored well
Nutritional/ technical	**Nutrient enhancement:** Processing strategies such as fermentation could enhance the protein content, digestibility and bioavailability of nutrients	**Nutritional imbalance:** Agri-wastes often have low protein/fat content and high fiber as compared to the protein from animal source as required by the carnivorous fish. It could impact growth adversely if not properly balanced
Operational	**Local sourcing/supply chain:** Use of local waste could stabilize the supply chain as compared to volatile global markets for the conventional ingredients	**Consistency and quality control:** Waste composition could vary, requiring rigorous quality control and optimized processing for consistency of the feed quality

and 42.53%) of crude protein. Alongside cod liver and vegetable oils and vitamin-mineral mix, the fish feed contained groundnut oil-cake, wheat bran, soybean meal, fish meal during formulation, and it was compared with a commercially available feed. The results revealed that the growth and the energy retention efficiency, feed conversion ratio, and protein retention efficiency (dietary utilization) were greater in the feed with 42.53% protein, clearly depicting the role protein–lipid combination in growth and feed utilization. They concluded that 40%–45% protein content in diet enhanced the growth and dietary utilization in young goldfish. Vitamin deficiencies are very prevalent in the unprocessed agri-residues, particularly vitamins A, E, B_{12} and D (Ugwuanyi, 2016). The cultured fish might exhibit stunted growth, weakened immunity, and low survival rates if these nutritional deficiencies are not addressed by supplementation or mixing with other suitable substances. Therefore, major nutritional imbalances in aquaculture diets may result from depending exclusively on agri-wastes without correcting these inadequacies.

3.4 Expected reduced growth and its implications

The combined issues of low digestibility, nutrient inadequacies, and anti-nutritional elements frequently result in decreased fish growth performance when agri-wastes are employed directly in fish feed without processing. Bitter substances such as tannins can make feed less palatable, which lowers feed intake and the total amount of nutrients available for growth. Due to the poor digestibility of the feed, many nutrients go unutilized through the digestive tract even though it was consumed by the fish. Thus, it leads to low feed conversion, increased production costs and extended farming times in order to reach market size. From the financials point of view, fish farming becomes less profitable as a result of low growth and thus slow expansion. In terms of environment, undigested plant residues and fibers pollute the water bodies, raise the ammonia (NH_3 and NH_4^+) levels, and upset the microbial equilibrium in the aquaculture system (Gnaneswari et al., 2022). Addressing these problems necessitates additional high-tech management strategies, including regular water exchange or greater aeration which increases the operational expenses even further. Therefore, the use of agri-wastes directly in fish feeds could not be recommended unless sufficiently treated for nutrient and digestibility improvement. Cutting-edge new innovative interventions of the new age need to be adopted to achieve the ambitious target.

Although agri-waste has a lot of potential as economical and sustainable aquaculture feed component, numerous issues with their direct incorporation are there that should be considered for investigation. While the high fiber content restricts digestion, the presence of ANFs dramatically reduces the digestibility and absorption of nutrients (Samtiya et al., 2020). Fish growth is further restricted and their health is impacted by the deficiencies of essential nutritional components. When these elements work together, the fish in an aquaculture system shall have decreased growth performance leading to financial loss and environmental damage. Appropriate processing techniques, such as fermentation, enzyme treatment, heat application, or chemical detoxification must be employed either as standalone or combined strategy to reduce the ANFs, enhance digestibility and balance the nutrient contents to ensure complete utilization of the agri-wastes as feed materials. With the right and optimal processing and altered formulation, agri-wastes can be useful ingredients of sustainable fish feed for reduced production cost and increased aquaculture productivity. Notwithstanding, environmental upkeep cannot be overlooked during the process optimization and strategy integration to achieve the desired results with sustainable growth and development of the sector.

4. Value-addition strategies to make agri-wastes amenable as fish feed ingredients

The objective of value-addition strategies to add agri-wastes to fish feed is digestibility, safety, and nutritional value enhancement before incorporating in the feed. Biological pretreatments to add value by improving protein content, breaking down the complex fibers, and reducing ANFs (anti-nutritional elements) that are frequently encountered in raw agricultural residues include ensiling and fermentation. Particle size, nutrient availability and water stability of the feed are further enhanced through additional mechanical processing strategies such as grinding, extrusion, and pelleting. Probiotics, enzymes, or microbial inoculants could be employed as biochemical processes to transform the low-value wastes into superior feed ingredients with improved nutritional profiles. Some agricultural wastes could be employed to produce single-cell proteins that would provide aquaculture with a reliable and sustainable protein. Eliminating the obnoxious substances is ensured through detoxification techniques such as thermal treatment and chemical neutralization. Solid state fermentation (SSF) is a quite popular and effective biotechnological strategy that turns low-value agri-residues to high-protein and readily digestible fish feed with high PER, by employing microbes (such as fungi and bacteria) to grow on high-moisture agricultural waste and break down the complex fibers and toxins. Fig. 15.1 provides a schematic presentation about employing SSF to biologically convert and add value to agri-waste before incorporating in fish feed formulation.

4.1 SSF and detoxification of anti-nutritional factors (ANFs) in fish feed

SSF reduces ANFs levels, thereby enhance the bioavailability and digestibility of nutrients in agri-industrial residues (Ajila et al., 2012). ANFs are bioactive compounds that hinder the bioavailability and digestibility of food and feed nutrients. ANFs such as hydrogen cyanide, caffeine, oxalates, phytates, tannins, polyphenols, and saponins adversely impact fish health. For instance, antigenic proteins (β-conglycinin and glycinin) and glucosinolates (mustard oil glycosides) found in agri-industrial wastes impact fish growth performance and weaken the immune

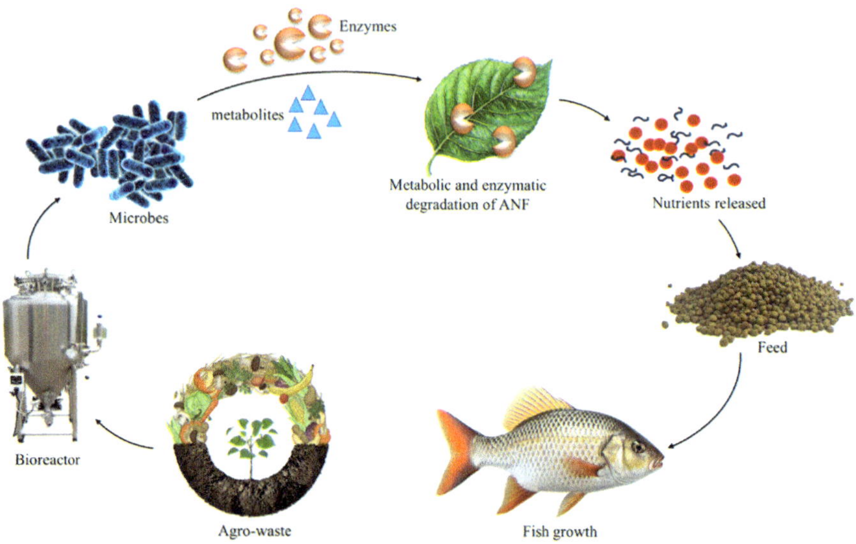

FIGURE 15.1

Employing SSF to convert agri-waste and add value to incorporate in fish feed.

system by reducing the nutrient quality of feeds (Yafetto et al., 2023). The intermediates that are generated through the metabolism of ANF also reduce the amount of one or more vital nutrients in feed. Treatment with polyethene glycol or microbes (through SSF) could reduce the harmful effects of ANFs. Trypsin inhibitor-related protein bands were removed by *Aspergillus oryzae*-mediated SSF of soybean meal, indicating their degradation (Kalaiselvan et al., 2025). Numerous investigations using bacteria, yeasts and filamentous fungi showed that SSF might actually biodetoxify and reduce ANFs in agri-industrial wastes (Yafetto, 2022). Fermented cassava peels had much lower phytate and cyanide levels than in unfermented peels that had higher levels of these anti-nutritional elements (Jumare et al., 2024).

These findings implied that cassava peel substrates could be detoxified by a mixed culture of *Saccharomyces cerevisiae* and *Lactobacillus* spp. through SSF. De-oiled rice bran fermented with *Rhizopus oryzae* showed a notable decrease in phytate and trypsin inhibition activities, indicating that the fungus actively metabolized these substances (Ranjan et al., 2019). Co-fermentation with *Aspergillus niger*, *Candida utilis,* and *Bacillus subtilis* reduced tannins and phytic acid more in *Moringa oleifera* leaf meal than the fermentation involving a single microbe (Shi et al., 2020). Nutritional quality of corn–soybean meal was improved by fermenting it with *B. subtilis* and *Enterococcus faecium*, which reduced allergenic proteins such as glycinin and β-conglycinin (Kalaiselvan et al., 2025). Another study found that the degradative activity of *B. subtilis*, *C. utilis,* and *Enterococcus faecalis* reduced the phytic acid, glucosinolate, and tannin in rapeseed meal by 20%, 96%, and 36%, respectively when subjected to SSF (Hu et al., 2016). Thus, SSF is a great way to reduce ANFs beyond doubt, that make the fish feed made from fermented agri-industrial wastes safe, nourishing, and healthful for aquacultured fish. The microbial fermentation and metabolic processes that affect the nutritional quality of agri-feedstuff is provided in Fig. 15.2 for ready reference.

4.2 Use of enzymes to enhance feed conversion of fish feeds

Compared to fish fed with commercial diet, feed conversion ratio, and total immunoglobulin of cultured fish (grass carp and gray mullet) increased after the enzymes-mix (1% and 2% weight by weight) were added to agri-wastes. Papain pretreatment of soybean residue might improve common carp feed digestibility and conversion while further enhancing the water quality in culture tanks (Choi et al., 2016). Feed proteins could be hydrolyzed into smaller and more digestible peptides by proteases such as papain (found in papaya) and bromelain (found in pineapple). Additionally, it was observed that a commercial enzyme Ronozyme VP improved growth performance and net protein consumption in tilapia fed diet based on palm kernel meal (Wattanakul et al., 2021). Fish protein hydrolyzate, which is produced as a fish byproduct (wastes include heads, viscera, skin, and skeleton) processing with enzyme, is also a possible component of aquaculture diets (Bhaskar and Mahendrakar, 2008). It would be a good idea to incorporate fruit processing residues from papaya or pineapple into food waste pellets to reduce the feed production cost.

4.3 Using vitamin–mineral premix for enhancing nutritional values of fish feed

Minerals and vitamins are frequently added to fish feed to improve nutritional value of animal feeds, particularly vegetable-based diets. It was demonstrated that the addition of a vitamin-mineral premix to broadhead catfish (*Clarias microcephalus*) meal improved the weight gain and survival rate as compared to those who did not (Wong et al., 2016). A different study found that phosphorus in mineral premix might increase the weight gain, feed conversion ratio (PCR) and PER in tilapia (Hangsapreurke et al., 2020). It is expected that incorporating the full range of vitamins and minerals would also be important for fish growth. Tang et al. (2017) revealed that hybrid tilapia (*O. niloticus* X *Oreochromis aureus*) fed diets supplemented with copper, iron and zinc significantly raised amylase activity in hepatopancreas homogenates, and amylase and lipase in intestine. Huang et al. (2011) demonstrated that thiamin included in diet significantly enhanced the activities of various enzymes (such as lipase, trypsin, alkaline phosphatase, etc.) in the intestine of juvenile Jian carp (*Cyprinus carpio* var. Jian). Addition of the premix enhanced protein digestibility of food waste-based pellets from 65% to 80% that led to better growth of fish (Mo et al., 2019). Thus, addition of vitamin-mineral premix in feed pellets containing agri-wastes will be beneficial for fish growth.

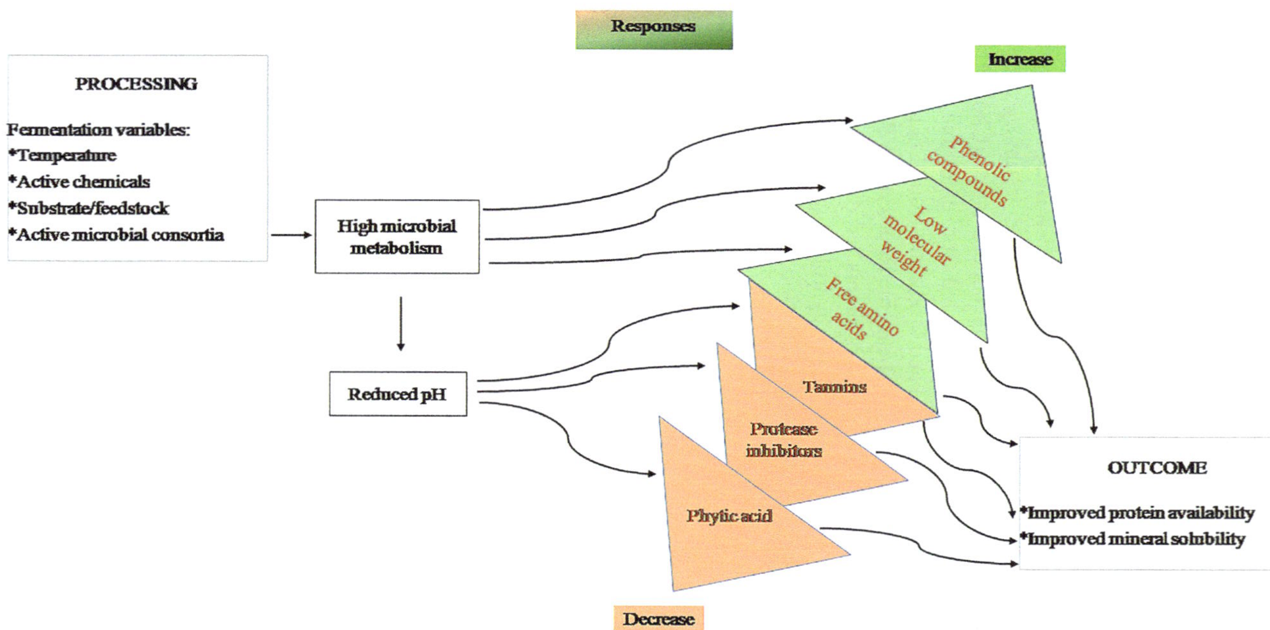

FIGURE 15.2

The interactions between the fermentation variables and microbial metabolism to alter nutritional availability in agri-waste making it amenable to include in aquafeed.

4.4 Use of probiotics for enhanced fish health and wellbeing

Probiotics are live microbial feed additives that improve intestinal balance of the fish. Bandyopadhyay et al. (2015) reported that dietary *S. cerevisiae* boosted growth and immunity in the Indian major carp *Labeo rohita* (Ham.) juveniles. By incorporating 2% baker's yeast (*S. cerevisiae*) into the feed pellets (Choi et al., 2016) aimed to improve the quality of the food-waste-based diets. They found higher growth rates in terms of FCR and PER in grass carp fed these diets, which improved their specific growth rate (SGR) and relative weight gain (RGW). Yeast cells attached to the intestinal wall enhanced the amylase enzyme activity thereby boosting the digestibility of the food-waste-based pellets, the primary phenomenon behind the beneficial effects of yeasts supplementation (Rahayu et al., 2024). Also, the yeast cells attached to the intestinal wall could stimulate innate immune responses and protect fish against infections. This was possible because yeasts contained β-glucans, mannan oligosaccharides, and nucleic acid which act as immunostimulants.

Feeding trials showed that adding these active components to feeds increased the disease resistance against *Aeromonas hydrophila* in a variety of fish species, including the common carp (*C. carpio*) and Asian catfish (*Clarias batrachus*), in controlled environment (Banu et al., 2020; Mahanty et al., 2013). Another investigation on the immunological response of seabream (*Sparus aurata*) and tilapia (*Oreochromis niloticus*) using whole yeast also showed positive effects (Abu-Elala et al., 2013). Yet another study revealed that whole baker's yeast was also a potential alternative to replace fish meal protein to rear seabass (*Dicentrarchus labrax*) juveniles that had higher nutritional requirements (Xie et al., 2022). Therefore, addition of whole yeast into food-waste based feed would be more convenient than the extracted and purified constituents, with associated lower cost and uncomplicated techniques that necessitate low-skill-set requirements. Byproducts generated from brewing industry contain appreciably high number of yeast cells that could be collected easily and used for the purpose, thereby rationalizing the feed production cost.

5. Gut microbiota and functional response on fish health

Gut microbiota plays a determining role in effective utilization of the nutrient in a feed. These microbes act like natural helpers in the digestive system, breaking down complex substance (such as proteins, fats, and fibers) in the diet into simpler forms. It makes it easier for fish to absorb the nutrient and utilize it for growth and energy. The microbes also help to produce certain useful compounds such as short-chain fatty acids and vitamins, antioxidants that further aid in good metabolism and wellbeing. Overall, a healthy gut microbiota ameliorates the digestion, enhances nutrient assimilation and helps the aquacultured fish to use different types of feeds better, especially the phyto-based unconventional diets.

Nutrient metabolism in fish is associated with diversity and function of the gut microbiota. Carbohydrate digestion, for instance, is largely dependent on the quality and quantity of carbohydrates in the diet. As aquaculture industry heads toward greener strategy l such as the use of phyto-based feeds due to the soaring feed cost and restricted supply of fishmeal and fish oil, understanding how gut microbes helped fish utilize these ingredients has become increasingly important. Herbivorous and omnivorous fish usually have more diverse gut microflora that are capable of breaking down the NSPs of plants. Some beneficial microbes, such as marine *Bacillus* sp. found in European seabass, have demonstrated their ability to hydrolyze NSPs and help in carbohydrate metabolism. Likewise, research on Nile tilapia and yellowtail showed that the gut microbiota actively participated in carbohydrate digestion and helped regulate the related metabolic genes. Study on longfin yellowtail juvenile concluded that the intestinal microbiota was mostly dominated by Proteobacteria, Firmicutes, Bacteroidetes, Cyanobacteria, and Actinobacteria. These bacterial groups mostly contributed to carbohydrate metabolism.

Protein metabolism also was significantly influenced by fish gut bacteria. Gut microbes could either utilize amino acids for their own growth, or break down the host-derived proteins and even synthesize essential amino acids for metabolism when high-protein feed was consumed (Dai et al., 2011). Microbial activity helped maintain overall amino acid profile within the intestine. Amino acids such as threonine are particularly significant as they help in mucin production and protect the gut barrier (Dai et al., 2011). Microbial enzymes break down the proteins into peptides and free amino acids. These amino acids help in various metabolic pathways such as deamination or decarboxylation. This produces several byproducts such as secondary metabolites, few of which are beneficial (some other could be harmful). For instance, high protein metabolism accumulates compounds such as ammonia, hydrogen sulfide, indoles, and phenols that could irritate or damage gut lining and increase the risk of digestive disorders (Macfarlane and Macfarlaneet, 2012).

The fish intestinal microbiota again has a vital role to play in lipid metabolism. The ability of fish to consume and store fats is frequently correlated with altered microbial population, which can vary rapidly in response to dietary or environmental changes. Studies on zebrafish (*Danio rerio*) and grass carp (*Ctenopharyngodon idella*) demonstrated that fish had increased lipid absorption and expression of lipid-related metabolic genes as Bacillota (Firmicutes) predominated in the gut microbiome (Ni et al., 2014; Sheng et al., 2018). However, sudden alteration in the diet altered the intestinal microbiota sometimes that affected lipid metabolism, even before a noticeable change in the genetic response. Feed additives such as α-lipoic acid were reported to significantly improve lipid metabolic pathways in tilapia, though their impact on gut microbial structure was little known (Navari-Izzo et al., 2002), which needed further investigation. Additionally, differences in the microbial composition as observed between the wild and aquacultured fish often mirror the differences in fatty acid synthesis pathways. Environmental stress such as heavy metal (such as copper) exposure could also disrupt both the microbiota and lipid-processing mechanisms. Another important lipid metabolism regulation factor is bile acids, which interact with the gut microbiota and activate specific receptors such as TGR5 involved in lipid metabolism. While the relationship is well studied in mammals, the precise mechanism in fish remains to be explored in detail. All in all, research in recent times has strongly vindicated the digestion and absorption roles of gut microbiota in fish, and overall carbohydrate, protein, and lipid metabolism. As aquaculture continues to grow and adopt more sustainable novel feeds, leveraging the beneficial roles of intestinal microbes will be essential to improve fish nutrition, health, and productivity.

5.1 Fish immunity and antioxidant defense

Diets play a pivotal role in providing immunity in fish. Fermented fish feed is assumed to maintain high lactic acid bacteria (LAB) population in fish gut. LAB, especially *Lactobacillus* sp., is known to activate immune cells to stimulate the production of cytokine that are helpful in functional modulation of immunological response. *Lactobacillus* sp. is necessary for cytokine production in helper T-cells, interleukin (IL-4, IL-10) production, which is responsible for growth of B-cell and isotype switching. Functional modulation of the cascade helps in antibody formation.

The commensal *Enterococcus casseliflavus* (EC-001) demonstrated to significantly boost host immunity by elevating total protein of serum, immunoglobulin (Ig), complement component C3, and protease activity of skin mucus (Akbari et al., 2021). Host-derived *Lactobacillus plantarum* enhanced immune performance in African catfish (*C. gariepinus*) by elevating the hematological parameters such as hemoglobin concentration, packed cell volume, and WBC (white blood cell) counts (Nwanna and Tope-Jegede, 2017). Similarly, host-derived *L. plantarum* N11 and *Bacillus velezensis* H3.1 in Nile tilapia (*Oreochromis niloticus*) strengthened both the mucosal and systemic immune responses such as higher serum lysozyme activity (Van Doan et al., 2018). Several *Bacillus* sp. (such as *B. velezensis* TPS3N, *Bacillus amyloliquefaciens* TPS17 and *B. subtilis* TPS4) also significantly enhanced the intestinal and skin mucosal immunity in tilapia. Moreover, *B. subtilis* AB1 strengthened both humoral and cellular immune responses by increasing the leukocyte counts and serum phagocytic and lysozyme activities.

Host-associated microbial strains showed strong immunotherapeutic potential. For instance, *Bacillus pumilus* A97 isolated from *Trachinotus ovatus* strengthened the host nonspecific immunity, upregulated kidney TLR8 and TLR9 expression, and boosted disease resistance (Liu et al., 2020). The commensal *Psychrobacter* sp. SE6 regulated intestinal immunity and maintained microbiota homeostasis (Sun et al., 2014). Similarly, *Lactococcus petauri* LF3 and *Bacillus siamensis* LF4 significantly boosted serum lysozyme activity and enhanced gut immune responses and microbial composition in *Lateolabrax japonicus*, the Japanese seabass (Yang et al., 2021, 2022a, 2022b). Host-associated beneficial microbes also helped to functionally modulate the key inflammatory pathways by regulating the expression of pro-inflammatory cytokines (e.g., ILs, IFNs, TNFs) and anti-inflammatory cytokines (e.g., TGFs). For instance, *Rummeliibacillus stabekisii* isolated from tilapia gut significantly increased the expression of cytokine-related genes such as the heat-shock protein 70, TNF-α, IL-1β, and TGF-β in the head kidney in teleosts (Tan et al., 2019). Additionally, *B. pumilus* SE5 and its immunogenic components LTA and PGN were found to upregulate TLR1, TLR2, TLR5, and MyD88, activate downstream immune effector molecules, and beneficially modify the gut microbiota by increasing the abundance of *Lactobacillus* and reducing *Vibrio* population (Yang et al., 2019).

5.2 Omics-based assessment of gut and systemic responses

SSF could play a role in nutritional value-addition, supplying bioactive chemicals, modifying gut microbiota and controlling the immune function modifications in fish (Fig. 15.3). It was crucial to comprehend these interactions to maximize the use of fermented substances in formulated aquafeed for improved growth, sustainability, and health of aquatic animals. The expression of intestinal inflammatory genes varied significantly in juvenile turbot fed diets containing various quantity of fermented soybean meal for 79 days (Li et al., 2019). Fish fed with 60% fermented soybean meal had higher TNF-α and IL-1β levels, while the TGF-β1 levels remained constant (Li et al., 2019). Enzymes related genes including SOD, CAT, and GSH-Px levels were significantly upregulated in *H. molitrix* when fermented rice bran was added to experimental diets at a level of up to 20% (Lin et al., 2022). Rice bran was an excellent protein, fatty acids, amino acids, and antioxidants source (Omarini et al., 2019). Juvenile Gibel carp (*Carrassius gibelio*; also known as Prussian carp and silver Prussian carp) fed fermented moringa leaves showed significant decrease in spleen immune-related gene expression (TLR2, IL-8, IL-1β, and TNF-α) level when challenged with *Aeromonas hydrophila* as compared to the pre-challenge levels (Zhang et al., 2020). Similarly, stress-related antioxidant genes such as SOD, CAT, MDA, and GSH-Px were significantly impacted in crayfish *Procambarus clarkii* by adding varying quantity (0.5%, 1%, and 2%) of moringa leaves in diets (Li et al., 2024). Fish could enjoy better health and a robust immune system due to the vital minerals, β-carotene, phytochemicals, and natural antioxidants present in moringa leaves. Immune-related gene expressions were considerably impacted when Golden pompano was fed fermented cotton seed meal (FCSM) at 0%, 12.5%, 25%, 50%, and 100% concentrations. Notably, the 12.5% FCSM cohort had higher IL-10 expression, whereas the 12.5%, 50%, and 100% FCSM cohorts had higher TNF-α and IL-1β levels. ZO-1, Claudin-3 and Claudin-15 were upregulated in the 25% FCSM cohort, and IL-8 and NF-kβ levels were markedly elevated in fish fed 100% FCSM diet (Wang et al., 2023). Protein content of cottonseed meal increased through microbial fermentation that also produced tiny peptides and growth-modulating substances that might help enhance the nutritional value and activated fish metabolism.

Effects of fermented unconventional feed additives with olive cake on fish gene and protein expression were investigated. TNF-α and IL-1β gene expression in Nile

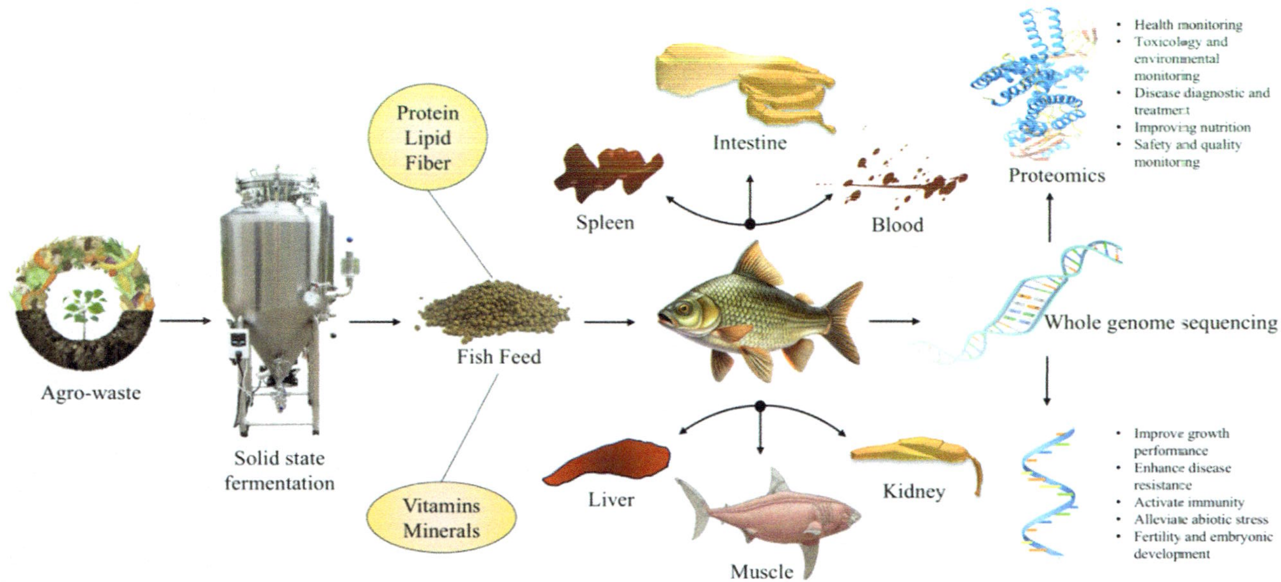

FIGURE 15.3

An overview of application of modified agri-waste fish-feed in modulating transcriptomics and proteomics in aquaculture.

tilapia (*O. niloticus*) intestine fed diets supplemented with *A. oryzae* fermented 0%, 5%, 10%, 15%, and 20% olive oil-cake (FOC) for 90 days significantly increased in 10% FOC cohort as compared to others (Ismail et al., 2021). Also, phenolic compounds with antioxidant, antibacterial, and anti-inflammatory properties were found in olive oil-cake include tyrosol, oleuropein, and hydroxytyrosol (Abdallah et al., 2023). In addition to modifying the immunological response and gene expression linked to inflammation and immune function in fish, fermentation might increase the bioavailability and activity of these beneficial substances.

6. Environmental impact assessment

Aquaculture has strong and multitude environmental implications (Mishra, 1998b; Mishra and Rath, 1999). Sustainable aquaculture development depends on evidence-based environmental assessments, structured monitoring methods, proficient modeling tools, and application of optimized standard management practices. Strict regulations and compliance to international codes of conduct are important to reduce environmental impacts while allowing aquaculture to contribute safely and sustainably to global aquaculture food production. Treweek (1995) argued that although ecology was central to understanding of environmental change, ecological input within EIA has been poorly designed and underused. Ecological assessments included in Environmental Statements (ESs) are often criticized for weak scientific precision, limited baseline data and inability to predict or evaluate ecological impacts.

Environmental Impact Assessment (EIA) is a tool that is intended to accesses sustainable development by recognizing and evaluating the environmental consequences of proposed anthropogenic activities. EIA in a fishery ecosystem is primarily measured through approaches that consider chemical, ecological and nutritional parameters. Chemical assessment strategies mainly focus on measuring physical and chemical parameters such as nutrients, oxygen levels and organic matter in water and sediments (ICES report, 1989). Similarly, ecological assessment strategies involve altered community structure, biodiversity, and indicator species, especially those organisms present in the benthic region and are sensitive to organic enrichment (Weston and Gowen, 1988). Nutritional assessment strategies measure the waste outputs based on the feed intake, digestibility, fish growth, and nutrient uptake, and the quantity of available nitrogen and phosphorus released into the environment (Cho, 1991). Each of these approaches has its own benefits and limitations, and combining them often yields more reliable assessments to formulate the environmental impact statement (EIS).

Some data highlight the shortcomings in ecological practice within EIA, including inability to quantify habitat loss, restricted field surveys, periodic data collection, and over dependance on descriptive rather than analytical methods. Many EISs failed to take into an account the indirect, cumulative, and long-term ecological impacts instead of focusing on easily observed and localized effects.

EIA of ecology concludes that anthropogenic activities significantly affect the ecosystem balance, biodiversity loss, and disruption in ecological functions. Figuring out the shortcomings at early stage could help minimize the impact on the ecosystem and the ecology.

6.1 EIAs related to feeding patterns and nutrient release to the environment

EIA for feeding patterns and the environmental impact typically employs methodologies such as the life cycle assessment (LCA) to quantify environmental impacts from "cradle-to-gate" (from raw material extraction to the farm gate) or throughout life cycle. Key assessment indicators include: carbon footprint (kg CO_2-equivalent), eutrophication potential (kg $PO4^{3-}$-equivalent or kg N-equivalent), acidification potential (g SO_2-equivalent), water footprint (litters of blue and green water consumed), and land use (m^2 X year).

6.2 Mitigation strategies

While EIA is the **process** to overall predict and evaluate the environmental effects of an envisaged project, EIS is a **document** that details the possible impacts, suggested mitigation strategies and eco-friendly alternatives, and formally reports the EIA process findings for the decision (or policy) makers as also for consumption of the public. Thus, the EIA is effectively about the "how" (the nitty-gritties of the study, the detailed analysis, and the public review, etc.), and the EIS is about the "what" (the final output or report). The specific mitigation strategies that could be undertaken to address the environmental impact in line with the parametric requirements under the EIS are as below:

Diet/ration balancing: Ensuring that the fish diet is nutritionally balanced to precisely meet the nutritional requirements while reducing the waste nutrient in the form of undesired environmental manure.

Feed efficiency improvement: Enzymes (such as phytase to release P) in the formulated feed, and smart feeding strategies would increase the digestibility and assimilation of the nutrients, which would mean less release of N and P into the environment per unit product.

Manure management: Treating the unavoidable manure that is generated as a valuable resource and applying it judiciously to the landmass as per the nutrient requirements of crops could help address runoff issues and reduce emissions.

Adopting sustainable practices: State-of-the-art agricultural practices such as rotational grazing (regenerative agriculture) and integration of crop and livestock systems could enhance the soil quality, increase carbon sequestration, and minimize nutrient loss to the environment.

Dietary shifts: A cultural shift of the human consumption patterns from high-impact animal produce (such as beef) to the low-impact alternatives such as shrimp, fish, or plant-based alternatives could potentially reduce the overall environmental footprints significantly in numerous impact categories.

The accumulation of excess reactive nitrogen (N) and phosphorus (P) in the environment is a key sustainability concern. Incessant use of synthetic fertilisers and unscientific manure management in livestock and crop production systems are the primary sources of N and P loss to the environment.

6.2.1 Aquatic ecosystem concerns

The agriculture land and effluent runoffs from aquaculture facilities add excess N and P into the contiguous freshwater and marine environments. This phenomenon leads to eutrophication that leads to excess algal growth, depletion of oxygen (hypoxia), and formation of "dead zones" that are unproductive and harm the fish and other aquatic lives.

6.2.2 Terrestrial and atmospheric concerns

The emission of Ammonia (NH_3 or NH_4^+) due to the fertilizer and manure application increases soil acidification and atmospheric aerosol loading. These could harm the vegetation, alter the ecosystem functions in the long run, and ultimately impact human health and wellbeing.

6.2.3 Greenhouse gas (GHG) emission concerns

The production, processing and transport of feed account for a significant portion of greenhouse gases (GHGs_ emissions from food systems, especially the methane (CH_4) generated through enteric fermentation in ruminants and the nitrous oxide (N_2O) generated during soil and manure management.

6.2.4 Land and water use concerns

The quantum of land and water resources use vary as the feeding patterns differ. For instance, animal-based formulated diets usually need significantly more land and water masses than plant-based diets. The production of feed crops for large-scale livestock operations could trigger deforestation in regional pockets, thereby impacting the biodiversity and the carbon sequestration patterns.

7. Economic feasibility assessments

Massive population explosion over the past two decades, alongside the ensuing widespread usage of nonrenewable resources, has detrimentally impacted on the environmental quality and forced the adoption of sustainable practices. Future generations depend on developing sustainable renewable processes and utilization of alternative resources that could take the place of fossil fuels. Since the idea of sustainability is entirely predicated on the circularity of all required materials, the shift from a linear to a circular economy is currently an essential component to manage resources in an eco-efficient manner.

A key component of the circular economy, the circular bioeconomy is essential in attaining resource and environmental sustainability. Bioeconomy utilizes biologically derived materials and mimics or employs natural processes to accomplish an effective and cyclic use of resources (Venkata Mohan et al., 2020). The European Commission defines bioeconomy as "the production of renewable biological resources and the conversion of these resources and waste streams into value-added products, including food, feed, biobased products, and bioenergy" (Coppola et al., 2021). Environmental impact is minimized in this manner, influx of renewable resources is encouraged and anthropogenic intervention and dependency in the materials derived from fossil fuels is decreased. In this line, interest in using marine bioresource refuse such as the massive amount of waste generated by fisheries and aquaculture has increased, which is a very promising resource for materials with high market value with the growing awareness and particular attention to advancements in more environmentally friendly greener practices (Shavandi et al., 2019; Coppola et al., 2021). Fish waste is mostly used to prepare fish meal as it contains nearly the same protein profile as fish meat (Mo et al., 2018). However, a significant amount of the biomass is either discarded (usually incinerated that increases energy consumption, involves financial implications, and is environmentally detrimental; Coppola et al., 2021), or is used to generate low-value products. Additionally, the nutrient make up in fish waste makes it a good plant nutrient and even easy to compost. A variety of fertilizers may be made from fish waste (Ahuja et al., 2020), and commercial fish-based fertilizers are currently applied to horticultural and agricultural crops (Coppola et al., 2021).

Additionally, fish waste has high percent of biodegradable organics that might be recycled and utilized as a desirable cosubstrate for waste activated sludge to boost methane production during anaerobic codigestion (Coppola et al., 2021). Fish farming is also the main driver of 91%–93% eutrophication, while just 5%–7% of the total feed contributes to fish diet due to direct nutrient release. Energy carriers were noticeable only in turbot recirculating system. Seabass cage culture showed 65% and 41% more eutrophication than turbot and trout, respectively. Feed production dominated climate impacts in seabass (86%) and trout (73%) aquaculture systems, while turbot (61%) had the highest (6017 kg CO_2-eq) emissions because of the farm energy consumption (Aubin et al., 2009).

Just 4%–8% came from infrastructure compared to trout flow-through system (15%) in seabass cages. Energy transporters to farm-level operations made a smaller contribution to climate change in seabass cages (5%). In both the seabass and trout systems, feed production was the primary source of emission acidification (68% and 80%, respectively). The energy need to manufacture pellets and crop ingredients production were linked to acidity. Energy carriers employed at farm level were the primary cause of acidity in the turbot recirculating system (52%). The acidity of turbot recirculating system (14%) was significantly influenced by equipments. Additionally, compared to the trout flow-through or seabass cage systems (19 and 25 kg SO_2-eq/tonne, respectively), the turbot recirculating system had much higher (48 kg SO_2-eq/tonne) acidifying emissions (Aubin et al., 2009). Feed production was the only source of net primary production utilization. Turbot recirculating system and seabass cages showed large differences (17%). The variations in FCR between systems (1.21 in turbot and trout, 1.77 in seabass) accounted for these levels that was somewhat offset by the total amount of fish oil and fish meal included in the meals (69% in turbot and trout, 50% in seabass).

The energy use fluctuated between the systems. Seabass systems were dominated by feed production (72%), turbot recirculating systems were primarily driven by farm energy use (86%), and trout farms relied primarily on farm energy (47%) and feed (40%) with modest chemical inputs. Compared to seabass (54,656 MJ/tonne) and trout (78,229 MJ/tonne), turbot had the highest (290,986 MJ/tonne) energy consumption (Ghamkhar et al., 2021). In contrast to its low water dependency, the system was distinguished by its high level of EU and the associated impact categories of acidification and climate change. Due to the very low feed efficiency of the system, seabass cage system was typified by significant eutrophication and a greater need of circulating water.

With somewhat minimal effects on each of the selected categories, trout flow-through system seemed to be a decent middle ground (Briones-Hidrovo et al., 2023). To create production systems that are socially, ecologically, and economically sustainable, production methods in integrated and multitrophic systems have been enhanced. Numerous factors, including the species being cultured, the production techniques, applied aquaculture technology, the purpose, aquafeed used, other allochthonous inputs, and so in, must be considered when it comes to sustainable consumption in aquaculture. The demand for vital components used in fish feed, particularly fish meal and fish oil, has peaked due to the rise in aquaculture production, particularly the intensive systems (Zlaugotne et al., 2022). Nonetheless, numerous attempts have been undertaken to substitute fish oil and fish meal in fish-feeds to ensure environmental sustainability without compromising the productivity. Research has demonstrated that substituting marine microalgae for wild fish could cut the catch by as much as 30%. Use of linseed oil and microalgae *Spirulina* sp. up to 50% in mullet diets is an example of fishmeal and fish oil substitution (Hussain et al., 2024). Soybean products such as fermented soybean and bioprocessed soybean meal could replace fish meal in rainbow trout diet (Zhang et al., 2023). Using agri-wastes in the feeds of carnivorous fish presents a techno-economic opportunity to significantly reduce feed costs, which typically account for 60%–70% of operational expenses in aquaculture. However, success is highly dependent on addressing specific nutritional and processing challenges to ensure commercial viability.

7.1 Cost-benefit analysis

No literature and scientific analyses could be complete in itself to lead to the level of technology transfer (translational research and development) stage without a proper cost-benefit analysis done. This is because of the fact that a commercializable techno-business package must meet the basic economic feasibility parameters alongside the technical feasibility tests, together known as the techno-economic feasibility analysis. Below is provided a gist of a representative techno-economic feasibility in general and the cost-benefit analyses in particular that could provide better clarity on the underlying principles (Table 15.3).

7.2 Techno-economic feasibility factors

The economic feasibility (viability) of strategies to incorporate agri-wastes into the formulated feeds designed for carnivorous fish would depend heavily on several critical factors as below:

***The processing technology*:** Advanced processing methods such as SSF or bioconversion into microbial proteins (SCPs) are crucial to enhance the nutrient profiles and neutralize ANFs present in the phyto-materials in their raw forms. The capital and operating costs of the material requirements are major aspects to consider in the overall economic analysis.

***Optimal inclusion levels*:** Studies have demonstrated that moderate levels (e.g., 20%–40% of diet) of inclusion often provided the best balance of cost reduction without compromising the growth performance in carnivorous species such as *C. gariepinus*, while an inclusion level that is higher than this might impact the growth adversely.

***Species specificity*:** As compared to the omnivorous, carnivorous fish species specifically has a high protein requirement and a limited ability to utilize carbohydrates. Therefore, the feed formulation strategy must meet this specific consideration, possibly limiting the percent inclusion of certain selectively recalcitrant carbohydrate-rich agri-wastes.

***Market price of conventional feed*:** Economic incentives to incorporate agri-wastes in formulated feeds are strongly linked to the prevailing high market costs of fishmeal and other conventional feed ingredients. If the prices of conventional feed drops significantly, it will have an economic impact on the agri-waste based feed business, thereby weakening the market link.

***Policies and incentives*:** Government waste management and sustainable practices policies and incentives could make agri-waste valorization initiatives more profitable, potentially through subsidies or carbon credits.

While significant technical challenges exist, particularly with the processing and nutrient balancing, the use of agri-wastes as feed ingredient is a promising, cost-effective and sustainable strategy. This may, however, entail process optimization through rigorous small-scale (prototype) testing before adoption of the appropriate technology for on-field replication.

8. The Timor–Leste case study

The Timor–Leste National Aquaculture Development Strategy (NADS) aimed to significantly boost the national fish production through aquaculture and per capita fish consumption by 2030. Aquaculture required sustainable source of feed and locally available ingredients in order to achieve high fish output. Crude protein could make up at least 20% of the fish diet. Soybean or animal-origin ingredients could be added to plant-based ingredients to increase the protein level; the latter required import of feed or feed materials (Briones-Hidrovo et al., 2023). However, any feed mill that relied on imported materials would not be financially feasible in Timor–Leste where aquaculture industry is just picking up. Nile tilapia aquaculture fish production and consumption is an ideal species to meet the goals of Timor–Leste as it uses green water and relies primarily on the feeds composed of plant-based materials. A survey of farmers and other stakeholders was held to ascertain possible and potential local fish feed ingredients, followed by a workshop on feed formulation. Based on the seasonal or year-round availability of on-farm feed production, the survey identified 26 types of ingredients form plant and seven ingredients from animal sources (Pant et al., 2024).

Rice bran, maize bran, leucaena leaf meal, taro/cocoyam leaves, fresh cassava, and blood meal were among the potential materials that could be utilised to formulate fish feeds. However, the latter had appreciably high fiber content and low protein values. Four sets of commercial feed formulation alternatives with 26%–27% specified protein content using local and/or imported (fishmeal and soybean meal) ingredients were recommended during the workshop. The feed formulation exercise showed that these substances could be used to produce feed with only about 15% crude protein that could be increased to 18% if blood meal was also added. To meet the fish production and consumption targets of National Aquaculture Development (NAD) through aquaculture, the workshop also determined that support for the private sector was essential, either for the import of high-quality feed itself or for the import of high-quality feed ingredients in bulk for to produce the recommended feeds in-house (Pant et al., 2024).

9. Knowledge gaps and future prospects

Use of agri-wastes in carnivorous fish feed presents a sustainable alternative to the conventional fishmeal and soybean meal as feed ingredients. However, the key knowledge gaps, and the technological and operational challenges that exist must be addressed before recommending its widespread field-scale adoption. The future prospects would be more oriented toward advanced processing technologies and circular bioeconomy approaches.

9.1 Knowledge gaps and challenges

***ANFs*:** A primary challenge is the presence of ANFs (e.g., phytate, trypsin inhibitors, gossypol and cyanide) in plant-based agri-wastes, which affects nutrient utilization, growth performance and fish health adversely at high concentrations. Research has not yet fully optimized the processing methods to consistently eliminate these across waste types.

***Nutritional imbalance*:** Carnivorous fish has specific high protein, essential amino acids (such as methionine) and fatty acids (Omega-3) requirements that agri-wastes often lack or present in insufficient quantity. Blending various plant proteins is suggested, but a precise understanding of the optimal combination to meet the species-specific requirements is a knowledge gap.

***Species-specific requirement*:** Most existing studies generalize the efficacy of plant-based ingredients. The optimal inclusion levels and processing methods vary considerably among carnivorous fish species (e.g., salmonids exhibit reduced growth at lower replacement levels than some omnivores), necessitating more targeted research.

***Mechanism of action*:** While studies showed positive effects on growth or immune response, the underlying biological and molecular mechanisms are often not fully understood. This lack of mechanistic data compromises the ability to formulate reliable, effective, and universally acceptable diets.

***Toxicity and optimal concentration*:** Conflicting results are revealed in existing literature, as beneficial effects are often observed only at very low inclusion rates (e.g., 1%–4% w/w); higher concentrations can lead to toxic effects. Rigorous small-scale testing is needed to determine the optimal, nontoxic inclusion level for each waste type and fish species.

***Pesticide residues*:** Agri-wastes could contain pesticide residues or heavy metals which might pose environmental and fish safety concerns (as well as bioaccumulation and biomagnification issues in humans) that require further assessment including bioassays.

9.2 Future prospects and opportunities

***Bioprocessing and fermentation*:** The future would lay in advanced processing technologies such as SSF and microbial fermentation (using fungi, yeasts or bacteria). These methods could degrade fiber, detoxify ANFs, and enrich the agri-wastes with digestible protein, making them more suitable for inclusion in the formulated feeds meant for carnivorous fish.

***Insect bioconversion*:** Using insects (e.g., black soldier flies, mealworms) for bioconversion of agri-wastes into high-protein insect meal is a promising strategy. Insect meal has an amino acid profile comparable to fishmeal and mitigates the disadvantages in directly using crop wastes.

***Nutraceutical and immunostimulant potential*:** Specific agri-wastes (such as grape seeds or certain medicinal herbs) may contain compounds with nutraceutical properties that can boost fish immunity and antioxidant defenses, potentially replacing synthetic chemical stimulants.

***Circular economy integration*:** Repurposing agricultural waste into high-value aquafeed ingredients aligns with circular economy principles, reducing environmental pollution from waste disposal and lowering feed production costs.

***Advanced research tools*:** Emerging and cutting-edge techniques and technologies such as genomics, bioinformatics and encapsulation could help researchers to judiciously formulate and understand the nutrient utilization patterns better, and improve the functionality of the agri-waste-based feeds.

***Regional models and policies*:** Developing region-specific models and policies will facilitate adoption by farmers and ensure that local agro-climatic conditions and available agri-waste streams are leveraged effectively.

10. Conclusion

Transforming environmental liabilities into valuable nutritional resources, incorporating agri-waste into carnivorous fish feed could foster sustainable and circular blue economy. While addressing the significant waste disposal challenge it also mitigates the heavy reliance of the aquaculture industry on marine resources for fishmeal and fish oil that are expensive and finite. While there is a significant opportunity, however, navigating the complex issues related to nutritional balance and fish health is required to realize the full potential. Although the shift to agri-waste-based feeds is challenging, innovative processing and strategic nutritional management provide novel vistas and a clear pathway to overcome it. Such transition is an alternative and necessary evolution toward a resilient, cost-effective and environmentally responsible aquaculture practices that align well with the global circular bioeconomy principles in line with SDG-14 of the Global Sustainable Goals (SDGs). To ensure sustainable aquaculture growth while safeguarding economic health and the environment, research and collaborative efforts should consider these in future endeavor that would further refine these strategies.

References

Abdallah, M., Ben Aoun, R., Ben Amor, M., Touhami, I., Habibi, M., Trabelsi, N., 2023. Useful bioactive compounds from olive tree by-products (leaves, branches, fruits). In: Agricultural Waste: Environmental Impact, Useful Metabolites and Energy Production, pp. 331–358. https://doi.org/10.1007/978-981-19-8774-8_13.

Abu-Elala, N., Marzouk, M., Moustafa, M., 2013. Use of different *Saccharomyces cerevisiae* biotic forms as immune-modulator and growth promoter for *Oreochromis niloticus* challenged with some fish pathogens. Int. J. Vet. Sci. Med. 1 (1), 21–29. https://doi.org/10.1016/j.ijvsm.2013.05.001.

Ahuja, I., Dauksas, E., Remme, J.F., Richardsen, R., Løes, A.K., 2020. Fish and fish waste-based fertilizers in organic farming—with status in Norway: a review. Waste Manag. 115, 95–112. https://doi.org/10.1016/j.wasman.2020.07.025.

Ajila, C.M., Brar, S.K., Verma, M., Tyagi, R.D., Godbout, S., Valéro, J.R., 2012. Bio-processing of agro-byproducts to animal feed. Crit. Rev. Biotechnol. 32 (4), 382–400. https://doi.org/10.3109/07388551.2012.659172.

Akbari, H., Shekrabi, S.P.H., Soltani, M., Mehrgan, M.S., 2021. Effects of potential probiotic *Enterococcus casseliflavus* (EC-001) on growth performance, immunity, and resistance to

Aeromonas hydrophila infection in common carp (*Cyprinus carpio*). Probiotics Antimicrob. Proteins 13 (5), 1316–1325. https://doi.org/10.1007/s12602-021-09771-x.

Aubin, J., Papatryphon, E., Van der Werf, H.M.G., Chatzifotis, S., 2009. Assessment of the environmental impact of carnivorous finfish production systems using life cycle assessment. J. Clean. Prod. 17 (3), 354–361. https://doi.org/10.1016/j.jclepro.2008.08.008.

Bala, S., Garg, D., Sridhar, K., Inbaraj, B.S., Singh, R., Kamma, S., Tripathi, M., Sharma, M., 2023. Transformation of agro-waste into value-added bioproducts and bioactive compounds: micro/nano formulations and application in the agri-food-pharma sector. Bioengineering 10 (2), 152. https://doi.org/10.3390/bioengineering10020152.

Bandyopadhyay, P., Swain, S.K., Mishra, S., 2005. Growth and dietary utilisation in goldfish (*Carassius auratus* Linn.) fed diets formulated with various local agro-produces. Bioresour. Technol. 96, 731–740. https://doi.org/10.1016/j.biortech.2004.06.018.

Bandyopadhyay, P., Mishra, S., Sarkar, B., Swain, S.K., Pal, A., Tripathy, P.P., Ojha, S.K., 2015. Dietary *Saccharomyces cerevisiae* boosts growth and immunity of IMC *Labeo rohita* (Ham.) juveniles. Indian J. Microbiol. 55 (1), 81–87. https://doi.org/10.1007/s12088-014-0500-x.

Banu, M.R., Akter, S., Islam, M.R., Mondol, M.N., Hossain, M.A., 2020. Probiotic yeast enhanced growth performance and disease resistance in freshwater catfish gulsatengra, *Mystus cavasius*. Aquac. Rep. 16, 100237. https://doi.org/10.1016/j.aqrep.2019.100237.

Barik, S.K., Mishra, S., Ayyappan, S., 2000. Decomposition patterns of unprocessed and processed lignocellulosics in a freshwater fish pond. Aquat. Ecol. 34 (2), 185–204. https://doi.org/10.1023/A:1009981319515.

Bhaskar, N.B.N.S., Mahendrakar, N.S., 2008. Protein hydrolysate from visceral waste proteins of Catla (*Catla catla*): Optimization of hydrolysis conditions for a commercial neutral protease. Bioresour. Technol. 99 (10), 4105–4111. https://doi.org/10.1016/j.biortech.2007.09.006.

Boyd, C.E., 2017. General relationship between water quality and aquaculture performance in ponds. In: Fish Diseases. Academic Press, pp. 147–166. https://doi.org/10.1016/B978-0-12-804564-0.00006-5.

Briones-Hidrovo, A., Quinteiro, P., Dias, A.C., 2023. Investigating the environmental sustainability of a seabass and seabream aquaculture system in Portugal based on life cycle and nexus approaches. Sci. Total Environ. 890, 164195. https://doi.org/10.1016/j.scitotenv.2023.164195.

Chapman, F.A., 2000. Ornamental fish culture, freshwater. Encyclopedia of Aquaculture, 3 (1). John Wiley & Sons, New York, USA, pp. 602–610.

Chattaraj, S., Behera, B.K., Das Mohapatra, P.K., 2024a. Effective valorization of Chlorella biomass and Brewers spent grain substituting fish meal and soybean meal in the diet of herbivorous fish Cirrhinus reba for higher growth, digestibility and sustainable cultivation. J. Appl. Phycol. 36 (1), 217–232.

Chattaraj, S., Mitra, D., Chattaraj, M., Ganguly, A., Thatoi, H., Mohapatra, P.K.D., 2024b. Brewers' spent grain as fish feed ingredient: evaluation of bio-safety and analysis of its impact on gut bacteria of Cirrhinus reba by 16S Metagenomic sequencing. Curr. Res. Microbial Sci. 7, 100286.

Chattaraj, S., Mitra, D., Ganguly, A., Thatoi, H., Mohapatra, P.K.D., 2024c. A critical review on the biotechnological potential of Brewers' waste: Challenges and future alternatives. Curr. Res. Microbial Sci. 6, 100228.

Chattaraj, S., Chattaraj, M., Mitra, D., Ganguly, A., Thatoi, H., Das Mohapatra, P.K., 2025. 16S amplicon sequencing of the gastrointestinal microbiota of *Cirrhinus reba* and isolation of an autochthonous probiotic using culture based approaches. Syst. Microbiol. Biomanufacturing 5 (1), 156–170.

Chattaraj, S., Ganguly, A., Mitra, D., et al., 2026a. Sustainable valorization of algal biomass in fish nutrition and dietetics: a critical review and scientometric study. Waste Biomass Valor. https://doi.org/10.1007/s12649-026-03619-7.

Chattaraj, S., Sahoo, S.S., Behera, S., Thatoi, H., 2026b. Sustainable aquafeed development using Spirulina: Effects on growth performance of Catla catla. Algal Res. 104447.

Cho, C.Y., 1991. Quantitation of fish culture wastes by biological (nutritional) and chemical (limnological) methods; the development of high nutrient dense (HND) diets. In: Nutritional Strategies And Aquaculture Waste. Proceedings of the 1st International Symposium on Nutritional Strategies and Manage. Fish Nutrition Research Laboratory, Department of Nutritional Sciences, College of Biological Science, University of Guelph, pp. 37–50.

Choi, W.M., Lam, C.L., Mo, W.Y., Wong, M.H., 2016. Upgrading food wastes by means of bromelain and papain to enhance growth and immunity of grass carp (*Ctenopharyngodon idella*). Environ. Sci. Pollut. Control Ser. 23 (8), 7186–7194. https://doi.org/10.1007/s11356-015-4863-2.

Coppola, D., Lauritano, C., Palma Esposito, F., Riccio, G., Rizzo, C., de Pascale, D., 2021. Fish waste: from problem to valuable resource. Mar. Drugs 19 (2), 116. https://doi.org/10.3390/md19020116.

Dai, Z.L., Wu, G., Zhu, W.Y., 2011. Amino acid metabolism in intestinal bacteria: links between gut ecology and host health. Front. Biosci. 16 (1), 1768–1786. https://doi.org/10.2741/3820.

De Boni, A., Ottomano Palmisano, G., De Angelis, M., Minervini, F., 2022. Challenges for a sustainable food supply chain: a review on food losses and waste. Sustainability 14 (24), 16764. https://doi.org/10.3390/su142416764.

Drosou, F., Kekes, T., Boukouvalas, C., 2023. Life cycle assessment of the canned fruits industry: sustainability through waste valorization and implementation of innovative techniques. AgriEngineering 5 (1), 395–412. https://doi.org/10.3390/agriengineering5010026.

Ganguly, A., Chattaraj, S., Ganguly, M., Chattaraj, M., Banerjee, A., Mandal, A., Ali Khan, M., Sen, S.K., Das Mohapatra, P.K., 2025. Effect of three probiotic *Bacillus* strains supplemented feeds on growth, carcass composition and blood parameters of *Clarias magur* (Hamilton). J. Appl. Aquac. 37 (1), 19–44.

Ghamkhar, R., Boxman, S.E., Main, K.L., Zhang, Q., Trotz, M.A., Hicks, A., 2021. Life cycle assessment of aquaculture systems: does burden shifting occur with an increase in production intensity? Aquac. Eng. 92, 102130. https://doi.org/10.1016/j.aquaeng.2020.102130.

Gnaneswari, M.D., Narala, V.R., Subramani, P.A., 2022. Components of aquaculture as sources of environmental pollution and available remedial measures. In: Impact of COVID-19 on emerging contaminants: one health framework for risk assessment and remediation. Springer Nature, Singapore, pp. 305–325. https://doi.org/10.1007/978-981-19-1847-6_13.

Gupta, N., Mahur, B.K., Izrayeel, A.M.D., Ahuja, A., Rastogi, V.K., 2022. Biomass conversion of agricultural waste residues for different applications: a comprehensive review. Environ. Sci. Pollut. Control Ser. 29 (49), 73622–73647. https://doi.org/10.1007/s11356-022-22802-6.

Hangsapreurke, K., Pratoomchat, B., Seetaikum, J., Kitcharoen, N., 2020. Effect of dietary supplementation of minerals premix on growth performances and return in red tilapia (*Oreochromis nilloticus* x *Oreochromis mossambicus*). Maejo Int. J. Ener. Environ. Commun. 2 (1), 23–28. https://doi.org/10.54279/mijeec.v2i1.244948.

Henkin, J.M., Mainali, K., Sharma, B.K., Yadav, M.P., Ngo, H., Sarker, M.I., 2025. A review of chemical and physical analysis, processing, and repurposing of brewers' spent grain. Biomass 5 (3), 42. https://doi.org/10.3390/biomass5030042.

Hu, Y., Wang, Y., Li, A., Wang, Z., Zhang, X., Yun, T., Qiu, L., Yin, Y., 2016. Effects of fermented rapeseed meal on antioxidant functions, serum biochemical parameters and intestinal morphology in broilers. Food Agric. Immunol. 27 (2), 182 193. https://doi.org/10.1080/09540105.2015.1079592.

Huang, H.H., Feng, L., Liu, Y., Jiang, J., Jiang, W.D., Hu, K., Li, S.H., Zhou, X.Q., 2011. Effects of dietary thiamin supplement on growth, body composition and intestinal enzyme activities of juvenile Jian carp (*Cyprinus carpio* var. Jian). Aquac. Nutr. 17 (2), e233–e240. https://doi.org/10.1111/j.1365-2095.2010.00756.x.

Hussain, S.M., Bano, A.A., Ali, S., Rizwan, M., Adrees, M., Zahoor, A.F., Sarker, P.K., Hussain, M., Arsalan, M.Z.U.H., Yong, J.W.H., Naeem, A., 2024. Substitution of fishmeal: highlights of potential plant protein sources for aquaculture sustainability. Heliyon 10 (4). https://doi.org/10.1016/j.heliyon.2024.e26573.

Ismail, T., Hegazi, E., Nassef, E., El-Din, M.T.S., Dawood, M.A., Abdo, S.E., Gewaily, M.S., 2021. Gut immune-related gene expression, histomorphometry and hematoimmunological assays in Nile tilapia (*Oreochromis niloticus*) fed *Aspergillus oryzae* fermented olive cake. Fish Shellfish Immunol. 117, 299–310. https://doi.org/10.1016/j.fsi.2021.07.006.

Jan, K., Riar, C.S., Saxena, D.C., 2017. Characterization of agro-industrial byproducts and wastes for sustainable industrial application. J. Food Meas. Char. 11 (3), 1254–1265. https://doi.org/10.1007/s11694-017-9503-8.

Jumare, F.I., Salleh, M.M., Ihsan, N., Hussin, H., 2024. Cassava waste as an animal feed treatment: past and future. Rev. Environ. Sci. Biotechnol. 23 (3), 839–868. https://doi.org/10.1007/s11157-024-09701-7.

Kalaiselvan, P., Devi, N.C., Deepti, M., Devi, A.A., Akamad, K., Dheeran, P., Debbarma, S., Vadivel, D., Rajesh, D., 2025. Solid-state fermentation (SSF)-a sustainable future technology in aquafeeds? Front. Mar. Sci. 12, 1669719. https://doi.org/10.3389/fmars.2025.1669719.

Kari, Z.A., Sukri, S.A.M., Rusli, N.D., Mat, K., Mahmud, M.B., Zakaria, N.N.A., Wee, W., Hamid, N.K.A., Kabir, M.A., Ariff, N.S.N.A., Abidin, S.Z., 2023. Recent advances, challenges, opportunities, product development and sustainability of main agricultural wastes for the aquaculture feed industry–a review. Ann. Anim. Sci. 23 (1), 25–38. https://doi.org/10.2478/aoas-2022-0082.

Li, Z, Luo, W., Zhou, Q., Sun, C., Zheng, X., Liu, B., Mpange, K., Zhu, A., Wang, A., 2024. Investigation of the fermentation process of Moringa oleifera leaves and its effects on the growth performance, antioxidant capacity, and intestinal microbiome of *Procambarus clarkii*. Antioxidants 13 (11), 1355. https://doi.org/10.3390/antiox13111355.

Li, C., Zhang, B., Wang, X., Pi, X., Wang, X., Zhou, H., Mai, K., He, G., 2019. Improved utilization of soybean meal through fermentation with commensal *Shewanella* sp. MR-7 in turbot (*Scophthalmus maximus* L.). Microb. Cell Fact. 18 (1), 214. https://doi.org/10.1186/s12934-019-1265-z.

Lin, H., He, S., Tan, B., Zhang, X., Lin, Y., Yang, Q., 2022. Effect of rice protein meal replacement of fish meal on growth, anti-oxidation capacity, and non-specific immunity for juvenile shrimp *Litopenaeus vannamei*. Animals 12 (24), 3579. https://doi.org/10.3390/ani12243579.

Liu, S., Wang, S., Cai, Y., Li, E., Ren, Z., Wu, Y., Guo, W., Sun, Y., Zhou, Y., 2020. Beneficial effects of a host gut-derived probiotic, *Bacillus pumilus*, on the growth, non-specific immune response and disease resistance of juvenile golden pompano, *Trachinotus ovatus*. Aquaculture 514, 734446. https://doi.org/10.1016/j.aquaculture.2019.734446.

Mahanty, A., Mishra, S., Bosu, R., Maurya, U.K., Netam, S.P., Sarkar, B., 2013. Phytoextracts-synthesized silver nanoparticles inhibit bacterial fish pathogen *Aeromonas hydrophila*. Indian J. Microbiol. 53 (4), 438–446. https://doi.org/10.1007/s12088-013-0409-9.

Macfarlane, G.T., Macfarlane, S., 2012. Bacteria, colonic fermentation, and gastrointestinal health. J. AOAC Int. 95 (1), 50–56. https://doi.org/10.5740/jaoacint.SGE_Macfarlane.

Mirzaei-Aghsaghali, A., Maheri-Sis, N., 2008. Nutritive value of some agro-industrial by-products for ruminants-A review. World J. Zool., 3 (2), 40–46 (ISSN 1817-3098).

Mishra, S., Rath, S.C., 1999. Environmental implications of aquaculture. Aquaculture 4 (4), 33–36.

Mishra, S., 1998a. Perspective of aquaculture development. Yojana 42 (3), 37–41 (ISSN: 0971-8400).

Mishra, S., 1998b. Aquaculture and environment. Yojana 42 (9), 31–32 (ISSN: 0971-8400).

Mishra, S., 2008. Limnology research. In: Applied Bioinformatics, Statistics & Economics in Fisheries Research, CIFA. ICAR, p. 447.

Mishra, S., Barik, S.K., Ayyappan, S., Mohapatra, B.C., 2000. Fish bioassays for evaluation of raw and bioremediated dairy effluent. Bioresour. Technol. 72 (3), 213–218. https://doi.org/10.1016/S0960-8524(99)00124-8.

Mo, W.Y., Man, Y.B., Wong, M.H., 2018. Use of food waste, fish waste and food processing waste for China's aquaculture industry: needs and challenge. Sci. Total Environ. 613, 635–643. https://doi.org/10.1016/j.scitotenv.2017.08.321.

Mo, W.Y., Man, Y.B., Wong, M.H., 2019. Evaluation of nutritional values of food waste-based feed pellets and common feeding materials for culturing freshwater fish. In: Environmental Sustainability and Education for Waste Management: Implications for Policy and Practice. Springer Singapore, Singapore, pp. 305–321.

Moraes, G., de Almeida, L.C., 2020. Nutrition and functional aspects of digestion in fish. In: Biology and Physiology of Freshwater Neotropical Fish. Academic Press, pp. 251–271. https://doi.org/10.1016/B978-0-12-815872-2.00011-7.

Mutwedu, V.B., Bacigale, S.B., Mugumaarhahama, Y., Muhimuzi, F.L., Munganga, B., Ayagirwe, R.B., Nguezet, P.M.D., Manyawu, G., 2022. Smallholder farmers' perception and challenges toward the use of crop residues and agro-industrial byproducts in livestock feeding systems in Eastern DR Congo. Sci. Afr. 16, e01239. https://doi.org/10.1016/j.sciaf.2022.e01239.

Nwanna, L.C., Tope-Jegede, H., 2017. Effects of dietary *Lactobicillus plantarum* on the growth, carcass quality, and immune response of African catfish (*Clarias gariepinus*) challenged with *Salmonella typhi*. J. Appl. Aquacult. 29 (1), 62–80. https://doi.org/10.1080/10454438.2016.1209713.

Navari-Izzo, F., Quartacci, M.F., Sgherri, C., 2002. Lipoic acid: a unique antioxidant in the detoxification of activated oxygen species. Plant Physiol. Biochem. 40 (6–8), 463–470. https://doi.org/10.1016/S0981-9428(02)01407-9.

Ni, J., Yan, Q., Yu, Y., Zhang, T., 2014. Factors influencing the grass carp gut microbiome and its effect on metabolism. FEMS (Fed. Eur. Microbiol. Soc.) Microbiol. Ecol. 87 (3), 704–714. https://doi.org/10.1111/1574-6941.12256.

Obosi, P.I., Enyidi, U.D., 2025. Solid state fermentation of palm kernel cake and brewers' spent grain elevates amino acid profile and proximate compositions of aqua-feed and feed amino acid fluxes. J. Aquat. Sci. 40 (1), 87–99.

Omarini, A.B., Labuckas, D., Zunino, M.P., Pizzolitto, R., Fernández-Lahore, M., Barrionuevo, D., Zygadlo, J.A., 2019. Upgrading the nutritional value of rice bran by solid-state fermentation with *Pleurotus sapidus*. Fermentation 5 (2), 44. https://doi.org/10.3390/fermentation5020044.

Pant, J., Teoh, S.J., Gomes, S., Pereira, A., Pereira, M., De Jesus, L.S., Du Carmu, A.D.F., Bhujel, R.C., 2024. Sustainable intensification of genetically improved farmed Tilapia (GIFT) in Timor-Leste's farming systems: challenges and opportunities. Agric. Syst. 216, 103874. https://doi.org/10.1016/j.agsy.2024.103874.

Rahayu, S., Amoah, K., Huang, Y., Cai, J., Wang, B., Shija, V.M., Jin, X., Anokyewaa, M.A., Jiang, M., 2024. Probiotics application in aquaculture: its potential effects, current status in China and future prospects. Front. Mar. Sci. 11, 1455905. https://doi.org/10.3389/fmars.2024.1455905.

Ram, S., Narwal, S., Gupta, O.P., Pandey, V., Singh, G.P., 2020. Anti-nutritional factors and bioavailability: approaches, challenges, and opportunities. In: Wheat and Barley Grain Biofortification, pp. 101–128. https://doi.org/10.1016/B978-0-12-818444-8.00004-3.

Ranjan, A., Sahu, N.P., Deo, A.D., Kumar, S., 2019. Solid state fermentation of de-oiled rice bran: effect on in vitro protein digestibility, fatty acid profile and anti-nutritional factors. Food Res. Int. 119, 1–5. https://doi.org/10.1016/j.foodres.2019.01.054.

Samtiya, M., Aluko, R.E., Dhewa, T., 2020. Plant food anti-nutritional factors and their reduction strategies: an overview. Food Prod. Process. Nutr. 2 (1), 6. https://doi.org/10.1186/s43014-020-0020-5.

Sarkar, A., Chattaraj, S., Chattaraj, M., Maity, S., Ganguly, A., Nandi, B., Pahari, P.R., Das Mohapatra, P.K., 2022. Ichthyodiversity and water quality parameters of Uttar Dinajpur, West Bengal, India. North-West. J. Zool. 18 (2).

Shah, A.M., Zhang, H., Shahid, M., Ghazal, H., Shah, A.R., Niaz, M., Naz, T., Ghimire, K., Goswami, N., Shi, W., Xia, D., 2025. The vital roles of agricultural crop residues and agro-industrial by-products to support sustainable livestock productivity in subtropical regions. Animals 15 (8), 1184. https://doi.org/10.3390/ani15081184.

Shavandi, A., Hou, Y., Carne, A., McConnell, M., Bekhit, A.E.D.A., 2019. Marine waste utilization as a source of functional and health compounds. Adv. Food Nutr. Res. 87, 187–254. https://doi.org/10.1016/bs.afnr.2018.08.001.

Sheng, Y., Ren, H., Limbu, S.M., Sun, Y., Qiao, F., Zhai, W., Du, Z.Y., Zhang, M., 2018. The presence or absence of intestinal microbiota affects lipid deposition and related genes expression in zebrafish (*Danio rerio*). Front. Microbiol. 9, 1124. https://doi.org/10.3389/fmicb.2018.01124.

Shi, H., Su, B., Chen, X., Pian, R., 2020. Solid state fermentation of *Moringa oleifera* leaf meal by mixed strains for the protein enrichment and the improvement of nutritional value. PeerJ 8, e10358. https://doi.org/10.7717/peerj.10358.

Singh, P.K., Mohanty, P., Mishra, S., Adhya, T.K., 2022a. Food waste valorisation for biogas-based bioenergy production in circular bioeconomy: opportunities, challenges, and future developments. Front. Energy Res. 10, 903775. https://doi.org/10.3389/fenrg.2022.903775.

Singh, P.K., Shukla, P., Verma, S.K., Mishra, S., Parhi, P.K., 2022b. Zero-waste biorefineries for circular economy. In: Biotechnology for Zero Waste: Emerging Waste Management Techniques, pp. 439–456. https://doi.org/10.1002/9783527832064.ch29.

Sun, Y.Z., Xia, H.Q., Yang, H.L., Wang, Y.L., Zou, W.C., 2014. TLR2 signaling may play a key role in the probiotic modulation of intestinal microbiota in grouper *Epinephelus coioides*. Aquaculture 430, 50–56. https://doi.org/10.1016/j.aquaculture.2014.03.042.

Swidiq, M., Jolly, K., Emmanuel, Z., George, L., 2012. Utilization of crop residues and agro-industrial by-products in livestock feeds and feeding systems of Uganda. Int. J. Biosci. 2 (4), 82–89.

Tan, H.Y., Chen, S.W., Hu, S.Y., 2019. Improvements in the growth performance, immunity, disease resistance, and gut microbiota by the probiotic *Rummeliibacillus stabekisii* in Nile tilapia (*Oreochromis niloticus*). Fish Shellfish Immunol. 92, 265–275. https://doi.org/10.1016/j.fsi.2019.06.027.

Tang, L., Huang, K., Xie, J., Yu, D., Sun, L., Huang, Q., Bi, Y., 2017. Dietary copper affects antioxidant and immune activity in hybrid tilapia (*Oreochromis niloticus* × *Oreochromis aureus*). Aquac. Nutr. 23 (5), 1003–1015. https://doi.org/10.1111/anu.12468.

Treweek, J., 1995. Ecological impact assessment. Impact Assessment 13 (3), 289–315. https://doi.org/10.1080/07349165.1995.9726099.

Ugwuanyi, J.O., 2016. Enzymes for nutritional enrichment of agro-residues as livestock feed. In: Agro-industrial Wastes as Feedstock for Enzyme Production. Academic Press, pp. 233–260. https://doi.org/10.1016/B978-0-12-802392-1.00010-1.

Van Doan, H., Hoseinifar, S.H., Khanongnuch, C., Kanpiengjai, A., Unban, K., Van Kim, V., Srichaiyo, S., 2018. Host-associated probiotics boosted mucosal and serum immunity, disease resistance and growth performance of Nile tilapia (*Oreochromis niloticus*). Aquaculture 491, 94–100. https://doi.org/10.1016/j.aquaculture.2018.03.019.

Venkata Mohan, S., Varjani, S., Pant, D., Sauer, M., Chang, J.S., 2020. Circular bioeconomy approaches for sustainability. Bioresour. Technol. 318, 124084. https://doi.org/10.1016/j.biortech.2020.124084.

Wang, S., Xu, G., Zou, J., 2024. Soluble non-starch polysaccharides in fish feed: implications for fish metabolism. Fish Physiol. Biochem. 50 (1), 1–22. https://doi.org/10.1007/s10695-022-01131-y.

Wang, Z., Liao, S., Wang, J., Wang, Y., Huang, Z., Yu, W., Huang, X., Lin, H., Luo, M., Cheng, Z., Zhou, C., 2023. Effects of fermented cottonseed meal substitution for fish meal on intestinal enzymatic activity, inflammatory and physical-barrier-related gene expression, and intestinal microflora of juvenile golden pompano (*Trachinotus ovatus*). Fishes 8 (9), 466. https://doi.org/10.3390/fishes8090466.

Wattanakul, W., Thongprajukaew, K., Hahor, W., Suanyuk, N., 2021. Optimal replacement of soybean meal with fermented palm kernel meal as protein source in a fish meal-soybean meal-based diet of sex reversed red tilapia (*Oreochromis niloticus* × *O. mossambicus*). Animals 11 (8), 2287. https://doi.org/10.3390/ani11082287.

Weston, D.P., Gowen, R.J., 1988. Assessment and Prediction of the Effects of Salmon Net-Pen Culture on the Benthic Environment. School of Oceanography, College of Ocean and Fishery Sciences, University of Washington. Top of Form.

Wong, M.H., Mo, W.Y., Choi, W.M., Cheng, Z., Man, Y.B., 2016. Recycle food wastes into high quality fish feeds for safe and quality fish production. Environ. Pollut. 219, 631–638. https://doi.org/10.1016/j.envpol.2016.06.035.

Xie, X., Wang, J., Guan, Y., Xing, S., Liang, X., Xue, M., Wang, J., Chang, Y., Leclercq, E., 2022. Cottonseed protein concentrate as fishmeal alternative for largemouth bass (*Micropterus salmoides*) supplemented a yeast-based paraprobiotic: effects on growth performance, gut health and microbiome. Aquaculture 551, 737898. https://doi.org/10.1016/j.aquaculture.2022.737898.

Yafetto, L., 2022. Application of solid-state fermentation by microbial biotechnology for bioprocessing of agro-industrial wastes from 1970 to 2020: a review and bibliometric analysis. Heliyon 8 (3), e09173. https://doi.org/10.1016/j.heliyon.2022.e09173.

Yafetto, L., Odamtten, G.T., Wiafe-Kwagyan, M., 2023. Valorization of agro-industrial wastes into animal feed through microbial fermentation: a review of the global and Ghanaian case. Heliyon 9 (4), e14814. https://doi.org/10.1016/j.heliyon.2023.e14814.

Yang, H.L., Liu, Z.Y., Jian, J.T., Liu, Z.X., Ye, J.D., Sun, Y.Z., 2022a. Host-derived probiotics shape the intestinal microbial composition, but not putative function in juvenile Japanese seabass (*Lateolabrax japonicus*). Front. Mar. Sci. 9, 878633. https://doi.org/10.3389/fmars.2022.878633.

Yang, H.L, Liu, Z.Y., Jin, Y.M., Liu, Z.X., Zhang, B.Y., Yuan, Z.H., Ye, J.D., Sun, Y.Z, 2022b. Preventive and reparative functions of host-associated probiotics against soybean meal induced growth, immune suppression and gut injury in Japanese seabass (*Lateolabrax japonicus*). Fish Shellfish Immunol. 128, 651–663. https://doi.org/10.1016/j.fsi.2022.08.034.

Yang, H.L, Liu, Z.Y., Jian, J.T., Ye, J.D., Sun, Y.Z., 2021. Host-associated *Bacillus siamensis* and *Lactococcus petauri* improved growth performance, innate immunity, antioxidant activity and ammonia tolerance in juvenile Japanese seabass (*Lateolabrax japonicus*). Aquac. Nutr. 27 (6), 2739–2748. https://doi.org/10.1111/anu.13399.

Yang, H.L, Sun, Y.Z., Hu, X., Ye, J.D., Lu, K.L., Hu, L.H., Zhang, J.J., 2019. Bacillus pumilus SE5 originated PG and LTA tuned the intestinal TLRs/MyD88 signaling and microbiota in grouper (*Epinephelus coioides*). Fish Shellfish Immunol. 88, 266–271. https://doi.org/10.1016/j.fsi.2019.03.005.

Zhang, X., Sun, Z., Cai, J., Wang, J., Wang, G., Zhu, Z., Cao, F., 2020. Effects of dietary fish meal replacement by fermented moringa (*Moringa oleifera* Lam.) leaves on growth performance, nonspecific immunity and disease resistance against *Aeromonas hydrophila* in juvenile gibel carp (*Carassius auratus* gibelio var. CAS III). Fish Shellfish Immunol. 102, 430–439. https://doi.org/10.1016/j.fsi.2020.04.051.

Zhang, Q., Yang, Q., Guo, M., Li, F., Qin, M., Xie, Y., Xu, J., Liu, Y., Tong, T., 2023. The effects of dietary fermented soybean meal supplementation on the growth, antioxidation, immunity, and mTOR signaling pathway of juvenile coho salmon (*Oncorhynchus kisutch*). Fishes 8 (9), 448. https://doi.org/10.3390/fishes8090448.

Zlaugotne, B., Pubule, J., Blumberga, D., 2022. Advantages and disadvantages of using more sustainable ingredients in fish feed. Heliyon 8 (9), e10527. https://doi.org/10.1016/j.heliyon.2022.e10527.

Further reading

Wu, Y., Song, K., 2021. Anaerobic co-digestion of waste activated sludge and fish waste: methane production performance and mechanism analysis. J. Clean. Prod. 279, 123678. https://doi.org/10.1016/j.jclepro.2020.123678.

Zhou, J.S., Chen, H.J., Ji, H., Shi, X.C., Li, X.X., Chen, L.Q., Du, Z.Y., Yu, H.B., 2018. Effect of dietary bile acids on growth, body composition, lipid metabolism and microbiota in grass carp (*Ctenopharyngodon idella*). Aquac. Nutr. 24 (2), 802–813. https://doi.org/10.1111/anu.12609.

CHAPTER

16

Nonconventional feed ingredients in aquaculture: Effects on fish health, gene expression, and metagenomic insights into the alteration of fish gut microbiome

Koushik Ghosh[1], Dibyendu Acharya[1], Tanaya Sukul[1], Abhisek Mukhopadhyay[1] and Seyed Hossein Hoseinifar[2]

[1]*Aquaculture Laboratory, The University of Burdwan, Department of Zoology, Burdwan, West Bengal, India;* [2]*Department of Fisheries, Gorgan University of Agricultural Sciences and Natural Resources, Gorgan, Golestan, Iran*

1. Introduction

Aquaculture has emerged as one of the fastest-growing food production sectors worldwide, playing a vital role in meeting the global demand for high-quality animal protein. The intensification of aquaculture practices, however, has created a pressing need for nutritionally balanced, cost-effective, and environmentally sustainable feed formulations (Chattaraj et al., 2026a; Ganguly et al., 2025). Traditionally, fish meal and fish oil have served as the primary protein and lipid sources in aquafeeds due to their optimal amino acid profile, high digestibility, and presence of essential fatty acids (Hardy, 2010). Nevertheless, the continuous exploitation of wild fish stocks for animal feed production has resulted in overfishing, rising prices, and ecological imbalances (Tacon and Metian, 2015). Consequently, the aquaculture industry faces an urgent challenge to identify sustainable, nonconventional feed (NCF) alternatives that can support efficient fish growth without compromising health or environmental integrity.

NCFs—encompassing plant-derived meals (e.g., oilseed cakes, legume by-products, and leaf meals), agro-industrial residues, insect meals, microbial and algal biomasses, and fermented substrates—are increasingly being evaluated as substitutes for fish meal (Lourenço et al., 2025; (Das and Ghosh, 2023; Chattaraj et al., 2024a; Acharya et al., 2025). These feedstuffs not only reduce the dependency on marine resources but also promote circular bioeconomy principles by valorizing

Fish Feed Formulation with Agroindustrial Wastes and Byproducts. https://doi.org/10.1016/B978-0-443-41728-3.00005-8

agricultural and food wastes (Nath et al., 2023). However, the incorporation of such unconventional ingredients often introduces challenges related to nutrient bioavailability, interference caused by antinutritional factors (ANFs), fiber complexity, and variations in lipid and amino acid composition (Gatlin III et al., 2007; Ghosh et al., 2019). These nutritional and biochemical disparities directly influence the digestive physiology of fish and indirectly modulate the gut microbial ecosystem, which plays a crucial role in maintaining host health and metabolic homeostasis (Siddik et al., 2023).

The inclusion of NCFs exerts multifaceted effects on fish physiology, encompassing alterations in metabolism, immunity, oxidative stress response, and gene expression patterns (Hafeez et al., 2024). Many plant-based ingredients contain ANFs such as tannins, saponins, phytic acid, and protease inhibitors, which can interfere with protein digestibility and mineral absorption, subsequently triggering metabolic and inflammatory stress (Ghosh et al., 2019). Dietary exposure to such compounds often modulates the expression of genes related to digestive enzymes (amylase, lipase, trypsin), antioxidant defense (superoxide dismutase, catalase, glutathione peroxidase), and immune signaling pathways (IL-1β, TNF-α, NF-κB) (Huang et al., 2022; Mihailović et al., 2021; Chattaraj and Das Mohapatra, 2024). Conversely, fermentation, enzymatic hydrolysis, or microbial bioprocessing of these ingredients can substantially reduce ANFs while generating bioactive peptides, single-cell protein, and prebiotic oligosaccharides that enhance gut health and modulate gene expression beneficially (Weng et al., 2023). Similarly, NCFs such as insect meals and algal biomasses have been reported to enhance metabolic and antioxidant-related gene expression, indicating improved physiological resilience (Ardó et al., 2025). The incorporation of probiotics, prebiotics, and synbiotics also improves growth, digestion, metabolic and antioxidant parameters, gut health, and survivability, altering different growth and immune-related genes (Amenyogbe et al., 2020; Ganguly et al., 2025). Thus, the molecular responses induced by these alternative ingredients are crucial indicators of their nutritional compatibility and biological safety, providing deeper insight into how diet-driven gene regulation supports sustainable fish production.

The fish gastrointestinal (GI) tract harbors a complex and dynamic microbial community that performs essential functions, including nutrient digestion, short-chain fatty acid (SCFA) production, vitamin synthesis, immune regulation, and pathogen inhibition (Egerton et al., 2018; Chattaraj et al., 2025). The composition and functionality of this microbiome are strongly influenced by host species, life stage, environmental conditions, and—most importantly—dietary composition (Llewellyn et al., 2014). Even minor alterations in feed formulations can lead to marked shifts in microbial community structure, metabolic activity, and symbiotic interactions within the gut. Such diet-induced microbial perturbations can affect the host's nutrient utilization efficiency, growth performance, and disease resistance. Therefore, understanding how alternative feed resources reshape gut microbial ecology is vital for developing nutritionally optimized and health-promoting aquafeeds.

In recent years, the advent of next-generation sequencing and metagenomic approaches has revolutionized our understanding of microbial ecology in aquatic organisms (Satya et al., 2024). Unlike traditional culture-based methods, metagenomics allows comprehensive analysis of both culturable and unculturable microorganisms, enabling the identification of entire microbial consortia and their functional potential (Clagnan et al., 2024). Through whole-genome shotgun sequencing and 16S rRNA amplicon sequencing, researchers can profile the taxonomic diversity, metabolic pathways, and functional gene repertoires of gut microbiota under different feeding regimes (Tyagi et al., 2019; Chattaraj et al., 2024b). This has provided unprecedented insights into how NCFs alter microbial community assembly, enzymatic functions related to carbohydrate and protein metabolism, and the biosynthesis of bioactive metabolites such as SCFAs, bile acids, and antimicrobial peptides (Predescu et al., 2024).

The use of metagenomics has revealed that diets containing plant-based or fermented feed ingredients often promote an increase in bacterial taxa capable of fiber degradation and fermentation, such as *Lactobacillus*, *Bacillus*, *Clostridium*, *Prevotella*, and *Bacteroides* (Tawfik et al., 2024). These microbes contribute to the production of SCFAs like acetate, propionate, and butyrate, which serve as energy sources for intestinal epithelial cells and play roles in modulating immune responses (Ringø et al., 2022). Conversely, NCFs high in indigestible polysaccharides or phytochemicals can sometimes disrupt microbial equilibrium, leading to reduced diversity or enrichment of potentially opportunistic species (Sharma and Singh, 2025). Thus, balancing the nutritional benefits of alternative feeds with their microbiome-mediated effects is essential for sustainable aquaculture nutrition.

Furthermore, the integration of metagenomic data with metatranscriptomic, metabolomic, and metaproteomic analyses offers a multidimensional perspective on host–microbe–diet interactions. Such systems biology approaches enable the elucidation of metabolic cross-talk between gut microbes and the host, highlighting how dietary inputs modulate gene expression and functional outputs within the microbiome (Satya et al., 2024; Egerton et al., 2018). These insights are critical not only for understanding nutrient assimilation and immune modulation but also for developing next-generation functional feeds incorporating probiotics, prebiotics, and synbiotics that enhance gut health and resilience against environmental stressors (Mafe et al., 2025).

In the context of NCF utilization, solid-state fermentation (SSF) and enzymatic pretreatment technologies have gained significant attention (Ghosh et al., 2019). Fermentation of plant and agro-industrial substrates by beneficial microorganisms such as *Pichia*, *Aspergillus*, or *Bacillus* spp. has been shown to improve protein digestibility, degrade ANFs like tannins, phytic acid, and trypsin inhibitors, and enrich bioactive compounds that promote beneficial gut microbes (Ghosh et al., 2019). Metagenomic studies following the inclusion of such fermented feeds often demonstrate enhanced microbial diversity and an increased abundance of beneficial taxa associated with carbohydrate metabolism, SCFA production, and immune enhancement (Deka et al., 2021). Thus, SSF-based feed processing represents a promising strategy to mitigate the negative impacts of raw plant materials while simultaneously shaping a healthier gut microbiome.

From a broader ecological and sustainability standpoint, studying gut microbiome alterations through metagenomic tools also provides indicators of how dietary strategies affect nutrient cycling and waste output in aquaculture systems. Microbial shifts influencing nitrogen, phosphorus, and carbon metabolism can directly affect feed conversion ratios (FCR) and the environmental footprint of aquaculture operations (Tawfik et al., 2024). Therefore, understanding these microbial dynamics contributes not only to improving fish health and productivity but also to promoting environmentally responsible aquaculture practices aligned with the principles of the blue bioeconomy (Nath et al., 2023).

In summary, investigations on physiological parameters, gene expressions, and metagenomics have opened new avenues for exploring the complex interplay between diet composition, nutrient utilization, and microbial ecology in fish. As aquaculture transitions toward more sustainable and resource-efficient feeding strategies, it becomes imperative to integrate microbial insights into feed formulation research. This chapter aims to provide a comprehensive overview of NCF resources, the challenges associated with their use in feed formulations, and the metagenomic alterations that occur in fish gut microbiota. It will also address the gene expressions and health issues resulting from the inclusion of NCFs. It will discuss the functional implications of these microbial shifts on digestion, immunity, and overall host performance while emphasizing the role of metagenomic technologies as powerful tools for feed evaluation and optimization. Ultimately, such knowledge will help to bridge the gap between nutritional innovation and microbial ecology, guiding the design of future aquafeeds that are both sustainable and biologically compatible with the host gut ecosystem (Fig. 16.1).

2. NCFs in aquaculture

2.1 Need for NCFs

The growing demand for aquaculture products, coupled with the escalating cost and limited availability of conventional feed ingredients such as fish meal and soybean meal, has intensified the search for NCF resources (Acharya et al., 2025). These alternative ingredients—sourced from underutilized plants, agro-industrial by-products, aquatic macrophytes, insects, algae, and microbial biomass—offer a promising solution to reduce feed costs, minimize dependence on imported raw materials, and promote environmental sustainability through waste valorization and circular economy practices (Chattaraj et al., 2026b). Conventional feeds often face supply chain fluctuations, nutrient imbalances, and ecological concerns, such as overexploitation of fish stocks and deforestation linked to crop cultivation, making diversification imperative (Saini et al., 2024). NCFs, when processed to remove anti-nutritional factors and balanced nutritionally, can maintain or even enhance growth

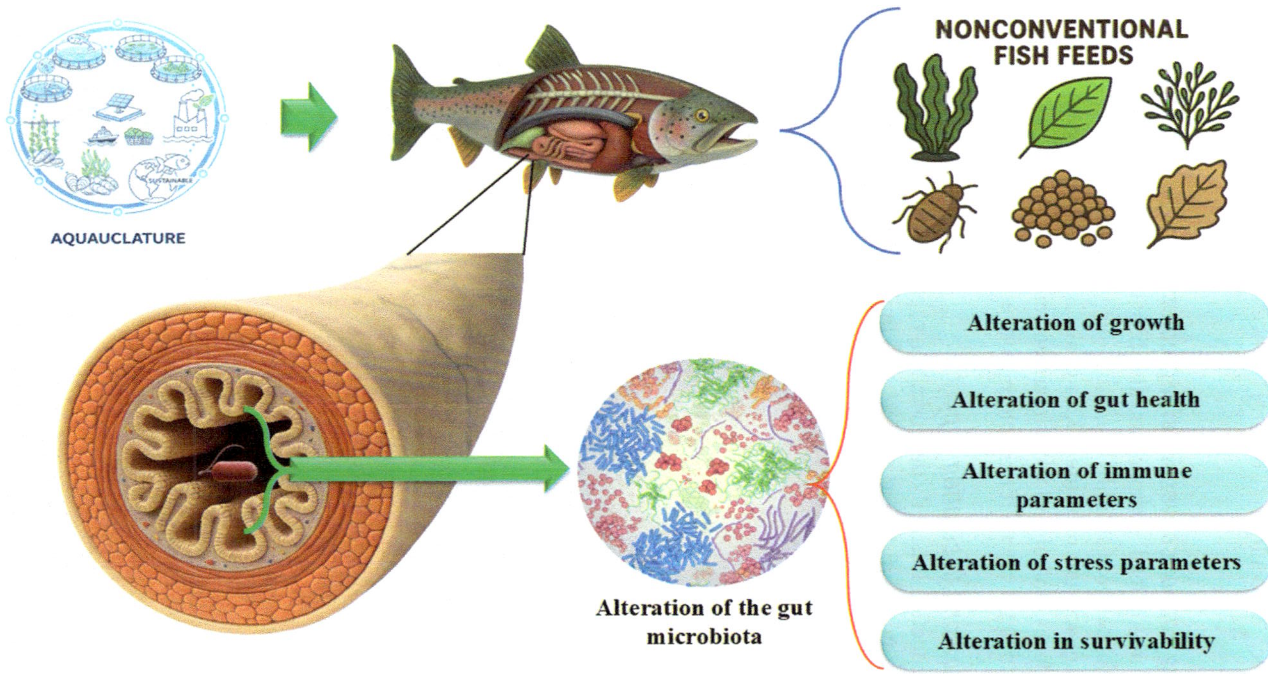

FIGURE 16.1

Use of NCF stuffs in the aquaculture industry and their effect on the alteration of fish health.

performance, immunity, and product quality in cultured species. Their strategic adoption not only supports farmer profitability but also aligns with the broader goals of sustainable aquaculture, resource efficiency, and climate-resilient food production.

2.1.1 Rising cost of conventional ingredients

The cost of major feed components like fish meal, soybean meal (SBM), and groundnut cake has been steadily increasing due to growing demand from livestock, poultry, and aquaculture industries (Kari et al., 2023). This escalation in raw material prices reduces profitability for fish farmers, especially small- and medium-scale producers, making it essential to explore more affordable feed resources.

2.1.2 Limited and fluctuating supply

Fish meal production is heavily dependent on capture fisheries, which are subject to seasonal variation, overfishing concerns, and climate change impacts (Soliman et al., 2017). Similarly, the supply of soybean meal is linked to agricultural productivity, which can be affected by land-use change, drought, and global market competition. This unpredictability in availability encourages the search for more stable, locally available alternatives.

2.1.3 Environmental sustainability

Overreliance on conventional feed sources exerts ecological pressure, such as depletion of wild fish stocks for fish meal production and deforestation for soybean cultivation. NCFs, especially those derived from agricultural by-products, aquatic plants, insects, and microbial biomass, offer more sustainable options that can reduce the environmental footprint of aquaculture (Hodar et al., 2020).

2.1.4 Utilization of local and underused resources

Many regions have an abundance of underutilized plants (e.g., duckweed, *Moringa* leaves, *Azolla*), agro-industrial residues (e.g., fruit pomace, oilseed cakes), and aquatic weeds (e.g., water hyacinth, water lettuce) that can be processed into nutritionally valuable feeds (Ghosh et al., 2019). Incorporating these into diets reduces dependence on imported ingredients and supports rural economies.

2.1.5 Waste reduction and circular economy

The aquaculture sector plays a significant role in waste valorization by using NCFs derived from food processing wastes, crop residues, or insect larvae reared on organic waste (Kari et al., 2023). This not only provides an economical protein source but also contributes to circular bioeconomy principles, reducing landfill pressure and environmental pollution.

2.1.6 Nutritional diversity and functional benefits

Certain NCF ingredients, such as microalgae, fermented plant products, and insect meals, contain bioactive compounds, pigments, essential fatty acids, and antioxidants that can enhance fish growth, health, and product quality (Bacchetti et al., 2020).

They can also improve gut microbiota balance and disease resistance, offering added value beyond basic nutrition. This diversity of nutrients ensures the health and quality of the final product, instilling confidence in the industry's practices.

2.1.7 Reducing dependence on imports

Many aquaculture-producing countries import a large share of their feed ingredients, making them vulnerable to international price volatility and supply chain disruptions. Locally sourced NCFs improve feed security and resilience.

2.2 Utilization of different nonconventional plant-derived materials in aquafeed formulation

The relatively lower cost of plant-based feedstuffs compared to fish meal has encouraged their exploration in aquafeed formulation. However, nonconventional plant sources often require additional processing to enhance their nutritive value and reduce antinutritional factors, which can add to their overall cost. Consequently, extensive research has been directed toward evaluating the nutritional potential of various plant-derived ingredients for aquaculture (Edwards et al., 1985; Ray and Das, 1996; Bairagi et al., 2002a). Several plant materials have demonstrated considerable promise as substitutes for conventional feedstuffs in formulated diets for fish. Different plant parts—when appropriately processed—can serve as practical and economical alternatives in aquafeed preparation.

Broadly, plant ingredients used in aquafeed can be classified into five categories: (1) seed meals, (2) oil cakes, (3) algae, (4) agricultural by-products, and (5) aquatic or terrestrial leaf meals. Certain alternative protein sources, such as oil cakes (Ramachandran et al., 2007) and sugarcane bagasse (Holanda and Ramos, 2016), have already found substantial applications in recycling and food-processing industries over recent decades. The main challenge lies in identifying plant materials that do not compete with human food requirements, yet provide adequate nutritional quality to match the feed efficiency of animal protein–based diets. The increasing commercial production of microalgae for biofuels and human nutrition has also stimulated interest in their use in aquafeeds (Gouveia et al., 2009).

Aquatic weeds present another potential resource; however, their relatively low protein content, along with high collection and labor costs, can limit their large-scale use. Nonetheless, these plants have been increasingly utilized as green manure (Vidya and Girish, 2014) and as livestock fodder (Iqbal, 1999). Some terrestrial plant leaf meals—due to their minimal industrial applications—can be incorporated into carp diets at low cost, with expenses generally restricted to collection and labor. Optimized use of such leaf meals is crucial not only for economic feasibility but also for nutrient recycling and overall ecosystem sustainability.

2.2.1 Seed meals

Seed meals are plant-derived products obtained after oil extraction from seeds, making them an important source of protein, energy, and essential amino acids for aquafeed formulation. Common examples include soybean meal, groundnut meal, sunflower meal, and cottonseed meal. These materials are often valued for their relatively high crude protein content, balanced amino acid profile, and ease of availability (Hardy, 2010; Oso and Ashafa, 2021). However, their direct inclusion in fish diets can be constrained by the presence of ANFs such as tannins, phytic acid, trypsin inhibitors, and saponins, which can impair nutrient digestibility and fish health (Francis et al., 2001; Ghosh et al., 2019).

To improve their nutritional value, various processing techniques such as heat treatment, fermentation, soaking, and enzyme supplementation have been employed (El-Saidy and Gaber, 2002; Ramachandran et al., 2007). Fermentation, in particular, has gained attention due to its dual role in degrading ANFs and enhancing the bioavailability of amino acids and minerals (Bairagi et al., 2002a; Feng et al., 2018). Additionally, SSF using microbial strains such as *Bacillus pumilus* and *Pichia kudriavzevii* has been shown to improve feed palatability and nutrient assimilation significantly (Banerjee et al., 2023; Mandal and Ghosh, 2013b).

From an economic perspective, seed meals are attractive as partial replacements for fish meal, especially in semi-intensive and intensive aquaculture systems where cost control is critical (Tacon and Metian, 2015). However, excessive reliance on certain seed meals—particularly soybean meal—may lead to supply competition with livestock and human consumption sectors. Therefore, diversification toward locally available, underutilized seed meals could offer a sustainable solution for reducing feed costs while maintaining fish growth performance and health.

2.2.2 Oil cakes

Oil cakes are by-products obtained after the mechanical pressing or solvent extraction of oil from seeds such as groundnut, mustard, sunflower, cottonseed, linseed, and palm kernel. These cakes are nutrient-rich, containing substantial levels of crude protein (typically 20%–45%), residual oil (5%–10%), and essential minerals, making them attractive alternatives or supplements to conventional fish meal in aquafeeds (Tacon and Metian, 2015; Tyapkova et al., 2016). The residual oil contributes energy, while the protein fraction provides a cost-effective plant-derived amino acid source, particularly for omnivorous and herbivorous fish species. However, oil cakes may also contain various antinutritional compounds such as gossypol (in cottonseed cake), phytic acid, tannins, and glucosinolates, which can negatively affect growth performance, nutrient absorption, and fish health (Francis et al., 2001; Zhou et al., 2012). Processing methods like solvent extraction, heat treatment, fermentation, or enzymatic degradation have been shown to reduce these ANFs significantly, improving digestibility and nutrient availability (Ghosh et al., 2019). For example, fermentation with specific yeast or bacterial strains can enhance the crude protein content while breaking down fibrous components and phytic acid, thus improving the bioavailability of phosphorus and other minerals (Adebo et al., 2022).

Oil cakes are generally more suitable for partial replacement of fish meal rather than complete substitution, with recommended inclusion levels varying between

10% and 40% depending on the fish species, processing technique, and amino acid supplementation strategy (Tyapkova et al., 2016; Hodar et al., 2020). Notably, oil cakes such as groundnut cake and sunflower cake have demonstrated positive growth performance and FCR improvements in carps and tilapia when incorporated at optimal levels (Zehra, 2020; Ayoola et al., 2023).

2.2.3 Algae

Algae, encompassing both microalgae (e.g., *Chlorella*, *Spirulina*, *Nannochloropsis*) and macroalgae (seaweeds such as *Ulva*, *Gracilaria*, *Sargassum*), are increasingly recognized as high-potential NCF resources for aquaculture. They are rich in proteins, polyunsaturated fatty acids (PUFAs), pigments (carotenoids, chlorophylls, phycobiliproteins), vitamins, and minerals (Gajendiran and Narayanan, 2025; Tibbetts et al., 2020). Microalgae, in particular, offer an excellent amino acid profile, omega-3 fatty acids (EPA, DHA), and bioactive compounds with immunostimulatory and antioxidant properties (Gao et al., 2024; González-Meza et al., 2023). Macroalgae, although typically lower in protein than microalgae, provide polysaccharides such as laminarin, alginate, and carrageenan, which may act as prebiotics, enhancing gut microbiota balance and immune function in fish (Patel et al., 2021). Furthermore, algal production can be carried out using nonarable land, saline or wastewater, and integrated multi-trophic aquaculture systems, thus aligning with circular economy principles (Tibbetts et al., 2020).

Despite these advantages, the large-scale inclusion of algae in aquafeeds faces challenges related to production cost, variability in nutrient composition, and the presence of antinutritional polysaccharides in certain species. Processing techniques such as enzymatic hydrolysis, fermentation, and mechanical cell wall disruption can enhance digestibility and nutrient release (González-Meza et al., 2023). In recent years, dried microalgal biomass has successfully replaced 10%–20% of fish meal protein in species such as tilapia and carp without adverse effects on growth (Gao et al., 2024).

2.2.4 Agricultural by-products

Agricultural by-products such as rice bran, wheat bran, maize gluten feed, sugarcane bagasse, fruit pomace, and vegetable trimmings are valuable low-cost feed resources that can reduce feed formulation expenses while promoting waste valorization. These materials, often generated in significant quantities by the crop processing industry, provide carbohydrates, fiber, and moderate protein levels (Ragaza et al., 2023).

In aquaculture, rice bran and wheat bran have been widely used as supplemental feeds for carp polyculture systems. In contrast, maize gluten feed offers higher protein and energy density for formulated diets. Additionally, fruit and vegetable processing residues such as citrus pulp and tomato pomace can contribute bioactive compounds, pigments, and vitamins (Sagar et al., 2018). However, their nutritional value is highly variable depending on crop variety, processing method, and storage conditions, necessitating compositional analysis before feed incorporation.

A limitation of many agricultural by-products is their high fiber and lignin content, which may reduce digestibility in fish. Processing techniques like pelleting, extrusion, fermentation, or enzyme supplementation (cellulase, xylanase) can help improve digestibility and nutrient release (Ragaza et al., 2023; Ghosh et al., 2019). When properly processed, these by-products can replace traditional ingredients in omnivorous fish diets, supporting cost-effective and environmentally sustainable aquaculture practices (Mahmud et al., 2025; Gao et al., 2024).

2.2.5 Aquatic and terrestrial leaf meals

Leaf meals derived from both aquatic plants (e.g., *Azolla*, duckweed, water hyacinth) and terrestrial plants (e.g., *Moringa oleifera*, mulberry, cassava leaves, *Leucaena*) represent another important category of NCF resources. These materials are typically rich in minerals, carotenoids, and other phytochemicals with potential health benefits (Leitzmann, 2016). Duckweed (*Lemna* spp.), for example, grows rapidly in nutrient-rich waters and can be harvested year-round, providing a sustainable protein source with balanced amino acids comparable to soybean meal (Ragaza et al., 2023).

Aquatic macrophytes like *Azolla* are also rich in vitamins and can fix atmospheric nitrogen, enabling cost-effective production in small-scale aquaculture systems. Terrestrial leaves such as *Moringa* contain significant levels of vitamins A and C, calcium, and antioxidants, making them valuable for improving fish health and coloration.

The main challenges in using leaf meals include high fiber content, the presence of secondary metabolites (tannins, oxalates, cyanogenic glycosides), and seasonal availability. Processing methods such as drying, ensiling, fermentation, and enzyme supplementation can improve nutrient digestibility and reduce ANFs levels (Asghar et al., 2024; Taye and Etefa, 2020). Inclusion levels typically range from 5% to 15% in aquafeeds, depending on the plant source and target species. Recent studies have reported that partial substitution of fish meal with *Moringa* or duckweed meal enhanced growth performance, immune responses, and pigmentation in species such as tilapia and ornamental fishes (Ayoola et al., 2023).

3. Challenges of using NCF ingredients as fish feed

The incorporation of NCF ingredients in aquaculture diets is increasingly explored as a means to reduce dependence on costly and limited conventional feed resources such as fish meal and soybean meal (Hussain et al., 2024). However, their utilization presents a range of nutritional, technical, and economic challenges that must be addressed before achieving large-scale application. One of the primary limitations is the presence of ANFs such as tannins, phytates, saponins, oxalates, and trypsin inhibitors, which interfere with nutrient digestibility and absorption (Ghosh et al., 2019). These compounds can bind with essential nutrients like proteins and minerals, reducing their bioavailability and ultimately hampering fish growth and feed efficiency. Although processing methods such as soaking, fermentation, heat treatment, and enzyme supplementation can reduce these ANFs, they also increase feed production costs and may not eliminate the adverse effects completely.

Another vital issue is the variability in nutrient composition among different batches of nonconventional ingredients. Factors such as plant species, geographical origin, seasonal variation, and postharvest handling greatly influence the nutritional quality of raw materials (Acharya et al., 2025). This inconsistency makes it difficult for feed formulators to design balanced diets that consistently meet the nutrient requirements of target fish species. Moreover, imbalances in essential amino acids—especially lysine, methionine, and threonine—are commonly found in plant-derived ingredients (Ghosh et al., 2019). These deficiencies can impair protein synthesis and growth unless they are supplemented with synthetic amino acids or other protein sources, which again raises formulation costs.

Digestibility and palatability also pose significant challenges. Many plant-based and agro-industrial by-products contain high levels of crude fiber, lignin, or non-starch polysaccharides that are difficult for fish to digest (Ghosh et al., 2019). This leads to reduced feed conversion efficiency and increased waste output, which can deteriorate water quality and environmental conditions in aquaculture systems (Chattaraj et al., 2024a). Additionally, some ingredients impart bitter flavors or strong odors that reduce feed intake, affecting overall growth performance.

From a safety perspective, contamination risks associated with nonconventional ingredients cannot be ignored. Ingredients sourced from waste streams or agricultural residues may contain pesticide residues, mycotoxins, or heavy metals that pose health risks to fish and consumers alike (Van Asselt et al., 2023). Ensuring the microbiological and chemical safety of such materials requires stringent quality control measures that add further costs and logistical complexities.

Finally, there are economic and logistical constraints in sourcing, processing, and storing nonconventional ingredients. Many potential feed resources are available only seasonally or in specific regions, leading to irregular supply chains. The lack of established processing infrastructure, storage facilities, and transport networks further limits their consistent availability for feed production.

4. Functional roles of the gut microbiome in nutrient acquisition

The role of the gut microbiota in fish nutrition goes well beyond mere colonization of the GI tract. It encompasses a range of enzymatic and metabolic functions that contribute to the breakdown of dietary components, thereby improving nutrient availability, absorption, and overall digestive efficiency.

4.1 Enhanced digestive capacity

Fish generally do not possess the full range of digestive enzymes needed to effectively break down complex macromolecules, particularly those present in plant-based feed ingredients. Therefore, autochthonous gut microbiota is increasingly recognized as a key secondary factor that positively impacts the host's digestion and nutrient absorption.

4.1.1 Enzyme production

Unlike in ruminants and higher vertebrates, the potential role of endosymbionts in the nutritional physiology of fish has only been acknowledged in recent years (Ray et al., 2012). In addition to the fish's own digestive enzymes, exogenous enzymes synthesized and supplied by the resident gut microbiota are increasingly recognized as significant secondary contributors to nutrition and feed efficiency in fish. Notably, gut-associated bacteria have been found to produce a variety of enzymes, including carbohydrases, phosphatases, esterases, lipases, and peptidases, which may play a vital role in enhancing digestive processes (Ghosh et al., 2019). Extensive research on Indian major carps (e.g., Ray et al., 2012; Mandal and Ghosh, 2013a; Das and Ghosh, 2014; Dutta et al., 2015; Dutta and Ghosh 2015; Banerjee et al., 2016; Mukherjee and Ghosh, 2016; Mukherjee et al., 2016, 2017) as well as other teleost species (e.g., Cahill 1990; Ringø et al., 2010, 2016; Llewellyn et al., 2014; Al-Hisnawi et al., 2015; Hoseinifar et al., 2016; Ringø and Song, 2016) has highlighted the presence of indigenous gut-associated microbiota in fish and their positive role in supporting nutritional functions. Microbes secrete a diverse array of extracellular enzymes (e.g., cellulases, xylanases, amylases, proteases, lipases) that break down complex macromolecules into simple, absorbable compounds.

4.1.2 Carbohydrate metabolism

While early studies on microbial amylase activity in the fish gut detected the presence of bacterial amylase, detailed characterization and identification of the specific amylase-producing bacterial strains were seldom performed (Lesel et al., 1986; Das and Tripathi, 1991; Bairagi et al., 2002b). Ghosh et al. (2002) were the first to report the presence of proteolytic and amylolytic *Bacillus* species—*B. circulans*, *B. cereus*, and *B. pumilus*, in the gut of the Indian major carp, rohu.

4.1.3 Protein and lipid metabolism

Certain bacterial taxa are known to produce extracellular proteases (Dutta and Ghosh, 2015; Dutta et al., 2015; Mukherjee and Ghosh, 2016; Mukherjee et al., 2016, 2017; Kavitha et al., 2018a) and lipases (Dutta and Ghosh, 2015; Dutta et al., 2015; Mukherjee and Ghosh, 2016; Mukherjee et al., 2016, 2017) in Indian major carps, aiding the host's endogenous enzymes in breaking down dietary protein and lipids into easily absorbed amino acids and fatty acids. For instance, in some omnivorous species, the functional genes associated with lipid transport and metabolism are enriched in the gut microbial community, indicating a key role in host energy balance (Zhang et al., 2020).

4.2 Synthesis of essential metabolites

The symbiotic microbes function as a highly specialized and efficient biological factory, continuously producing a diverse array of essential compounds that play a critical role in supporting the host organism's growth, development, metabolic regulation, and overall physiological maintenance.

4.2.1 Short-chain fatty acids

The anaerobic fermentation of nondigestible carbohydrates by bacteria (e.g., those from the phyla Firmicutes and Fusobacteria) yields SCFAs (acetate, propionate, and butyrate) (Xu et al., 2023; Jin et al., 2019a). Butyrate is especially significant, acting as the main energy source for colonocytes, supporting the integrity of the intestinal epithelium, and contributing to the regulation of immune responses in European sea bass (*Dicentrarchus labrax*) fed a plant-based diet (Gliozheni, 2018). The production of SCFAs is closely associated with increased energy extraction from the diet and better overall gut health.

4.2.2 Vitamin biosynthesis

Key bacterial groups, notably certain species of Cetobacterium (a common member of the Fusobacteriota) and Lactobacillus, are known to synthesize essential B vitamins (Qi et al., 2023). These microbially derived vitamins supplement the host's dietary intake and are vital cofactors in numerous host metabolic processes, including hematopoiesis and energy metabolism.

4.3 Gut microbiome and metabolic regulation

The influence of the gut microbiome on metabolism has been explored in several fish species, revealing notable impacts. In grass carp (*Ctenopharyngodon idella*), shifts in microbiota composition have been linked to alterations in various biosynthetic and metabolic pathways, including those related to carbohydrates, amino acids, and lipids (Ni et al., 2014). In zebrafish, microbial colonization of the gut enhances epithelial absorption of fatty acids (Semova et al., 2012), and fish with a normal microbiota show greater lipid accumulation in the intestinal lining, along with upregulated expression of lipid metabolism genes, compared to germ-free counterparts lacking microbiota (Sheng et al., 2018). Similarly, Japanese flounder (*Paralichthys olivaceus*) fed diets supplemented with *Bacillus clausii* exhibit improved weight gain, feed efficiency, and overall growth performance relative to fish on unsupplemented control diets, likely due to enhanced nutrient digestibility and increased food intake (Ye et al., 2011). Collectively, these findings highlight the significant role of gut microbiota in regulating metabolic processes in fish.

4.4 Intestinal morphophysiology

A well-balanced and diverse gut microbiota is crucial for the healthy development of intestinal architecture. Zhang et al. (2021) showed that dietary supplementation of *Bacillus licheniformis* not only increased the growth and influenced the intestinal development and disease resistance of common carp but also altered the intestine microbiota structure at 10^8 CFU/g concentration. Commensal bacteria support the development of gut epithelial cells, promote greater villus height and density, and strengthen the mucosal barrier that collectively enhance the absorptive surface area and help suppress inflammation that could impair nutrient absorption.

5. Alteration of microbial community due to nonconventional feeding

Diets have been shown to influence the gut microbiome and its relationship to health in humans and diverse animal systems (Nayak, 2010; Rinninella et al., 2019). Aquaculture research has increasingly focused on diet that replace fish meal and fish oil, as sourcing these ingredients contributes to the depletion of wild fish population globally (Naylor et al., 2009). Diet replacement study usually assesses whether a nonconventional ingredient can match or exceed the productivity of diets based on conventional fish meal and fish oil levels however, few studies examine how such dietary components influence microbiome diversity or how these changes relate to animal performance.

Regarding protein alternatives, black soldier fly (*Hermetia illucens*) meal has been evaluated as a partial replacement for fish meal in rainbow trout (*Oncorhynchus mykiss*) diets. Positive results have generally been observed when using the early developmental stages of this insect, with no adverse effect on growth or survival and positively influencing the gut microbiome (Bruni et al., 2018; Rimoldi et al., 2019; Terova et al., 2019). Partial replacement of fish meal (10%–30%) in rainbow trout diets has been found to increase bacterial diversity in both fecal (Terova et al., 2019) and mucosal microbiomes (Bruni et al., 2018; Rimoldi et al., 2019). In addition to modulating overall microbiome composition, this insect meal specifically promoted lactic acid bacteria (LAB) within the fish intestinal contents (*Carnobacterium divergeni*) (Bruni et al., 2018) and feces (*Facklamia*, *Enterococcus*, *Lactobacillus*, and *Pediococcus*) (Terova et al., 2019). Rimoldi et al. (2019) found that an insect-based diet decreased the abundance of potentially pathogenic Gram-negative bacteria, including *Aeromonas* and *Citrobacter* genera, while increasing the prevalence of *Mycoplasma* spp., which are considered potentially beneficial for the gut health of rainbow trout. Soybean meal is the most commonly used plant-based protein substitute in aqua-feeds (Gatlin III et al., 2007), although its inclusion has been linked to adverse effects on the gut health of farmed fish (Francis et al., 2001; Krogdahl et al., 2003, 2010) is well established. Several ANFs, such as protease inhibitor, agglutinin, and allergenic proteins, have been identified in soybean (Wang et al., 2014). For example, feeding Atlantic salmon with soy protein concentrate at seawater temperatures above 17°C induce diarrhea, vomiting, and reduction in growth, while no such symptoms were found in case of fish meal diet under same conditions (Green et al., 2013). Green et al. (2013) found that incorporating just 10% soybean protein concentrate into the diet of *Salmo salar* induced diarrhea and vomiting, while also altering the gut microbiota, as analyzed by terminal restriction fragment length polymorphism and 16S rRNA clone library techniques. In *Penaeus vannamei*, the partial replacement of fish meal with dehulled, oil-extracted soybean meal was shown to impair growth, feed efficiency, and innate immunity, and to induce an unfavorable gut microbiome (Hu et al., 2019).

Additives including fructooligosaccharides (FOS) and sodium butyrate have been shown to reduce the detrimental effects of soybean meal on animal growth and gut microbiome health (Hu et al., 2019; Han et al., 2020). The combination of FOS and soybean meal positively influenced the bacterial microbiome, growth rate, survival, feed utilization, digestive enzyme activity, and nonspecific immune responses in *P. vannamei* (Hu et al., 2019). The inclusion of the prebiotic mannan-oligosaccharide, derived from yeast cell walls, in *S. salar* diets containing soy protein concentrate has been shown to prevent disease development, likely through modulation of the intestinal microbiome (Green et al., 2013).

Mixtures of plant- and animal-derived ingredients have also been investigated for their effects on the gut microbiota. Rimoldi et al. (2018) found that feeding rainbow trout a high-protein diet (80%) containing a substantial proportion of poultry by-product meal, a reduced level of fish meal (11%), and a plant protein component (20%) resulted in growth performance comparable to that of fish fed a diet rich in fish meal (37.3%) and devoid of poultry by-products. Although no significant differences in microbial alpha or beta-diversity were observed between the two diets, the gut microbiota in both cases was dominated by γ-proteobacteria and exhibited low levels of Lactobacillales, a profile not considered beneficial for overall fish health by the authors. In contrast, fish fed a lower-protein diet (56%) that also contained low levels of fish meal (11.2%) but higher proportions of vegetable proteins and poultry by-products developed a microbiota enriched with potentially beneficial bacteria, such as *Fusobacterium*, known for SCFA production. However, this diet was associated with reduced growth performance.

The proportions of proteins, lipids, and carbohydrates in the diet have a significant influence on both the gut microbiome composition and overall animal performance (Qiao et al., 2017; Sun et al., 2018). For instance, elevated dietary lipid levels have been associated with negative alterations in the gut microbiota, including an increase in potentially pathogenic bacterial populations. In *Portunus trituberculatus*, a diet high in crude fat (15.1%) resulted in reduced weight gain and survival rates compared to a low-fat diet (5.8%) (Sun et al., 2018). In the high-fat diet group, Proteobacteria dominated the gut microbiota (~78%), with Vibrionaceae accounting for approximately 42% of the total bacterial community, this family constituted only about 7% in the low-fat diet group. In contrast, Leptotrichiaceae were found to comprise around 61% of the bacterial community in the low-fat diet group, but only about 1% in those fed the high-fat diet (Sun et al., 2018).

Plant-based oils are the most commonly used alternatives to fish oil, primarily due to their wide availability and cost-effectiveness (Ringø et al., 2016). Similar to soybean meal, alternatives to soybean oil are among the most extensively studied, and as will be further discussed, they have been shown to impact both animal performance and the associated microbial community (Zhang et al., 2014; You et al., 2019). Although oilseeds, cereal grains, and legumes have been explored as alternatives to fish oil (as reviewed by Gatlin III et al., 2007), there remains a lack of microbiome-focused studies involving these ingredients. Further research is therefore needed to understand how alternative lipid sources influence the intestinal microbial community, especially given emerging evidence that dietary lipids can shape the long-term composition of the gut microbiota. For example, in juvenile Atlantic salmon, Jin et al. (2019b) reported that fish fed a diet low in long-chain n-3 polyunsaturated fatty acids (LC-n-3 PUFA)- using linseed and palm oils, had a higher relative abundance of pathogenic *Mycoplasma* spp. compared to those fed a diet rich in LC-n-3 PUFA from North Atlantic fish oil. This effect was observed after the fish were transferred from freshwater tanks to a shared marine environment (Jin et al., 2019b). Likewise, the complete substitution of fish oil with soybean oil in the diet of golden pompano (*Trachinotus ovatus*) resulted in adverse effects on intestinal health (You et al., 2019). In *P. vannamei*, a diet containing an equal mixture of soybean oil, beef tallow, and linseed oil (SBL), with approximately 44% PUFAs, led to greater weight gain and survival compared to diets based solely on soybean oil (50.15% PUFAs) or beef tallow (9.21% PUFAs) (Zhang et al., 2014). The SBL diet (comprising soybean oil, beef tallow, and linseed oil) was associated with a higher abundance of Rhizobiaceae in the gut microbiota, a trend that diminished in animals fed diets containing only soybean oil or beef tallow. Conversely, Aeromonadaceae and Enterobacteriaceae were slightly more abundant in the guts of animals receiving the soybean oil and beef tallow diets. In golden pompano (*Trachinotus ovatus*), linseed oil, with a PUFA content of 66%, was found to be more suitable than soybean oil, which contained 55% PUFA (You et al., 2019). Soybean oil had a more pronounced negative impact on both intestinal health and microbial composition compared to linseed oil. LC-PUFAs appeared to exert beneficial effects on the gut microbiota, with specific fatty acids (C20:4n-6, C20:5n-3, and C22:6n) showing significant positive correlations with *Lactococcus* spp. and *Bacillus* spp., and a negative correlation with *Mycoplasma* spp. (You et al., 2019).

Carbohydrate source also plays a critical role in shaping both animal performance and the gut microbiome. In *P. vannamei* cultured under low salinity conditions, shrimp fed a cornstarch-based diet exhibited poorer growth performance compared to those fed diets containing sucrose or glucose (Qiao et al., 2017). The gut bacterial community in shrimp receiving cornstarch was marked by reduced abundances of *Chloroflexi*, *Verrucomicrobia*, and TM7, all of which were present at similar levels in the sucrose and glucose-fed groups. Additionally, the cornstarch diet led to an increased abundance of opportunistic pathogens, including *Desulfovibrio* spp. and *Aeromonas* spp. The authors suggested that these microbiome alterations, induced by the carbohydrate source, were likely responsible for the observed decline in shrimp growth performance (Qiao et al., 2017).

In addition to research focused on replacing fish meal and fish oil, a growing number of studies investigate functional nutritional strategies designed to promote animal growth and health through the inclusion of additives such as probiotics, prebiotics, and synbiotics. Probiotics are live microorganisms administered in specific quantities with the aim of colonizing the GI tract, modulating the microbiome, and providing health benefits to the host. In contrast, prebiotics are nondigestible dietary components that promote host health by selectively stimulating the growth or activity of beneficial gut microorganisms (Schrezenmeir and De Vrese, 2001). Synbiotics, which combine both probiotics and prebiotics, are intended to synergistically enhance these effects (Schrezenmeir and De Vrese, 2001). Although numerous studies have demonstrated that such feed additives can improve growth performance and disease resistance in aquaculture species (Irianto and Austin, 2002), relatively few have investigated their specific impacts on the gut microbiome. From an aquaculture perspective, feed additives capable of positively modulating the microbiome are of particular interest due to their potential to enhance productivity, increase yield, and reduce susceptibility to infection. For instance, in *Apostichopus japonicus*, dietary supplementation with *Bacillus licheniformis*, either as a probiotic or in combination with alginate oligosaccharides as a symbiotic, led to enhanced growth performance, improved innate immune responses, or increased survival following a *Vibrio splendidus* challenge (Wang et al., 2017). Additionally, both diets promoted an increase in the relative abundance of potentially beneficial bacteria, such as *Bacillus* spp. and *Lactococcus* spp., while slightly reducing the abundance of *Vibrio* spp. (Wang et al., 2017).

Poly-β-hydroxybutyrate (PHB) is a microbial polymer synthesized by various bacteria and other organisms as a carbon and energy storage compound (Luongo et al., 2017). Its production typically occurs under nutrient-limited but carbon-rich conditions, such as when nitrogen, phosphorus, sulfur, or oxygen is scarce (Bonartseva et al., 2002; Kavitha et al., 2018b). Several studies have demonstrated that dietary supplementation with low concentrations of PHB can enhance animal performance and modulate the intestinal microbiome (De Schryver et al., 2010; Najdegerami et al., 2012; Sui et al., 2016). For instance, the inclusion of 2% and 5% PHB in the diet of juvenile European sea bass (*Dicentrarchus labrax*) led to increased body weight, lowered intestinal pH, likely promoting SCFA production and altered the composition of the GI microbiota, as observed using denaturing gradient gel electrophoresis (DGGE) analysis (De Schryver et al., 2010). Similarly, dietary PHB at 1% and 3% improved growth performance in Chinese mitten crab (*Eriocheir sinensis*), enhanced digestive enzyme activities (pepsin, trypsin, and lipase in the hepatopancreas), and modulated immune responses (increased superoxide dismutase and decreased alkaline and acid phosphatase levels in haemolymph serum). PHB supplementation also increased bacterial richness and diversity in the gut microbiome (Sui et al., 2016). In Siberian sturgeon (*Acipenser baerii*) fingerlings, dietary supplementation with 2% PHB resulted in a slight improvement in weight gain and survival. Additionally, it modulated the intestinal microbiota by increasing microbial diversity, promoting aerobic metabolism in culturable bacteria, and enriching beneficial taxa such as *Bacillus* spp. and members of the Ruminococcaceae family (Najdegerami et al., 2012).

Various probiotics have been shown to influence the gut microbiome of key aquaculture species. Al-Hisnawi et al. (2019) demonstrated that the probiotic *Pediococcus acidilactici* modulates the intestinal microbiome of *Oncorhynchus mykiss* and enhances the innate immune response by upregulating the expression of proinflammatory cytokines and immunoglobulin T genes. Although the probiotic had no significant effect on alpha-diversity, it influenced the composition of specific microbial taxa. In the posterior intestine, it increased the relative abundance of *Mycoplasma* spp., members of the order *Sphingomonadales*, *Leuconostoc* spp., *Sphingobacterium* spp., and *Lactococcus* spp. In the intestinal mucosa, higher abundances of *Mycoplasma* spp., *Pediococcus* spp., *Pannonibacter* spp., and *cc-115* were observed. Additionally, in the intestinal content, *Pediococcus* spp., *Yaniella* spp., *Leptotrichia* spp., *Paracoccus* spp., *Facklamia* spp., and unidentified members of the families Streptococcaceae and Peptostreptococcaceae were enriched.

Among other feed additives with beneficial effects, dietary garlic supplementation in *O. mykiss* at inclusion levels of 1.0%, 1.5%, and 2.0% improved growth performance and was positively correlated with an increased abundance of *Clostridium* spp. However, garlic also induced shifts in the gut microbial community, leading to a reduction in bacterial richness and diversity within the intestinal mucus (Büyükdeveci et al., 2018). In a separate study, dietary salt supplementation (5%) in Nile tilapia (*O. niloticus*) reared in freshwater enhanced the digestibility of lipids, protein, ash, and dry matter, while also promoting growth (Hallali et al., 2018). This salt-enriched diet significantly altered the gut microbiome by increasing microbial diversity in the posterior intestine and enriching specific taxa, including *Micrococcus* spp., *Succiniclasticum* spp., *Sediminibacterium* spp., and members of the Clostridiaceae family (Hallali et al., 2018).

In some cases, dietary trials have shown that changes in the bacterial microbiome do not necessarily align with performance indicators such as growth and survival. For example, Rimoldi et al. (2018) fed rainbow trout diets with varying proportions of fish meal, plant proteins, and poultry by-product meal, and observed that fish exhibiting weight loss, rather than weight gain, developed bacterial communities considered potentially beneficial. In gibel carp (*Carassius auratus gibelio*), supplementation of a commercial diet with 5% terrestrial medicinal plants *Pastinaca sativa*, *Astragalus membranaceus*, and *Atractylodes macrocephala* Koidz (commonly used in Chinese medicine) did not significantly affect growth or survival. However, this dietary treatment enhanced bacterial alpha-diversity in the gut and slightly reduced the relative abundance of potential fish pathogens, including *Aeromonas* spp., *Acinetobacter* spp., and *Shewanella* spp. (Wu et al., 2018). Therefore, diet formulation should take into account not only traditional performance metrics such as growth and survival but also the associated shifts in the gut bacterial community, which may have beneficial or harmful impacts on host health. Looking ahead, microbiome research is expected to play a key role in evaluating diet suitability by identifying microbial changes that correlate with improved or impaired animal performance, ultimately contributing to more sustainable and productive aquaculture practices.

6. Effect of NCFs on gene expression

Several recent studies have demonstrated that the inclusion of NCF ingredients and functional additives in aquaculture diets can significantly modulate gene expression patterns in fish, revealing intricate molecular adaptations in growth, metabolism, and immunity. For instance, Nishiguchi et al. (2024) reported that supplementing fish diets with black soldier fly larvae enhanced gut immune gene expression, suggesting improved mucosal defense mechanisms. Similarly, Tacchi et al. (2012) and Król et al. (2016) showed that partial substitution of fish meal with plant-based proteins in Atlantic salmon-induced substantial transcriptomic shifts, particularly in genes associated with digestion, inflammation, and nutrient transport. Kemski et al. (2020) found that juvenile yellow perch fed SBM-based diets upregulated cholesterol biosynthesis genes to maintain homeostasis, potentially reducing growth due to the high energy cost of this pathway. Yu et al. (2023) further demonstrated through transcriptome analysis that a low–fish meal diet in *Micropterus salmoides* led to differential expression of genes, characterized by the upregulation of orexigenic and inflammatory factors and the downregulation of antioxidant and anorexigenic genes.

Hoseinifar et al. (2017) demonstrated that dietary administration of sodium propionate (SP) significantly ($P < .05$) upregulated the expression of GH and IGF1, and downregulated the MYSTN gene. In another study, Safari et al. (2024) noticed that substituting fish meal with yellow mealworm meal in *Acipenser stellatus* significantly altered metabolism-related gene expression. Ahmadifar et al. (2021) reviewed the gene regulatory roles of herbal extracts in fish, emphasizing their effects on immunity, growth, and reproduction. Moreover, Hoseinifar et al. (2024) reported that host-associated probiotics modulated immune parameters and molecular markers in Caspian whitefish fry, reflecting the nutrigenomic impact of microbiome-targeted feeding. Collectively, these findings indicate that NCFs ingredients not only act as sustainable nutritional alternatives but also actively reprogram metabolic and immune gene networks, shaping fish physiology and health within modern aquaculture systems.

7. Effect of NCFs on fish health

Nonconventional fish feeds, often incorporating alternative protein sources like plant meals or insect proteins, can significantly impact fish health (Aragão et al., 2022). The presence of ANF in some nonconventional ingredients can potentially disrupt gut health, leading to inflammation or changes in the gut microbiota composition (dysbiosis). Conversely, processing methods like fermentation or the inclusion of certain novel ingredients can introduce beneficial compounds and probiotics, supporting a healthier gut and improved nutrient utilization. Effects on the immune system are variable; some ingredients may suppress immune function, while others, particularly those with functional properties, can be immunostimulatory, enhancing disease resistance. Regarding stress enzymes, changes in diet can influence the fish's ability to cope with stressors. Certain nonconventional feeds or feed additives have been shown to modulate the antioxidative system, potentially reducing oxidative stress and lowering levels of stress hormones like cortisol.

7.1 Effect on gut health

The replacement of fish meal with nonconventional feed ingredients, such as plant proteins, fermented meals, insect meals, and algal biomass, has shown diverse impacts on the intestinal histo-architecture of fish. Histological responses serve as sensitive indicators of gut health, nutrient absorption efficiency, and local immune modulation. Several studies have reported that unprocessed soybean meal and other plant-derived ingredients rich in ANFs may induce intestinal enteritis, villi atrophy, and epithelial disruption, particularly in carnivorous fish species. Conversely, bioprocessed or fermented plant meals, insect meals, and algal supplements often alleviate these adverse effects by improving mucosal integrity, increasing villi height, and enhancing enterocyte health. The following table (Table 16.1) summarizes representative studies demonstrating gut histological damage or improvement in different fish species fed nonconventional feed ingredients.

7.2 Effect on immunity

Beyond nutrient composition, NCF ingredients profoundly influence the innate and adaptive immune responses of fish. The immunomodulatory potential of these ingredients depends on their biochemical composition, presence of bioactive compounds, and degree of processing. Fermented soybean meal, and insect meals such as *Hermetia illucens* or *Tenebrio molitor*, and microalgal biomass (*Chlorella*, *Spirulina*, *Ulva*) are increasingly recognized for enhancing humoral and cellular immune parameters, including lysozyme activity, complement components, immunoglobulin levels, and cytokine expression. In contrast, unprocessed plant meals high in saponins or other ANFs may suppress immune function and induce inflammation. The studies compiled below (Table 16.2) will provide an overview of immune improvement or reduction in various fish species fed with nonconventional dietary ingredients.

7.3 Effect on antioxidant enzyme activity

Oxidative stress is a major physiological challenge during intensive aquaculture, and diet plays a pivotal role in modulating the antioxidant defense system of fish. NCFs, particularly those rich in phytochemicals, bioactive peptides, and microbial metabolites, can influence the activities of key antioxidant enzymes such as superoxide dismutase (SOD), catalase (CAT), and glutathione peroxidase (GPx). Fermented plant meals, insect-derived proteins, microalgae, and single-cell proteins have shown potential in enhancing these enzymatic defenses while reducing lipid peroxidation (MDA levels). Conversely, high inclusion levels of unprocessed or poorly balanced alternative ingredients may induce oxidative stress by impairing antioxidant responses. The following table (Table 16.3) compiles studies that report alterations in antioxidant enzyme activity in fish following the inclusion of various NCFs.

8. Future outlook

At present, the application of NCFs is inevitable for the formulation of animal feed, including fish. The incorporation of advanced biochemical, molecular, as well as

Table 16.1 Effects of NCFs ingredients on gut health in different fish species.

No.	References	Fish species	Nonconventional ingredient	Inclusion level	Feeding duration	Histological outcome	Key findings
1	Krogdahl et al. (2003)	Atlantic salmon (*Salmo salar*)	Soybean meal (raw)	10%–40%	8–12 weeks	Damage	Caused distal intestinal enteritis; reduced villi height, and increased inflammatory cell infiltration.
2	Krogdahl et al. (2015)	Atlantic salmon	Soy saponins	1–2 g/kg	8 weeks	Damage	Saponins identified as causal agents for SBM-induced enteritis; increased gut permeability.
3	Sahlmann et al. (2013)	Atlantic salmon	Soybean meal	20%	7–21 days	Damage (transient)	Early transcriptional and histological changes—mucosal lifting, lymphocyte infiltration.
4	Pang et al. (2023)	Tilapia (*Oreochromis niloticus*)	Fermented soybean meal (FSBM)	10%–30%	8 weeks	Improvement	FSBM reduced inflammation and improved villi height versus raw SBM; better gut morphology.
5	Zakaria et al. (2022)	African catfish (*Clarias gariepinus*)	FSBM	15%–45%	60 days	Improvement	Moderate FSBM inclusion enhanced mucosal integrity and nutrient absorption.
6	Di Rosa et al. (2023)	Rainbow trout (*Oncorhynchus mykiss*)	Black soldier fly (*Hermetia illucens*) meal	10%–50%	10 weeks	No damage/ Neutral	No major intestinal lesions; mild goblet cell increase at highest inclusion.

Continued

Table 16.1 Effects of NCFs ingredients on gut health in different fish species.—*cont'd*

No.	References	Fish species	Nonconventional ingredient	Inclusion level	Feeding duration	Histological outcome	Key findings
7	El-Desouky et al. (2024)	Nile tilapia (*O. niloticus*)	Yellow mealworm (*Tenebrio molitor*) meal	10%–40%	8 weeks	No damage/ Slight improvement	Normal intestinal histology; improved villi length at 20% inclusion.
8	Velichkova et al. (2024)	Rainbow trout	Microalgal meal (*Chlorella vulgaris*)	10%–30%	70 days	Improvement	Increased mucosal fold height; better epithelial integrity and microvilli density.
9	Huang et al. (2024)	Marbled eel (*Anguilla marmorata*)	Chlorella meal	5%–25%	60 days	Improvement	Moderate inclusion enhanced intestinal villi and reduced lipid deposition.
10	Samsing et al. (2024)	Barramundi (*Lates calcarifer*)	Single-cell protein (methanotrophic bacteria)	10%–60%	8 weeks	No damage	Gut histology comparable to fish meal control; no mucosal degeneration.

Table 16.2 Effects of NCFs ingredients on immunity in different fish species.

No.	References (Year)	Fish species	Nonconventional ingredient	Inclusion level	Feeding duration	Immune outcome	Key findings
1	Pang et al. (2023)	Nile tilapia (*O. niloticus*)	Fermented soybean meal (FSBM)	10%–30%	8 weeks	Improvement	FSBM enhanced lysozyme, complement (C3/C4) activity, and antioxidant enzymes versus raw SBM.
2	Van Tran et al. (2024)	Common carp (*Cyprinus carpio*)	Bacillus-fermented SBM + amino acids	15%–45%	60 days	Improvement	Fermentation plus amino acid blend improved innate immune parameters and gut health.
3	Rimoldi et al. (2024)	Gilthead seabream (*Sparus aurata*)	Black soldier fly (*Hermetia illucens*) meal	10%–50%	10 weeks	Improvement (moderate)	Moderate inclusion enhanced serum lysozyme and complement activity; excessive inclusion showed a neutral effect.
4	Lee et al. (2024)	Rainbow trout	Defatted BSF larvae meal	20%–60%	10 weeks	Reduction (high level)	Very high inclusion caused a mild reduction in serum lysozyme and SOD; moderate levels are safe.
5	Pradhan et al. (2023)	Nile tilapia (*O. niloticus*)	*Chlorella vulgaris* meal	5%–20%	60 days	Improvement	Chlorella supplementation increased respiratory burst, lysozyme and IgM titers.
6	Sattanathan et al. (2022)	Rohu (*Labeo rohita*)	Algal extract blend (Spirulina + Ulva)	1%–3%	45 days	Improvement	Enhanced innate immunity (NBT, MPO, lysozyme) and disease resistance to *Aeromonas hydrophila*.

Continued

Table 16.2 Effects of NCFs ingredients on immunity in different fish species.—*cont'd*

No.	References (Year)	Fish species	Nonconventional ingredient	Inclusion level	Feeding duration	Immune outcome	Key findings
7	Samsing et al. (2024)	Barramundi (*Lates calcarifer*)	Single-cell protein (methanotrophic bacteria)	10%–60%	8 weeks	Neutral/slight improvement	Comparable immune responses to fish meal control; no inflammatory suppression.
8	Calcagnile et al. (2024)	Various species	Probiotics (*Lactobacillus*, *Bacillus*)	—	—	Improvement	Probiotics consistently enhanced nonspecific immunity and cytokine expression in fish.
9	Hoseinifar et al. (2024)	Multiple species	Probiotic candidates	—	—	Improvement	Metaanalysis showed positive effects on lysozyme, complement, and mucosal immunity.
10	Tayyab et al. (2025)	Tilapia/ catfish	Probiotic–vaccine combination	—	—	Improvement	Probiotic prefeeding increased antibody titers and postvaccination survival rate.
11	Krogdahl et al. (2015)	Atlantic salmon	Soy saponins	1–2 g/kg	8 wks	Reduction	Saponin-induced enteritis suppressed local immune function and altered cytokine expression.
12	Sukul et al. (2023)	Rohu (*Labeo rohita*)	Autochthonous Bacilli and fructooligosaccharide	1%	90 days	Improvement	Improve growth, nonspecific immune parameters, and survivability.
13	Sukul and Ghosh (2025)	Rohu (*Labeo rohita*)	Pectin and *Bacillus* spp.	1%	90 days	Improvement	Improve growth, nonspecific immune parameters, and survivability.

Table 16.3 Effects of NCFs ingredients on antioxidant enzyme activity in different fish species.

No.	References (Year)	Fish species	Nonconventional ingredient (processing)	Inclusion/ Duration	Antioxidant outcome	Key finding (short)
1	Zhang et al. (2023)	Nile tilapia (*Oreochromis niloticus*)	Fermented soybean meal (FSBM) replacing unfermented SBM/ Fish meal	Various substitution levels; typical 6–8 weeks	Improved (↑ SOD, CAT, GPx; ↓ MDA) with FSBM versus raw SBM	Fermentation restored antioxidant enzyme activities that were reduced by unfermented SBM.
2	Abdel-Tawwab et al. (2022)	Nile tilapia	Chlorella vulgaris meal (algal meal)	~5–20%; 6–10 weeks	Improved (↑ SOD CAT, GPx; ↓ oxidative markers)	Chlorella supplementation upregulated antioxidant enzyme expression and reduced oxidative stress.
3	Sangsawang et al. (2024)	Nile tilapia/ tilapia-type studies	Defatted black soldier fly (*Hermetia illucens*) meal	10%–40% Typical; 8–10 weeks	Improved/Neutral (↑ SOD, GPx in many trials; reduced lipid peroxidation)	Defatted BSF diets often increased hepatic antioxidant enzymes and lowered MDA at moderate inclusion.
4	Yang et al. (2025)	Various (reported in BSF reviews)	BSF larvae/oil	Varied	Improved (SOD, GPx) in several trials	Multiple studies reported higher SOD and GPx after partial FM replacement with BSF.
5	Xie and Zhang (2024)	Hybrid grouper	Multi-strain probiotic mix (dietary additive)	Weeks (typical 6–8 weeks)	Improved (↑ T-AOC, SOD, CAT, GPx)	Probiotics increased antioxidant status along with improved growth and immune markers.
6	Chen et al. (2021)	–	*Chlorella*/CV meal replacement	For example, 10% FM replacement	Improved (↑ T-AOC, SOD, GPx)	Replacing a portion of FM with Chlorella elevated total antioxidant capacity and key enzymes.
7	Oliveira et al. (2024)	Various species	BSF meal replacing fish meal	Variable	Generally neutral to improved	Meta/experimental evidence: Moderate BSF inclusion tends not

Continued

Table 16.3 Effects of NCFs ingredients on antioxidant enzyme activity in different fish species.—*cont'd*

No.	References (Year)	Fish species	Nonconventional ingredient (processing)	Inclusion/ Duration	Antioxidant outcome	Key finding (short)
						to impair and often enhances antioxidant defences.
8	Tawfeek et al. (2024)	Nile tilapia (toxicity amelioration model)	*Chlorella* supplementation during pollutant exposure	Experimental exposure trials	Improved/ protective (restored SOD/ GPx; lowered MDA)	Chlorella diets attenuated pollutant-induced oxidative stress and normalized antioxidant enzyme levels.
9	Hu et al. (2025)	Various	Natural antioxidants and functional ingredients	—	Mechanistic overview (SOD, CAT, GPx central)	Summarizes enzymatic antioxidant responses to natural feed additives (algae, fermented meals, probiotics, plant extracts). Useful for background/mechanisms.

metagenomic tools into aquaculture nutrition research has profoundly expanded our understanding of how NCFs influence the physiology and gut microbial ecosystem of fish. As aquaculture continues to evolve toward sustainability and resource efficiency, metagenomics serves as a pivotal bridge connecting feed formulation, host physiology, and microbial ecology. The studies reviewed and discussed in this chapter demonstrate that dietary interventions—particularly those involving plant-based, microbial, algal, and fermented feed resources—significantly modulate the structure and functional potential of the gut microbiome (López Nadal et al., 2020; Yu and Yoshikuni, 2025). These microbial alterations are not merely reflective of compositional shifts but represent functional adaptations that influence nutrient utilization, metabolic efficiency, immune competence, and disease resilience in cultured fish species.

The findings from metagenomic analyses consistently emphasize that NCFs can either enhance or perturb the gut microbial balance depending on their composition, processing method, and inclusion level. When properly optimized, such feeds promote beneficial bacterial consortia such as *Lactobacillus*, *Bacillus*, and *Bacteroides*, which are associated with improved digestive enzyme activity, production of SCFAs, and immune modulation (Egerton et al., 2018). Conversely, unprocessed or poorly balanced feed formulations may increase microbial taxa linked to inflammation or suboptimal nutrient absorption. Therefore, a metagenomic framework provides essential guidance for fine-tuning feed formulations that foster a healthy gut microbiome and optimal fish performance.

Importantly, the incorporation of SSF and enzymatic pretreatment technologies represents a transformative step toward improving the nutritional quality and microbial compatibility of plant-based feeds (Ghosh et al., 2019). Fermented ingredients reduce ANFs, enhance amino acid availability, and introduce beneficial microbial metabolites that selectively stimulate commensal microbiota (Ghosh et al., 2019). Metagenomic evidence supports that such processing not only increases microbial diversity but also enriches functional genes associated with carbohydrate metabolism, SCFA synthesis, and immune-enhancing pathways. These advances align with the principles of a circular bioeconomy by promoting waste valorization and resource sustainability in aquafeed development (Nath et al., 2023).

From a broader perspective, the implications of metagenomic insights extend beyond feed evaluation to encompass ecosystem and environmental sustainability. Microbial communities play a vital role in nutrient cycling, nitrogen and phosphorus metabolism, and organic matter degradation, thereby influencing the overall efficiency and ecological footprint of aquaculture systems (Tacon and Metian, 2015; An et al., 2025). By integrating microbial community data with growth performance, FCR, and immune indices, aquaculture researchers can design predictive models for feed efficiency and disease resistance based on gut microbial signatures. Looking ahead, multi-omics integration—combining metagenomics, metatranscriptomics, metaproteomics, and metabolomics—will enable a holistic understanding of host–microbe–diet interactions at both molecular and ecological levels (Satya et al., 2024; Ferrocino et al., 2023). Such systems-level analyses will be instrumental in identifying core microbiome members and functional biomarkers that can serve as indicators of gut health and feed efficiency. Furthermore, advances in computational biology, machine learning, and network-based modelling will allow the prediction of microbial responses to novel feed components, thereby accelerating the design of next-generation aquafeeds.

In conclusion, metagenomic insights have redefined the paradigm of fish nutrition by uncovering the intricate microbial mechanisms that mediate dietary effects on host physiology. NCF ingredients, when strategically selected and biotechnologically optimized, hold immense potential to enhance fish growth, immunity, and environmental sustainability. The integration of metagenomic evidence into aquafeed research and formulation represents a transformative step toward precision aquaculture—an approach that aligns nutlritional innovation with microbial and ecological harmony. Continued exploration of this interdisciplinary frontier will ensure that aquaculture progresses not only as a source of food security but also as a model for sustainable and microbiome-conscious food production systems.

Disclosure statement

The authors hereby declare that they do not have any conflicts of interest. During the preparation of this manuscript, the authors used Grammarly to improve its readability and language. After using this tool, the authors carefully reviewed and edited the content, and take full responsibility for the content of the article.

References

Abdel-Tawwab, M., Mousa, M.A., Mamoon, A., Abdelghany, M.F., Abdel-Hamid, E.A., Abdel-Razek, N., Gewida, A.G., 2022. Dietary *Chlorella vulgaris* modulates the performance, antioxidant capacity, innate immunity, and disease resistance capability of Nile tilapia fingerlings fed on plant-based diets. Anim. Feed Sci. Technol. 283, 115181.

Acharya, D., Kari, Z.A., Wei, L.S., Ghosh, K., 2025. Nutrient profiling of some terrestrial plant leaves for their prospective use as nonconventional ingredients in carp diets. Aquac. Res. 2025 (1), 9989338.

Adebo, J.A., Njobeh, P.B., Gbashi, S., Oyedeji, A.B., Ogundele, O.M., Oyeyinka, S.A., Adebo, O.A., 2022. Fermentation of cereals and legumes: impact on nutritional constituents and nutrient bioavailability. Fermentation 8 (2), 63.

Ahmadifar, E., Pourmohammadi Fallah, H., Yousefi, M., Dawood, M.A., Hoseinifar, S.H., Adineh, H., Doan, H.V., 2021. The gene regulatory roles of herbal extracts on the growth, immune system, and reproduction of fish. Animals 11 (8), 2167.

Al-Hisnawi, A., Ringø, E., Davies, S.J., Waines, P., Bradley, G., Merrifield, D.L., 2015. First report on the autochthonous gut microbiota of brown trout (*Salmo trutta* Linnaeus). Aquac. Res. 46 (12), 2962–2971.

Al-Hisnawi, A., Rodiles, A., Rawling, M.D., Castex, M., Waines, P., Gioacchini, G., Merrifield, D.L., 2019. Dietary probiotic *Pediococcus acidilactici* MA18/5M modulates the intestinal microbiota and stimulates intestinal immunity in rainbow trout (*Oncorhynchus mykiss*). J. World Aquacult. Soc. 50 (6), 1133–1151.

Amenyogbe, E., Chen, G., Wang, Z., Huang, J., Huang, B., Li, H., 2020. The exploitation of probiotics, prebiotics and synbiotics in aquaculture: present study, limitations and future directions.: a review. Aquac. Int. 28 (3), 1017–1041.

An, S., Li, J., Du, J., Feng, L., Zhang, L., Zhang, X., Yang, G., 2025. Coupled nitrogen and phosphorus cycles mediated by coordinated variations of functional microbes in industrial recirculating aquaculture system. Water Res., 123726

Aragão, C., Gonçalves, A.T., Costas, B., Azeredo, R., Xavier, M.J., Engrola, S., 2022. Alternative proteins for fish diets: implications beyond growth. Animals 12 (9), 1211.

Ardó, L., Sándor, Z.J., Orbán, M., Szakáli, J., Biró, J., Szűcs, A.A., Békefi, E., 2025. Effects of feed additives (*Nannochloropsis gaditana* and *Hermetia illucens*) on growth and expression of antioxidant and cytokine genes in Nile tilapia (*Oreochromis niloticus*) subjected to air exposure stress. Animals 15 (12), 1776.

Asghar, M.U., Sajid, Q.U.A., Wilk, M., Konkol, D., Korczyński, M., 2024. Influence of various methods of processing soybeans on protein digestibility and reduction of nitrogen deposits in the natural environment–a review. Ann. Anim. Sci. 24 (4), 1037–1049.

Ayoola, S.O., Kolawole, A.S., Etim, F., 2023. Growth performance, haematological and biochemical responses of Nile tilapia *Oreochromis niloticus* fed graded levels of groundnut cake. Aceh J. Anim. Sci. 8 (3).

Bacchetti, T., Annibaldi, A., Comitini, F., Ciani, M., Damiani, E., Norici, A., Olivotto, I., 2020. Alternative ingredients for feed and food. In: The first outstanding 50 years of "Università Politecnica delle Marche" research achievements in life sciences. Springer International Publishing, Cham, pp. 529–545.

Bairagi, A., Ghosh, K.S., Sen, S.K., Ray, A.K., 2002a. Duckweed (*Lemna polyrhiza*) leaf meal as a source of feedstuff in formulated diets for rohu (*Labeo rohita* Ham.) fingerlings after fermentation with a fish intestinal bacterium. Bioresour. Technol. 85 (1), 17–24.

Bairagi, A., Ghosh, K.S., Sen, S.K., Ray, A.K., 2002b. Enzyme producing bacterial flora isolated from fish digestive tracts. Aquac. Int. 10 (2), 109–121.

Banerjee, S., Kari, Z.A., Téllez-Isaías, G., Ghosh, K., 2023. The use of linseed oil cake in the diets of rohu, *Labeo rohita* (Hamilton), after solid-state fermentation with a fish gut bacterium, *Bacillus pumilus* (KF640221): an appraisal on growth, digestibility, body composition, and hematobiochemical profile. Front. Mar. Sci. 10, 1278704.

Banerjee, S., Mukherjee, A., Dutta, D., Ghosh, K., 2016. Non-starch polysaccharide degrading gut bacteria in Indian major carps and exotic carps. Jordan J. Biol. Sci. 9 (1).

Bonartseva, G.A., Myshkina, V.L., Nikolaeva, D.A., Rebrov, A.V., Gerasin, V.A., Makhina, T. K., 2002. The biodegradation of poly-β-hydroxybutyrate (PHB) by a model soil community: the effect of cultivation conditions on the degradation rate and the physicochemical characteristics of PHB. Microbiology 71 (2), 221–226.

Bruni, L., Pastorelli, R., Viti, C., Gasco, L., Parisi, G., 2018. Characterisation of the intestinal microbial communities of rainbow trout (*Oncorhynchus mykiss*) fed with *Hermetia illucens* (black soldier fly) partially defatted larva meal as partial dietary protein source. Aquaculture 487, 56–63.

Büyükdeveci, M.E., Balcázar, J.L., Demirkale, İ., Dikel, S., 2018. Effects of garlic-supplemented diet on growth performance and intestinal microbiota of rainbow trout (*Oncorhynchus mykiss*). Aquaculture 486, 170–174.

Cahill, M.M., 1990. Bacterial flora of fishes: a review. Microb. Ecol. 19 (1), 21–41.

Calcagnile, M., Tredici, S.M., Alifano, P., 2024. A comprehensive review on probiotics and their use in aquaculture: biological control, efficacy, and safety through the genomics and wet methods. Heliyon 10 (24).

Chattaraj, S., Das Mohapatra, P.K., 2024. Immunostimulatory and antagonistic potential of the methanolic extract of Oedogonium intermedium SCB in *Cirrhinus reba* challenged with *Aeromonas hydrophila*. Aquacult. Int. 32 (3), 2523–2548.

Chattaraj, S., Behera, B.K., Das Mohapatra, P.K., 2024a. Effective valorization of Chlorella biomass and Brewers spent grain substituting fish meal and soybean meal in the diet of herbivorous fish *Cirrhinus reba* for higher growth, digestibility and sustainable cultivation. J. Appl. Phycol. 36 (1), 217–232.

Chattaraj, S., Mitra, D., Chattaraj, M., Ganguly, A., Thatoi, H., Mohapatra, P.K.D., 2024b. Brewers' spent grain as fish feed ingredient: evaluation of bio-safety and analysis of its impact on gut bacteria of Cirrhinus reba by 16S Metagenomic sequencing. Curr. Res. Microb. Sci. 7, 100286.

Chattaraj, S., Chattaraj, M., Mitra, D., Ganguly, A., Thatoi, H., Das Mohapatra, P.K., 2025. 16S amplicon sequencing of the gastrointestinal microbiota of *Cirrhinus reba* and isolation of an autochthonous probiotic using culture based approaches. Syst. Microbiol. Biomanufact. 5 (1), 156–170.

Chattaraj, S., Ganguly, A., Mitra, D., et al., 2026a. Sustainable valorization of algal biomass in fish nutrition and dietetics: a critical review and scientometric study. Waste Biomass Valor. https://doi.org/10.1007/s12649-026-03619-7.

Chattaraj, S., Sahoo, S.S., Behera, S., Thatoi, H., 2026b. Sustainable aquafeed development using Spirulina: effects on growth performance of Catla catla. Algal Res. 104447.

Chen, W., Luo, L., Han, D., Long, F., Chi, Q., Hu, Q., 2021. Effect of dietary supplementation with *Chlorella sorokiniana* meal on the growth performance, antioxidant status, and immune response of rainbow trout (*Oncorhynchus mykiss*). J. Appl. Phycol. 33 (5), 3113–3122.

Clagnan, E., Costanzo, M., Visca, A., Di Gregorio, L., Tabacchioni, S., Colantoni, E., Bevivino, A., 2024. Culturomics-and metagenomics-based insights into the soil microbiome preservation and application for sustainable agriculture. Front. Microbiol. 15, 1473666.

Das, J., Ghosh, K., 2023. Nutrient profiling of five freshwater algae for their prospective use as fish feed ingredients. Algal Res. 74, 103173.

Das, K.M., Tripathi, S.D., 1991. Studies on the digestive enzymes of grass carp, *Ctenopharyngodon idella* (Val.). Aquaculture 92, 21–32.

Das, P., Ghosh, K., 2014. The presence of phytase in yeasts isolated from the gastrointestinal tract of four major carps [*Labeo rohita* (Hamilton, 1822), *Catla catla* (Hamilton, 1822), *Cirrhinus mrigala* (Hamilton, 1822), *Hypophthalmichthys molitrix* (Valenciennes, 1844)], climbing perch [*Anabas testudineus* (Bloch, 1792)] and Mozambique tilapia [*Oreochromis mossambicus* (Linnaeus, 1758)]. J. Appl. Ichthyol. 30 (2), 403–407.

De Schryver, P., Sinha, A.K., Kunwar, P.S., Baruah, K., Verstraete, W., Boon, N., Bossier, P., 2010. Poly-β-hydroxybutyrate (PHB) increases growth performance and intestinal bacterial range-weighted richness in juvenile European sea bass, *Dicentrarchus labrax*. Appl. Microbiol. Biotechnol. 86 (5), 1535–1541.

Deka, P., Mehetre, G.T., Lalnunmawii, E., Upadhyaya, K., Singh, G., Hashem, A., Singh, B.P., 2021. Metagenomic analysis of bacterial diversity in traditional fermented foods reveals food-specific dominance of specific bacterial taxa. Fermentation 7 (3), 167.

Di Rosa, A.R., Caccamo, L., Pansera, L., Oteri, M., Chiofalo, B., Maricchiolo, G., 2023. Influence of *Hermetia illucens* larvae meal dietary inclusion on growth performance, gut histological traits and stress parameters in *Sparus aurata*. Animals 13 (3), 339.

Dutta, D., Ghosh, K., 2015. Screening of extracellular enzyme-producing and pathogen inhibitory gut bacteria as putative probiotics in mrigal, *Cirrhinus mrigala* (Hamilton, 1822). Int. J. Fish. Aquat. Stud. 2 (4), 310–318.

Dutta, D., Banerjee, S., Mukherjee, A., Ghosh, K., 2015. Selection and probiotic characterization of exoenzyme-producing bacteria isolated from the gut of *Catla catla* (Actinopterygii: Cypriniformes: Cyprinidae). Acta Ichthyol. Piscatoria (4).

Edwards, P., Kamal, M., Wee, K.L., 1985. Incorporation of composted and dried water hyacinth in pelleted feed for the tilapia *Oreochromis niloticus* (Peters). Aquac. Res. 16 (3), 233–248.

Egerton, S., Culloty, S., Whooley, J., Stanton, C., Ross, R.P., 2018. The gut microbiota of marine fish. Front. Microbiol. 9, 873.

El-Desouky, F.F., Ibrahim, M.A., Abd El-Razek, I.M., El-Nabawy, E.S.M., Amer, A.A., Zaineldin, A.I., Dawood, M.A., 2024. Improving yellow mealworm (*Tenebrio molitor*) utilization with sodium butyrate in Nile tilapia diets: effects on growth performance, intestinal histology, antioxidative response, and blood biomarkers. Aquac. Nutr. 2024 (1), 2442308.

El-Saidy, D.M., Gaber, M.M., 2002. Complete replacement of fish meal by soybean meal with dietary L-lysine supplementation for Nile tilapia *Oreochromis niloticus* (L.) fingerlings. J. World Aquacult. Soc. 33 (3), 297–306.

Feng, R., Chen, L., Chen, K., 2018. Fermentation trip: amazing microbes, amazing metabolisms. Ann. Microbiol. 68 (11), 717–729.

Ferrocino, I., Rantsiou, K., McClure, R., Kostic, T., De Souza, R.S.C., Lange, L., MicrobiomeSupport Consortium, 2023. The need for an integrated multi-OMICs approach in microbiome science in the food system. Compr. Rev. Food Sci. Food Saf. 22 (2), 1082–1103.

Francis, G., Makkar, H.P., Becker, K., 2001. Antinutritional factors present in plant-derived alternate fish feed ingredients and their effects in fish. Aquaculture 199 (3–4), 197–227.

Gatlin III, D.M., Barrows, F.T., Brown, P., Dabrowski, K., Gaylord, T.G., Hardy, R.W., Wurtele, E., 2007. Expanding the utilization of sustainable plant products in aquafeeds: a review. Aquac. Res. 38 (6), 551–579.

Gajendiran, K., Narayanan, M., 2025. Algae as a promising source of vitamins and protein: current prospects and challenges. In: Algae-Derived Biochemicals of Industrial Importance, pp. 163–173.

Ganguly, A., Chattaraj, S., Ganguly, M., Chattaraj, M., Banerjee, A., Mandal, A., Ali Khan, M., Sen, S.K., Das Mohapatra, P.K., 2025. Effect of three probiotic Bacillus strains supplemented feeds on growth, carcass composition and blood parameters of *Clarias magur* (Hamilton). J. Appl. Aquacult. 37 (1), 19–44.

Gao, S., Chen, W., Cao, S., Sun, P., Gao, X., 2024. Microalgae as fishmeal alternatives in aquaculture: current status, existing problems, and possible solutions. Environ. Sci. Pollut. Control Ser. 31 (11), 16113–16130.

Ghosh, K., Ray, A.K., Ringø, E., 2019. Applications of plant ingredients for tropical and subtropical freshwater finfish: possibilities and challenges. Rev. Aquacult. 11 (3), 793–815.

Ghosh, K., Sen, S.K., Ray, A.K., 2002. Characterization of Bacilli isolated from the gut of rohu, *Labeo rohita*, fingerlings and its significance in digestion. J. Appl. Aquacult. 12 (3), 33–42.

Gliozheni, E., 2018. Effects of Some Stable Derivatives of Butyrate used as feed additives on fish intestinal microbiota, Histone Modifications and the Expression of Genes Related to Epigenetic Regulatory Mechanisms and Immune Response.

González-Meza, G.M., Elizondo-Luevano, J.H., Cuellar-Bermudez, S.P., Sosa-Hernández, J. E., Iqbal, H.M., Melchor-Martínez, E.M., Parra-Saldívar, R., 2023. New perspective for macroalgae-based animal feeding in the context of challenging sustainable food production. Plants 12 (20), 3609.

Gouveia, L., Batista, A.P., Sousa, I., Raymundo, A., Bandarra, N.M., 2009. Microalgae in novel food products. In: Algae: Nutrition, Pollution control and Energy Sources, pp. 265–300.

Green, T.J., Smullen, R., Barnes, A.C., 2013. Dietary soybean protein concentrate-induced intestinal disorder in marine farmed Atlantic salmon, *Salmo salar* is associated with alterations in gut microbiota. Vet. Microbiol. 166 (1–2), 286–292.

Hafeez, M., Gogi, M.D., Imran, M., Abbas, A., Saeed, K., Arif, A., Ramzan, R., 2024. Additives in fish feeds: non-conventional way for improving fish health. In: Complementary and alternative medicine: feed additives, 385.

Hallali, E., Kokou, F., Chourasia, T.K., Nitzan, T., Con, P., Harpaz, S., Cnaani, A., 2018. Dietary salt levels affect digestibility, intestinal gene expression, and the microbiome, in Nile tilapia (*Oreochromis niloticus*). PLoS One 13 (8), e0202351.

Han, F., Xu, C., Qi, C., Lin, Z., Li, E., Wang, C., Chen, L., 2020. Sodium butyrate can improve intestinal integrity and immunity in juvenile Chinese mitten crab (*Eriocheir sinensis*) fed glycinin. Fish Shellfish Immunol. 102, 400–411.

Hardy, R.W., 2010. Utilization of plant proteins in fish diets: effects of global demand and supplies of fishmeal. Aquac. Res. 41 (5), 770–776.

Hodar, A.R., Vasava, R.J., Mahavadiya, D.R., Joshi, N.H., 2020. Fish meal and fish oil replacement for aqua feed formulation by using alternative sources: a review. J. Exp. Zool. India 23 (1).

Holanda, L.R., Ramos, F.S., 2016. Reuse of waste sugarcane agribusiness and green power generation. J. Clean Energy Technol. 4 (5), 341–345.

Hoseinifar, S.H., Eshaghzadeh, H., Vahabzadeh, H., Peykaran Mana, N., 2016. Modulation of growth performances, survival, digestive enzyme activities and intestinal microbiota in common carp (*Cyprinus carpio*) larvae using short chain fructooligosaccharide. Aquac. Res. 47 (10), 3246–3253.

Hoseinifar, S.H., Safari, R., Dadar, M., 2017. Dietary sodium propionate affects mucosal immune parameters, growth and appetite related genes expression: insights from zebrafish model. Gen. Comp. Endocrinol. 243, 78–83.

Hoseinifar, S.H., Faheem, M., Liaqat, I., Van Doan, H., Ghosh, K., Ringø, E., 2024. Promising probiotic candidates for sustainable aquaculture: an updated review. Animals 14 (24), 3644.

Hu, X., Yang, H.L., Yan, Y.Y., Zhang, C.X., Ye, J.D., Lu, K.L., Sun, Y.Z., 2019. Effects of fructooligosaccharide on growth, immunity and intestinal microbiota of shrimp (*Litopenaeus vannamei*) fed diets with fish meal partially replaced by soybean meal. Aquac. Nutr. 25 (1), 194–204.

Hu, Z., Yang, K., Yu, A., 2025. The extraction technologies, nutritional compositions, and health benefits of *Camellia oleifera* seed oil: a review. J. Oleo Sci. 74 (10), 859–878.

Huang, K., Liu, X., Ma, R., Wang, B., Ho, S.H., Chen, J., Xie, Y., 2024. Effects of substituting fish meal with *Chlorella* meal on growth performance, whole-body composition, pigmentation, and physiological health of marbled eel (*Anguilla marmorata*). Algal Res. 80, 103523.

Huang, W., Yao, C., Liu, Y., Xu, N., Yin, Z., Xu, W., Ai, Q., 2022. Effects of dietary eucommia ulmoides leaf extract on growth performance, expression of feeding-related genes, activities of digestive enzymes, antioxidant capacity, immunity and cytokines expression of large yellow croaker (*Larimichthys crocea*) larvae. Br. J. Nutr. 128 (9), 1711–1719.

Hussain, S.M., Bano, A.A., Ali, S., Rizwan, M., Adrees, M., Zahoor, A.F., Naeem, A., 2024. Substitution of fishmeal: highlights of potential plant protein sources for aquaculture sustainability. Heliyon 10 (4).

Iqbal, S., 1999. Duckweed aquaculture: potentials, possibilities and limitations for combined wastewater treatment and animal feed production in developing countries. In: SANDEC Report No. 6/99. EAWAG-SANDEC, Duebendorf, Switzerland, pp. 1–99.

Irianto, A., Austin, B., 2002. Probiotics in aquaculture. J. Fish. Dis. 25 (11), 633–642.

Jin, M., Kalainy, S., Baskota, N., Chiang, D., Deehan, E.C., McDougall, C., Abraldes, J.G., 2019a. Faecal microbiota from patients with cirrhosis has a low capacity to ferment non-digestible carbohydrates into short-chain fatty acids. Liver Int. 39 (8), 1437–1447.

Jin, Y., Angell, I.L., Sandve, S.R., Snipen, L.G., Olsen, Y., Rudi, K., 2019b. Atlantic salmon raised with diets low in long-chain polyunsaturated n-3 fatty acids in freshwater have a *Mycoplasma*-dominated gut microbiota at sea. Aquac. Environ. Interact. 11, 31–39.

Kari, Z.A., Sukri, S.A.M., Rusli, N.D., Mat, K., Mahmud, M.B., Zakaria, N.N.A., Wei, L.S., 2023. Recent advances, challenges, opportunities, product development and sustainability of main agricultural wastes for the aquaculture feed industry–a review. Ann. Anim. Sci. 23 (1), 25–38.

Kavitha, G., Rengasamy, R., Inbakandan, D., 2018a. Polyhydroxybutyrate production from marine source and its application. Int. J. Biol. Macromol. 111, 102–108.

Kavitha, M., Raja, M., Perumal, P., 2018b. Evaluation of probiotic potential of *Bacillus* spp. isolated from the digestive tract of freshwater fish *Labeo calbasu* (Hamilton, 1822). Aquac. Rep. 11, 59–69.

Kemski, M.M., Rappleye, C.A., Dabrowski, K., Bruno, R.S., Wick, M., 2020. Transcriptomic response to soybean meal-based diets as the first formulated feed in juvenile yellow perch (*Perca flavescens*). Sci. Rep. 10 (1), 3998.

Krogdahl, Å., Bakke-McKellep, A.M., Baeverfjord, G., 2003. Effects of graded levels of standard soybean meal on intestinal structure, mucosal enzyme activities, and pancreatic response in Atlantic salmon (*Salmo salar* L.). Aquac. Nutr. 9 (6), 361–371.

Krogdahl, Å., Gajardo, K., Kortner, T.M., Penn, M., Gu, M., Berge, G.M., Bakke, A.M., 2015. Soya saponins induce enteritis in Atlantic salmon (*Salmo salar* L.). J. Agric. Food Chem. 63 (15), 3887–3902.

Krogdahl, Å., Penn, M., Thorsen, J., Refstie, S., Bakke, A.M., 2010. Important antinutrients in plant feedstuffs for aquaculture: an update on recent findings regarding responses in salmonids. Aquac. Res. 41 (3), 333–344.

Król, E., Douglas, A., Tocher, D.R., Crampton, V.O., Speakman, J.R., Secombes, C.J., Martin, S.A., 2016. Differential responses of the gut transcriptome to plant protein diets in farmed Atlantic salmon. BMC Genom. 17 (1), 156.

Lee, J., Park, Y., Song, D., Chang, S., Cho, J., 2024. Effects of defatted and hydrolyzed black soldier fly larvae meal as an alternative fish meal in weaning pigs. Animals 14 (11), 1692.

Leitzmann, C., 2016. Characteristics and health benefits of phytochemicals. Forsch. Komplementmed. Res. Complement. Med. 23 (2), 69–74.

Lesel, R., Fromageot, C., Lesel, M., 1986. Cellulose digestibility in grass carp, *Ctenopharyngodon idella* and in goldfish, *Carassius auratus*. Aquaculture 54 (1–2), 11–17.

Llewellyn, M.S., Boutin, S., Hoseinifar, S.H., Derome, N., 2014. Teleost microbiomes: the state of the art in their characterization, manipulation and importance in aquaculture and fisheries. Front. Microbiol. 5, 207.

López Nadal, A., Ikeda-Ohtsubo, W., Sipkema, D., Peggs, D., McGurk, C., Forlenza, M., Brugman, S., 2020. Feed, microbiota, and gut immunity: using the zebrafish model to understand fish health. Front. Immunol. 11, 114.

Lourenço, S., Neves, M., Gonçalves, E.M., Roseiro, C., Pombo, A., Raimundo, D., Pinheiro, J., 2025. Application of fruit and vegetable processing by-products as ingredients in aquafeed. Discov. Food 5 (1), 45.

Luongo, V., Ghimire, A., Frunzo, L., Fabbricino, M., d'Antonio, G., Pirozzi, F., Esposito, G., 2017. Photofermentative production of hydrogen and poly-β-hydroxybutyrate from dark fermentation products. Bioresour. Technol. 228, 171–175.

Mafe, A.N., Smart, O.O., Edo, G.I., Akpoghelie, P.O., Gaaz, T.S., Yousif, E., Umar, H., 2025. Genesis, health benefits, and future perspectives of probiotics: exploring endogenous and exogenous classes, innovations, and research gaps. Probiotics Antimicrob. Proteins 1–34.

Mahmud, M.N., Ritu, F.Y., Ansary, A.A., Haque, M.M., 2025. Exploring protein-based fishmeal alternatives for aquaculture feeds in Bangladesh. Aquac. Nutr. 2025 (1), 3198303.

Mandal, S., Ghosh, K., 2013a. Isolation of tannase-producing microbiota from the gastrointestinal tracts of some freshwater fish. J. Appl. Ichthyol. 29 (1), 145–153.

Mandal, S., Ghosh, K., 2013b. Optimization of tannase production and improvement of nutritional quality of two potential low-priced plant feedstuffs under solid state fermentation by *Pichia kudriavzevii* isolated from fish gut. Food Biotechnol. 27 (1), 86–103.

Mihailović, M., Dinić, S., Arambašić Jovanović, J., Uskoković, A., Grdović, N., Vidaković, M., 2021. The influence of plant extracts and phytoconstituents on antioxidant enzymes activity and gene expression in the prevention and treatment of impaired glucose homeostasis and diabetes complications. Antioxidants 10 (3), 480.

Mukherjee, A., Ghosh, K., 2016. Antagonism against fish pathogens by cellular components and verification of probiotic properties in autochthonous bacteria isolated from the gut of an Indian major carp, *Catla catla* (H amilton). Aquac. Res. 47 (7), 2243–2255.

Mukherjee, A., Dutta, D., Banerjee, S., Ringø, E., Breines, E.M., Hareide, E., Ghosh, K., 2016. Potential probiotics from Indian major carp, *Cirrhinus mrigala*. characterization, pathogen inhibitory activity, partial characterization of bacteriocin and production of exoenzymes. Res. Vet. Sci. 108, 76–84.

Mukherjee, A., Dutta, D., Banerjee, S., Ringø, E., Breines, E.M., Hareide, E., Ghosh, K., 2017. Culturable autochthonous gut bacteria in rohu, Labeo rohita. In vitro growth inhibition against pathogenic *Aeromonas* spp., stability in gut, bio-safety and identification by 16S rRNA gene sequencing. Symbiosis 73 (3), 165–177.

Najdegerami, E.H., Tran, T.N., Defoirdt, T., Marzorati, M., Sorgeloos, P., Boon, N., Bossier, P., 2012. Effects of poly-β-hydroxybutyrate (PHB) on Siberian sturgeon (*Acipenser baerii*) fingerlings performance and its gastrointestinal tract microbial community. FEMS Microbiol. Ecol. 79 (1), 25–33.

Nath, P.C., Ojha, A., Debnath, S., Sharma, M., Nayak, P.K., Sridhar, K., Inbaraj, B.S., 2023. Valorization of food waste as animal feed: a step towards sustainable food waste management and circular bioeconomy. Animals 13 (8), 1366.

Nayak, S.K., 2010. Role of gastrointestinal microbiota in fish. Aquac. Res. 41 (11), 1553–1573.

Naylor, R.L., Hardy, R.W., Bureau, D.P., Chiu, A., Elliott, M., Farrell, A.P., Nichols, P.D., 2009. Feeding aquaculture in an era of finite resources. Proc. Natl. Acad. Sci. 106 (36), 15103–15110.

Ni, J., Yan, Q., Yu, Y., Zhang, T., 2014. Factors influencing the grass carp gut microbiome and its effect on metabolism. FEMS Microbiol. Ecol. 87 (3), 704–714.

Nishiguchi, H., Suryadi, I.B.B., Ali, M.F.Z., Miura, C., Miura, T., 2024. Dietary black soldier fly (*Hermetia illucens*)—Dipterose-BSF—enhanced zebrafish innate immunity gene expression and resistance to *Edwardsiella tarda* infection. Insects 15 (5), 326.

Oliveira, C.G.D., Freitas, D.D.A., Ribeiro, P.A.P., Teixeira, R.R.C., Silva, R.F.D., Gamarano, P.G., Costa, L.S., 2024. Impact of replacing fish meal with black soldier fly (*Hermetia illucens*) meal on diet acceptability in juvenile Nile tilapia: palatability and nutritional and health considerations for dietary preference. Aquac. Res. 2024 (1), 3409955.

Osu, A.A., Ashafa, A.O., 2021. Nutritional composition of grain and seed proteins. In: Grain and seed proteins functionality. IntechOpen.

Pang, A., Peng, C., Xie, R., Wang, Z., Tan, B., Wang, T., Zhang, W., 2023. Effects of fermented soybean meal substitution for fish meal on intestinal flora and intestinal health in pearl gentian grouper. Front. Physiol. 14, 1194071.

Patel, A.K., Singhania, R.R., Awasthi, M.K., Varjani, S., Bhatia, S.K., Tsai, M.L., Dong, C.D., 2021. Emerging prospects of macro-and microalgae as prebiotic. Microb. Cell Fact. 20 (1), 112.

Pradhan, J., Sahu, S., Das, B.K., 2023. Protective effects of *Chlorella vulgaris* supplemented diet on antibacterial activity and immune responses in rohu fingerlings, *Labeo rohita* (Hamilton), subjected to *Aeromonas hydrophila* infection. Life 13 (4), 1028.

Predescu, N.C., Stefan, G., Rosu, M.P., Papuc, C., 2024. Fermented feed in broiler diets reduces the antinutritional factors, improves productive performances and modulates gut microbiome—a review. Agriculture 14 (10), 1752.

Qi, X., Zhang, Y., Zhang, Y., Luo, F., Song, K., Wang, G., Ling, F., 2023. Vitamin B12 produced by *Cetobacterium somerae* improves host resistance against pathogen infection through strengthening the interactions within gut microbiota. Microbiome 11 (1), 135.

Qiao, F., Liu, Y.K., Sun, Y.H., Wang, X.D., Chen, K., Li, T.Y., Zhang, M.L., 2017. Influence of different dietary carbohydrate sources on the growth and intestinal microbiota of *Litopenaeus vannamei* at low salinity. Aquac. Nutr. 23 (3), 444–452.

Ragaza, J.A., Kumar, V., Traifalgar, R.F.M., 2023. Innovative, non-conventional ingredients for sustainable aquafeeds. Front. Sustain. Food Syst. 7, 1256460.

Ramachandran, S., Singh, S.K., Larroche, C., Soccol, C.R., Pandey, A., 2007. Oil cakes and their biotechnological applications–a review. Bioresour. Technol. 98 (10), 2000–2009.

Ray, A.K., Das, I., 1996. Evaluation of dried aquatic weed, *Pistia stratiotes*, meal as a feedstuff in pelleted feed for rohu, *Labeo rohita*, fingerlings. J. Appl. Aquacult. 5 (4), 35–44.

Ray, A.K., Ghosh, K., Ringø, E.J.A.N., 2012. Enzyme-producing bacteria isolated from fish gut: a review. Aquac. Nutr. 18 (5), 465–492.

Rimoldi, S., Di Rosa, A.R., Oteri, M., Chiofalo, B., Hasan, I., Saroglia, M., Terova, G., 2024. The impact of diets containing *Hermetia illucens* meal on the growth, intestinal health, and microbiota of gilthead seabream (*Sparus aurata*). Fish Physiol. Biochem. 50 (3), 1003–1024.

Rimoldi, S., Gini, E., Iannini, F., Gasco, L., Terova, G., 2019. The effects of dietary insect meal from *Hermetia illucens* prepupae on autochthonous gut microbiota of rainbow trout (*Oncorhynchus mykiss*). Animals 9 (4), 143.

Rimoldi, S., Terova, G., Ascione, C., Giannico, R., Brambilla, F., 2018. Next generation sequencing for gut microbiome characterization in rainbow trout (*Oncorhynchus mykiss*) fed animal by-product meals as an alternative to fishmeal protein sources. PLoS One 13 (3), e0193652.

Ringø, E.Z.Z.V., Zhou, Z., Vecino, J.G., Wadsworth, S., Romero, J., Krogdahl, Å., Merrifield, D.L., 2016. Effect of dietary components on the gut microbiota of aquatic animals. A never-ending story? Aquac. Nutr. 22 (2), 219–282.

Ringø, E., Song, S.K., 2016. Application of dietary supplements (synbiotics and probiotics in combination with plant products and β-glucans) in aquaculture. Aquac. Nutr. 22 (1), 4–24.

Ringø, E., Harikrishnan, R., Soltani, M., Ghosh, K., 2022. The effect of gut microbiota and probiotics on metabolism in fish and shrimp. Animals 12 (21), 3016.

Ringø, E., Løvmo, L., Kristiansen, M., Bakken, Y., Salinas, I., Myklebust, R., Mayhew, T.M., 2010. Lactic acid bacteria vs. pathogens in the gastrointestinal tract of fish: a review. Aquac. Res. 41 (4), 451–467.

Rinninella, E., Raoul, P., Cintoni, M., Franceschi, F., Miggiano, G.A.D., Gasbarrini, A., Mele, M.C., 2019. What is the healthy gut microbiota composition? A changing ecosystem across age, environment, diet, and diseases. Microorganisms 7 (1), 14.

Safari, R., Hoseinifar, S.H., Azadi, H., Delavar, F.H., Pakzad, M., Jafari, V., Sahebi, H.P., 2024. Dietary substitution of fishmeal by yellow mealworm and its effects on growth and metabolism-related gene expressions in *Acipenser stellatus*. J. Insects Food Feed 11 (8), 1499–1516.

Sagar, N.A., Pareek, S., Sharma, S., Yahia, E.M., Lobo, M.G., 2018. Fruit and vegetable waste: bioactive compounds, their extraction, and possible utilization. Compr. Rev. Food Sci. Food Saf. 17 (3), 512–531.

Sahlmann, C., Sutherland, B.J., Kortner, T.M., Koop, B.F., Krogdahl, Å., Bakke, A.M., 2013. Early response of gene expression in the distal intestine of Atlantic salmon (*Salmo salar* L.) during the development of soybean meal induced enteritis. Fish Shellfish Immunol. 34 (2), 599–609.

Saini, V.P., Singh, A.K., Paul, T., Thakur, A., Singh, M., 2024. Advances in Environment Management for Sustainable Fisheries & Livestock Production.

Samsing, F., Sullivan, R., Truong, H., Rombenso, A., Sangster, C.R., Bannister, J., Becker, J. A., 2024. Replacement of fishmeal with a microbial single-cell protein induced enteropathy and poor growth outcomes in barramundi (*Lates calcarifer*) fry. J. Fish. Dis. 47 (9), e13985.

Sangsawang, A., Kovitvadhi, S., Pewhom, A., Kovitvadhi, U., Kovitvadhi, A., Wongoutong, C., Paankhao, N., 2024. Impacts of substituting fish meal with full-fat or defatted black soldier fly (*Hermetia illucens*) larvae on growth, quality, and health of Nile tilapia (*Oreochromis niloticus*) fingerlings. Aquac. Rep. 38, 102348.

Sattanathan, G., Liu, W.C., Padmapriya, S., Pushparaj, K., Sureshkumar, S., Lee, J.W., Kim, I. H., 2022. Effects of dietary blend of algae extract supplementation on growth, biochemical, haemato-immunological response, and immune gene expression in *Labeo rohita* with *Aeromonas hydrophila* post-challenges. Fishes 8 (1), 7.

Satya, S., Sharma, S., Choudhary, G., Kaushik, G., 2024. Advances in environmental microbiology: a multi-omic perspective. In: Microbial Omics in Environment and Health. Springer Nature Singapore, pp. 175–204.

Schrezenmeir, J., De Vrese, M., 2001. Probiotics, prebiotics, and synbiotics—approaching a definition. Am. J. Clin. Nutr. 73 (2), 361s–364s.

Semova, I., Carten, J.D., Stombaugh, J., Mackey, L.C., Knight, R., Farber, S.A., Rawls, J.F., 2012. Microbiota regulate intestinal absorption and metabolism of fatty acids in the zebrafish. Cell Host Microbe 12 (3), 277–288.

Sharma, S., Singh, A.P. (Eds.), 2025. Traditional Foods: Impact on Gut Health. CRC Press.

Sheng, Y., Ren, H., Limbu, S.M., Sun, Y., Qiao, F., Zhai, W., Zhang, M., 2018. The presence or absence of intestinal microbiota affects lipid deposition and related genes expression in zebrafish (*Danio rerio*). Front. Microbiol. 9, 1124.

Siddik, M.A., Francis, P., Rohani, M.F., Azam, M.S., Mock, T.S., Francis, D.S., 2023. Seaweed and seaweed-based functional metabolites as potential modulators of growth, immune and antioxidant responses, and gut microbiota in fish. Antioxidants 12 (12), 2066.

Soliman, N.F., Yacout, D.M., Hassaan, M.A., 2017. Responsible fishmeal consumption and alternatives in the face of climate changes. Int. J. Mar. Sci. 7.

Sui, L., Ma, G., Lu, W., Deng, Y., Bossier, P., De Schryver, P., 2016. Effect of poly-β-hydroxybutyrate on growth, enzyme activity and intestinal microbial community of Chinese mitten crab, *Eriocheir sinensis* (Milne-Edwards) juveniles. Aquac. Res. 47 (11), 3644–3652.

Sukul, T., Ghosh, K., 2025. Combined application of pectin and Bacillus spp. in the diets of rohu, *Labeo rohita* (Hamilton): effects on growth, feed utilization, immunity, haemato-biochemical profile and pathogen resistance. Aquac. Rep. 43, 102885.

Sukul, T., Kari, Z.A., Téllez-Isaías, G., Ghosh, K., 2023. Autochthonous Bacilli and fructooligosaccharide as functional feed additives improve growth, feed utilisation, haemato-immunological parameters and disease resistance in rohu, *Labeo rohita* (Hamilton). Animals 13 (16), 2631.

Sun, P., Jin, M., Ding, L., Lu, Y., Ma, H., Yuan, Y., Zhou, Q., 2018. Dietary lipid levels could improve growth and intestinal microbiota of juvenile swimming crab, *Portunus trituberculatus*. Aquaculture 490, 208–216.

Tacchi, L., Secombes, C.J., Bickerdike, R., Adler, M.A., Venegas, C., Takle, H., Martin, S.A., 2012. Transcriptomic and physiological responses to fishmeal substitution with plant proteins in formulated feed in farmed Atlantic salmon (*Salmo salar*). BMC Genom. 13 (1), 363.

Tacon, A.G., Metian, M., 2015. Feed matters: satisfying the feed demand of aquaculture. Rev. Fish. Sci. Aquacult. 23 (1), 1–10.

Tawfeek, W.S., Kassab, A.S., Al-Sokary, E.T., Abass, M.E., Sherif, A.H., 2024. Chlorella vulgaris algae ameliorates chlorpyrifos toxicity in Nile tilapia with special reference to antioxidant enzymes and *Streptococcus agalactiae* infection. Mol. Biol. Rep. 51 (1), 616.

Tawfik, M.M., Lorgen-Ritchie, M., Król, E., McMillan, S., Norambuena, F., Bolnick, D.I., Martin, S.A., 2024. Modulation of gut microbiota composition and predicted metabolic capacity after nutritional programming with a plant-rich diet in Atlantic salmon (*Salmo salar*): insights across developmental stages. Anim. Microbiome 6 (1), 38.

Taye, D., Etefa, M., 2020. Review on improving nutritive value of forage by applying exogenous enzymes. Int. J. Vet. Sci. Anim. Husb. 5, 72–79.

Tayyab, M., Islam, W., Waqas, W., Zhang, Y., 2025. Probiotic–vaccine synergy in fish aquaculture: exploring microbiome-immune interactions for enhanced vaccine efficacy. Biology 14 (6), 629.

Terova, G., Rimoldi, S., Ascione, C., Gini, E., Ceccotti, C., Gasco, L., 2019. Rainbow trout (*Oncorhynchus mykiss*) gut microbiota is modulated by insect meal from *Hermetia illucens* prepupae in the diet. Rev. Fish Biol. Fish. 29 (2), 465–486.

Tibbetts, S.M., Patelakis, S.J., Whitney-Lalonde, C.G., Garrison, L.L., Wall, C.L., MacQuarrie, S.P., 2020. Nutrient composition and protein quality of microalgae meals produced from the marine prymnesiophyte *Pavlova* sp. 459 mass-cultivated in enclosed photobioreactors for potential use in salmonid aquafeeds. J. Appl. Phycol. 32 (1), 299–318.

Tyagi, A., Singh, B., Billekallu Thammegowda, N.K., Singh, N.K., 2019. Shotgun metagenomics offers novel insights into taxonomic compositions, metabolic pathways and antibiotic resistance genes in fish gut microbiome. Arch. Microbiol. 201 (3), 295–303.

Tyapkova, O., Osen, R., Wagenstaller, M., Baier, B., Specht, F., Zacherl, C., 2016. Replacing fishmeal with oilseed cakes in fish feed–a study on the influence of processing parameters on the extrusion behavior and quality properties of the feed pellets. J. Food Eng. 191, 28–36.

Van Asselt, E.D., Arrizabalaga-Larrañaga, A., Focker, M., Berendsen, B.J.A., Van de Schans, M.G.M., Van der Fels-Klerx, H.J., 2023. Chemical food safety hazards in circular food systems: a review. Crit. Rev. Food Sci. Nutr. 63 (30), 10319–10331.

Van Tran, T., Kim, Y.S., Yun, H.H., Nguyen, D.H., Bui, T.T., Van Tran, P., 2024. A blend of bacillus-fermented soybean meal, functional amino acids, and nucleotides improves nutrient digestibility, bolsters immune response, reduces diarrhea, and enhances growth performance in weaned piglets. J. Anim. Sci. 102, skae293.

Velichkova, K., Sirakov, I., Stoyanova, S., Simitchiev, A., Yovchev, D., Stamatova-Yovcheva, K., 2024. Effect of replacing fishmeal with algal meal on growth parameters and meat composition in rainbow trout (*Oncorhynchus mykiss* W.). Fishes 9 (7), 249.

Vidya, S., Girish, L., 2014. Water hyacinth as a green manure for organic farming. Int. J. Res. Appl. Nat. Soc. Sci. 2 (6), 65–72.

Wang, T., Qin, G.X., Sun, Z.W., Zhao, Y., 2014. Advances of research on glycinin and β-conglycinin: a review of two major soybean allergenic proteins. Crit. Rev. Food Sci. Nutr. 54 (7), 850–862.

Wang, X., Sun, Y., Wang, L., Li, X., Qu, K., Xu, Y., 2017. Synbiotic dietary supplement affects growth, immune responses and intestinal microbiota of *Apostichopus japonicus*. Fish Shellfish Immunol. 68, 232–242.

Weng, L., Wang, Z., Zhuang, W., Yang, T., Xu, X., Liu, J., Ying, H., 2023. Effect of replacing fish meal using fermented soybean meal on growth performance, intestine bacterial diversity, and key gene expression of largemouth bass (*Micropterus salmoides*). Fermentation 9 (6), 520.

Wu, Z.B., Gatesoupe, F.J., Li, T.T., Wang, X.H., Zhang, Q.Q., Feng, D.Y., Li, A.H., 2018. Significant improvement of intestinal microbiota of gibel carp (*Carassius auratus gibelio*) after traditional Chinese medicine feeding. J. Appl. Microbiol. 124 (3), 829–841.

Xie, G., Zhang, M., 2024. Effects of dietary multi-strain probiotics on growth, immunity, and antioxidant enzyme activity of hybrid grouper (*Epinephelus fuscoguttatus*♀ × *E. lanceolatus*♂). Fish Shellfish Immunol. 136, 108700.

Xu, J., Wang, R., Liu, W., Yin, Z., Wu, J., Yu, X., Zhan, X., 2023. The specificity of ten non-digestible carbohydrates to enhance butyrate-producing bacteria and butyrate production in vitro fermentation. Food Sci. Hum. Wellness 12 (6), 2344–2354.

Yang, Y., Zhu, T., Jin, M., Li, X., Xie, S., Cui, Y., Zhou, Q., 2025. Black soldier fly larvae oil can partially replace fish oil in the diet of the juvenile mud crab (*Scylla paramamosain*). Anim. Nutr. 20, 469–486.

Ye, J.D., Wang, K., Li, F.D., Sun, Y.Z., 2011. Single or combined effects of fructo-and mannan oligosaccharide supplements and *Bacillus clausii* on the growth, feed utilization, body composition, digestive enzyme activity, innate immune response and lipid metabolism of the Japanese flounder *Paralichthys olivaceus*. Aquac. Nutr. 17 (4), e902–e911.

You, C., Chen, B., Wang, M., Wang, S., Zhang, M., Sun, Z., Li, Y., 2019. Effects of dietary lipid sources on the intestinal microbiome and health of golden pompano (*Trachinotus ovatus*). Fish Shellfish Immunol. 89, 187–197.

Yu, H., Ge, X., Zhang, L., Chen, X., Ren, M., Liang, H., 2023. Transcriptome analysis reveals the feeding response and oxidative stress in juvenile *Micropterus salmoides* fed a low-fish-meal diet with enzyme-hydrolysed intestinal mucosa protein substitution. Aquaculture 570, 739441.

Yu, S., Yoshikuni, Y., 2025. Biotechnological advances in algae-based foods: applications in nutrition and microbiome health. Curr. Opin. Biotechnol. 94, 103335.

Zakaria, M.K., Kari, Z.A., Van Doan, H., Kabir, M.A., Che Harun, H., Mohamad Sukri, S.A., Wei, L.S., 2022. Fermented soybean meal (FSBM) in African catfish (*Clarias gariepinus*) diets: effects on growth performance, fish gut microbiota analysis, blood haematology, and liver morphology. Life 12 (11), 1851.

Zhang, J., Huang, M., Feng, J., Chen, Y., Li, M., Chang, X., 2021. Effects of dietary *Bacillus licheniformis* on growth performance, intestinal morphology, intestinal microbiome, and disease resistance in common carp (*Cyprinus carpio* L.). Aquac. Int. 29 (3), 1343–1358.

Zhang, M.L., Li, M., Sheng, Y., Tan, F., Chen, L., Cann, I., Du, Z.Y., 2020. *Citrobacter* species increase energy harvest by modulating intestinal microbiota in fish: nondominant species play important functions. mSystems 5 (3), 10–1128.

Zhang, M., Sun, Y., Chen, K., Yu, N., Zhou, Z., Chen, L., Li, E., 2014. Characterization of the intestinal microbiota in Pacific white shrimp, *Litopenaeus vannamei*, fed diets with different lipid sources. Aquaculture 434, 449–455.

Zhang, Q., Guo, M., Li, F., Qin, M., Yang, Q., Yu, H., Tong, T., 2023. Evaluation of fermented soybean meal to replace a portion fish meal on growth performance, antioxidant capacity, immunity, and mTOR signaling pathway of coho salmon (*Oncorhynchus kisutch*). Aquac. Nutr. 2023 (1), 2558173.

Zhou, Q.C., Zeng, W.P., Wang, H.L., Wang, T., Wang, Y.L., Xie, F.J., 2012. Dietary arginine requirement of juvenile Pacific white shrimp, *Litopenaeus vannamei*. Aquaculture 364, 252–258.

Zehra, S., 2020. Effect of replacing fishmeal with ground nut cake on growth, feed conversion and carcass composition of fingerling *Oreochromis niloticus* (Nile Tilapia). SUST J. Agric. Vet. Sci. 21 (2).

CHAPTER

17 Concerns due to the use of alternative sources of protein in feed on the physiology and well-being of fish (for example agricultural waste)

Rahul Kumar[1], B.E. Guerra Sierra[2], Rachan Karmakar[3], Bahman Khoshru[4], Arindam Ganguly[5], Tanmoy Sarkar[6], Addisu Assefa[7], Navneet Joshi[8], Shraddha Bhaskar Sawant[9], Mona F.A. Dawood[10] and Debasis Mitra[1]

[1]*Department of Microbiology, Graphic Era (Deemed to be University), Dehradun, Uttarakhand, India;* [2]*Universidad de Santander-Facultad de Ciencias Exactas Naturales Y Agropecuarias, Bucaramanga, Colombia;* [3]*Department of Environmental Science, Graphic Era (Deemed to be University), Dehradun, Uttarakhand, India;* [4]*Soil and Water Research Institute (SWRI), Agricultural Research, Education and Extension Organization (AREEO), Karaj, Iran;* [5]*Department of Microbiology, Bankura Sammilani College, Bankura, West Bengal, India;* [6]*School of Agriculture, Swami Vivekananda University, Kolkata, West Bengal, India;* [7]*Applied Microbiology, Madda Walabu University, Bale Robe, Ethiopia;* [8]*Department of Biosciences, School of Liberal Arts and Sciences, Mody University of Science and Technology, Lakshmangarh, Rajasthan, India;* [9]*Bihar Agricultural University, Bhagalpur, Bihar, India;* [10]*Botany and Microbiology Department, Faculty of Science, Assiut University, Assiut, Egypt*

1. Introduction

The growing international market for aquaculture commodities has put immense pressure on conventional feed materials, most importantly fishmeal, which is the major protein source in aquafeeds owing to its good amino acid balance, high digestibility, and palatability (Hua et al., 2019). However, sustainability issues related to fishmeal manufacturing, such as overfishing, environmental degradation, and rising costs, have encouraged the aquaculture sector to investigate alternative protein sources. These alternatives, including insect meal, microbial proteins, single-cell proteins, and plant-based proteins, have promising prospects for minimizing dependence on sea resources and enhancing the economic feasibility of fish culture (Hadi and Brightwell, 2021). Notwithstanding their potential contributions, the use of alternative protein ingredients has identified several issues concerning their effects on the physiology and general health of cultured fish stocks (Aragão et al., 2022). One of the most important

Fish Feed Formulation with Agroindustrial Wastes and Byproducts. https://doi.org/10.1016/B978-0-443-41728-3.00001-0

issues is the nutritional quality of such alternative proteins, as most plant-derived foods possess anti-nutritional factors (ANFs), such as protease inhibitors, tannins, and phytic acid, which reduce nutrient digestibility and absorption (Salim et al., 2023). The incorporation of these substances usually results in lower feed efficiency and growth rate, which degrades the physiological growth of fish. Additionally, alternative proteins often have imbalanced amino acid compositions, missing necessary amino acids such as methionine and lysine, which are important for sustaining protein synthesis, immune status, and tissue repair in fish (Hissen et al., 2023). In addition to nutritional imbalances, the well-being of the gastrointestinal tract of fish is compromised by alternative protein sources. Vegetarian diets have also been associated with enteritis and other inflammatory diseases in species such as salmon, leading to morphological alterations such as villi shortening and enhanced inflammatory cell infiltration (Aidos et al., 2023). Such damage to the intestine not only compromises nutrient absorption but also puts undue stress on the immune system and, in turn, may enhance vulnerability to pathogens. Furthermore, dietary changes excluding fishmeal may reduce the availability of bioactive compounds naturally present in marine proteins, including specific peptides, omega-3 fatty acids, and micronutrients that support immune competence.

This deficiency can weaken the immune responses of fish, making them more vulnerable to disease outbreaks and diminishing vaccine effectiveness. Oxidative stress and metabolic disturbances are additional physiological concerns associated with alternative protein sources (Fig. 17.1) (Singh et al., 2021). Dietary lipid and fatty acid imbalances and antioxidant imbalances, such as vitamin E and selenium, can interfere with redox homeostasis, causing cellular damage and impaired energy metabolism, which impairs growth performance and fish health. Palatability is another critical parameter that determines the success of alternative proteins in fish diets (Afrin et al., 2023). A few fish species exhibit strong taste and texture preferences when it comes to feed, and the use of new protein sources can decrease feed intake and growth rate, and increase waste production with the potential to reduce water quality. Chronic nutritional and gastrointestinal disorders resulting from suboptimally formulated feeds over time can affect the behavior and welfare of fish, and documented effects include altered levels of activity, aggression, and stress resilience. Although alternative protein sources are vital to the development of sustainable aquaculture, these physiological and welfare issues require prudent formulation, processing methods to reduce anti-nutritional factors, and supplementation to compensate for nutritional deficiencies (Fig. 17.1) (Thazeem et al., 2022). Extensive research and development activities are paramount for maximizing the use of alternative proteins, such that the health and quality of life of cultured fish are not compromised in the pursuit of more sustainable and ethically desirable aquaculture (Table 17.1).

2. Nutritional challenges of agro-waste in fish feed

The quest for low-cost and sustainable feed ingredients in aquaculture has spurred growing interest in the use of agro-waste as a novel protein and nutrient source

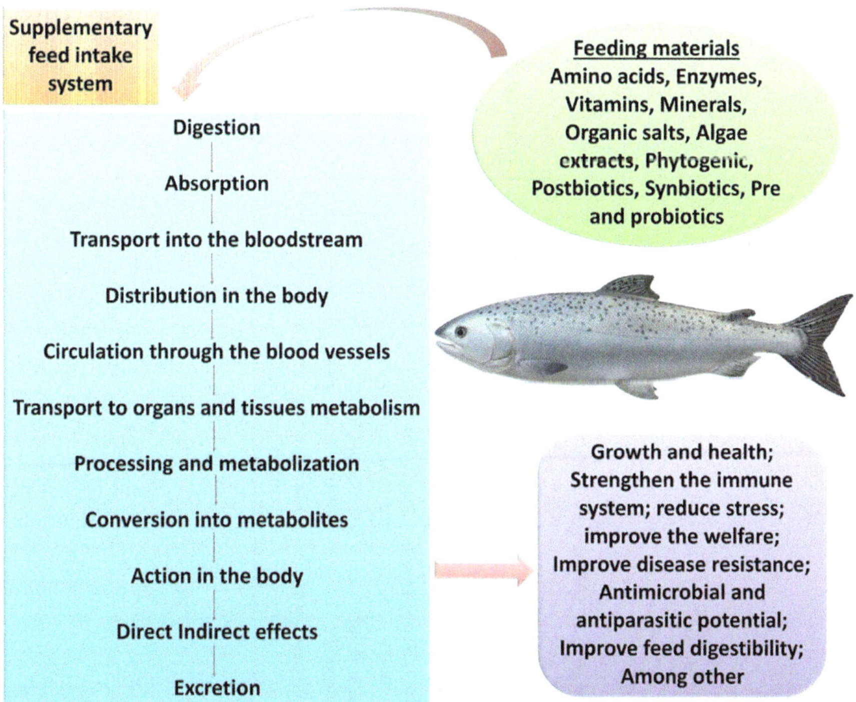

FIGURE 17.1 Physiological effects of agricultural waste-based diets on fish health.
Impact of agricultural waste-based diets on fish health and physiology.

for fish feed (Bala et al., 2023). Argo-waste, which consists of by-products of crop processing, such as rice bran, wheat bran, corn husk, fruit and vegetable skins, and oilseed cakes, is a vast and largely untapped resource with great potential for lowering the cost of feeds and the environmental footprint of aquaculture. Nevertheless, despite these benefits, the use of agro-waste in fish feed carries substantial nutritional difficulties that can be detrimental to fish growth, health, and performance (Amjad et al., 2025). These difficulties are mainly due to the inherent nature of agro-waste materials, which tend to possess inferior protein quality, imbalanced amino acid contents, and excessive quantities of anti-nutritional components, fiber, and indigestible matter, all of which render them less favorable for use as major feed ingredients. One of the key nutritional deficits of agro-waste is its generally low protein and defective amino acid balance compared to normal fishmeal or formulated diets (Akpoilih et al., 2024). Although certain oilseed cakes from crops such as soybean and sunflower provide moderate protein content, most agro-waste products lack adequate essential amino acids, such as lysine, methionine, and tryptophan, which play a central role in protein synthesis and metabolic processes in fish. This limitation leads to impaired protein utilization and lower growth rates when fish are fed diets highly dependent on these products (Aragão et al., 2022).

Table 17.1 Impact of agro-waste-based diets on fish immunity and disease resistance.

Ingredient	Fish species	Immunity markers affected	Impact on disease resistance	Observations	References
Rice bran	Nile tilapia	Lysozyme activity, complement system	Moderate enhancement	Improved innate immunity; needs balanced inclusion	Fath El-Bab et al. (2022)
Soybean cake	Common carp	Phagocytic activity, respiratory burst	Variable; can suppress at high levels	Heat treatment improves immune response	Song et al. (2022)
Mustard cake	Catfish	Serum immunoglobulins, oxidative stress markers	Reduced resistance due to toxicity	Detoxification required to avoid immune suppression	Reda et al. (2021)
Legume meal (pea)	Rainbow trout	Cytokine expression, leukocyte count	Mild improvement	Fermentation improves immunomodulatory effects	Shea et al. (2024)

Additionally, the protein within agro-waste is usually less bioavailable and digestible because of the complex structural matrices within which it is contained, which reduces its availability and nutritional value. Another major issue lies in the fact that most agro-waste products contain high fiber content and anti-nutritional factors (ANFs). Fibers, particularly non-starch polysaccharides, are highly resistant to digestion by most fish species, especially carnivorous and omnivorous fish, which do not possess the enzymatic equipment to effectively degrade complex carbohydrates effectively (Wang et al., 2024).

High dietary fiber can diminish nutrient uptake by accelerating the digesta passage rate and physically obstructing the digestive processes (Zhuang et al., 2024). Similarly, ANFs such as tannins, phytic acid, saponins, and enzyme inhibitors in plant-based agro-waste may bind nutrients or inhibit enzymes involved in digestion, thereby reducing feed efficiency. Such substances tend to trigger gut inflammation, disrupt intestinal morphology, and impair immune competence, eventually compromising fish welfare and health (De Marco et al., 2023). The diversity in the nutritional content of agro-waste makes it challenging to formulate food. Variations in crop type, growing conditions, processing, and storage may result in inconsistent nutrient content and the occurrence of contaminants, such as mycotoxins or pesticide residues. These variations create challenges in providing consistent quality feed and expected fish performance (Munguti et al., 2021). Moreover, the occurrence of lignin and other sophisticated phenolic compounds in agro-waste restricts microbial breakdown, thereby challenging nutrient recovery without specific processing methods, such as fermentation, enzymatic treatment, or thermal treatment (Bibi et al., 2023; Kumar et al., 2024). However, agro-waste is a viable alternative ingredient when it is well processed and supplemented to suit the nutritional needs of fish. Technological innovations involving pretreatment to neutralize ANFs and fiber levels, amino acid supplementation, and mixing with other proteins have achieved some success in circumventing most of the nutritional challenges facing agro-waste (Balakrishnan et al., 2025; Kumar et al., 2025a,b). However, proper assessment of the digestive ability and nutritional requirements of farmed fish is necessary to optimize the utilization of agro-waste and ensure minimal negative impacts on growth and health. Although agro-waste presents a sustainable avenue for cutting fees and enhancing resource efficiency in aquaculture, its nutritional limitations must be carefully examined and managed strategically (Sinha et al., 2022; Kumar et al., 2025a). Minimizing the problems of low protein quality, antinutritional compounds, high fiber levels, and compositional heterogeneity is essential for the formulation of balanced, digestible, and health-enhancing diets that enhance maximum fish growth and well-being in a sustainable manner (Table 17.2).

3. Incomplete amino acid profile

The theory of an incomplete amino acid profile is a paramount nutritional issue when formulating aquaculture feed, particularly when alternative protein sources are used

Table 17.2 Growth performance and feed conversion in fish fed agricultural waste-based diets.

Ingredient	Fish species	Inclusion level (%)	Growth rate (g/day)	Feed conversion ratio (FCR)	Observations
Rice bran	Tilapia	20	1.2	1.8	Moderate growth; slight decrease at higher levels
Soybean cake	Common carp	25	1.5	1.6	Good growth; needs enzyme supplementation
Mustard cake	Catla	15	1.1	2.0	Reduced feed intake; signs of mild toxicity
Fruit by-products	Rainbow trout	10	1.3	1.7	Acceptable growth; improved palatability

as substitutes for conventional fishmeal. Proteins consist of amino acids, which are some of the essential amino acids that cannot be produced by fish and must be nutritionally provided in food. Incomplete amino acid profiles represent a situation in which one or more of the essential amino acids are in inadequate quantities compared to the nutritional needs of fish, reducing the overall applicability of the diet protein. This deficiency has a significant impact on fish growth, metabolism, immune responses, and physiological well-being. Fishmeal, the traditional protein source of aquafeeds, is virtually ideal as it contains all essential amino acids in proportions that are very close to those required by fish. However, most plant proteins and other alternative ingredients have unbalanced or incomplete amino acid profiles. For instance, soybean meal, which is the most common plant protein, tends not to provide sufficient methionine and cysteine sulfur amino acids, which are crucial for protein formation and antioxidant protection. Similarly, by-products of cereals also lack lysine, a critical amino acid crucial for tissue repair as well as enzyme activity. These shortcomings have the effect that even when the overall protein level in a diet is sufficient, the lack of limiting amino acids hinders the ability of fish to use all the proteins for growth and maintenance (Li et al., 2021). The effects of diets with limited amino acid profiles are complex. Growth retardation is the most common outcome because without a balanced supply of essential amino acids, fish cannot synthesize proteins that support muscle development and other physiological functions. In many cases, this inefficiency results in poor feed conversion ratios and increases feed costs.

Suboptimal amino acid intake impairs immune responses and reduces the ability of fish to resist diseases and recover from stress (Hissen et al., 2023; Kamboj et al., 2025). Amino acids, such as arginine and histidine, are involved in immune cell proliferation and regulation of inflammation, and their shortage is therefore particularly detrimental to fish health. In addition, insufficient essential amino acids can affect nitrogen metabolism, causing increased ammonia excretion and water quality problems in aquaculture systems (Yavuzcan Yildiz et al., 2017). The limited amino acid profiles are compounded by the presence of anti-nutritional factors (ANFs) in most alternative protein feeds. ANFs can interfere with amino acid absorption and protein digestion, thereby further decreasing the bioavailability of essential amino acids. Such double action not only impairs nutrient utilization, but might also induce physiological stress, gut inflammation, and decreased appetite, all of which are detrimental to fish well-being (Eed Desouky et al., 2025). Complementing the deficient amino acid profiles in aquafeeds involves tactical strategies. The most widespread tactic is to add crystalline or synthetic amino acids to the feed to improve the diet. Methionine, lysine, and threonine are added to plant diets because of their natural deficiency. This addition promotes protein efficiency, growth, and immune status. A second strategy involves mixing several protein sources in a manner that optimizes a more complete amino acid profile by utilizing complementary profiles of the ingredients (Dimina et al., 2022). For example, mixing legume meals with high lysine levels with cereal meals with high methionine levels can be used to produce a diet with a better amino acid spectrum. Advances in biotechnology have also offered the potential to enhance amino acid profiles of non-traditional feed ingredients. Genetic engineering of crops to enhance the content of critical amino acids or through the utilization of fermentation processes to enhance the content of amino acids available in plant-based proteins is a new approach for improving the nutritional value of plant-based proteins (Langyan et al., 2022). In addition, the utilization of new protein sources, such as insect meals or microbial proteins, is being explored based on their more balanced amino acid profiles, which tend to be closer to fishmeal. The incomplete amino acid profiles of alternative proteins are a major constraint to their efficient utilization of alternative proteins in aquafeeds. Maintaining an equilibrated supply of all essential amino acids is crucial to maximize fish growth, health, and feed efficacy (Manam, 2023). Through targeted supplementation, blending of ingredients, and novel feed processing technologies, these challenges can be overcome and sustainable, nutritionally balanced diets that enhance productive and healthy aquaculture systems can be developed.

4. Imbalance in lipid composition

Lipids are a vital component of fish nutrition and act as energy-dense nutrients, cell membrane structural elements, and precursors for essential bioactive molecules, such as hormones and eicosanoids. They are critical in determining the physiology, growth performance, and health of fish, and are dependent on the food composition of lipids, specifically the proportion of saturated, monounsaturated, and polyunsaturated fatty acids (Trushenski and Rombenso, 2020). Nonetheless, novel feed ingredients bring about a distortion in lipid composition with respect to conventional fishmeal and fish oil-based diets, which presents various challenges to aquaculture nutrition. One of the significant problems related to lipid distortion is the shortage or surplus of essential fatty acids, particularly omega-3 long-chain polyunsaturated fatty acids (LC-PUFAs) such as eicosapentaenoic acid (EPA) and docosahexaenoic acid (DHA) (Mariamenatu and Abdu, 2021). These fatty acids are essential for fish membrane fluidity, brain development, vision, reproduction, and the modulation of inflammatory reactions. Most vegetable oils that are alternatively used in place of fish oil are high in omega-6 fatty acids but fall short of the necessary EPA and DHA content (Petersen et al., 2024). This imbalance affects the omega-6 to omega-3 ratio and may induce pro-inflammatory states, compromise immune responses, and lower disease resistance. In addition, insufficient LC-PUFA consumption can impair typical growth and development processes, thereby hindering overall fish performance. Imbalances may also cause changes in the lipid metabolism. Fish have a limited ability to elongate and desaturate the shorter-chain fatty acids present in vegetable oils into longer-chain fatty acids, with the accumulation of less desirable fatty acids and decreased incorporation of essential LC-PUFAs into tissues (Glencross et al., 2025). This metabolic constraint frequently leads to lipid deposition irregularities, including the undue accumulation of fat within the liver or visceral cavity, which is described as fatty liver syndrome. These conditions impair organ function, lower feeding efficiency, and make organisms more susceptible to stress and infection (Inbaraj et al., 2022).

The structural integrity and fluidity of cellular membranes are influenced by an improper balance between saturated and unsaturated fatty acids. Membranes that are high in saturated fats become more brittle, potentially impairing membrane-bound enzyme function, nutrient transport, and receptor activity (Ali and Szabó, 2023). This brittleness can interfere with cellular signaling and responsiveness, which are essential for establishing and maintaining homeostasis and responding to changes in the environment. Sensory and product quality-wise, lipid imbalance affects the fatty acid profile of fish flesh, which affects fillet texture, flavor, and storage stability. Decreased levels of healthy omega-3 fatty acids decrease the nutritional quality of fish for human consumption, thereby eroding one of the primary benefits (Yan et al., 2024). In addition, excessive contents of unsaturated fatty acids, especially polyunsaturated fatty acids, contribute to increased susceptibility to lipid oxidation, which causes rancidity and off-flavors if not regulated by antioxidants or processing methods.

Lipid imbalances are related to feed stability and storage. Some alternative lipids are more susceptible to oxidation during feed manufacturing and storage, thereby reducing feed palatability and nutritional value (Jackson and Penumetcha, 2019). Oxidized lipids produce free radicals and reactive species that can cause oxidative stress in fish, injure tissues and immune cells, and further weaken health. To overcome these challenges, aquafeed formulas tend to be formulated using well-balanced lipid mixtures of plant oils supplemented with oils derived from marine sources to obtain a more balanced fatty acid profile (Colombo et al., 2023). Supplementation with purified EPA and DHA concentrates or algal oils with high LC-PUFA content is now becoming widespread to address the fatty acid needs of cultured fish species. Improvements in feed technology, such as microencapsulation of fragile lipids and application of natural antioxidants, have preserved the integrity of lipids during storage and digestion (Soukoulis and Bohn, 2018). An imbalance in the lipid composition of fish feed, especially a lack of important omega-3 fatty acids and uneven fatty acid ratios, is a major nutritional and physiological problem. These imbalances hamper growth, health, immune responses, and product quality, underlining the need for accurate formulation techniques and novel supplementation strategies (Puri et al., 2022). Maintaining the correct lipid profile in aquafeeds is critical for maximizing fish welfare, facilitating sustainable production, and providing high-quality seafood to consumers.

5. Increased fiber and anti-nutritional factors (ANFs)

The use of alternative protein sources and agro-industrial by-products in fish diets frequently leads to increased concentrations of dietary fiber and anti-nutritional factors (ANFs), which pose serious challenges to aquaculture nutrition and fish health (Ababor et al., 2023). In contrast to conventional fishmeal, most plant ingredients and agro-waste have high concentrations of indigestible carbohydrates and bioactive compounds that interfere with nutrient use, digestion, and the physiological health of fish. The effects of elevated fiber and ANFs are multifaceted and involve digestive efficiency, growth performance, immune status, and overall fish well-being (Amenyogbe et al., 2024). Fish diet fiber is mainly composed of non-starch polysaccharides, cellulose, hemicellulose, and lignin, which are poorly digested by many fish species, notably carnivorous and omnivorous species. The absence of endogenous enzymes to digest such complex carbohydrates implies that a high fiber content tends to be associated with low nutrient digestibility and energy availability.

High fiber content increases gastrointestinal transit time and reduces the contact time between digestive enzymes and nutrients, leading to decreased feed efficiency (Bashar et al., 2021). In addition, the large fiber component can physically block nutrient absorption surfaces within the intestine and interfere with the absorption of proteins, fats, vitamins, and minerals essential for maximum growth and metabolism. Anti-nutritional factors are a broad category of naturally occurring phytochemicals present in plant-based feed ingredients that interfere with digestion and metabolism (Malik et al., 2025). Tannins, phytates, saponins, protease inhibitors, lectins, and alkaloids are anti-nutritional factors. Phytates, for instance, chelate minerals such as calcium, zinc, and iron, which precipitate as insoluble complexes, lowering their availability and inducing mineral deficiency. Protease inhibitors disrupt digestive enzymes such as trypsin and chymotrypsin, resulting in poor protein digestion and enhanced nitrogen loss (Gopalraaj et al., 2024). Tannins and saponins cause intestinal mucosa damage, which triggers inflammation and disrupts gut morphology, lowering the absorption of nutrients and impairing barrier functions against pathogens. The presence of ANFs is also likely to induce oxidative stress through the production of reactive oxygen species (ROS), which are harmful to cellular components including lipids, proteins, and DNA (Venditti and Di Meo, 2020). Oxidative damage compromises immune function, making fish more susceptible to infection and environmental stress.

Additionally, certain ANFs can exert direct toxicity or modify hormone metabolism, further compromising physiological homeostasis. The joint action of augmented fiber and ANFs manifests in delayed rates of growth, lower feed conversion ratios, and increased mortality in the worst instances (Chuang et al., 2021). In carnivorous fish, such as salmonids, high-fiber and ANF diets are especially detrimental because salmonids have evolved to eat low-fiber, animal-based diets and have limited enzymatic equipment for fiber breakdown. Herbivorous and omnivorous fish will better tolerate more fiber, but overconsumption of fiber and ANFs will still compromise their health and performance (Hofmann et al., 2025). Solutions to counter these dietary challenges include the physical, chemical, and biological processing of feedstuffs to minimize fiber and ANF concentrations. Heat treatment, extrusion, and fermentation can hydrolyze resistant carbohydrates and inactivate most ANFs, thereby enhancing digestibility and feed quality. Adding cellulases, xylanases, and phytases to enzyme supplementation improves the availability of fiber and minerals in the gut (Moita and Kim, 2022). In addition, selecting low-ANF crop varieties and blending multiple sources to dilute the concentrations of ANFs in the formulation will improve feed formulation. The substantial challenges posed by increased fiber and anti-nutritional factors in alternative feed ingredients do not contribute to the nutrition challenges in aquaculture, such as digestibility, nutrient absorption, and health conditions in fish (Siddik et al., 2024). Resolution of these problems using novel processing methods, enzyme addition, and judicious ingredient choice is critical for maximizing feed efficiency, sustainable aquaculture, and fish welfare. An ongoing investigation into the pathways through which fiber and ANFs influence fish physiology will aid in the creation of better dietary means to tap the potential of alternative protein sources without hampering fish performance (Table 17.3).

6. Physiological and health concerns

6.1 Digestive health

Digestive health is a key foundation for fish welfare and production, directly affecting nutrient uptake, growth, and disease resistance. Optimal digestive

Table 17.3 Anti-nutritional factors (ANFs) in common agricultural waste ingredients and their effects on fish health.

Ingredient	Common ANFs	Impact on fish physiology	Mitigation strategies	References
Rice bran	Phytic acid, tannins, fiber	Reduced mineral absorption, digestive disturbances, lower growth	Phytase enzyme supplementation, fermentation, heat treatment	Irakli et al. (2020)
Soybean cake	Trypsin inhibitors, lectins, saponins, phytic acid	Inhibited protein digestion, gut inflammation, reduced feed efficiency	Heat treatment, fermentation, enzyme supplementation	Cao et al. (2024)
Mustard cake	Glucosinolates, tannins, and phytic acid	Liver toxicity, reduced feeding intake, impaired metabolism	Heat processing, detoxification, blending with other ingredients	Vujetić et al. (2025)
Legume meal (pea, lentil)	Protease inhibitors, tannins, oligosaccharides	Reduced protein digestibility, intestinal irritation, flatulence	Soaking, fermentation, enzyme addition	Rudra et al. (2023)

performance is particularly difficult to sustain in aquaculture when alternative feed materials or agro-waste products with complex compositions are used (Abasubong et al., 2024). These ingredients tend to have high fiber, anti-nutritional components, and indigestible material contents that negatively impact the structure and function of the gastrointestinal tract of fish and consequently compromise digestive health. The digestive system of fish includes the mouth, esophagus, stomach, intestine, liver, and related glands, which cooperate to digest feed and absorb nutrients (Zhukova and Stroganov, 2022). The effectiveness of this system is based on the biochemical and physical compatibility of the diet with digestive enzymes and gut morphology of the fish. Most carnivorous fish species that dominate aquaculture have comparatively straightforward digestive systems developed for protein- and lipid-rich foods, with little or no ability to process high fiber or complex carbohydrates (Steinberg, 2022). On feeding fish diets with high concentrations of plant-based ingredients or agro-wastes, excess fiber tends to retard digestion by prolonging gut transit time or leading to physical irritation, destroying the soft mucosal lining. Anti-nutritional factors (ANFs) in other feedstuffs can cause digestive disorders. Protease inhibitors inhibit important digestive enzymes such as trypsin and chymotrypsin, resulting in impaired protein digestion (Solovyev et al., 2023). Lectins and tannins can adhere to the gut mucosa, impairing nutrient uptake and inducing inflammation or morphological alterations, such as villus shortening and damage to epithelial cells. Phytates bind essential minerals, making them unavailable for absorption, and leading to indirect deficiency effects on the function of digestive enzymes and cellular metabolism. Such dietary insults frequently result in morphological changes in the intestines (Lall and Kaushik, 2021). Histological investigations showed that high-fiber and ANF-containing diets may lead to increased gut wall thickness, inflammation, and villi reduction, all of which decrease the absorptive surface area. This injury not only restricts nutrient absorption, but also compromises the integrity of the gut barrier, enhancing susceptibility to pathogen invasion and systemic infection (Zhang et al., 2020). Intestinal mucosal health is important for digestion and immune protection because it contains a significant percentage of immune cells tasked with monitoring microbial numbers and avoiding disease. The gut microbiota is instrumental in the maintenance of digestive health.

Healthy microbial populations facilitate the fermentation of complex dietary carbohydrates, vitamin synthesis, and competitive inhibition of pathogens. However, high-fiber or ANF diets can cause microbial imbalances, support the overgrowth of pathogenic bacteria, and induce dysbiosis. This imbalance can lead to inefficient nutrient use, inflammation, and an increased vulnerability to enteric diseases. Digestive health management in fish that consume alternative diets requires multifaceted approaches. Processing techniques, such as extrusion, fermentation, and heat treatment, can break down the fiber content and inactivate most ANFs, enhancing digestibility (Maksimenko et al., 2024). Supplementation with exogenous enzymes (e.g., cellulases, proteases, and phytases) improves the hydrolysis of complex feed ingredients and enhances nutrient availability. Prebiotic and probiotic inclusion also supports healthy gut microflora, restores balance, and enhances gut integrity. Digestive health is a key component of successful aquaculture, especially when utilizing nontraditional feed sources (Doolam et al., 2025). Matching diets to fish digestive physiology, reducing anti-nutritional effects, and facilitating a healthy gut microbiota is crucial for maximal nutrient uptake, growth, and disease resistance. Ongoing research on feed formulation and gut health will allow for more sustainable and efficient aquaculture production (Boyd et al., 2020).

6.2 Stress response and behavior

Stress response and behavior are critical indicators of fish health and welfare, reflecting how environmental factors, including diet composition, impact physiological and psychological well-being. In aquaculture, the shift toward alternative feed ingredients and agro-waste utilization often introduces nutritional imbalances, anti-nutritional factors, and indigestible compounds that can provoke stress responses and alter normal fish behavior (Lackner and Besharati, 2025). Understanding these effects is vital for optimizing fish welfare, growth, and overall production success. Fish are stressed when environmental or internal conditions complicate homeostasis. Nutritional deficiencies, including deficiencies in essential amino acids, fatty acids, vitamins, and minerals, serve as chronic stressors that interfere with metabolic functions (Ciji and Akhtar, 2021). Diets with large amounts of fiber and anti-nutritional factors lead to gastrointestinal discomfort, malabsorption, and inflammation, stimulating physiological stress responses. These involve stimulation of the hypothalamic-pituitary-interrenal (HPI) axis, resulting in increased cortisol secretion, the major stress hormone in fish (Peixoto et al., 2024). Increased cortisol affects energy metabolism, inhibits immune response, and interferes with growth, thus decreasing fish performance. Behavioral responses usually follow physiological stress. Nutritionally stressed fish present with decreased feeding activity, lethargy, aberrant swimming, and increased aggression or social isolation (Dara et al., 2023). Decreased appetite depresses feed consumption, aggravates nutritional deficiencies, and delays growth. Shifts in schooling behavior or enhanced surface respiration may indicate suboptimal water quality or internal morbidity. These behavioral changes also affect the stocking density control and make harvesting timing more difficult. Anti-nutritional compounds (ANFs) as tannins, lectins, and saponins, directly cause gastrointestinal irritation and systemic inflammatory reactions, contributing to enhanced stress (Yan et al., 2025). Chronic inflammation drains energy from growth and reproduction into immune defenses, resulting in a self-perpetuating cycle that compromises fish well-being.

In addition, oxidative stress introduced by pro-oxidant feed components in low-quality diets further injures tissues and cells, compromising their resilience to environmental stress. Stress and behavioral modifications have far-reaching consequences for fish welfare and sustainability in aquaculture (Sharma et al., 2024). Chronic cortisol elevation results in immunosuppression, enhancing susceptibility to infectious illness and diminishing the effect of vaccines. Behavioral stress can compromise reproductive success, induce developmental disorders, and increase mortality. In commercial contexts, these problems manifest as economic losses due to reduced production efficiency and augmented management expenses (Shastak et al., 2023). Several approaches are required to reduce stress and behavioral disorders associated with nutrition. Supplying balanced, digestible, and acceptable diets according to species requirements is the foundation of this study. Processing techniques such as fermentation and extrusion minimize ANFs and enhance nutrient accessibility. Feed supplementation with antioxidants, essential fatty acids, and immunostimulants enhances stress resistance and immune competence (Ndudzo et al., 2023). Environmental enrichment and good husbandry techniques also reduce stressors and enhance natural behavior and welfare. Biomarkers of stress, such as cortisol measurements, heat shock proteins, and behavioral inspection, allow identification of early signs of nutritional stress and provide an opportunity for timely intervention (Dhama et al., 2019). Improved knowledge of neuroendocrine control of stress and behavior in fish allows for the design of precision nutrition approaches to minimize negative effects. Fish stress responses and behavior are inextricably linked to dietary quality and composition. Nutritional stress disrupts physiological processes, compromises immunity, and modifies behavior, with negative consequences on fish welfare and production performance (Dara et al., 2023). These issues should be addressed using optimized feed processing, formulation, and management strategies to facilitate sustainable and responsible aquaculture production.

6.3 Liver and kidney function

The liver and kidneys are essential organs in fish that play pivotal roles in metabolism, detoxification, osmoregulation, and overall physiological homeostasis. The functioning of these organs is critical for health maintenance and growth promotion, particularly in aquaculture, where the diet composition directly determines organ health directly (Dawood, 2021). The greater incorporation of alternate protein sources and agro-industrial by-products in fish diets poses challenging nutrient problems, which may have negative impacts on liver and kidney functions, leading to reduced health and productivity. The liver is the central metabolic center for processing, storage, synthesis of key biomolecules, and detoxification of toxic substances (Topić Popović et al., 2023). It is a key site for carbohydrate, lipid, and protein metabolisms. Substitution of feed ingredients may also vary largely in terms of nutrient composition with respect to conventional fishmeal, with occasionally imbalanced lipid levels, high carbohydrate content, or anti-nutritional components (Jannathulla et al., 2019). These differences may lead to hepatic stress as exemplified by lipid accumulation (hepatic steatosis), inflammation, and oxidative damage. Fatty liver disease due to excessive fat accumulation in hepatocytes compromises liver function by lowering its ability to control metabolism and detoxify xenobiotics, which can increase the risk of sensitivity to environmental toxins and pathogens (Naiel et al., 2023). Antinutrients that occur in plant or agro-waste feed ingredients, such as tannins, saponins, and protease inhibitors, have the potential to cause liver inflammation and oxidative stress. Oxidative stress arises when the formation of reactive oxygen species (ROS) surpasses antioxidant defense mechanisms, causing cellular injury (Chowdhury and Saikia, 2020). Not only does this impair hepatocyte function, but it also interferes with the production and excretion of bile necessary for lipid digestion and waste removal. Chronic liver injury can decrease feed efficiency, inhibit growth, and increase mortality.

Fish kidneys play an essential role in nitrogenous waste excretion, osmoregulation, and acid-base balance. Dietary content affects the kidney load and health (Takvam et al., 2023). Low-digestible protein foods or those with toxins enhance the excretion of nitrogenous waste products, such as ammonia and urea, adding to the renal burden. Overloading with metabolic waste can cause nephrotoxicity, manifested by the degeneration of tubules, damage to glomeruli, and compromised filtration function (Javed et al., 2025). These effects compromise the ability of fish to maintain internal homeostasis, especially in challenging environmental conditions such as varying salinity or temperature. Moreover, contaminants and heavy metals present in alternative feed ingredients can accumulate in the kidney tissues, exacerbating their toxic effects. Kidney dysfunction may manifest as reduced growth, abnormal swimming behavior, and increased mortality (Ghafarifarsani et al., 2024). As the kidneys are also involved in immune functions by producing immune cells, compromised kidney function can compromise overall fish immunity. Histopathological analysis of liver and kidney tissues frequently indicates morphological alterations, such as vacuolation, necrosis, inflammatory cell infiltration, and fibrosis, in fish fed diets rich in anti-nutritional factors or unbalanced nutrients (Kakwi et al., 2021). These structural alterations are associated with biochemical markers, such as elevated liver enzymes (ALT and AST), elevated blood urea nitrogen (BUN), and creatinine, reflecting organ stress and injury. Modification of the unfavorable effects on liver and renal function includes judicious feed formulation to control nutrients and limit anti-nutritional compounds (Marimuthu et al., 2022). Processing techniques, such as heat processing, fermentation, and the addition of enzymes, reduce toxic substances and enhance digestibility. The incorporation of antioxidants, such as vitamin E, selenium, and polyphenols, enhances organ well-being through the neutralization of oxidative stress. Ongoing assessment of organ function via biochemical analyses and histology is critical for identifying early indicators of dysfunction and aligning feeding plans accordingly (Belenguer et al., 2024). Liver and kidney functions are essential to fish physiology and are extremely sensitive to dietary quality. Alternative protein sources and fish feed agro-waste pose potential risks to these organs in terms of nutritional imbalance, toxicants, and oxidative stress (Bertocci and Mannino, 2022). Balancing nutrition, reducing harmful substances, and enhancing antioxidant defenses are essential for maintaining organ function, fish health, and the sustainability of aquaculture.

7. Long-term health risks

7.1 Chronic inflammation and immune system suppression

Chronic immune suppression and inflammation are major physiological issues in aquaculture, particularly with the feeding of fish diets that include alternative protein sources and agro-industrial by-products. Although these foodstuffs have cost and sustainability benefits, their intricate structures, such as anti-nutritional factors (ANFs) and unbalanced nutrients, can induce long-term inflammatory reactions and impair immune function, thereby influencing fish health, growth, and survival (Albrektsen et al., 2022). Inflammation is a normal defense response triggered by injury, infection, or toxic stimuli. However, chronic inflammation occurs when the stimulus is present for more than a few days or if controlling mechanisms are ineffective, causing sustained tissue damage and impaired function. In fish, foodstuffs such as high dietary fiber, tannins, lectins, saponins, and protease inhibitors, which are characteristic of most plant-derived feedstuffs, irritate the gastrointestinal tract (Francis et al., 2001). This inflammation leads to mucosal injury and persistent stimulation of the immune system within the gut, the largest immune organ in fish. The intestinal mucosa reacts to irritants by invoking immune cells and secreting pro-inflammatory cytokines, inducing a state of chronic inflammation. This repeated activation of the immune system injures the epithelial lining, lowers villus height, and violates tight junctions, which compromises the gut barrier function (Kucuksezer et al., 2023). A weakened barrier increases the threat of invasion by pathogens and systemic infections. The diminished capacity of the gut for absorption leads to uptake of fewer nutrients, compounding nutritional stress, and inflammation. Systemic effects also occur in chronic gut inflammation. The resources and energy allocated to sustain the inflammatory process decrease availability for growth, reproduction, and other processes (Naya-Català et al., 2021). Pro-inflammatory cytokines in the bloodstream can also act on other organs, such as the liver and kidneys, leading to metabolic imbalances and organ dysfunction. Immunosuppression usually results from chronic inflammation.

Immune exhaustion and dysregulation results from long-term exposure to ANFs and continuous inflammatory stimuli. Such immune suppression is reflected in reduced phagocytic ability, antibody and antimicrobial peptide production, and impairment of cellular immunity. Fish are increasingly prone to bacterial, viral, and parasitic infections, which increase morbidity and mortality rates (Grayfer et al., 2018). Oxidative stress is known to be strongly associated with chronic inflammation. Chronic inflammation favors the production of reactive oxygen species (ROS), which can lead to cellular structural components such as lipids, proteins, and DNA damage. Oxidative damage inactivates immune cell functions and compromises tissue integrity. In the absence of sufficient antioxidant defenses, usually compromised in nutritionally unbalanced diets, oxidative stress enhances tissue damage and impairs recovery (Hissen et al., 2023). Gut microbiota also serves as a critical regulator of inflammation and immunity. Healthy microbiomes maintain homeostasis of the immune system and resistance to pathogens. High-fiber and ANF-containing diets can disrupt homeostasis and cause dysbiosis, which fuels inflammation and immune pathology. Restoration of microbial equilibrium using probiotics and prebiotics can counteract these effects, promoting gut health and immune resilience (Amenyogbe et al., 2024). The neutralization of chronic inflammation and immune suppression involves a multi-pronged approach. Feed formulations must be designed to reduce ANFs by selecting appropriate ingredients and processing protocols, such as fermentation, heat treatment, or enzyme supplementation (Siddik et al., 2024). Nutritional balance, with sufficient essential amino acids, fatty acids, vitamins, and minerals, ensures immune competence. Antioxidants such as vitamin E, selenium, and polyphenols can counteract oxidative stress. Immunostimulants and functional feed additives have the potential to boost immune responses and alleviate inflammation. Chronic inflammation and immunosuppression are also inter-related problems in fish fed alternative protein-based diets. Resolving these conditions is essential for ensuring the health of fish, maximizing growth, and facilitating sustainable aquaculture production. Ongoing research into feed formulation, gut health, and immune modulation will also advance measures to counteract these negative effects.

7.2 Altered reproductive performance

Reproductive performance is a core element of fish biology and has a direct impact on aquaculture productivity and sustainability (Mitra et al., 2023). The quality and composition of fish diets play key roles in controlling reproductive physiology, such as gametogenesis, spawning behavior, and viability of offspring. The growing replacement of conventional fishmeal with substitute protein sources such as plant meals and agro-industrial by-products in aquafeeds, which are economically and environmentally sound, has been concerning in view of their possible negative impact on reproductive outcomes in a negative way (Debbarma et al., 2025). Nutrition directly influences the endocrine system that governs reproduction. Important fatty acids, such as long-chain polyunsaturated fatty acids (LC-PUFAs), EPA and DHA, vitamins (A, E, C), minerals (zinc and selenium), and amino acids, play crucial roles in hormone production, gamete development, and embryonic development. Other feedstuffs contain unbalanced lipid profiles, reduced levels of essential nutrients, and increased amounts of anti-nutritional factors (ANFs), which hamper nutrient utilization and uptake (Salim et al., 2023). These dietary deficiencies have the potential to interfere with the synthesis and secretion of major reproductive hormones, such as gonadotropins, estrogens, and androgens, and consequently hamper gonadal development and function (Babaniyi et al., 2024). Among the most frequently observed consequences of alternative protein-rich diets is a decrease in fecundity, the number of eggs laid by females, and compromised sperm quality among males. An inadequate dietary lipid profile, particularly a lack of omega-3 fatty acids, impairs membrane fluidity and gamete function, resulting in diminished fertilization and viability of embryos (Bera et al., 2025). Moreover, dietary deficiencies in minerals and vitamins essential for antioxidant protection augment oxidative stress in reproductive tissues and induce injury to the oocyte and spermatozoon. Endocrine-disrupting effects can be caused by anti-nutritional components, such as phytates, tannins, and saponins, found in most vegetable-based feedstuffs.

They disrupt hormone receptor binding, enzyme function, and gene expression during the reproductive processes. Repeated exposure to these substances can retard sexual development, decrease spawning intervals, and modify the secondary sexual traits necessary for mating (Langston et al., 2020). Dietary composition also affects broodstock health. Metabolic stress and chronic inflammation caused by low-quality diets can channel energy from reproduction to immune defense and maintenance. This energy compromise lowers the total reproductive investment, with negative consequences on gamete quality and spawning success. In addition, disrupted gut microbiota due to non-conventional feed ingredients may affect nutrient uptake and hormone metabolism, consequently indirectly affecting reproductive performance indirectly (Kopylov et al., 2021). In operational aquaculture, these phenomena manifest as irregular spawning cycles, low egg viability, and hatchery success, compromising production efficiency and profitability. Poor reproductive performance also calls for more broodstock replacement, making selective breeding programs difficult. Reducing the negative impact on reproduction requires preparing balanced diets with sufficient levels of essential nutrients, particularly LC-PUFAs, vitamins, and trace minerals (Mititelu et al., 2024). Ingredient processing to minimize anti-nutritional factors and promote ingredient digestibility is vital. Functional additive supplementation with antioxidants, precursors to hormones, and probiotics facilitates reproductive health by increasing nutritional availability, minimizing oxidative impairment, and ensuring hormonal stability. Continuous monitoring of reproductive parameters, such as gonadosomatic index (GSI), hormone concentration, sperm mobility, and quality of eggs, is necessary for the early identification of dietary influences and in advising nutritional modifications (Manibusan and Touart, 2017). Advances in endocrinology and molecular biology provide a basis for understanding the mechanisms by which diet affects reproduction so that precise feeding strategies can be developed to meet species and life stage needs (Bertucci, 2023). Changes in reproductive function are a substantial long-term issue to be considered with the use of alternative protein feed ingredients in fish diets. Correcting nutritional imbalances, reducing anti-nutritional factors, and maintaining reproductive physiology through specially formulated feeds is responsible for maintaining broodstock health and successful aquaculture reproduction (Engdaw and Geremew, 2024).

8. Mitigation strategies

8.1 Improving digestibility

Aquaculture nutrition is a growing concern, particularly with the addition of alternative protein sources such as plant meals and agro-industrial by-products. These materials are less digestible than traditional fishmeal because of structural elements, such as high fiber levels, complex carbohydrates, and anti-nutritional factors (ANFs), which inhibit nutrient utilization and absorption (Ma et al., 2022). Increasing digestibility not only optimizes feed efficiency and growth performance but also minimizes waste production and enhances environmental friendliness. Sophisticated feed processing methods can enhance digestibility by breaking down plant cell walls and denaturing ANFs such as protease inhibitors and lectins. Fermentation is another promising method, as it degrades complex polysaccharides and ANFs, thereby improving protein bioavailability and micronutrients (Adebo et al., 2022). Fermented soybean meal has been reported to enhance growth and minimize intestinal inflammation in several species compared with unfermented alternatives. Enzyme supplementation is common to cover the limited endogenous enzyme capacity in most fish species, especially carnivores that are poorly adapted for digesting plant substances. Supplementation with exogenous enzymes, such as phytase, cellulase, xylanase, and protease promotes the hydrolysis of phytate, non-starch polysaccharides, and proteins, respectively (Singh et al., 2019). Reducing the particle size through feed milling and fine grinding of feed materials maximizes mixing, consistent pellet quality, and effective nutrient extraction in the gut. Optimizing gut health is central to digestibility, as high-fiber or ANF alternative feed ingredients can interfere with intestinal morphology, lowering absorptive surface area and enzyme secretion. Functional feed additives, such as organic acids, can reduce intestinal pH, stimulate digestive enzymes, and suppress pathogenic bacteria (Dittoe et al., 2018). Water temperature and feeding frequency also affect digestive efficiency, as fish are ectothermic, and enzymatic activity and gut motility are temperature-dependent in the environment. Feeding practices based on species-specific metabolic rates and life stages maximize digestion and minimize feed waste.

8.2 Balancing amino acid and fatty acid profiles

Amino acid and fatty acid balances are crucial in fish nutrition, especially when substituting conventional fishmeal with other protein sources such as plant meals, insect meals, and agro-industrial by-products. These novel ingredients often carry incomplete or unbalanced profiles of essential amino acids and fatty acids, which can impair fish growth, health, reproduction, and general physiological processes (Hăbeanu et al., 2024). A balanced nutrient composition of aquafeeds is essential to satisfy species-specific nutritional needs and to maintain optimum performance and sustainability in aquaculture. Amino acids are protein building blocks that are essential for tissue building, metabolism, and immune functions. However, sustainable vegetable protein sources often lack certain essential amino acids, leading to inferior protein synthesis, slower growth rates, and lower immune status (Salter and Lopez-Viso, 2021). To correct these deficiencies, accurate formulation processes are undertaken, such as the complementary blending of various plant protein sources to offer a more balanced amino acid supply. Synthetic amino acids are bioavailable and enable the diet to be tailored to the needs of fish without surplus protein load, thereby minimizing nitrogenous waste production and environmental influence (Kim et al., 2019). Fatty acid profiles, particularly those of essential fatty acids, are vital for fish growth and health. Long-chain polyunsaturated fatty acids (LC-PUFAs), such as eicosapentaenoic acid (EPA) and docosahexaenoic acid (DHA), are necessary for cell membrane structure, neural function, inflammation control, and reproduction. Fishmeal and fish oil naturally contain these fatty acids, but many plant oils contain shorter-chain fatty acids and omega-6 fatty acids, which cannot be efficiently converted into EPA and DHA by fish (Kitessa et al., 2014). Maintaining fatty acid balance in fish feeds involves supplementing other oils with marine-sourced omega-3 sources or incorporating new ingredients, such as microalgae rich in EPA and DHA. Nutritional balancing also includes antioxidant protection, as unsaturated fatty acids are susceptible to oxidation, which affects cell membranes and feed quality (Ponnampalam et al., 2022). Formulation strategies involving complementary ingredient mixing, synthetic supplementation, and new sources of lipids are central to overcoming the disadvantages of non-conventional feedstuffs and driving aquaculture productivity and environmental sustainability.

8.3 Optimizing feed formulation

Feed formulation optimization is crucial for sustainable aquaculture, balancing fish diets against their nutritional needs, while supporting growth, health, and feed efficiency. As the world adopts fishmeal alternatives such as plant meals, insect proteins, and agro-industrial by-products, feed formulation has become more challenging owing to the variable nutrient composition and presence of antinutritional factors (ANFs) (Ravi et al., 2020). Comprehensive ingredient characterization is necessary to optimize the feed formulation, including proximate analysis, amino acid profiles, lipid content, and digestibility coefficients. The least-cost formulation algorithm is commonly used, involving nutrient needs, cost of ingredients, and availability to arrive at cost-efficient and nutritionally complete feeds. Trial feeding studies have been conducted to evaluate growth performance, feed conversion ratio (FCR), health indicators, and product quality (Davison et al., 2023). Repeated adjustments ensured that nutrient content and ingredient inclusion percentages maximized fish performance without impairing welfare or the environment. In addition to enzymes such as phytase, protease, and carbohydrases, probiotics and prebiotics maintain gut health and modulate immune responses, leading to improved nutrient assimilation. Antioxidants defend the elements in the feed against oxidation and ensure fish health under stressful conditions (Hoseinifar et al., 2020). Processing methods are combined with formulation practices to improve feed quality, such as extrusion, which enhances pellet stability and acceptability and improves nutrient availability through protein denaturation and starch gelatinization. Ecological sustainability is a major concern in feed formulation, with optimization of nutrient content to decrease surplus protein and phosphorus release into aquatic ecosystems and reduce the environmental footprint (Wilfart et al., 2023). The addition of local sources of alternative ingredients encourages the principles of circular economy.

Advances in precision nutrition and modeling technology are improving feed formulation, allowing species-specific databases, growth modeling, and digestibility tests to tailor nutritional requirements. Emerging strategies include metabolomics and gut microbiome evaluation to enhance health and resilience (Fassarella et al., 2021). Overall, optimizing feed formulations in aquaculture requires a multidisciplinary approach that integrates ingredient assessment, nutritional equilibration, additive addition, process technologies, and environmental factors.

9. Conclusion

Transitioning to alternative protein sources for fish diets is crucial for sustainable aquaculture; however, it presents significant nutritional and physiological challenges. These alternatives often contain incomplete amino acid profiles, unbalanced fatty acids, excessive fiber, and anti-nutritional compounds that impair nutrient digestibility and utilization. These issues can compromise growth, immune function, liver and kidney performance, and reproductive capability, ultimately affecting fish health and farm yields. To avoid these issues, an integrated strategy involving an accurate feed formulation, ingredient processing, and dietary supplementation is required. Processing technologies, such as extrusion, fermentation, and enzyme treatment, reduce anti-nutritional factors and increase nutrient availability. Supplementing feeds with additives, such as probiotics, enzymes, and antioxidants, enhances digestion, gut health, and immunity. Future feed-development technologies include nanoencapsulation and selective breeding. Balancing nutritional science with effective feed management and disease monitoring is essential for sustainable aquaculture. Ongoing research and development efforts are crucial for creating efficient, cost-saving, and environment-friendly aquafeeds to sustain seafood demand in the global market.

References

Ababor, S.A., Tamiru, M., Alkhtib, A., Wamatu, J., Kuyu, C.G., Teka, T.A., Arega Terefe, L., Burton, E.J., 2023. The use of biologically converted agricultural byproducts in chicken nutrition. Sustainability 15 (19), 14562. https://doi.org/10.3390/su151914562.

Abasubong, K.P., Gabriel, N.N., Adjoumani, J.J.Y., 2024. The application of fermented rice bran and its influence on aquatic species: a dynamic study. In: sustainable feed ingredients and additives for aquaculture farming: perspectives from Africa and Asia, pp. 243–270.

Adebo, J.A., Njobeh, P.B., Gbashi, S., Oyedeji, A.B., Ogundele, O.M., Oyeyinka, S.A., Adebo, O.A., 2022. Fermentation of cereals and legumes: impact on nutritional constituents and nutrient bioavailability. Fermentation 8 (2). https://doi.org/10.3390/fermentation8020063.

Afrin, S., Shimul, T.J.T., Sezu, N.H., Nandi, S.K., Suma, A.Y., Alam, M.T., Kabir, M.A., 2023. Attractability and palatability of formulated diets incorporated with fermented aquatic weeds meal (FAWM) for Asian catfish Clarias batrachus fingerling. Agric. Rep. 2, 25–40.

Aidos, L., Mirra, G., Pallaoro, M., Herrera Millar, V.R., Radaelli, G., Bazzocchi, C., Modina, S.C., Di Giancamillo, A., 2023. How do alternative protein resources affect the intestine morphology and microbiota of Atlantic salmon? Animals 13 (12). https://doi.org/10.3390/ani13121922.

Akpoilih, B.U., Nwafili, S.A., Erondu, E.S., 2024. Microbial phytases as functional feed additives in aquaculture: impact, challenges, recent developments and future opportunities. In: Sustainable feed ingredients and additives for aquaculture farming: perspectives from Africa and Asia, pp. 521–563.

Albrektsen, S., Kortet, R., Skov, P.V., Ytteborg, E., Gitlesen, S., Kleinegris, D., Mydland, L.T., Hansen, J.Ø., Lock, E.J., Mørkøre, T., James, P., Wang, X., Whitaker, R.D., Vang, B., Hatlen, B., Daneshvar, E., Bhatnagar, A., Jensen, L.B., Øverland, M., 2022. Future feed resources in sustainable salmonid production: a review. Rev. Aquacult. 14 (4), 1790–1812. https://doi.org/10.1111/raq.12673.

Ali, O., Szabó, A., 2023. Review of eukaryote cellular membrane lipid composition, with special attention to the fatty acids. Int. J. Mol. Sci. 24 (21), 15693. https://doi.org/10.3390/ijms242115693.

Amenyogbe, E., Droepenu, E.K., Ayisi, C.L., Boamah, G.A., Duker, R.Q., Abarike, E.D., Huang, J.S., 2024. Impact of probiotics, prebiotics, and synbiotics on digestive enzymes, oxidative stress, and antioxidant defense in fish farming: current insights and future perspectives. Front. Mar. Sci. 11. https://doi.org/10.3389/fmars.2024.1368436.

Amjad, M., Hussain, S.M., Ali, S., Rizwan, M., Ahmad, A., Yong, J.W.H., 2025. Agro-waste biochar conversion into a fish feed additive: assessing its effects on the health and performance of Cyprinus carpio. Waste Biomass Valoriz. 16 (2), 901–910. https://doi.org/10.1007/s12649-024-02714-x.

Aragão, C., Gonçalves, A.T., Costas, B., Azeredo, R., Xavier, M.J., Engrola, S., 2022. Alternative proteins for fish diets: implications beyond growth. Animals 12 (9), 1211.

Babaniyi, G.G., Ajao, B.H., Akor, U.J., Babaniyi, E., 2024. Reproductive endocrinology drug development: hormones, metabolism, and fertility in female reproductive health. In: Perspectives of Quorum Quenching in New Drug Development. CRC Press, pp. 187–206.

Bala, S., Garg, D., Sridhar, K., Inbaraj, B.S., Singh, R., Kamma, S., Tripathi, M., Sharma, M., 2023. Transformation of agro-waste into value-added bioproducts and bioactive compounds: micro/nano formulations and application in the agri-food-pharma sector. Bioengineering 10 (2). https://doi.org/10.3390/bioengineering10020152.

Balakrishnan, S., Chandrasekaran, I.R., Jegadeesan, R., Ramachandran, K., Govindasamy, A. K., Kandasamy, G., 2025. Eco-innovation in action: valorizing vegetable waste into high-value ingredients for sustainable applications. Front. Sustain. Food Syst. 9. https://doi.org/10.3389/fsufs.2025.1507120.

Bashar, A., Hasan, N.A., Haque, M.M., Rohani, M.F., Hossain, M.S., 2021. Effects of dietary silica nanoparticle on growth performance, protein digestibility, hematology, digestive morphology, and muscle composition of Nile tilapia. Front. Mar. Sci. 8. https://doi.org/10.3389/fmars.2021.706179.

Belenguer, Á., Naya-Català, F., Calduch-Giner, J.À., Pérez-Sánchez, J., 2024. Exploring multifunctional markers of biological age in farmed gilthead sea bream (Sparus aurata): a transcriptomic and epigenetic interplay for an improved fish welfare assessment approach. Int. J. Mol. Sci. 25 (18), 9836. https://doi.org/10.3390/ijms25189836.

Bera, P., Dhara, H., Mandal, S., Sau, S.K., Das, S., 2025. Nutritional Strategies for Improving Reproductive Performance in Fishes. Springer Science and Business Media LLC, pp. 409–430. https://doi.org/10.1007/978-3-031-92858-1_20.

Bertocci, F., Mannino, G., 2022. Can agri-food waste Be a sustainable alternative in aquaculture? A bibliometric and meta-analytic study on growth performance, innate immune system, and antioxidant defenses. Foods 11 (13), 1861. https://doi.org/10.3390/foods11131861.

Bertucci, J.I., 2023. Environmental regulation of feeding, growth, and reproduction in fish: influence of nutrition and physical parameters on the endocrine system. Front. Endocrinol. 14, 1281959.

Bibi, F., Ilyas, N., Saeed, M., Shabir, S., Shati, A.A., Alfaifi, M.Y., Amesho, K.T.T., Chowdhury, S., Sayyed, R.Z., 2023. Innovative production of value-added products using agro-industrial wastes via solid-state fermentation. Environ. Sci. Pollut. Control Ser. 30 (60), 125197–125213. https://doi.org/10.1007/s11356-023-28765-6.

Boyd, C.E., D'Abramo, L.R., Glencross, B.D., Huyben, D.C., Juarez, L.M., Lockwood, G.S., McNevin, A.A., Tacon, A.G.J., Teletchea, F., Tomasso, J.R., Tucker, C.S., Valenti, W.C., 2020. Achieving sustainable aquaculture: historical and current perspectives and future needs and challenges. J. World Aquacult. Soc. 51 (3), 578–633. https://doi.org/10.1111/jwas.12714.

Cao, Y., Xu, M., Lu, J., Cai, G., 2024. Simultaneous microbial fermentation and enzymolysis: a biotechnology strategy to improve the nutritional and functional quality of soybean meal. Food Rev. Int. 40 (5), 1296–1311. https://doi.org/10.1080/87559129.2023.2212048.

Chowdhury, S., Saikia, S.K., 2020. Oxidative stress in fish: a review. J. Sci. Res. 12 (1), 145–160. https://doi.org/10.3329/jsr.v12i1.41716.

Chuang, W.Y., Lin, L.J., Shih, H.D., Shy, Y.M., Chang, S.C., Lee, T.T., 2021. The potential utilization of high-fiber agricultural by-products as monogastric animal feed and feed additives: a review. Animals 11 (7). https://doi.org/10.3390/ani11072098.

Ciji, A., Akhtar, M.S., 2021. Stress management in aquaculture: a review of dietary interventions. Rev. Aquacult. 13 (4), 2190–2247. https://doi.org/10.1111/raq.12565.

Colombo, S.M., Roy, K., Mraz, J., Wan, A.H.L., Davies, S.J., Tibbetts, S.M., Øverland, M., Francis, D.S., Rocker, M.M., Gasco, L., Spencer, E., Metian, M., Trushenski, J.T., Turchini, G.M., 2023. Towards achieving circularity and sustainability in feeds for farmed blue foods. Rev. Aquacult. 15 (3), 1115–1141. https://doi.org/10.1111/raq.12766.

Dara, M., Carbonara, P., La Corte, C., Parrinello, D., Cammarata, M., Parisi, M.G., 2023. Fish welfare in aquaculture: physiological and immunological activities for diets, social and spatial stress on mediterranean aqua cultured species. Fishes 8 (8). https://doi.org/10.3390/fishes8080414.

Davison, C., Michie, C., Tachtatzis, C., Andonovic, I., Bowen, J., Duthie, C.A., 2023. Feed conversion ratio (FCR) and performance group estimation based on predicted feed intake for the optimisation of beef production. Sensors 23 (10). https://doi.org/10.3390/s23104621.

Dawood, M.A.O., 2021. Nutritional immunity of fish intestines: important insights for sustainable aquaculture. Rev. Aquacult. 13 (1), 642–663. https://doi.org/10.1111/raq.12492.

De Marco, G., Cappello, T., Maisano, M., 2023. Histomorphological changes in fish gut in response to prebiotics and probiotics treatment to improve their health status: a review. Animals 13 (18), 2860.

Debbarma, S., Deb, S., Yadav, N.K., Kashyap, S., Debbarma, P., Chrisolite, B., Yirang, Y., Meena, D.K., Singh, S.K., Choudhury, T.G., Patel, A.B., 2025. Waste not, want not: unlocking the innovative potential of organic and eco-friendly insect and algal resources for future aquaculture. Aquac. Int. 33 (2). https://doi.org/10.1007/s10499-024-01814-8.

Dhama, K., Latheef, S.K., Dadar, M., Samad, H.A., Munjal, A., Khandia, R., Karthik, K., Tiwari, R., Yatoo, M.I., Bhatt, P., Chakraborty, S., Singh, K.P., Iqbal, H.M.N., Chaicumpa, W., Joshi, S.K., 2019. Biomarkers in stress related diseases/disorders: diagnostic, prognostic, and therapeutic values. Front. Mol. Biosci. 6. https://doi.org/10.3389/fmolb.2019.00091.

Dimina, L., Rémond, D., Huneau, J.F., Mariotti, F., 2022. Combining plant proteins to achieve amino acid profiles adapted to various nutritional objectives—an exploratory analysis using linear programming. Front. Nutr. 8. https://doi.org/10.3389/fnut.2021.809685.

Dittoe, D.K., Ricke, S.C., Kiess, A.S., 2018. Organic acids and potential for modifying the avian gastrointestinal tract and reducing pathogens and disease. Front. Vet. Sci. 5, 216.

Doolam, B., Mishra, B., Surabhi, D., Mandal, S.K., Sada, S., Reddy, N.R., Panda, J., Rustagi, S., Mishra, A.K., Mohanta, Y.K., 2025. A systematic review of potential bioactive compounds from Saccharomyces cerevisiae: exploring their applications in health promotion and food development. Environ. Dev. Sustain. 27 (2), 2945–2982. https://doi.org/10.1007/s10668-024-04969-9.

Eed Desouky, H., Mahmoud Sayed, N., Prudence Abasubong, K., Zhang, Z., 2025. Nutritional and physiological effects of high-fat diets in finfish: effects on growth, immunity, lipid metabolism, and intestinal health: a review. J. Comp. Physiol. B. https://doi.org/10.1007/s00360-025-01626-z.

Engdaw, F., Geremew, A., 2024. Broodstock nutrition in Nile tilapia and its implications on reproductive efficiency. Front. Aquac. 3. https://doi.org/10.3389/faquc.2024.1281640.

Fassarella, M., Blaak, E.E., Penders, J., Nauta, A., Smidt, H., Zoetendal, E.G., 2021. Gut microbiome stability and resilience: elucidating the response to perturbations in order to modulate gut health. Gut 70 (3), 595–605. https://doi.org/10.1136/gutjnl-2020-321747.

Fath El-Bab, A.F., Majrashi, K.A., Sheikh, H.M., Shafi, M.E., El-Ratel, I.T., Neamat-Allah, A.N.F., El-Raghi, A.A., Elazem, A.Y.A., Abd-Elghany, M.F., Abdelnour, S.A., Abduh, M.S., Jaremko, M., Naiel, M.A.E., 2022. Dietary supplementation of Nile tilapia (Oreochromis niloticus) with β-glucan and/or Bacillus coagulans: synergistic impacts on performance, immune responses, redox status and expression of some related genes. Front. Vet. Sci. 9. https://doi.org/10.3389/fvets.2022.1011715.

Francis, G., Makkar, H.P., Becker, K., 2001. Antinutritional factors present in plant-derived alternate fish feed ingredients and their effects in fish. Aquaculture 199 (3–4), 197–227.

Ghafarifarsani, H., Fazle Rohani, M., Raeeszadeh, M., Ahani, S., Yousefi, M., Talebi, M., Sazzad Hossain, M., 2024. pesticides and heavy metal toxicity in fish and possible remediation—a review. Ann. Anim. Sci. 24 (4), 1007–1024. https://doi.org/10.2478/aoas-2024-0012.

Glencross, B.D., Bachis, E., Betancor, M.B., Calder, P., Liland, N., Newton, R., Ruyter, B., 2025. Omega-3 futures in aquaculture: exploring the supply and demands for long-chain omega-3 essential fatty acids by aquaculture species. Rev. Fish. Sci. Aquacult. 33 (2), 167–216. https://doi.org/10.1080/23308249.2024.2388563.

Gopalraaj, J., Velayudhannair, K., Arockiasamy, J.P., Radhakrishnan, D.K., 2024. The effect of dietary supplementation of proteases on growth, digestive enzymes, oxidative stress, and intestinal morphology in fishes—a review. Aquac. Int. 32 (1), 745–765. https://doi.org/10.1007/s10499-023-01191-8.

Grayfer, L., Kerimoglu, B., Yaparla, A., Hodgkinson, J.W., Xie, J., Belosevic, M., 2018. Mechanisms of fish macrophage antimicrobial immunity. Front. Immunol. 9. https://doi.org/10.3389/fimmu.2018.01105.

Hăbeanu, M., Gheorghe, A., Dinita, G., Mihalcea, T., 2024. An in-depth insight into the profile, mechanisms, functions, and transfer of essential amino acids from mulberry leaves to silkworm Bombyx mori L. pupae and fish. Insects 15 (5), 332. https://doi.org/10.3390/insects15050332.

Hadi, J., Brightwell, G., 2021. Safety of alternative proteins: technological, environmental and regulatory aspects of cultured meat, plant based meat, insect protein and single-cell protein. Foods 10 (6), 1226. https://doi.org/10.3390/foods10061226.

Hissen, K.L., He, W., Wu, G., Criscitiello, M.F., 2023. Immunonutrition: facilitating mucosal immune response in teleost intestine with amino acids through oxidant-antioxidant balance. Front. Immunol. 14. https://doi.org/10.3389/fimmu.2023.1241615.

Hofmann, L.C., Strauss, S., Shpigel, M., Guttman, L., Stengel, D.B., Rebours, C., Gjorgovska, N., Turan, G., Balina, K., Zammit, G., Adams, J.M.M., Ahsan, U., Bartolo, A.G., Bolton, J.J., Domingues, R., Dürrani, Ö., Eroldogan, O.T., Freitas, A., Golberg, A., Kremer, K.I., Marques, F., Milia, M., Steinhagen, S., Sucu, E., Vargas-Murga, L., Zemah-Shamir, S., Zemah-Shamir, Z., Meléndez-Martínez, A.J., 2025. The green seaweed Ulva: tomorrow's "wheat of the sea" in foods, feeds, nutrition, and biomaterials. Crit. Rev. Food Sci. Nutr. 65 (19), 3728–3763. https://doi.org/10.1080/10408398.2024.2370489.

Hoseinifar, S.H., Yousefi, S., Van Doan, H., Ashouri, G., Gioacchini, G., Maradonna, F., Carnevali, O., 2020. Oxidative stress and antioxidant defense in fish: the implications of probiotic, prebiotic, and synbiotics. Rev. Fish. Sci. Aquacult. 29 (2), 198–217. https://doi.org/10.1080/23308249.2020.1795616.

Hua, K., Cobcroft, J.M., Cole, A., Condon, K., Jerry, D.R., Mangott, A., Praeger, C., Vucko, M.J., Zeng, C., Zenger, K., Strugnell, J.M., 2019. The future of aquatic protein: implications for protein sources in aquaculture diets. One Earth 1 (3), 316–329. https://doi.org/10.1016/j.oneear.2019.10.018.

Inbaraj, S., Sejian, V., Ramasamy, S., 2022. Role of environmental stressor-host immune system–pathogen interactions in development of infectious disease in farm animals. Biol. Rhythm Res. 53 (5), 707–724.

Irakli, M., Lazaridou, A., Biliaderis, C.G., 2020. Comparative evaluation of the nutritional, antinutritional, functional, and bioactivity attributes of rice bran stabilized by different heat treatments. Foods 10 (1), 57.

Jackson, V., Penumetcha, M., 2019. Dietary oxidised lipids, health consequences and novel food technologies that thwart food lipid oxidation: an update. Int. J. Food Sci. Technol. 54 (6), 1981–1988. https://doi.org/10.1111/ijfs.14058.

Jannathulla, R., Rajaram, V., Kalanjiam, R., Ambasankar, K., Muralidhar, M., Dayal, J.S., 2019. Fishmeal availability in the scenarios of climate change: inevitability of fishmeal replacement in aquafeeds and approaches for the utilization of plant protein sources. Aquac. Res. 50 (12), 3493–3506. https://doi.org/10.1111/are.14324.

Javed, A., Mahmood, T., Tiwari, R., Ahsan, F., Ansari, V.A., Bano, S., Uzzam Khan, M.M., Khan, A., 2025. Navigating nephropathy and nephrotoxicity: understanding pathophysiology unveiling clinical manifestations, and exploring treatment approaches. J. Basic Clin. Physiol. Pharmacol. 36 (2–3), 69–93. https://doi.org/10.1515/jbcpp-2024-0220.

Kakwi, D.G., Awolumate, S., Audu, B.S., Oshinbajo, D.O., Oyedapo, F.A., Mohammed, M.Y., Ositimehin, K.M., 2021. Some aspects of histology of liver and intestine of Cyprinus carpio fingerlings fed raw and processed Mucuna pruriens diets. Niger. J. Anim. Prod. 48 (3), 50–62. https://doi.org/10.51791/njap.v48i3.2968.

Kamboj, N., Kumar, R., Mitra, D., 2025. Immunomodulation in autoimmune disorders: harnessing the power of stem cell. Stem Cell Ther. 409–462.

Kim, S.W., Less, J.F., Wang, L., Yan, T., Kiron, V., Kaushik, S.J., Lei, X.G., 2019. Meeting global feed protein demand: challenge, opportunity, and strategy. Ann. Rev. Anim. Biosci. 7, 221–243. https://doi.org/10.1146/annurev-animal-030117-014838.

Kitessa, S.M., Abeywardena, M., Wijesundera, C., Nichols, P.D., 2014. DHA-containing oilseed: a timely solution for the sustainability issues surrounding fish oil sources of the health-benefitting long-chain omega-3 oils. Nutrients 6 (5), 2035–2058. https://doi.org/10.3390/nu6052035.

Kopylov, A.T., Malsagova, K.A., Stepanov, A.A., Kaysheva, A.L., 2021. Diversity of plant sterols metabolism: the impact on human health, sport, and accumulation of contaminating sterols. Nutrients 13 (5). https://doi.org/10.3390/nu13051623.

Kucuksezer, U.C., Ozdemir, C., Yazici, D., Pat, Y., Mitamura, Y., Li, M., Sun, N., D'Avino, P., Bu, X., Zhu, X., Akdis, M., Nadeau, K., Ogulur, I., Akdis, C.A., 2023. The epithelial barrier theory: development and exacerbation of allergic and other chronic inflammatory diseases. Asia Pac. Allergy 13 (1), 28–39. https://doi.org/10.5415/apallergy.0000000000000005.

Kumar, R., Kamboj, N., Mitra, D., 2024. Microbial innovations in agriculture: interdisciplinary approaches to leveraging microbes for food sustainability and security. Indian J. Microbiol. Res. 11 (3), 129–139.

Kumar, R., Kamboj, N., Mitra, D., Ray, A., 2025a. Strengthening local communities through advanced recycling technology for sustainable growth. In: Mitra, D., Choudhury, T., Madan, A., Chattaraj, S., Pellegrini, M. (Eds.), AI technologies for enhancing recycling processes. IGI Global Scientific Publishing, pp. 393–420.

Kumar, R., Kamboj, N., Mitra, D., Verma, D., Choudhury, T., 2025b. Bacterial strategies: novel pathways for sustainable development through wastewater. In: Mitra, D., Choudhury, T., Madan, A., Chattaraj, S., Pellegrini, M. (Eds.), AI technologies for enhancing recycling processes. IGI Global Scientific Publishing, pp. 421–440.

Lackner, M., Besharati, M., 2025. Agricultural waste: challenges and solutions, a review. Waste 3 (2), 18. https://doi.org/10.3390/waste3020018.

Lall, S.P., Kaushik, S.J., 2021. Nutrition and metabolism of minerals in fish. Animals 11 (9). https://doi.org/10.3390/ani11092711.

Langston, W.J., 2020. Endocrine disruption and altered sexual development in aquatic organisms: an invertebrate perspective. J. Mar. Biol. Assoc. U. K. 100 (4), 495–515.

Langyan, S., Yadava, P., Khan, F.N., Dar, Z.A., Singh, R., Kumar, A., 2022. Sustaining protein nutrition through plant-based foods. Front. Nutr. 8. https://doi.org/10.3389/fnut.2021.772573.

Li, X., Zheng, S., Wu, G., 2021. Nutrition and functions of amino acids in fish. In: Amino acids in nutrition and health: amino acids in the nutrition of companion, zoo and farm animals. Springer International Publishing, pp. 133–168.

Ma, S., Wang, H., Dou, Y., Liang, X., Zheng, Y., Wu, X., Xue, M., 2022. Anti-nutritional factors and protein dispersibility index as principal quality indicators for soybean meal in diet of Nile tilapia (Oreochromis niloticus GIFT), a meta-analysis. Animals 12 (14), 1831. https://doi.org/10.3390/ani12141831.

Maksimenko, A., Belyi, L., Podvolotskaya, A., Son, O., Tekutyeva, L., 2024. Exploring sustainable aquafeed alternatives with a specific focus on the ensilaging technology of fish waste. Fermentation 10 (5), 258. https://doi.org/10.3390/fermentation10050258.

Malik, M.A., Sardar, P., Munilkumar, S., Varghese, T., Bhat, N.M., Dar, S., Kumar, S., 2025. Comparative evaluation of microbial fermentation on the nutritional and anti-nutritional

profile of Eichhornia crassipes leaf meal. Aquac. Int. 33 (4). https://doi.org/10.1007/s10499-025-01990-1.
Manam, D.V.K., 2023. Fish feed nutrition and its management in aquaculture. Int. J. Fish. Aquat. Stud. 11 (2), 58–61. https://doi.org/10.22271/fish.2023.v11.i2a.2791.
Manibusan, M.K., Touart, L.W., 2017. A comprehensive review of regulatory test methods for endocrine adverse health effects. Crit. Rev. Toxicol. 47 (6), 440–488. https://doi.org/10.1080/10408444.2016.1272095.
Mariamenatu, A.H., Abdu, E.M., 2021. Overconsumption of omega-6 polyunsaturated fatty acids (PUFAs) versus deficiency of Omega-3 PUFAs in modern-day diets: the disturbing factor for their "balanced antagonistic metabolic functions" in the human body. J. Lipids 2021 (1), 8848161.
Marimuthu, V., Sarawagi, A.D., Kumar, A., Paul, S., Sampath, V., Issara, U., Al-Dhabi, N.A., Arasu, M.V., Balasubramanian, B., Sureshkumar, S., 2022. Glimpse of Feed and Feed Additive Necessity and Mycotoxin Challenges in Aquaculture Aquaculture Science and Engineering. Springer Nature, pp. 401–430. https://doi.org/10.1007/978-981-19-0817-0_15.
Mititelu, M., Lupuliasa, D., Neacsu, S.M., Olteanu, G., Busnatu, Ș.S., Mihai, A., Scafa-Udriste, A., 2024. Polyunsaturated fatty acids and human health: a key to modern nutritional balance in association with polyphenolic compounds from food sources. Foods 14 (1), 46.
Mitra, A., Abdel-Gawad, F.K., Bassem, S., Barua, P., Assisi, L., Parisi, C., Temraz, T.A., Vangone, R., Kajbaf, K., Kumar, V., Guerriero, G., 2023. Climate change and reproductive biocomplexity in fishes: innovative management approaches towards sustainability of fisheries and aquaculture. Water 15 (4), 725. https://doi.org/10.3390/w15040725.
Moita, V.H.C., Kim, S.W., 2022. Nutritional and functional roles of phytase and xylanase enhancing the intestinal health and growth of nursery pigs and broiler chickens. Animals 12 (23). https://doi.org/10.3390/ani12233322.
Munguti, J., Odame, H., Kirimi, J., Obiero, K., Ogello, E., Liti, D., 2021. Fish feeds and feed management practices in the Kenyan aquaculture sector: challenges and opportunities. Aquat. Ecosyst. Health Manag. 1, 82–89. https://doi.org/10.14321/aehm.024.01.12.
Naiel, M.A.E., Negm, S.S., Ghazanfar, S., Shukry, M., Abdelnour, S.A., 2023. The risk assessment of high-fat diet in farmed fish and its mitigation approaches: a review. J. Anim. Physiol. Anim. Nutr. 107 (3), 948–969. https://doi.org/10.1111/jpn.13759.
Naya-Català, F., do Vale Pereira, G., Piazzon, M.C., Fernandes, A.M., Calduch-Giner, J.A., Sitjà-Bobadilla, A., Conceição, L.E.C., Pérez-Sánchez, J., 2021. Cross-talk between intestinal microbiota and host gene expression in gilthead sea bream (Sparus aurata) juveniles: insights in fish feeds for increased circularity and resource utilization. Front. Physiol. 12. https://doi.org/10.3389/fphys.2021.748265.
Ndudzo, A., Pullen, J., Magwaba, T., Ndlovu, S., Moyo, M., Sibanda, S., Mugoti, A., 2023. Incorporation of functional feed ingredients to substitute antimicrobials in animal nutrition: opportunities for livestock production in developing countries. Int. J. Livest. Prod. 14, 44–57.
Peixoto, D., Carvalho, I., Machado, M., Aragão, C., Costas, B., Azeredo, R., 2024. Dietary tryptophan intervention counteracts stress-induced transcriptional changes in a teleost fish HPI axis during inflammation. Sci. Rep. 14 (1). https://doi.org/10.1038/s41598-024-57761-0.
Petersen, K.S., Maki, K.C., Calder, P.C., Belury, M.A., Messina, M., Kirkpatrick, C.F., Harris, W.S., 2024. Perspective on the health effects of unsaturated fatty acids and

commonly consumed plant oils high in unsaturated fat. Br. J. Nutr. https://doi.org/10.1017/S0007114524002459.

Ponnampalam, E.N., Kiani, A., Santhiravel, S., Holman, B.W.B., Lauridsen, C., Dunshea, F. R., 2022. The importance of dietary antioxidants on oxidative stress, meat and milk production, and their preservative aspects in farm animals: antioxidant action, animal health, and product quality—invited review. Animals 12 (23). https://doi.org/10.3390/ani12233279.

Puri, V., Nagpal, M., Singh, I., Singh, M., Dhingra, G.A., Huanbutta, K., Dheer, D., Sharma, A., Sangnim, T., 2022. A comprehensive review on nutraceuticals: therapy support and formulation challenges. Nutrients 14 (21). https://doi.org/10.3390/nu14214637.

Ravi, H.K., Degrou, A., Costil, J., Trespeuch, C., Chemat, F., Vian, M.A., 2020. Larvae mediated valorization of industrial, agriculture and food wastes: biorefinery concept through bioconversion, processes, procedures, and products. Processes 8 (7). https://doi.org/10.3390/PR8070857.

Reda, R.M., Nasr, M.A.F., Ismail, T.A., Moustafa, A., 2021. Immunological responses and the antioxidant status in african catfish (Clarias gariepinus) following replacement of dietary fish meal with plant protein. Animals 11 (5). https://doi.org/10.3390/ani11051223.

Rudra, S.G., Singh, A., Pal, P., Thakur, R.K., 2023. Antinutritional Factors in Lentils: Their Effect on Bioavailability of Nutrients and Significance in Human Health Lentils: Production, Processing Technologies, Products, and Nutritional Profile. wiley, pp. 339–364.

Salim, R., Nehvi, I.B., Mir, R.A., Tyagi, A., Ali, S., Bhat, O.M., 2023. A review on antinutritional factors: unraveling the natural gateways to human health. Front. Nutr. 10. https://doi.org/10.3389/fnut.2023.1215873.

Salter, A.M., Lopez-Viso, C., 2021. Role of novel protein sources in sustainably meeting future global requirements. Proc. Nutr. Soc. 80 (2), 186–194. https://doi.org/10.1017/S0029665121000513.

Sharma, A., Goel, H., Sharma, S., Rathore, H.S., Jamir, I., Kumar, A., Thimmappa, S.C., Kesari, K.K., Kashyap, B.K., 2024. Cutting edge technology for wastewater treatment using smart nanomaterials: recent trends and futuristic advancements. Environ. Sci. Pollut. Control Ser. 31 (48), 58263–58293. https://doi.org/10.1007/s11356-024-34977-1.

Shastak, Y., Pelletier, W., Colors, C., Roles, C., 2023. Astaxanthin's antioxidant impact on fish oxidative stress and reproductive performance. Animals 13 (21), 3357. https://doi.org/10.3390/ani13213357.

Shea, Z., Ogando do Granja, M., Fletcher, E.B., Zheng, Y., Bewick, P., Wang, Z., Singer, W. M., Zhang, B., 2024. A review of bioactive compound effects from primary legume protein sources in human and animal health. Curr. Issues Mol. Biol. 46 (5), 4203–4233. https://doi.org/10.3390/cimb46050257.

Siddik, M.A.B., Julien, B.B., Islam, S.M.M., Francis, D.S., 2024. Fermentation in aquafeed processing: achieving sustainability in feeds for global aquaculture production. Rev. Aquacult. 16 (3), 1244–1265. https://doi.org/10.1111/raq.12894.

Singh, A.K., Tiwari, U.P., Berrocoso, J.D., Dersjant-Li, Y., Awati, A., Jha, R., 2019. Effects of a combination of xylanase, amylase and protease, and probiotics on major nutrients including amino acids and non-starch polysaccharides utilization in broilers fed different level of fibers. Poult. Sci. 98 (11), 5571–5581. https://doi.org/10.3382/ps/pez310.

Singh, S., Kola, P., Kaur, D., Singla, G., Mishra, V., Panesar, P.S., Mallikarjunan, K., Krishania, M., 2021. Therapeutic potential of nutraceuticals and dietary supplements in the prevention of viral diseases: a review. Front. Nutr. 8. https://doi.org/10.3389/fnut.2021.679312.

Sinha, A.K., Rakesh, S., Mitra, B., Roy, N., Sahoo, S., Saha, B.N., Dutta, S., Bhattacharya, P. M., 2022. Agricultural Waste Management Policies and Programme for Environment and Nutritional Security Input Use Efficiency for Food and Environmental Security. Wiley, pp. 627–664. https://doi.org/10.1007/978-981-16-5199-1_21.

Solovyev, M., Kashinskaya, E., Gisbert, E., 2023. A meta-analysis for assessing the contributions of trypsin and chymotrypsin as the two major endoproteases in protein hydrolysis in fish intestine. Comp. Biochem. Physiol. Mol. Integr. Physiol. 278, 111372.

Song, H., Zhang, S., Yang, B., Liu, Y., Kang, Y., Li, Y., Qian, A., Yuan, Z., Cong, B., Shan, X., 2022. Effects of four different adjuvants separately combined with Aeromonas veronii inactivated vaccine on haematoimmunological state, enzymatic activity, inflammatory response and disease resistance in crucian carp. Fish Shellfish Immunol. 120, 658–673. https://doi.org/10.1016/j.fsi.2021.09.003.

Soukoulis, C., Bohn, T., 2018. A comprehensive overview on the micro- and nano-technological encapsulation advances for enhancing the chemical stability and bioavailability of carotenoids. Crit. Rev. Food Sci. Nutr. 58 (1), 1–36. https://doi.org/10.1080/10408398.2014.971353.

Steinberg, C.E.W., 2022. Carbohydrate Preference and Metabolism—"Life's Little Luxury Digested. Springer Science and Business Media LLC, pp. 393–434. https://doi.org/10.1007/978-3-030-87227-4_18.

Takvam, M., Wood, C.M., Kryvi, H., Nilsen, T.O., 2023. Role of the kidneys in acid-base regulation and ammonia excretion in freshwater and seawater fish: implications for nephrocalcinosis. Front. Physiol. 14. https://doi.org/10.3389/fphys.2023.1226068.

Thazeem, B., Umesh, M., Sarojini, S., Allwyn Vyas, G., Adhithya Sankar, S., Sapthami, K., Suresh, S., Stanly, L.M., 2022. Developments in Feeds in Aquaculture Sector: Contemporary Aspects Aquaculture Science and Engineering. Springer Nature, pp. 35–78. https://doi.org/10.1007/978-981-19-0817-0_2.

Topić Popović, N., Čižmek, L., Babić, S., Strunjak-Perović, I., Čož-Rakovac, R., 2023. Fish liver damage related to the wastewater treatment plant effluents. Environ. Sci. Pollut. Control Ser. 30 (17), 48739–48768. https://doi.org/10.1007/s11356-023-26187-y.

Trushenski, J.T., Rombenso, A.N., 2020. Trophic levels predict the nutritional essentiality of polyunsaturated fatty acids in fish—introduction to a special section and a brief synthesis. N. Am. J. Aquac. 82 (3), 241–250. https://doi.org/10.1002/naaq.10137.

Venditti, P., Di Meo, S., 2020. The role of reactive oxygen species in the life cycle of the mitochondrion. Int. J. Mol. Sci. 21 (6). https://doi.org/10.3390/ijms21062173.

Vujetić, J., Spasevski, N., Dragojlović, D., 2025. Processing techniques of removing antinutrients from oilseed cakes as by-products intended for animal feeding. Food Feed Res. 52 (1), 37–51.

Wang, S., Xu, G., Zou, J., 2024. Soluble non-starch polysaccharides in fish feed: implications for fish metabolism. Fish Physiol. Biochem. 50 (1), 1–22.

Wilfart, A., Garcia-Launay, F., Terrier, F., Soudé, E., Aguirre, P., Skiba-Cassy, S., 2023. A step towards sustainable aquaculture: multiobjective feed formulation reduces environmental impacts at feed and farm levels for rainbow trout. Aquaculture 562, 738826. https://doi.org/10.1016/j.aquaculture.2022.738826.

Yan, K., Guo, F., Kainz, M.J., Li, F., Gao, W., Bunn, S.E., Zhang, Y., 2024. The importance of omega-3 polyunsaturated fatty acids as high-quality food in freshwater ecosystems with implications of global change. Biol. Rev. 99 (1), 200–218. https://doi.org/10.1111/brv.13017.

Yan, Z., Liu, Z., Zhou, C., Tan, Z., 2025. Anti-nutritional factors of plant protein feeds for ruminants and methods for their elimination. Animals 15 (8), 1107. https://doi.org/10.3390/ani15081107.

Yavuzcan Yildiz, H., Robaina, L., Pirhonen, J., Mente, E., Domínguez, D., Parisi, G., 2017. Fish welfare in aquaponic systems: its relation to water quality with an emphasis on feed and faeces—a review. Water 9 (1), 13. https://doi.org/10.3390/w9010013.

Zhang, H., Ran, C., Teame, T., Ding, Q., Hoseinifar, S.H., Xie, M., Zhang, Z., Yang, Y., Olsen, R.E., Gatlin, D.M., Ringø, E., Duan, M., Zhou, Z., 2020. Research progress on gut health of farmers teleost fish: a viewpoint concerning the intestinal mucosal barrier and the impact of its damage. Rev. Fish Biol. Fish. 30 (4), 569–586. https://doi.org/10.1007/s11160-020-09614-y.

Zhuang, J., Zhu, J., Cheung, P.C.K., Li, C., 2024. The physical and chemical interactions between starch and dietary fiber: their impact on the physicochemical and nutritional properties of starch. Trends Food Sci. Technol. 149, 104566. https://doi.org/10.1016/j.tifs.2024.104566.

Zhukova, K., Stroganov, A.N., 2022. Anatomy of the digestive system of lumpfish (Cyclopterus lumpus) as an adaptation to puffing behavior. Anat. Rec. 305 (7), 1732–1738. https://doi.org/10.1002/ar.24799.

CHAPTER

18 Bioeconomic and environmental implications of utilizing nonconventional feed ingredients in aquaculture

Manisha Dash[1], Lopamudra Subudhi[2], Abhisek Das[1], Prasanta Rath[1], Alok Kumar Panda[1] and Shibani Mohapatra[2]

[1]*Environmental Science Laboratory, School of Applied Sciences, Kalinga Institute of Industrial Technology (Deemed to Be University), Bhubaneswar, Odisha, India;* [2]*Center for Biotechnology, Siksha 'O' Anusandhan (Deemed to Be University), Bhubaneswar, Odisha, India*

1. Introduction

Aquaculture is the largest food production sector, and could play an important role in future sustainable food production. The aquaculture industry grew in 2018 by 114.5 million metric tons of live weight production. Aquaculture is critical for food security, defined here as providing quality food for the increasing global population. The consumer market for seafood is growing at a rate of more than 3.1% per year, even though the global population is growing at a rate of 1.6% per year. The aquaculture sector is increasing and growing faster at a rate of 2.1% per year when compared to other animal producing sectors, like livestock. This might be due to the fact that production is more profitable for fishers and aquaculture than marketing or selling (Azra et al., 2021). Although there are undoubtedly a variety of beneficial opportunities for aquaculture to supply food, its sustainable development opportunities depend on many variables, including knowledge, experience, regulations, ecological conditions for cultured organisms growth, and the presence of large marketplaces for the sale of aquaculture products (Tacon, 2020). At present, 50% of world fish supplies are produced through aquaculture and number one as a high-quality source of protein supply and the number two source of protein by weight. Terrestrial and freshwater organisms supply approximately 43% of protein in total proliferation of proteins and prevent undernourishment and malnutrition in varying populations, especially in developing countries. Aquaculture supplies, like other forms of terrestrial agriculture, are completely reliant on nutrient availability to support feed requirements (FAO, 2020). The situation has also been made worse by the expanding human population and rising demand for animal products. Given the circumstances described above, it is critical to identify and utilize nonconventional feed resources (NCFRs)

Fish Feed Formulation with Agroindustrial Wastes and Byproducts. https://doi.org/10.1016/B978-0-443-41728-3.00004-6

in the future in order to feed millions of animals and ensure their food and nutritional security (Chattaraj et al., 2024, 2026a). By encouraging the use of alternate fodder sources, traditional agricultural practices would be diversified and sustainable use of natural resources would help conserve biodiversity. To date the use of nonconventional forage resources in livestock feeding has been limited due to poor information pertaining to their nutrient potential (Katoch et al., 2018).

2. The role of feed in aquaculture sustainability

2.1 Conventional feed ingredients

2.1.1 Fish meal

Feed is the largest input in aquaculture production. As such, an expansion of aquaculture practices to achieve the expected fish production must be supported by the expansion of the aqua-feed industry. Fish meal, a traditional source of protein in aqua feeds, has become widely regarded as the main dietary protein source in aqua feed production (Chattaraj and Das Mohapatra, 2024; Ganguly et al., 2025). However, the recent decrease in supply and high demand for fish meal have resulted in increased prices, and total reliance on fishmeal should not be assumed. The higher price and limited supply of fish meal have motivated researchers and feed manufacturers to look for protein replacements that would allow for more sustainable aqua feeds. In this context, plant-protein sources are widely available and are generally available at lower market prices than fish meal (Agboola et al., 2021; Yakubu et al., 2021). While plant-protein ingredients are high in protein, the antinutritional factors (ANFs) in the plant-protein ingredients reduce their use in aqua feed. Anti-nutritional components have an impact on the digestion and utilization of nutrients from cultured species. In general, aqua feed suppliers are most concerned with the objective of neutralizing or destroying the anti-nutritional substances phytic acid and trypsin inhibitors to effectively feed the ingredients to fish. More specifically, some of the proteins derived from plants, such as soybean meal, are already being used in aqua feeds to partially replace fish meal protein sources. In addition to soybean meal, all of these have a large potential to replace either a part of all traditional fish meal protein sources for fish. Traditional or conventional fish meal protein sources in aqua feeds are typically replaced with corn gluten meal, palm kernel meal, cottonseed meal, canola meal, jatropha kernel meal, and peanut meal (Chakraborty et al., 2019).

2.1.2 Soybean meal

A major purpose behind opting for processed soy products, such as soy protein concentrate, is the presence of saponins and other ANFs that would be found in conventional soybean meal, which can lead to distal intestine enteritis and slow growth rates in fish like Atlantic salmon and rainbow trout. The move toward plant-based ingredients provokes ethical and sustainability discussions (Agboola et al., 2021; Yakubu et al., 2021). The enhancement of plant-based materials in aquafeeds will likely put pressure on land and freshwater use, increased energy, decreased ecosystem and forest biodiversity, and potentially in crop production as an issue. The most important point is that the increase in soy protein and other plant materials to aquaculture reduces levels of direct human consumption. Thus, we see an increased desire for alternate and sustainable novel feed ingredients for aquaculture (Fry et al., 2016).

3. Nonconventional feed ingredients

3.1 Agro-industrial by products

Several factors have influenced some researchers to test the efficiency of various plant products treated as partial and total replacers of Fish meal in aquaculture feed, namely, their easy accessibility, availability, and what is perceived to be protein quality. Typical plant by-products used as a contributor of protein to fish feed are mulberry leaves, water hyacinth, duck weed, legumes, cassava leaves, barn yard grass, and so on (Mondal et al., 2012). Among these, nutrient abundant mulberry leaves are being used recently as part of a fish feed due to substantial amounts produced as waste from sericulture industries. Also, cauliflower leaves, another nutrient rich product generated in retail markets by the Indian subcontinent, are largely ignored and disposed of with little implications for recycling. However, the use of this plant resource in fish diet has not been undertaken to a large extent. Although intense research effort during the last two decades has well-documented the feasibility to replace FM protein with plant protein (PP) in fish feed formulation, and thus potentially improve sustainability, predictability and access, with cost effectiveness in the feed provision (Samaddar, 2018).

4. Insect meals

The approximate one million insect species in the world are important for the stability of food chains and ecosystems (de Castro et al., 2018). The potential application of insect meal (IM) in fish feed for aquaculture has grown significantly. Since they are a rich supply of both macro and micronutrients, insects can be a substantial source of energy. As a result, they are the perfect meal for fish. When we look at the nutritional value of different insect orders in comparison to fishmeal and soybean meal, we find that insects have less crude protein than fishmeal but the right amount of crude protein. Fishmeal has a good amino acid profile, making it a desirable choice for aquatic feed. The quantity of essential amino acids is a crucial factor in evaluating the quality of proteins (Hameed et al., 2022). Because fishmeal contains large levels of easily digested necessary amino acids, particularly methionine and lysine, which are thought to be lacking in grains, the primary source of animal feed, it is a highly suitable protein for aquaculture. Less research has been done on the amino acid profile of several insect groups. A normal amino acid profile for Diptera, for instance, contains enough lysine and methionine, making it comparable to fishmeal. at comparison to fishmeal, many insect groups contain EAA at quantities that are equal to or higher. Leucine in Coleoptera and Orthopterans and histidine and threonine in Diptera are more abundant than in fishmeal. Insects also have high amounts of polyunsaturated fatty acid *n*-6, which are lower than soybean meal but higher than fishmeal (Shah et al., 2022).

5. Single-cell proteins

Several types of microbial sources can be used to generate single-cell protein (SCP) products, including microalgae, yeast, and bacteria. They are all under investigation and being commercially produced with unique advantages and disadvantages (Table 18.1) (Jones et al., 2020).

5.1 Microalgae

Microalgae may be good sources of food due to their chemical composition which can include proteins, essential fatty acids (eicosapentaenoic acid and docosahexaenoic acid [DHA]), and other bioactive compounds (phenolic compounds and pigments) (Pereira et al., 2020; Chattaraj et al., 2026b). In regards to proteins, total protein weight differs greatly among 60% of microalgal species with an average total protein weight of 34%. Selecting suitable species with the appropriate protein characteristics is also important to ensure it is viable and has a specialization in "filling the protein gap" as a part of excess demand by supply of protein around the world (Jones et al., 2020). Marine species represent the highest number of documented species, in terms of both availability and compositional ranges. *Isochrysis galbana*, *Tetraselmis suecica*, *Dunaliella tertiolecta*, and *Chlorella stigmatophora* are the species most often categorized as "marine." Their amino acid profiles are similar to, if not the same, as those for the reference egg or human milk protein provides very limited amounts of cystine and methionine, but a relatively large amount of lysine. One caveat to keep in mind with all these species is that their total nucleic acid content is lower, at less than 7% of dry weight, than that of either the bacteria or yeasts (Pereira et al., 2022).

5.2 Yeast

Yeasts are valuable supplements in aqua feeds because of their greater level of protein and amino acids, energy, and micronutrients in comparison to grain and oilseed meal. Yeasts have a thick cell wall that resists digestive enzymes, reducing digestibility, so some researchers have homogenized the yeast cells (e.g., by centrifugation) in order to overcome the digestibility issue with yeast extracts (Shurson, 2018). The only drawback associated with yeast cell extracts is that, due to the

Table 18.1 An overview of single cell protein sources.

Sources of single cell proteins	Examples of specific organisms	Special characteristics	Range of protein content (%)	Challenges	References
Microalgae	*Chlorella vulgaris* *Desmodesmus* sp.	Production of omega-3 fatty acids	60–70	Rupturing cells to liberate nutrients	Ritala et al. (2017)
Yeast	*Saccharomyces cerevisiae* *Candida utilis*	Manufacturing of micronutrients and vitamins Utilizing a range of feedstocks	30–50	Boost the amount of EAA and protein	Jones et al. (2020)
Bacteria	*Methylococcus capsulatus* *Cupravidus nectar*	Content of high protein	50–80	Concerns with palatability	Ritala et al. (2017)

presence of the cell wall, they are rich in mannanoligosaccharides (MOS), beta glucans, chitin, and nucleotides, all of which serve as prebiotic fibers and immune stimulants for fish species. As previously mentioned with regard to the microbial and algal types of SCP, the relative amounts of protein, lipid, and other components, will also differ depending on the production applications (e.g., temperature, drying time, organic substrates used, the timing of harvest, storage, and yeast strains) (Nalage et al., 2016).

5.3 Bacteria

Bacteria can reproduce in relatively low time (compared to cell masses) along with growing in a variety of raw materials and edible substrates, that is, starches and sugars. Bacteria can grow quite readily on wastes of organic materials, petrochemicals, that is, ethanol, methanol, and nitrogen. They can also grow in natural water with the addition of minerals and nutrients, and also reduce any levels of nutrient deficiencies required for bacterial growth. *Methylophilus* spp. can achieve 2 h generation time and has been established as a proven useful component of animal feed with protein content chemically superior to any yeast or fungi as a potential source of protein (Sharif et al., 2021).

6. Bioeconomic implications

Based on the ongoing demand for the fish and shellfish by keeping its economic viability and maintaining its environmental sustainability, world aquaculture industry have faced a number of challenges. Due to the finite resource supply, traditional aquafeed ingredients such as fishmeal and fish oil produced from wild caught forage fish, its cost increases drastically which becomes a major-concerns to the environment. In order to find out a promising alternative to the conventional feed materials, extensive research is going on to develop a nonconventional feed ingredients (NCFIs) such as SCP, microalgae product, IMs and other agricultural waste materials. which not only addresses the bioeconomic concern but also the environmental issue that threaten the industries long-term sustainability (Yakubu et al., 2021).

6.1 Cost-effectiveness of NCFIs

6.1.1 Comparative analysis of ingredient costs

The cost and availability of fish feed ingredient varies from one another which can facilitates the adoption of these alternative protein sources that directly influence the adoption rates and its commercial feasibility. Recent aquafeed market has establish the price point for soybean meal approximately $570 per metric ton whereas for fishmeal it is roughly $1870 per metric ton. One of the most promising insect-based alternatives Black soldier fly (BSF) have ranges from $2000 to 5000 per ton depending on its production scale and its processing methods. Industry projection suggests that as the production scale increases, it can significantly decrease its cost which at commercial scale targets around $2500 per ton (Naylor et al., 2021).

The Italian economic case study examining sea bass farming provides concrete insights into cost implications at the farm level. In a research Pulina et al. reported that the exceeding costs comprise around 63% of total farm operational expenses. From their study they found that introducing *Tenebrio molitor* meal at 50% inclusion rates with prices of €10 per kg would increase the feeding cost share to approximately 87% of the total expenses (Pulina et al., 2018).

6.1.2 Impact on feed conversion ratio (FCR) and production efficiency

In aquaculture the utilization of processed animal proteins has shown excellent result. In gilthead seabream growth performance, processed animal proteins shows better fish growth performance, high resistance for nutrient and energy, with lesser feed conversion ratio (FCR) as compared to the control treatments. The significant improvement in FCR observed in fish feed diets along with high protein digestibility leads to the reduction in the harmful environmental effect of seabream production (Aragão et al., 2020).

Research on alternative feed sources for fish production, especially concerning nontraditional plant classifications, have shown variations in feed conversion efficiency in relation to species across aquaculture development. Research conducted on gilthead sea bream fed diets formulated using plant by-products, glutens, and concentrates demonstrated better fish growth performance with a lower FCR than fish fed diets based on traditional feed ingredient(s) in the controls. The plant based treatments remarkably showed a better FCR value and influenced nutrient use and retention of energy than diets based on traditional fishmeal (Aragão et al., 2020).

Nawaz et al. observed an enhancement in the production of biomass with the incorporation of 10% *Moringa oleifera* leaves power in conventional fishmeal achieved 65.16 g body weight whereas 63.20 g increase is observed with conventional diet. Their study revealed that the aggregate could save around 13% when fish were fed with *M. oleifera* leaves power diets (Nawaz et al., 2023).

BSF larvae have emerged as a particularly promising NCFI with favorable FCR outcomes. Research has demonstrated that fish feed can be replaced with BSF meal for African catfish and rainbow trout without harming results. For rainbow trout, 100% of the feed can be replaced and African catfish, 66% can be swapped out. The study confirmed that at a 66% replacement rate, African catfish grew just as well and survived at the same rate as the fish eating the regular feed (Bartucz et al., 2023).

6.1.3 Economic feasibility for small- and large-scale farmers

Shivashankar et al. has studied on economic analyses of NCFIs adoption which demonstrates significant potential for improving profitability among small scale farmers. Their investigation into the economics of composite fish culture in 0.2 ha tanks, has indicated economic feasibility per farm, while the study's fisher farmers themselves were able to achieved net profits of ₹60,816 with surviving rate of 42.6%. Their research illustrate that composite fish culture can potentially improve local livelihoods through low input costs and an adaptive scale of operation (Shivashankar et al., 2024). In Nigeria, local feed production using plant-based NCFIs shows profitability indicators with 3.24% net returns on investment. The use of plant-based sources of ingredients resulted in a 50% reduction in the use of fishmeal, and years of experience and the initial capital investment were statistically determined the net income (Omeje et al., 2023).

Small scale BSF farming in Malawi demonstrates economic potential where small scale farmer could generate around MK507,100 gross revenue for one larval cycle leading to a potential yearly income of MK2,535,500 (over $2500). The research on the benefit cost ratio showed that BSF maintains its profitability even when facing variations of up to 10% in its costs, output levels, and the prices it can sell its products (Munthali et al., 2023).

For large scale operations, economic modeling indicates substantial potential returns from NCFIs adoption. Commercial BSF farmers in Malawi could generate MK1,536,100 for one larval cycle and MK7,680,750 per annum. The economic surplus analysis for insect-based feed (IBF) in Uganda revealed potential net economic benefit of USD 0.73 billion over 20 years with a benefit cost ratio of 28:1 and an internal rate of return of 138% (Munthali et al., 2023).

An organic aquaculture system that uses NCFIs has excellent economic performance in comparison to conventional aquaculture. Research conducted on Indian major carps reported that the organic aquaculture system had the highest net present value of 1.06 million USD, and an internal rate of return of 51% per hectare of fish culture pond. Additionally, the organic culture system achieved a production of 19 tons per hectare as compared to the conventional system (Beg et al., 2024).

6.2 Market dynamics and adoption barriers

6.2.1 Consumer perceptions and market acceptance

Consumers requirement of the aquaculture products from fish feed with NCFIs varies significantly based on the ingredient type and its processing methods. Research on organic fish production indicates that organically grown fish sold for USD 2.75 kg^{-1} that can command premium prices around 30% higher than the conventionally grown fish (Beg et al., 2024). An application of theory of planned behavior (TPB) to aquaculture research indicates that fish farmers accustomed to manufactured feed are more inclined to incorporate nonconventional ingredients. The acceptance of this feed innovation tends to be dependent on a few key elements: expected social pressure, the perceived importance and benefits of the new ingredient, the perceived relative advantage, and the uncertainty of its outcomes. Studies looking at Indian carp farmers perceptions also suggest that the same aspects influence the adoption of the NCFIs. The research emphasizes the importance of addressing farmer concerns and providing adequate information to facilitate the adoption of innovation feed alternatives (Brugere et al., 2021) (Figs. 18.1 and 18.2).

6.2.2 Regulatory challenges and quality standards

Regulatory frameworks governing NCFIs use in aquaculture vary significantly across the regions, creating challenges for its widespread adoption. The ASC feed

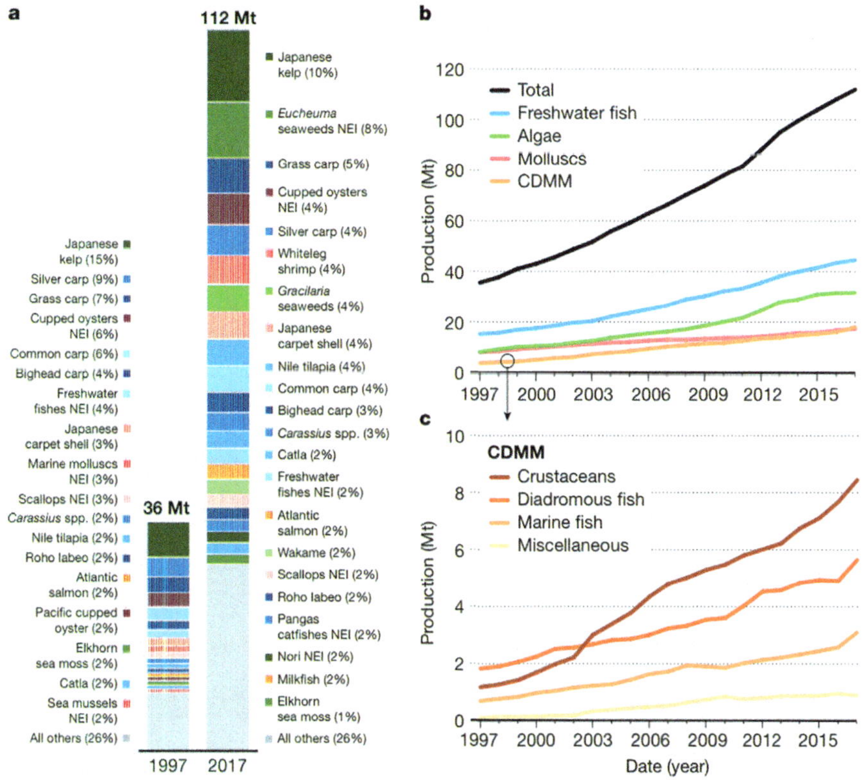

FIGURE 18.1

Global aquaculture production. (A) The species composition is shown for 1997 and 2017. Green, plants and algae; blue, freshwater fish; pink, shellfish; orange, diadromous fish. (B) Total, freshwater fish, algae, molluscs, and CDMM, which comprises crustaceans, diadromous fish, marine fish, and miscellaneous species and is expanded in C. Algae comprised more than 99% of the production weight of "algae and aquatic plants" production in 2017 (Naylor et al., 2021).

standard represents one of the most comprehensive approaches to deal the challenges in aquafeed sustainability. The standard addresses the interconnected challenges faced by the aquafeed industry feed ingredients have different impacts depending on ingredient type, location and the method of production. Feed performance criteria implementation varies among the countries with Denmark being noted as the only EU country using the feed performance criteria as a component of aquaculture regulatory system. Danish regulation require farmers to demonstrate FCR not exceeding 1.2:1–1.5:1 with minimum digestibility of 70% and fines content less than 1% of total feed (Shipton et al., 2013) (Tables 18.2 and 18.3).

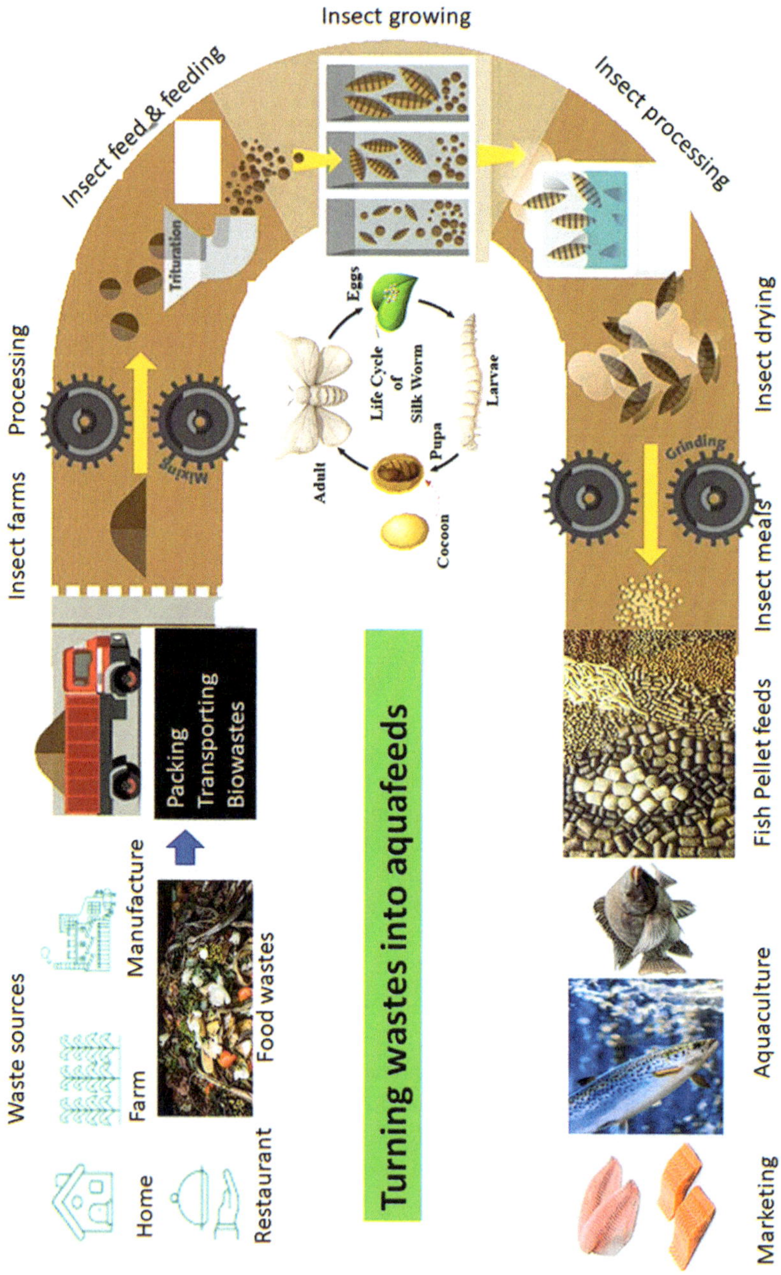

FIGURE 18.2

Conversion of wastes into proteins using insects to produce insect meals to replace fishmeal in aquafeeds for fish culture (Alfiko et al., 2022).

Table 18.2 Major ingredients and typical feed conversion ratios (FCR) in feeds for common aquaculture species (Boyd et al., 2007).

	Ingredient content (%)				
	Atlantic salmon	***Trout***	***Shrimp***	***Tilapia***	***Channel catfish***
Soybean meal	14.0	15.0	24.5	38.3	34.5
Corn meal	10			48.8	20.4
Wheat middlings	18	27	27	4	20
Fish meal	30	25	19	6	2
Oil	24	16	1.5	1.5	2
Feed conversion ratio	1	1.2	2	1.8	2.2

Table 18.3 Major anti-nutrients present in some commonly used alternative ingredients for fish feed (Gopan et al., 2020).

Plant-derived nutrient source	Anti-nutrients present
Soybean meal	Protease inhibitors, phytic acid, allergens, lectins, saponins, phytoestrogens, antivitamins
Rapeseed meal	Protease inhibitors, phytic acid, tannins, glucosinolates
Lupin seed meal	Protease inhibitors, phytoestrogens, alkaloids, saponins
Pea seed meal	Protease inhibitors, lectins, tannins, phytic acid, saponins, antivitamins, cyanogens
Sunflower oil cake	Protease inhibitors, saponins, arginase inhibitor
Cotton seed meal	Phytic acid, phytoestrogens, gossypol, antivitamins, cyclopropenoic acid
Leucaena leaf meal	Mimosine
Alfalfa leaf meal	Protease inhibitors, phytoestrogens, antivitamins, saponins
Mustard oil cake	Glucosinolates, tannins
Sesame meal	Phytic acid, protease inhibitors

6.2.3 Scaling production and supply chain considerations

Scaling production of NCFIs presents both opportunities and challenges for the development of sustainable agriculture. The abundance and availability of the selected materials across various regions confirms sustainability of feed industry once feeds are formulated and customized to local farmers. Sources of plant-based materials produced by agricultural farmers significantly cut those from the processors which indicates the potential for working with small holder farmers using intrafirm resources Supply chain consideration for NCFIs scaling includes seasonal availability, processing requirements, and transportation costs. From the study, it indicates that most of the ingredients are generally available just after crop harvest, making timely access crucial for feed formulations. The development of the feed production units has been proposed to address its cost, various challenges during its production specifically for agricultural residues (Africa).

The potential to scale up BSF production has significant potential for addressing supply chain issues. Uganda, which has historically had a potential of producing from 3244 tons to 1.5 million tons of IBF according to the replacement rates, could also produce approximately 695–312,678 tons of NPK fertilizer from the biowaste recycling according to that same replacement rate. Approximately 0.09–41 million tons of biowaste could be recycled from biowaste in accordance and based on the conventional feed using the IBF replacement ratio (Abro et al., 2022).

6.3 Case studies

6.3.1 Success stories and pilot projects using NCFIs

Case study 1: microalgae-blend tilapia feed success.

The development of microalgae blend tilapia feed eliminates fishmeal and fish oil while improving growth and cost viability. A study created a high performance, fish free feed for Nile tilapia by combining two commercial microalgae. To replace fishmeal, the researchers used protein rich, defatted biomass from *Nannochloropsis oculate*. To replace fishoil, they used whole cells from a different microalga rich DHA. This formulation demonstrate a comparable growth performance to the conventional fed by achieving the cost challenges (Sarker et al., 2020).

Case study 2: seawater energy and agriculture system (SEAS).

The 2-hectare seawater energy and agriculture system (SEAS) pilot plant in Abu Dhabi demonstrates an innovative model that integrates aquaculture with the cultivation of Salicornia. In this system, the salicornia plants act as natural filters for fish and shrimp ponds. This approach allows for the production of both seafood and economically viable biofuels, all while utilizing nonarable land and nonpotable water. After harvesting salicornia seeds 2 t ha per year, the system produces vegetable oil 0.7 t ha per year which can be processed in to sustainable aviation fuel. The techno economic study shows that the biorefinery component becomes economically sustainable when reaching a production scale pf 900 barrels per day, that requires a SEAS surface of 20,000 hectares (Laveille et al., 2022)

Case study 3

Abdou et al. represents the life cycle assessment (LCA) approach and its environmental impacts generated by production of 1 ton of European seabass (*Dicentrarchus labrax*) and gilthead seabream (*Sparus aurata*) on a sea-cage aquaculture farm in Tunisia. They reported in their study that fish feed is primarily responsible for most of the environmental effects that are directly linked to production of fish meal and oil as feed components and the high concentrations of nitrogen and phosphorus discharged into the environment (Hamilton et al., 2016).

6.3.2 Economic outcomes and lessons learned

Analysis of small scale aquaponics systems reveals various economic outcomes based on its production ratios and its management practices. Research in south

Africa showed that conventional aquaponics cultural practices fish plant revenue model 59%–41% which is not economically viable with net profit of 42% and profitability index 0.35. However, the optimized revenue models of 42%–58% fish-to-plant ratio is achieved marginal feasibility with net profit of 59% with profitability index of 0.50 (Babatunde et al., 2023)

7. Environmental implications

7.1 Reducing pressure on wild fish stocks

7.1.1 Decreased reliance on fishmeal and fish oil

The aquaculture industries dependence on marine derived fishmeal and fish oil has created a significant environmental paradox. While aquaculture has developed to reduce pressure on wild fish stocks, the industry currently consumes approximately 71% of global fishmeal production and 73.9% of fish oil production. The decline in large predatory fish populations has forced the aquaculture industry ton become more dependent on smaller fish from lower trophic levels. This reliance has significant environmental consequences. In 2017, for instance, 18.3% of all fish caught from the ocean were used for aquaculture feed ingredients. Critically, the majority of these fish, which were processed into fishmeal and fish oil, were also suitable for direct human consumption. This induces a fundamental inefficiency in the global food systems nutritious fish that could feed coastal populations are instead processes into feed for farmed fish in developed countries (Boyd et al., 2022).

Similarly, microalgae-based alternatives have demonstrated significant potential for reducing fish meal dependency. A detailed study on *Nannochloropsis* sp. QH25 coproduct as a fishmeal replacement in rainbow trout feed found that 100% replacement of fishmeal. Do not significantly alter fish growth performance while significantly reducing biotic resource uses. The biotic resource use conversion ratio decreased from 17.69 kg C/kg fish in the reference diet to 7.15 kg C/kg fish in the complete replacement diet (Sarker et al., 2024).

Another study has been conducted to evaluate the cost effectiveness and utilization of maggot in the fish diets simultaneously to observe the rate of larval production from various culture enclosures and feeding trials. From their finding they found the aluminum box as the best choice that yields the highest maggots and found to be the most cost-effective out of all the culture enclosures tried. Their study proved that *Musca domestica* maggots could serve as a partial fish meal replacer in *Clarias gariepinus* fingerlings at 60%–70% inclusion level with good growth and nutrient utilization (Kolawole and Ugwumba, 2018).

7.1.2 Contribution to marine biodiversity conservation

The shift toward NCFIs represents a critical contribution to marine biodiversity conservation. The extraction of small pelagic fish for fishmeal production directly impacts marine food webs as these species serve as crucial forage fish for larger predators, seabirds and marine mammals. The removal of these foundational species can trigger cascading effects throughout marine ecosystem (Ticina et al., 2020). Research shows that small wild fish are vital for both healthy marine ecosystems and the food security of poor communities in many costal-regions. The limited availability of these fish species threatens not only marine biodiversity but also the livehoods and nutritional security of coastal communities that depend on these resources for subsistence and income (Boyd et al., 2022).

Study shows that complete substitution of plant-based alternatives can reduce the increasing demand for the marine derived ingredients, which is mainly occurs due to the inclusion of the resource intensive crops and their derived product to meet the nutritional requirements. Using plant ingredients instead of fish meal transfers demand from the sea to the land. This creates more competition for agricultural resources that are already strained, potentially affecting the environment, wildlife and the supply and cost of crops (Malcorps et al., 2019). A lifecycle assessment study of shrimp feeds reveals that its production have certain environmental impacts that includes terrestrial acidification, freshwater eutrophication, and global warming (Al Eissa et al., 2022; Chen et al., 2022). All these findings reveals the complexity of sustainability trade-off and the need of comprehensive environmental assessment. The development of an integrated multi-trophic agriculture (IMTA) system presents another innovative approach to reducing pressure on wild stocks. Research has demonstrated that marine periphyton from biofilters can effectively replace fish meal in fish diets, creating closed loop systems that utilize waste biomass from other aquaculture operations. This approach not only reduces fishmeal dependency but also address waste management challenges within aquaculture system (Hameed et al., 2022).

7.2 Resource use and circular economy

7.2.1 Utilization of agricultural and food waste

The concept of circular economy in aquaculture represent a paradigm shift from the traditional linear take-make-dispose model to a more sustainable approach that increases the biomass utilization that leads to reduction in waste generation and its impact on the environment. From the research it is observed that aquaculture sludge are rich in nutrients like nitrogen and phosphorous, that can be converted into biofertilizers through various processes including aerobic and anaerobic digestion, drying, composting, and vermicomposting. This approach not only addresses waste management challenges but also creates a valuable product for agricultural applications. This transition is particularly relevant in the context of feed production where agricultural and food processing waste can be transformed into valuable feed ingredients (Kurniawan et al., 2025). Research demonstrates that various agricultural by products that can serve as an effective feed ingredient for fish feed. This includes plant-based materials such as yam, plantain, banana, cowpeas, Mucuna, maize, cassava, millet, sorghum, groundnut, sun hemp seed, and brewery waste. The utilization of these locally available nonconventional feeds not only reduces its production cost but also contributes to waste reduction and resource efficiency (Audu and Yola,

2021). The BSF larvae production system is a successful example of valorization of waste in aquaculture feed production. The research demonstrates the potential of the BSF larvae to be reared efficiently on different plant byproducts such as cocoa bean shells, depectinized apple pomace, and potato peelings. Although larvae fed on plant by product recorded reduced growth performance than control feeds and the FCR for BSF shrimp feed larvae fed on plant by products was 0.8–1.0 that was reflecting an excellent feed efficiency (Barth et al., 2025).

7.2.2 Reduction in land, water, and energy footprints

The adoption of NCFIs has demonstrated various impact on land, water and energy foot prints with small alternatives which shows a significant improvement while others shows trade-offs that require careful consideration. Marine microalgae cultivation presents a particularly promising case for resource efficiency. Unlike terrestrial crops, marine microalgae do not require arable land and can achieve higher biomass yields as compared to conventional crops. Research on *Nannochloropsis* sp. Cultivation shows that while fresh water is required to maintain salinity due to evaporation, the harvesting process enables water recycling resulting in reduction in net water use. This Favors with terrestrial crops such as soybeans and corn which have comparatively higher water demands. A comprehensive study of microalgae-based feeds found that while water usage were comparable between reference and experimental diets, land use conversion ratio of the reference diet were significantly lower than all microalgae-based diets highlighting the need for the optimization of alternative ingredient production system (Sarker et al., 2024).

7.3 Life cycle assessment (LCA) of NCFIs

7.3.1 Greenhouse gas emissions comparison

LCA studies of NCFIs reveals complex patterns in greenhouse gas (GHG) emissions with various outcomes depending on the specific alternative ingredient and its production method. These assessments are very important for understanding the true environmental impact of converting conventional feed to NCFIs. Research by Sarkar et al. on microalgae-based feed alternatives demonstrates as a promising result for GHG emission. Using *Nannochloropsis* sp. QH25 to replace fishmeal does not create a major difference in GHGs, with the standard diet at 2.43 kg CO_2 e/kg fish and the replacement at 2.49 kg CO_2 e/kg fish. This is significant because feed for salmon farming is a top contributor to emissions. The way alternative ingredients are made greatly affects their carbon footprint. For microalgae, urea and carbon dioxide use during growth are the main sources of emissions. However, these emissions could be cut by using circular bioeconomy methods like capturing carbon naturally and recycling nitrate from waste. Their research suggests that these emissions could be substantially reduced through circular bioeconomy approaches including biological carbon sequestration and nitrate recycling from industrial or agricultural waste (Sarker et al., 2024).

Other comparative studies on aquaculture systems shows that this feed production can be the major cause for the GHG emission. In case of shrimp production chain, feed ingredients is identified as the main contributor to the global warming potential with poultry by product and fish meals Also the extensive feed product chain exhibit highest global warming potential because of land transformation mainly in mangrove areas. This research highlights the environmental impact with increasing the feed production intensity (Al Eissa et al., 2022; Chen et al., 2022). LCAs of rainbow trout production in recirculating aquaculture systems (RASs) indicates that the feed production is responsible for the major environmental impact between 67% and 98% across multiple categories including climate change. Further the research suggest that RASs plays an important role by improving the utilization of larger portion of plant-based feed ingredients as environmentally sustainable food system (Bergman et al., 2020).

7.3.2 Eutrophication and pollution potential

Eutrophication potential represents one of the most critical environmental concerns that is related with aquaculture feeds, as excess nutrients from fish feeds and waste can cause significant ecological damage to the aquatic ecosystem. The LCA of NCFIs reveals various environmental impact of both marine and freshwater eutrophication potential. Sarkar et al. has studied on the eutrophication potential of the microalgae-based feed ingredients. Their research on *Nannochloropsis* sp. QH25 replacement shows no significant difference in marine eutrophication potential between reference diets (0.00195 g N/kg fish) and complete replacement diets (0.00197 g N/kg fish). Similarly, freshwater eutrophication showed a comparable result between reference diet of about 0.000288 kg P/kg fish and replacement diets of about 0.000292 kg P/kg fish) (Sarker et al., 2024). The eutrophication potential of NCFIs demonstrates substantial improvements over conventional alternatives across both marine and freshwater environments. Marine microalgae protein production exhibits marine eutrophication potential of 0.00004 kg N per kg protein as compared to 0.0011 kg N per kg protein for fishmeal representing a 96% reduction in nitrogen related marine pollution potential. Similarly freshwater eutrophication potential shows even greater enhancement with microalgae achieving 0.0000652 kg P per kg protein as compared to 0.000132 kg P per kg protein for fishmeal that represents a reduction in phosphorous related fresh water pollution by 51% (McKuin et al., 2023). LCAs of shrimp production systems reveal that plant-based protein substitutes do not universally reduce eutrophication potential. Research found that soybean, wheat, and corn gluten meals that produce higher freshwater eutrophication potential than animal-based ingredients which demonstrates that substitutes plant-based proteins for animal-based feed ingredients that does not always yields positive environmental consequences (Al Eissa et al., 2022; Chen et al., 2022).

7.3.3 Waste management and nutrient recycling

The utilization of agricultural and food waste streams for NCFIs production exemplifies circular economy principles in aquaculture. May et al. successfully processed food waste into fish flakes containing protein ratios suitable for various fish species. In feeding trials with zebrafish (*Dani rerio*), diets formulated with up to 100% of the converted food waste resulted in high survivability and growth outcomes equivalent to standard commercial feeds. The research concludes that such recycled food waste holds significant potential as a feed source for omnivorous fish in small scale freshwater aquaculture (May et al., 2006). Leyva-López et al. (2020) found that agro-industrial waste utilization shows a significant potential as sources of bioactive compounds for aquaculture applications. The exploitation of agro industrial waste addresses multiple objectives providing added value to the production chain, reducing environmental waste burdens and supplying cost effective feed ingredients. Their study demonstrates that insects can effectively valorize byproducts by converting them into high value larval biomass that is suitable for feed production (Leyva-López et al., 2020). Biofloc technology (BFT) represents a paradigm shift in aquaculture waste management utilizing microbial ecosystem to convert organic waste into valuable biomass while maintaining water quality. Reports demonstrates that BFT can achieve minimal water exchange while maintaining high stocking densities through effective nutrient recycling. Studies shows that that BFT converts waste into Bioflocs that serve as natural food sources for fish and shrimp, reducing dependence on external feed inputs while improving nutrient utilization efficiency (Khanjani et al., 2024). Wang et al. also examine Biofloc integration with worm reactors observed that biofloc worm system increased fish production by 17.1% through protein recovery. The system effectively transformed aquaculture waste into Tubificidae biomass for direct feeding while improving aerobic denitrification processes. These integrated approaches demonstrate the potential for improvements in feed nitrogen use efficiency (Wang et al., 2024).

8. Nutritional and health considerations

8.1 Nutritional profiles of NCFIs

The quest for sustainable aquaculture has intensified the search for alternative protein sources to replace conventional fishmeal, which became increasingly expensive and environmentally unsustainable. NCFIs represents a diverse array of plant-based materials, animal byproducts and novel protein sources that offer different nutritional profiles and present novel challenges for aquafeed formulation (Eggink et al., 2025; Zuluaga-Hernandez et al., 2023).

8.1.1 Protein, lipid, and micronutrient content

Microalgae are an encouraging group of NCFIs due to their extremely high protein content and balanced amino acid composition. Chen et al. show that *Chlorella pyrenoidosa* contains a significant protein content (52.4%) with a total amino acid content of approximately 189.7 mg/g, making them a potentially protein rich ingredient for aquatic animals (Chen et al., 2022). The larvae meal of BSF (*Hermetia illucens*) has become a particularly important nonconventional feedstuff due to its high protein content and adequate amino acid profile. The protein content of BSF (*H. illucens*) larvae meal can vary quite significantly based on the substrate growing the larvae, where their work with maggot meal from layer manure substrate suggested protein content at the highest level of 45.36% protein while the lower level of protein from kitchen waste substrate had 30.91% (Widiyastuti et al., 2024). Another study mentions that black soldier fly larvae meal (BSFLM) is composed of protein that is from and ranging between 29.9% and 48.2%, which meets the protein needs of different fish species. The amino acid profile of BSFL meal has high levels of glutamic acid, aspartic acid, alanine, valine, leucine, and isoleucine. The even distribution of amino acids, especially branched-chain amino acids (BCAAs) in BSFL meals, can stimulate muscle protein synthesis and metabolic function in aquatic animals (Munguti et al., 2025).

The lipid content of microalgae varies significantly among the species and cultivation conditions. Study on *C. pyrenoidosa* demonstrates lipid content about 19.4% providing essential fatty acids for aquatic nutrition. However, the percentage of polyunsatufatty acids (PUFAs) in some microalgae preparations can be relatively low with *C. pyrenoidosa* showing only 1.2% PUFAs under certain cultivation conditions. This highlights the importance of optimizing cultivation parameters to enhance PUFAs production (Chen et al., 2022). *Chlamydomonas reinhardtii* exhibits particularly favorable fatty acids profiles with unsaturated fatty acids containing approximately 74% of total fatty acids out of which 48% are omega 3 fatty acids. The Alpha-linolenic acid (ALA) content in *C. reinhardtii* reaches 42.4% which is significantly higher than *Chlorella* (23.4) and *Spirulina* (0.12%) (Darwish et al., 2020). BSF (*H. illucens*) larvae exhibit highly variable lipid content based on the substrate types, composition, and processing methods. Maggot meal from kitchen waste exhibit the highest fat content at 31.97% while those grow on layer manure show the lowest about 18.44%. The fatty acid profile of BSF (*H. illucens*) larvae is dominated by lauric acid that represents the predominant fatty acid across all substrate types ranging from 22.94% to 33.79% (Widiyastuti et al., 2024).

8.1.2 Presence of antinutritional factors and strategies for mitigation

ANFs possess a significant challenge in applications of NCFIs that significantly limits its nutrient availability and compromising fish health and growth performance. Plant-based NCFIs commonly contain different ANFs including protease inhibitors, lectins, phytic acid, saponins, and phenolic compounds that can interfere with protein digestion mineral absorption and overall nutrient utilization. Although it has some antinutrients, its significant health advantages and rich nutritional profile make it a strong replacement for animal-based protein (Choręziak et al., 2025).

Quality control measures are essential for managing ANF risks in commercial applications. Regular analysis of ANFs concentrations, establishment of maximum allowable levels and development of standardized processing protocols that helps to ensure consistent feed quality and performance. Collaboration between feed manufactures, ingredient suppliers and aquaculture producers is necessary to develop effective management strategies that maintain both nutritional quality and economic viability while addressing the specific challenges posed by ANFs in NCFIs (Muntean et al., 2024).

Research on plant-based diets reveals complex interactions between ANFs and fish physiology that extend beyond simple growth performance impacts. Studies using 100% plant-based diets in rainbow trout demonstrate that high carbohydrate to protein ratios can induce beneficial changes in gut microbiota while decreasing expression of proinflammatory cytokines in liver tissue. These findings suggest that careful formulation of plant-based diets can achieve positive health outcomes despite the presence of ANFs, emphasizing the importance of holistic nutritional approaches rather than simple ingredient substitution strategies (Defaix et al., 2024).

8.1.3 Effects on fish growth and health

8.1.3.1 Growth performance and feed efficiency

The impact of various factors on the well-being and physical development of fish encompassing both positive and negative influences. Fish growth rate, general health, and reproductive success are influenced by a combination of environmental factors, diet, and genetics. In aquaculture, the PPs have been frequently used as a substitute for fishmeal. According to a reported study, plant-based diets have a detrimental effect on fish growth and health, possibly as a result of the presence of anti-nutritional substances (ANFs) (Magbanua and Ragaza, 2024). Fish growth depends on saturated and monounsaturated fatty acids because they are unable to synthesize PUFAs (Tocher, 2003). The fish's high tropic levels in aquatic habitats are explained by their ability to metabolize PUFA to produce LC-PUFA, which greatly facilitates the environment. Numerous studies have shown the importance of effective feed management strategies in reducing feed waste, lowering costs, promoting quick fish growth, and increasing profitability (Kébé et al., 2021).

8.1.4 Immune response and disease resistance

The microorganisms that produce chitinase or are chitinolytic can be used as supplements to remove chitin, which has a detrimental effect on fish growth. This method is very beneficial for reducing the adverse environmental effects of fish waste and inducing an immunological response to fish (Zhang et al., 2014). The goal of these feedings is to determine the fishing byproducts impact the overall health, FCRs, and growth rates of fish bred for food. Immune response and disease resistance are two indicators of these feed's nutritional sufficiency and health benefits. There are more sustainable and circular ways to create the microbial feedstock. Lipids, protein (SCP), and other particular nutritional components can be obtained from a wide variety of SCIs by processing, drying, and using them in aquaculture feed. They can also be used to improve immunological response and survival. The microorganisms can also be used as probiotics (Glencross et al., 2020; Chattaraj et al., 2025). Taurine availability in the food is linked to the main physiological and behavioral reactions of fish, including survival rate (Rotman et al., 2017).

8.1.5 Challenges and future perspectives

8.1.5.1 Technical and processing challenges

8.1.5.1.1 Palatability, digestibility, and feed formulation: Aquafeed's shelf life is limited, and the peroxidation of lipid makes it susceptible to rancidity (Kazuo, 2019). The feed's nutritional value and palatability are negatively impacted by the oxidized lipids (Phung et al., 2020). The path of effectively incorporating these micro- and macro-organisms as essential resources in aquaculture feeds may involve making additional efforts to understand them better, specifically their role as value-added products and their potential to enhance animal performance, nutrient availability, food palatability, and digestibility. Problems could arise from the marine microalgae species selected as fish feed have inefficient digestion and/or an accumulation of salt. In trial diets, algae provides only 10%–15% of the total protein needed for growth, without decreasing growth or feed intake. The realistic and useful assessment of possible fish feed ingredients is known as in vivo testing. An ideal feed composition and feeding for the selected fish species can be achieved by selecting the target cultured species. Climate adaptation, climate mitigation, and higher productivity can all be achieved through sustainable aquaculture production. Innovations in system design, feed management, and feed formulation can all be part of mitigation and adaptation initiatives. Fish waste can build up in the surrounding environment, causing dead zones and algal blooms oxygen depletion in the water (Mbaka et al., 2022).

8.1.6 Processing technologies for improving NCFI utilization

The growth of aquafeed through the circular bioeconomy is one of the more promising strategies. The utilization, reuse, and/or recycling of innovative fish feed ingredients that are nonconventional, such as byproducts, agricultural industry wastes, and natural raw materials that do not compete for human food, defines circular bioeconomy in aquaculture. However, there exists comparatively high levels of antinutrients in such innovative feed that limits its application in fish diets (Francis et al., 2001). The antinutrients in JCK cause more than only growth impairment; they also interfere with the production of vital blood cells like erythrocytes (Akinleye et al., 2012; Haghbayan and Shamsaie Mehrgan, 2015). One of the main drivers of global food security is the aquaculture industry (Pradeepkiran, 2019). The heightened demand for food has forced researchers to search for new and NCFIs, including insects, SCPs, algae, and land animal byproducts (Hameed et al., 2022). The current analysis concluded that, due to their high nutritional content, insects have a good chance of replacing fishmeal. Insects are a NCFI with the greatest potential to replace conventional feed ingredients. Insects nutritional value can be further enhanced by using appropriate processing technology.

8.1.7 Policy and regulatory framework

8.1.7.1 Current regulations on feed ingredients

Nontraditional feed ingredients are being investigated more and more by the aquaculture sector to lower expenses and increase sustainability. Implications for the environment, the economy, and regulations are important factors. The unconventional substances like microalgae, insects, and agricultural waste have potential; however, their effects on manufacturing costs, environmental impact, and regulatory compliance must be carefully considered. Food waste is a useful resource that can be applied to aquaculture species diets, significantly lowering the need for pricey feed materials (Cheng et al., 2014; Wong et al., 2016). Omnivorous aquaculture candidates like tilapia, can also benefit from food waste from restaurants and household leftovers intended for human consumption (Nasser et al., 2018). Feed ingredients have the ability to these issues by showing decreases in impact categories, such as the possibility of global warming at the level of seabream aquaculture production. There is potential for these issues, and promoting a more sustainable future for the aquaculture sector by investigating substitutes for feed ingredients.

8.1.8 Recommendations for policy support and standardization

There are substantial bioeconomic and environmental advantages to using unconventional feed ingredients in aquaculture; however, for broader acceptance and sustainable practices, legislative support and standardization are needed. These components, which include insects, microalgae, and agricultural waste, can enhance fish health and stress tolerance at lower feed costs. The environmental encouragement of research and development is recommended, with the establishment of precisely defined and quality requirements, as well as sustainable aquaculture methods. Aquaculture sustainability is negatively impacted by several variables and concerns about the exploitation of finite natural resources and environmental problems related to the pollution, conversion, and deterioration of aquatic and terrestrial environments (Boyd, et al. 2020). Public opinion, policy inequality, production costs, illnesses, and supply chain problems, on the other hand, provide challenges to social and economic sustainability (Peñalosa Martinell et al., 2020). The legal, policy, and regulatory frameworks that govern aquaculture and aquafeed use are provided by government agencies. The approach in environmental governance has emerged in recent years with the rise of certification bodies like the Global Aquaculture Alliance (GAA) and the Aquaculture Certification Council, and the World Wildlife Fund (WWF) aquaculture dialogs.

8.1.9 Research and innovation needs

8.1.9.1 Areas requiring further study (e.g., long-term impacts, species-specific responses)

Aquaculture can create a circular bioeconomy by using, reusing, and/or recycling unconventional but cutting-edge fish feed ingredients, such as by-products, agricultural industry processing wastes, and natural raw resources that do not compete with human food consumption. The FAO (2022) projects that by 2030, an estimated 202 million tonnes of aquatic animals will be produced, of which 96 million will come from capture fisheries and 106 million will come from aquaculture. It is anticipated that aquaculture production will increase by 26 million metric tons (MT) globally by 2030, requiring an additional 40 million MT of feed (Albrektsen et al., 2022).

Aquaculture faces a significant challenge in improving feed appropriateness and supporting the global increase in fish output. Feed formulation has been continuously improved over the last 20 years to boost feed efficiency and sustainability in aquaculture.

8.1.10 Role of biotechnology and genetic improvement

Aquaculture, or sustainable fish farming, is becoming more and more important to supplying the world's expanding seafood demand while lessening the strain on wild fish supplies. The use of conventional fish feed ingredients, including fishmeal and fish oil, which are obtained from wild-caught fish, is a major problem in aquaculture (Araujo et al., 2022). Despite their high nutritional content, these nutrients support overfishing and environmental damage. Innovative feed substitutes are being created to improve fish farming without sacrificing the production or health of the fish (Wohlfarth and Schroeder, 1979). To improve characteristics, including growth rate, feed conversion efficiency, and disease resistance, both conventional and cutting-edge breeding programs, such as genomic selection and marker-assisted selection, are being employed. The breeding techniques have been shown to increase feed efficiency in some fish species by up to 18%–19% over several generations.

9. Conclusion

The use of NCFIs in aquaculture can promote a more sustainable way to decrease reliance on fishmeal and soybean meal, decrease environmental impacts, and improve economic sustainability. IMs, SCPs, and agro-industrial byproducts improve feed conversion rates and nutrient recycling while enhancing circular economy and sustainability practices. Although NCFIs present challenges such as ANFs, digestibility, and regulatory hurdles, increases in biotechnology, processing methods, and supportive policies will assist with adoption. Ultimately, by improving bioeconomic efficiencies while ensuring ecological responsibility, NCFIs can be critical players in building a resilient aquaculture system that achieves global food and nutritional security.

References

Abro, Z., Macharia, I., Mulungu, K., Subramanian, S., Tanga, C.M., Kassie, M., 2022. The potential economic benefits of insect-based feed in Uganda. Front. Insect. Sci. 2, 968042.

Africa, S.-S. "*Development and Scaling of Sustainable Feeds for Resilient Aquatic Food Systems in Sub-Saharan Africa: Life Cycle Assessment*.".

Agboola, J.O., Øverland, M., Skrede, A., Hansen, J.Ø., 2021. Yeast as major protein-rich ingredient in aquafeeds: a review of the implications for aquaculture production. Rev. Aquacult. 13 (2), 949–970.

Akinleye, A., Kumar, V., Makkar, H., Angulo-Escalante, M., Becker, K., 2012. Jatropha platyphylla kernel meal as feed ingredient for Nile tilapia (*Oreochromis niloticus* L.): growth, nutrient utilization and blood parameters. J. Anim. Physiol. Anim. Nutr. 96 (1), 119–129.

Al Eissa, A., Chen, P., Brown, P.B., Huang, J.Y., 2022. Effects of feed formula and farming system on the environmental performance of shrimp production chain from a life cycle perspective. J. Ind. Ecol. 26 (6), 2006–2019.

Albrektsen, S., Kortet, R., Skov, P.V., Ytteborg, E., Gitlesen, S., Kleinegris, D., Mydland, L.T., Hansen, J.Ø., Lock, E.J., Mørkøre, T., 2022. Future feed resources in sustainable salmonid production: a review. Rev. Aquacult. 14 (4), 1790–1812.

Alfiko, Y., Xie, D., Astuti, R.T., Wong, J., Wang, L., 2022. Insects as a feed ingredient for fish culture: status and trends. Aquac. Fish. 7 (2), 166–178.

Aragão, C., Cabano, M., Colen, R., Fuentes, J., Dias, J., 2020. Alternative formulations for gilthead seabream diets: towards a more sustainable production. Aquac. Nutr. 26 (2), 444–455.

Araujo, G.S., Silva, J.W. A.d., Cotas, J., Pereira, L., 2022. Fish farming techniques: current situation and trends. J. Mar. Sci. Eng. 10 (11), 1598.

Audu, R., Yola, I.A., 2021. Contemporary issues in fisheries and aquaculture: a review on non—conventional feed ingredients for fish feed in Nigeria. Bayero J. Pure Appl. Sci. 13 (2), 22–28. https://doi.org/10.4314/bajopas.v13i2.3.

Azra, M.N., Okomoda, V.T., Tabatabaei, M., Hassan, M., Ikhwanuddin, M., 2021. The contributions of shellfish aquaculture to global food security: assessing its characteristics from a future food perspective. Front. Mar. Sci. 8, 654897.

Babatunde, A., Deborah, R.-A., Gan, M., Simon, T., 2023. Economic viability of a small scale low-cost aquaponic system in South Africa. J. Appl. Aquacult. 35 (2), 285–304.

Barth, A., Bendag, S., Klüber, P., Kreft, D., Schubert, P., Tegtmeier, D., Wilke, T., 2025. From plant by-products to insects to shrimp: a pathway to sustainable aquaculture feed in a circular economy. Aquac. Nutr. 2025 (1), 7288318.

Bartucz, T., Csókás, E., Nagy, B., Gyurcsák, M.P., Bokor, Z., Bernáth, G., Molnár, J., Urbányi, B., Csorbai, B., 2023. Black soldier fly (*Hermetia illucens*) meal as direct replacement of complex fish feed for rainbow trout (*Oncorhynchus mykiss*) and African catfish (*Clarias gariepinus*). Life 13 (10), 1978.

Beg, M.M., Roy, S.M., Moulick, S., Mandal, B., Kim, T., Mal, B.C., 2024. Economic feasibility study of organic and conventional fish farming systems of Indian major carps. Sci. Rep. 14 (1), 7001.

Bergman, K., Henriksson, P.J., Hornborg, S., Troell, M., Borthwick, L., Jonell, M., Philis, G., Ziegler, F., 2020. Recirculating aquaculture is possible without major energy tradeoff: life cycle assessment of warmwater fish farming in Sweden. Environ. Sci. Technol. 54 (24), 16062–16070.

Boyd, C.E., Tucker, C., McNevin, A., Bostick, K., Clay, J., 2007. Indicators of resource use efficiency and environmental performance in fish and crustacean aquaculture. Rev. Fish. Sci. 15 (4), 327–360.

Boyd, C.E., D'Abramo, L.R., Glencross, B.D., Huyben, D.C., Juarez, L.M., Lockwood, G.S., McNevin, A.A., Tacon, A.G., Teletchea, F., Tomasso Jr., J.R., 2020. Achieving sustainable aquaculture: historical and current perspectives and future needs and challenges. J. World Aquac. Soc. 51 (3), 578–633.

Boyd, C.E., McNevin, A.A., Davis, R.P., 2022. The contribution of fisheries and aquaculture to the global protein supply. Food Secur. 14 (3), 805–827.

Brugere, C., Padmakumar, K.P., Leschen, W., Tocher, D.R., 2021. What influences the intention to adopt aquaculture innovations? Concepts and empirical assessment of fish farmers' perceptions and beliefs about aquafeed containing non-conventional ingredients. Aquacult. Econ. Manag. 25 (3), 339–366.

Chakraborty, P., Mallik, A., Sarang, N., Lingam, S.S., 2019. A review on alternative plant protein sources available for future sustainable aqua feed production. Int. J. Chem. Stud. 7 (3), 1399–1404.

Chattaraj, S., Behera, B.K., Das Mohapatra, P.K., 2024. Effective valorization of Chlorella biomass and Brewers spent grain substituting fish meal and soybean meal in the diet of herbivorous fish Cirrhinus reba for higher growth, digestibility and sustainable cultivation. J. Appl. Phycol. 36 (1), 217–232.

Chattaraj, S., Chattaraj, M., Mitra, D., Ganguly, A., Thatoi, H., Das Mohapatra, P.K., 2025. 16S amplicon sequencing of the gastrointestinal microbiota of Cirrhinus reba and isolation of an autochthonous probiotic using culture based approaches. Syst. Microbiol. Biomanufact. 5 (1), 156–170.

Chattaraj, S., Das Mohapatra, P.K., 2024. Immunostimulatory and antagonistic potential of the methanolic extract of Oedogonium intermedium SCB in Cirrhinus reba challenged with Aeromonas hydrophila. Aquac. Int. 32 (3), 2523–2548.

Chattaraj, S., Sahoo, S.S., Behera, S., Thatoi, H., 2026a. Sustainable aquafeed development using Spirulina: effects on growth performance of Catla catla. Algal Res. 104447.

Chattaraj, S., Ganguly, A., Mitra, D., et al., 2026b. Sustainable valorization of algal biomass in fish nutrition and dietetics: a critical review and scientometric study. Waste Biomass Valor. https://doi.org/10.1007/s12649-026-03619-7.

Chen, F., Qian, J., He, Y., Leng, Y., Zhou, W., 2022. Could *Chlorella pyrenoidosa* be exploited as an alternative nutrition source in aquaculture feed? A study on the nutritional values and anti-nutritional factors. Front. Nutr. 9. https://doi.org/10.3389/fnut.2022.1069760.

Cheng, Z., Mo, W.-Y., Man, Y.-B., Nie, X.-P., Li, K.-B., Wong, M.-H., 2014. Replacing fish meal by food waste in feed pellets to culture lower trophic level fish containing acceptable levels of organochlorine pesticides: health risk assessments. Environ. Int. 73, 22–27.

Choręziak, A., Rosiejka, D., Michałowska, J., Bogdański, P., 2025. Nutritional quality, safety and environmental benefits of alternative protein sources—an overview. Nutrients 17 (7), 1148.

Darwish, R., Gedi, M.A., Akepach, P., Assaye, H., Zaky, A.S., Gray, D.A., 2020. Chlamydomonas reinhardtii is a potential food supplement with the capacity to outperform Chlorella and Spirulina. Appl. Sci. 10 (19), 6736.

de Castro, R.J.S., Ohara, A., dos Santos Aguilar, J.G., Domingues, M.A.F., 2018. Nutritional, functional and biological properties of insect proteins: processes for obtaining, consumption and future challenges. Trends Food Sci. Technol. 76, 82–89.

Defaix, R., Lokesh, J., Frohn, L., Le Bechec, M., Pigot, T., Véron, V., Surget, A., Biasutti, S., Terrier, F., Skiba-Cassy, S., 2024. Exploring the effects of dietary inulin in rainbow trout fed a high-starch, 100% plant-based diet. J. Anim. Sci. Biotechnol. 15 (1), 6.

Eggink, K.M., Gonçalves, R., Skov, P.V., 2025. Shrimp processing waste in aquaculture feed: nutritional value, applications, challenges, and prospects. Rev. Aquacult. 17 (1), e12975.

FAO, 2020. The state of World Fisheries and Aquaculture 2020: Sustainability in Action, pp. 1–244.

Francis, G., Makkar, H.P., Becker, K., 2001. Antinutritional factors present in plant-derived alternate fish feed ingredients and their effects in fish. Aquaculture 199 (3–4), 197–227.

Fry, J.P., Love, D.C., MacDonald, G.K., West, P.C., Engstrom, P.M., Nachman, K.E., Lawrence, R.S., 2016. Environmental health impacts of feeding crops to farmed fish. Environ. Int. 91, 201–214.

Ganguly, A., Chattaraj, S., Ganguly, M., Chattaraj, M., Banerjee, A., Mandal, A., Ali Khan, M., Sen, S.K., Das Mohapatra, P.K., 2025. Effect of three probiotic Bacillus strains supplemented feeds on growth, carcass composition and blood parameters of Clarias magur (Hamilton). J. Appl. Aquac. 37 (1), 19–44.

Glencross, B.D., Huyben, D., Schrama, J.W., 2020. The application of single-cell ingredients in aquaculture feeds—a review. Fishes 5 (3), 22.

Gopan, A., Lalappan, S., Varghese, T., Maiti, M., Peter, R., 2020. Anti-nutritional factors in plant-based aquafeed ingredients: effects on fish and amelioration strategies. Biosc. Biotech. Res. Comm. 13 (12), 01–09.

Haghbayan, S., Shamsaie Mehrgan, M., 2015. The effect of replacing fish meal in the diet with enzyme-treated soybean meal (HP310) on growth and body composition of rainbow trout Fry. Molecules 20 (12), 21058–21066. https://doi.org/10.3390/molecules201219751.

Hameed, A., Majeed, W., Naveed, M., Ramzan, U., Bordiga, M., Hameed, M., Ur Rehman, S., Rana, N., 2022. Success of aquaculture industry with new insights of using insects as feed: a review. Fishes 7 (6), 395.

Hamilton, H.A., Brod, E., Hanserud, O.S., Gracey, E.O., Vestrum, M.I., Bøen, A., Steinhoff, F. S., Müller, D.B., Brattebø, H., 2016. Investigating cross-sectoral synergies through integrated aquaculture, fisheries, and agriculture phosphorus assessments: a case study of Norway. J. Ind. Ecol. 20 (4), 867–881.

Jones, S.W., Karpol, A., Friedman, S., Maru, B.T., Tracy, B.P., 2020. Recent advances in single cell protein use as a feed ingredient in aquaculture. Curr. Opin. Biotechnol. 61, 189–197.

Katoch, R., Tripathi, A., Sood, S., 2018. Possibilities of non-conventional feed resources in livestock feeding—a review. Forage Res. 44, 141–151.

Kazuo, M., 2019. Prevention of fish oil oxidation. J. Oleo Sci. 68 (1), 1–11. https://doi.org/10.5650/jos.ess18144.

Kébé, N.N., Rieker, C., Fall, P.A., Diouf, D., Ndiaye, D., Mockenhaupt, T., Beuel, P., Bursche, J., 2021. Anaerobic co-digestion of fish processing waste with cow manure and waste of market (rests of fruits and vegetables): a lab scale batch test. J. Sustain. Bioenergy Syst. 11 (01), 45.

Khanjani, M.H., Sharifinia, M., Emerenciano, M.G.C., 2024. Biofloc technology (BFT) in aquaculture: what goes right, what goes wrong? A scientific-based snapshot. Aquac. Nutr. 2024 (1), 7496572.

Kolawole, A.A., Ugwumba, A.A.A., 2018. Economic evaluation of different culture enclosures for *Musca domestica* larval production and their utilization for *Clarias gariepinus* (Burchell, 1822) fingerlings diets. Not. Sci. Biol. 10 (4), 466–474.

Kurniawan, S.B., Ahmad, A., Imron, M.F., Abdullah, S.R.S., Othman, A.R., Hasan, H.A., 2025. Achieving a biocircular economy in the aquaculture sector through waste valorization. Toxics 13 (2), 131.

Laveille, P., Uratani, J., Barron, J.G., Brodeur-Campbell, M., Chandak, N.R., George, A., Morin, S., Galvan, A.R., Berthod, M., 2022. Sustainable pilot-scale production of a Salicornia oil, its conversion to certified aviation fuel, and techno-economic analysis of the related biorefinery. Biofuel Bioprod. Biorefin. 16 (1), 27–42.

Leyva-López, N., Lizárraga-Velázquez, C.E., Hernández, C., Sánchez-Gutiérrez, E.Y., 2020. Exploitation of agro-industrial waste as potential source of bioactive compounds for aquaculture. Foods 9 (7), 843.

Magbanua, T.O., Ragaza, J.A., 2024. Selected dietary plant-based proteins for growth and health response of Nile tilapia *Oreochromis niloticus*. Aquac. Fish. 9 (1), 3–19. https://doi.org/10.1016/j.aaf.2022.04.001.

Malcorps, W., Kok, B., van t Land, M., Fritz, M., van Doren, D., Servin, K., Van der Heijden, P., Palmer, R., Auchterlonie, N.A., Rietkerk, M., 2019. The sustainability conundrum of fishmeal substitution by plant ingredients in shrimp feeds. Sustainability 11 (4), 1212.

May, C., Shankar, P., Afriyie-Gyawu, E., Sittaramane, V., 2006. Recycling and repurposing food waste as feed for small-scale zebrafish (*danio rerio*) aquaculture. Zebrafish 23, 4.

Mbaka, M.M., Cox, J.H., Ronan, S., 2022. Aquaculture in Africa: Aquatic Animal Welfare, Impact on the Environment and the Sustainability of the Sector.

McKuin, B., Kapuscinski, A.R., Sarker, P.K., Cheek, N., Lim, J., Sabarsky, M., 2023. Comparative life cycle assessment of marine microalgae, Nannochloropsis sp. and fishmeal for sustainable protein ingredients in aquaculture feeds. Elem. Sci. Anth. 11 (1), 00083.

Mondal, K., Kaviraj, A., Mukhopadhyay, P.K., 2012. Effects of partial replacement of fishmeal in the diet by mulberry leaf meal on growth performance and digestive enzyme activities of Indian minor carp Labeo bata. Int. J. Aquat. Sci. 3 (1), 72–83.

Munguti, J., Muthoka, M., Mboya, J.B., Kyule, D., Meenakshisundaram, M., Tanga, C.M., 2025. Unraveling the potential of black soldier fly larvae as a sustainable protein source for Nile tilapia production in diverse aquaculture systems. Aquac. Nutr. 2025 (1), 3598843.

Muntean, G.-C., Simedru, D., Uiuiu, P., Tanaselia, C., Cadar, O., Becze, A., Coroian, A., 2024. Evaluation of alternative sources of proteins and other nutrients with potential applications in fish nutrition. Molecules 29 (10), 2332.

Munthali, M.G., Chilora, L., Goliath, M., Burke, W.J., Benbow, E., Kang'ombe, J., Safalaoh, A., 2023. The Economic Cost-Benefit Analysis of Black Soldier Fly as an Alternative Animal and Fish Feed ngredient in Malawi.

Nalage, D., Khedkar, G., Kalyankar, A., Sarkate, A., Ghodke, S., Bedre, V.B., Khedkar, C.D., 2016. Single cell proteins. Encyclopedia of food and health 4, 790–794.

Nasser, N., Abiad, M.G., Babikian, J., Monzer, S., Saoud, I.P., 2018. Using restaurant food waste as feed for Nile tilapia production. Aquac. Res. 49 (9), 3142–3150.

Nawaz, M., Tanvir, M.A., Ammar, M., Imran, M., Rehman, I., 2023. Growth enhancement in Labeo rohita by replacing conventional feed with Moringa leaves (Moringa oleifera) powder: growth enhancement in Labeo rohita. Markhor J. Zool. 27–32.

Naylor, R.L., Hardy, R.W., Buschmann, A.H., Bush, S.R., Cao, L., Klinger, D.H., Little, D.C., Lubchenco, J., Shumway, S.E., Troell, M., 2021. A 20-year retrospective review of global aquaculture. Nature 591 (7851), 551–563.

Omeje, J.E., Achike, A.I., Nwabeze, G.O., Ibiyo, L.M.O., Jimmy, S.P., 2023. Economic analysis of locally produced aquaculture feeds with complements of plant-based ingredients in Kainji Lake Basin, Nigeria. Res. World Agric. Econ. 4 (1), 54–61.

Peñalosa Martinell, D., Vergara-Solana, F.J., Almendarez-Hernández, L.C., Araneda-Padilla, M.E., 2020. Econometric models applied to aquaculture as tools for sustainable production. Rev. Aquacult. 12 (3), 1344–1359.

Pereira, A.G., Jimenez-Lopez, C., Fraga, M., Lourenço-Lopes, C., García-Oliveira, P., Lorenzo, J.M., Perez-Lamela, C., Prieto, M.A., Simal-Gandara, J., 2020. Extraction, properties, and applications of bioactive compounds obtained from microalgae. Curr. Pharm. Des. 26 (16), 1929–1950.

Pereira, A.G., Fraga-Corral, M., Garcia-Oliveira, P., Otero, P., Soria-Lopez, A., Cassani, L., Cao, H., Xiao, J., Prieto, M.A., Simal-Gandara, J., 2022. Single-cell proteins obtained by circular economy intended as a feed ingredient in aquaculture. Foods 11 (18), 2831.

Phung, N., Félix, R., Meggiolaro, D., Al-Ashouri, A., Sousa e Silva, G., Hartmann, C., Hidalgo, J., Köbler, H., Mosconi, E., Lai, B., 2020. The doping mechanism of halide perovskite unveiled by alkaline earth metals. J. Am. Chem. Soc. 142 (5), 2364–2374.

Pradeepkiran, J.A., 2019. Aquaculture role in global food security with nutritional value: a review. Transl. Anim. Sci. 3 (2), 903–910. https://doi.org/10.1093/tas/txz012.

Pulina, P., Arru, B., Madau, F., Furesi, R., Gasco, L., 2018. Insect Meal in the Fish Diet and Feeding Cost: First Economic Simulations on European Sea Bass Farming by a Case Study in Italy.

Ritala, A., Häkkinen, S.T., Toivari, M., Wiebe, M.G., 2017. Single cell protein—state-of-the-art, industrial landscape and patents 2001–2016. Front. Microbiol. 8, 2009.

Rotman, F., Stuart, K., Drawbridge, M., 2017. Effects of taurine supplementation in live feeds on larval rearing performance of California yellowtail *Seriola lalandi* and white seabass *Atractoscion nobilis*. Aquac. Res. 48 (3), 1232–1239.

Samaddar, A., 2018. A review of fish meal replacement with fermented biodegradable organic wastes in aquaculture. Int. J. Fish. Aquat. Stud. 6, 203–208.

Sarker, P.K., Kapuscinski, A.R., McKuin, B., Fitzgerald, D.S., Nash, H.M., Greenwood, C., 2020. Microalgae-blend tilapia feed eliminates fishmeal and fish oil, improves growth, and is cost viable. Sci. Rep. 10 (1), 19328.

Sarker, P.K., Figueroa, E., Kapuscinski, A.R., McKuin, B., Schoffstall, B.V., Fitzgerald, D., Greenwood, C., O'Shelski, K., Pasion, E.N., Gwynne, D., 2024. Towards cleaner environment: recycling microalgal co-product to reduce emissions and impacts while eliminating fishmeal in rainbow trout feed for sustainable aquaculture. Environ. Sci. Pollut. Control Ser. 31 (33), 46073–46086.

Shah, A.A., Totakul, P., Matra, M., Cherdthong, A., Hanboonsong, Y., Wanapat, M., 2022. Nutritional composition of various insects and potential uses as alternative protein sources in animal diets. Anim. Biosci. 35 (2), 317.

Sharif, M., Zafar, M.H., Aqib, A.I., Saeed, M., Farag, M.R., Alagawany, M., 2021. Single cell protein: sources, mechanism of production, nutritional value and its uses in aquaculture nutrition. Aquaculture 531, 735885.

Shipton, T., Hecht, T., Shipton, T., 2013. Economic, regulatory and legal review of feed management practices. In: On-Farm Feeding and Feed Management in Aquaculture, pp. 565–585.

Shivashankar, M., Prem, K.S.N., Sanketh, C.V., Rajegowda, Satish Gowda, C.S., 2024. Economic feasibility and trends in fish production: a case study of composite fish culture in Karnataka. Int. J. Res. Agron. 7 (10S), 614–617. https://doi.org/10.33545/2618060x.2024.v7.i10si.1855.

Shurson, G.C., 2018. Yeast and yeast derivatives in feed additives and ingredients: sources, characteristics, animal responses, and quantification methods. Anim. Feed Sci. Technol. 235, 60–76. https://doi.org/10.1016/j.anifeedsci.2017.11.010.

Tacon, A.G.J., 2020. Trends in global aquaculture and aquafeed production: 2000–2017. Rev. Fish. Sci. Aquacult. 28 (1), 43–56. https://doi.org/10.1080/23308249.2019.1649634.

Ticina, V., Katavic, I., Grubisic, L., 2020. Marine aquaculture impacts on marine biota in oligotrophic environments of the Mediterranean Sea-a review. Front. Mar. Sci. 7.

Tocher, D.R., 2003. Metabolism and functions of lipids and fatty acids in teleost fish. Rev. Fish. Sci. 11 (2), 107–184. https://doi.org/10.1080/713610925.

Wang, Y., Deng, M., Zhou, S., Li, L., Song, K., 2024. Increasing fish production in recirculating aquaculture system by integrating a biofloc-worm reactor for protein recovery. Water Res. X 24, 100246.

Widiyastuti, T., Rahayu, S., Suryapratama, W., Suhartati, F., 2024. Nutrient Profile, Protease and Cellulase Activities of Protein Extracted.

Wong, M.-H., Mo, W.-Y., Choi, W.-M., Cheng, Z., Man, Y.-B., 2016. Recycle food wastes into high quality fish feeds for safe and quality fish production. Environ. Pollut. 219, 631–638.

Wohlfarth, G.W., Schroeder, G.L., 1979. Use of manure in fish farming—a review. Agric. Wastes 1 (4), 279–299. https://doi.org/10.1016/0141-4607(79)90012-x.

Yakubu, C., Sharma, R., Sharma, S., Singh, B., 2021. Effects of Alkaline Fermentation Time on Anti-nutrients, Structural Interactions and Bioactive Profile of Locust Bean (Parkia Biglobosa).

Zhang, Y., Zhou, Z., Liu, Y., Cao, Y., He, S., Huo, F., Qin, C., Yao, B., Ringø, E., 2014. High-yield production of a chitinase from Aeromonas veronii B565 as a potential feed supplement for warm-water aquaculture. Appl. Microbiol. Biotechnol. 98 (4), 1651–1662.

Zuluaga-Hernandez, C.D., Hincapie, C.A., Osorio, M., 2023. Non-conventional ingredients for tilapia (*Oreochromis* spp.) feed: a systematic review. Fishes 8 (11), 556.

Further reading

Ndebele-Murisa, M., Mubaya, C., Kapute, F., Samundengo, A., Yossa, R., 2024. Development and Scaling of Sustainable Feeds for Resilient Aquatic Food Systems in Sub-Saharan Africa: Life Cycle Assessment. 1. https://cgspace.cgiar.org/items/b925a745-fdf2-4729-a95d-25916f45eb5d. (Accessed 27 August 2024).

CHAPTER

19 Future perspectives and research directions in fish feed formulations with non-conventional agricultural byproduct

Hrudayanath Thatoi[1], Mehak Kanwar[1], Pradeep K. Das Mohapatra[2], Partha Bandyopadhyay[3], Shabana Arjmand[4], Arindam Ganguly[5] and Sourav Chattaraj[1]

[1]*Centre for Industrial Biotechnology Research, School of Pharmaceutical Sciences, Siksha 'O' Anusandhan Deemed to be University, Bhubaneswar, Odisha, India;* [2]*Department of Microbiology, Raiganj University, Raiganj, West Bengal, India;* [3]*Finray Biotech Private Ltd, Ahmedabad, Gujarat, India;* [4]*Government Degree College, Srinagar, Kashmir, India;* [5]*Department of Microbiology, Bankura Sammilani College, Bankura, West Bengal, India*

1. Introduction

Aquaculture has developed into one of the fastest-developing segments of global food production systems and contributes significantly to food security, nutritional well-being, job creation, and economic development (Chattaraj et al., 2022). It appears that contemporary evaluations of the current state of aquaculture demonstrate that aquaculture currently provides a large portion of the aquatic food consumed globally and that it has expanded rapidly relative to capture fisheries (Henriksson et al., 2021; Costello et al., 2023). This growth is reflective of advancements in technology, increasing demand for seafood, and increased concern about the sustainability of production inputs such as feeds. Feed formulations are not only the greatest expense for the operation of aquaculture systems based on intensive production methods, but they also influence the level of environmental impact of the production process through greenhouse gas (GHG) emissions, nutrient runoff, and the overall efficiency of resource use (Henriksson et al., 2012; Agriculture Organization of the United Nations, 2018).

Fishmeal (FM) and fish oil (FO) have historically been two of the key ingredients in aquafeeds due to their high digestibility, favorable amino acid profile, and favorable long-chain omega-3 fatty acid profiles (Chattaraj et al., 2024a,b). The continued reliance on marine raw materials for aquafeeds has become increasingly constrained by ecological limitations, variability in supply, and economic uncertainty (Chattaraj and Das Mohapatra, 2024; Pelletier et al., 2009). As noted previously, environmental sustainability studies consistently show that feed production is one of the largest

Fish Feed Formulation with Agroindustrial Wastes and Byproducts. https://doi.org/10.1016/B978-0-443-41728-3.00018-6

contributors to the life cycle environmental impacts of aquaculture, which emphasizes the need for the identification of nutritionally equivalent and environmentally sustainable alternative feed ingredients (Henriksson et al., 2012). Therefore, the diversification of feed ingredients is not simply an option in terms of adjusting the formulation of aquafeeds to address issues associated with the sustainability of production inputs, but rather a structural imperative for the long-term sustainability of aquaculture.

Concurrent to the trends observed in the previous paragraph, agricultural and agroindustrial production generates massive amounts of byproducts and residue streams, many of which remain unused in spite of their rich composition. Global assessments of food losses and agricultural waste indicate the magnitude of the potential biomass streams available for the purpose of providing additional value-added applications in circular bioeconomy frameworks (Yadav et al., 2024). The development of new technologies in bioprocess engineering, including fermentation, enzymatic hydrolysis, microbial bioconversion, and precision drying technologies, has significantly enhanced the feasibility of converting oilseed cakes, cereal byproducts, fruit and vegetable residue, etc., into functional feed ingredients (Kusumowardani and Tjahjono, 2020). These approaches to revalue biomass streams are aligned with sustainable intensification principles by reducing waste streams, decreasing pressures on marine ecosystems, and improving resource efficiencies.

The use of plant-based proteins represents the area of research that has been explored the most as a substitute for fishmeal (FM). Soybean meal, pea products, oilseed cakes, and cereal co-products, for example, have the potential to provide scalability and economic feasibility; however, their inclusion in diets can be limited by antinutritional factors (ANF's), fiber content, and unbalanced amino acid compositions (Francis et al., 2001). More recent investigations using alternative plant residues including red vetchling, pea peels, pumpkin seed cake, and watermelon rind derivatives indicate that optimal inclusion rates can support growth performance, feed conversion ratios, and physiological stability in a variety of species (Mente et al., 2019; Gasco et al., 2019; Hobbi et al., 2022; Rangel, 2023). These findings emphasize the importance of the processing of ingredients, dietary supplementation, and species-specific evaluation in optimizing the functional benefits of plant-based feed ingredients.

Protein-rich meals produced from insects, specifically from black soldier fly larvae (BSFL) (Hermetiaillucens), have emerged as nutritionally robust and environmentally efficient protein sources. Feeding experiments and controlled field trials conducted in Atlantic salmon, tilapia, and carp indicate that partial substitution of FM with IM can maintain growth parameters and nutrient utilization and promote immune competence and gut health (Huyben et al., 2024; Albrektsen et al., 2022). In addition to protein sources, insect biomass contains chitin and bioactive peptides, which may also contribute to functional properties of insect-derived feed, although threshold values for inclusion must be established (Makkar et al., 2014). Although technical feasibility exists, consumer perceptions remain a critical factor determining market access as evidenced by recent studies examining the socio-demographic determinants of acceptability of fish fed insects (De Koning et al., 2020). These studies demonstrate that for the implementation of sustainable feed transition programs, there will be a need for parallel advances in both scientific validation and stakeholder engagement.

Microbial and microalgal proteins are becoming yet another area of innovation in the development of sustainable aquafeed options (Chattaraj and Das Mohapatra, 2024; Ganguly et al., 2025). Proteins derived from single-cell organisms, including bacteria, yeast, and microalgae, offer high-quality amino acid profiles and, in certain instances, contain high levels of DHA and other long-chain polyunsaturated fatty acids that were traditionally obtained from fish oil (Santos et al., 2025). Empirical evaluations of *Schizochytrium* sp. and other algal oils demonstrate the ability to replace FO in the diets of salmonid and marine fish without negatively impacting growth performance or tissue fatty acid composition (Sprague et al., 2016; Venkata et al., 2020). Other low-trophic resources, including polychete meal, shrimp processing waste, biofloc biomass, and integrated aquaponics-derived feed inputs, further increase the number of circular alternatives being investigated (Gasco et al., 2018; Medeiros et al., 2025; Avnimelech, 2009).

Together, these emerging innovations represent a fundamental shift toward the integration of circular bioeconomy practices within the context of aquaculture nutrition. The intentional incorporation of non-conventional agricultural byproducts into aquafeeds address several aspects of sustainability: the reduced dependence on finite marine resources, improvement in the efficiency of nutrient recycling, decreased environmental impacts, and improved economic resilience. However, the widespread adoption of these innovations in commercial aquaculture settings requires comprehensive evaluations addressing aspects including digestibility, metabolic effects, immune modulation, environmental performance, economic viability, and social/cultural acceptance (Chattaraj et al., 2023, 2024c).

Beginning with an overview of the background, we then explore recent advances in fish feed formulation using unconventional agricultural by-products. The synthesis of evidence from the use of plant residues, insect meals, microbial protein, and integrated production systems is intended to outline potential directions for sustainable development of aquaculture as it relates to nutritional sufficiency, environmental sustainability, and economic feasibility.

2. Classification and functional evaluation of non-conventional agricultural byproducts in aquafeed formulation

The diversification of aquafeed ingredients has led to systematic classification of non-conventional feed resources based on origin, biochemical composition, trophic level, and functional properties. Broadly, these alternatives can be categorized into (i) plant-derived agricultural residues and oilseed byproducts, (ii) insect-based proteins, (iii) microbial and single-cell proteins, (iv) microalgae and algae-derived lipids, and (v) other low-trophic or circular bioresource inputs such as polychetes, biofloc biomass, and seafood processing wastes. Each category presents distinct nutritional advantages, technological considerations, and sustainability implications.

2.1 Plant-derived agricultural residues and oilseed byproducts

Oilseed cakes from sunflowers, pumpkin seeds, soybeans, cereals, legume peas, and watermelon rinds, and legumes like red vetchling have been identified as having a high potential for use as partially replacing fishmeal in aquafeeds on account of their relatively well-established supply channels and wide geographic distribution. Inclusion of oilseed cakes in diets of Nile tilapia has been demonstrated through empirical assessments to improve growth performance, FCR, antioxidant status, and immune system parameters at appropriate inclusion levels (Hobbi et al., 2022). Similar results were observed with respect to the digestibility of pea peel meal and red vetchling derivatives included in diets of other fish species at nutritional thresholds specific to each species (Mente et al., 2019; Gasco et al., 2019). The incorporation of moderate amounts of watermelon rind into Labeorohita diets was also found to produce positive effects on hematological and biochemical response characteristics (Rangel, 2023).

Despite the apparent benefits associated with the inclusion of plant by-products in aquafeeds, they often contain ANFs such as phytates, protease inhibitors, and NSPs, which may negatively impact the bioavailability of nutrients and/or cause damage to intestinal tissue (Francis et al., 2001). Therefore, processing technologies such as fermentation, enzyme treatments, extrusion, and thermal treatments may be necessary to mitigate some of the negative impacts of ANFs. Formulation guidelines developed to optimize the inclusion of plant by-products in aquafeeds emphasize the importance of supplementing essential AA, mineral balancing, and enhancing digestibility to achieve maximum physiological compatibility across species.

2.2 Insect-based proteins

Alternative protein sources are now at the forefront of research for the next generation of aquaculture feed ingredients. One of these alternatives is the black soldier fly (*Hermetia illucens*) larva (BSFL). The nutritional profiles of BSFLs provide an optimal mix of essential amino acids with higher concentrations of protein than many other conventional alternative proteins, while their production systems allow for the utilization of organic waste streams thus providing a strong foundation for a circular bio-economy model (Huyben et al., 2024; Albrektsen et al., 2022).

In addition to the nutritional benefits of BSFL, several studies have shown that it is possible to replace significant portions of fishmeal with BSFL without negatively affecting the growth performance or the quality of the fillets when fed commercially (Huyben et al., 2024). In similar studies conducted using carp and tilapia, BSFL was shown to support the growth metrics of both fish and alter the expression of genes associated with the immune system (Albrektsen et al., 2022). Furthermore, the chitin fraction of insect frass has been shown to stimulate the immune response in fish, which could enhance the ability of fish to fight diseases (Makkar et al., 2014).

While there appears to be agreement regarding the potential use of BSFL as an alternative to fishmeal, the variability in the digestibility of BSFL and its lipid composition are influenced by the type of substrate used during rearing and the process methods used for the BSFL meal (Hobbi et al., 2022). Another variable that will influence the scalability of insect based feed is consumer acceptance. Recently, a study was completed that analyzed the behavior of European consumers toward sustainable seafood products including fish that were fed insects. The results showed that sustainability awareness significantly increases acceptance of fish that are fed insects (De Koning et al., 2020). Therefore, the scalability of insect-based feeds depends not only upon the nutritional validation of such products, but also on regulatory frameworks and market perceptions.

2.3 Microbial and single-cell proteins

Microbial proteins are usually single cell proteins (SCP) that have a higher yield of protein than other types of proteins and require relatively less land and water. Cultivation of microbial biomass is possible through the use of industrial by-products and renewable substrate to further improve on resource utilization (Santos et al., 2025). Improvements in fermentation technology has allowed for the production of increased amounts of biomass as well as to better meet nutritional requirements of animals by improving on the digestibility of the biomass and reducing nucleic acid content. Genetically engineered strains along with optimized fermentation conditions may help improve upon this digestibility and ultimately allow for SCPs to replace FM in feeds. There are studies showing compatibility of bacterial protein-based diets with European seabass without negative impacts on their health (Francis et al., 2001). The potential exists for SCP's to be included into commercially available aquafeeds; however they would need to be formulated in combination with strategies to optimize lipid inclusion in order to maximize benefits.

2.4 Microalgae and algae-derived lipids

Microalgae are an ingredient class which serve two functions: they are a source of protein, and they are also a source of lipids. Due to their high DHA (docosahexaenoic acid), species of microalgae, like *Schizochytrium* sp., can be used as a sustainable substitute for fish oil (Sprague et al., 2016). Studies on controlled feeding in sablefish and Atlantic salmon have shown that algal oils could completely or partially replace fish oil without affecting the fish's growth or fatty acid profile (Sprague et al., 2016; Venkata et al., 2020).

Microalgal biomass is also a source of bioactive, such as carotenoids, antioxidants, and polysaccharides; these can potentially provide benefits to gut health and oxidative stability (Chattaraj et al., 2026a,b). Therefore, the inclusion of algae-based products into aquafeed will help address both nutritional adequacy and added functional value. There still are, however, several remaining challenges to overcome prior to being able to make large scale production of algae-based aquafeeds economically competitive. Those challenges include the need for continued development of improved technology.

2.5 Low-trophic and circular bio-resource inputs

Additional feed alternatives to traditional fishmeal include polychete meal, shrimp waste and other derivatives from shrimp processing, insect exoskeleton (exuviae) and biofloc derived from recirculation or integrated aquaculture systems. A European seabass study found that feeding European seabass diets partially replaced with polychete meal was equivalent to those fed diets completely composed of fishmeal in terms of growth performance (Gasco et al., 2018). Shrimp processing wastes and insect exoskeleton contain chitin and can be used as an additional source of protein that could be incorporated into formulated feeds (Medeiros et al., 2025). Bioflocs represent a very unique form of aquaculture integration; they are aggregates of microorganisms grown in aquaculture systems that are then collected and converted into protein rich feed components (Avnimelech, 2009). Biofloc harvesting increases recycling of nutrients, decreases effluent and supports closed-loop aquaculture production.

These three categories illustrate the increasing list of agricultural by-products and alternative protein sources being developed for use in aquaculture feeds. While nutritional adequacy is still one of the first considerations when evaluating alternative ingredient systems for aquaculture, it is also important to evaluate digestibility, immunomodulatory activity, environmental footprint, cost-effectiveness, and social acceptability when determining whether alternative ingredient systems are viable over time (Table 19.1). Future interdisciplinary studies combining feed science, microbial biotechnology, environmental impact assessment, and socioeconomic assessment will be necessary to refine these novel ingredient systems.

3. Processing technologies and bioconversion strategies for enhancing the nutritional value of agroindustrial byproducts in aquafeeds

The integration of unconventional agro-industrial by-products into aqua-feed formulations is dependent upon both ingredient availability and technological advancement to improve the nutritional value, digestibility, safety, and functionality of the feed. Most raw agro-industrial by-products are complex, have varying levels of nutrients and may contain antinutrients that will require specific preprocessing strategies to render them suitable for aquaculture nutrition. As such, advances in the area of bio-process engineering (enzymatic modification, microbial fermentation and

Table 19.1 Multi-criteria sustainability matrix for alternative aquafeed ingredients.

Sustainability dimension	Insect-based proteins	Microalgal SCP	Yeast/Bacterial SCP	Agro-industrial by-products
Protein quality	High crude protein; generally balanced essential amino acids	Moderate–high protein; variable digestibility	High protein concentration; favorable amino acid profile	Moderate protein (oilseed meals); may require amino acid supplementation
Lipid contribution	Contains medium-chain fatty acids; limited long-chain omega-3	Can provide DHA/EPA depending on species	Minimal long-chain omega-3 contribution	Typically, low omega-3; depends on source
Digestibility considerations	Chitin may reduce digestibility at high inclusion levels	Cell wall rigidity may limit nutrient access without processing	Nucleic acid content requires reduction; digestibility generally high postprocessing	Fiber and antinutritional factors may reduce digestibility
Environmental resource use	Potentially low land footprint when reared on side streams; energy use depends on drying methods	Land-efficient cultivation; energy demand varies with system design	Fermentation systems scalable; carbon source determines footprint	Utilizes existing waste streams; minimal additional land requirement
Circular economy potential	Converts organic waste streams into protein	May integrate with industrial CO_2 capture systems	Can valorize industrial carbon substrates	Direct valorization of agroprocessing residues
Greenhouse gas profile	System-dependent; varies by substrate and energy inputs	Dependent on cultivation energy source	Dependent on fermentation energy and substrate	Generally low additional emissions when derived from existing waste
Functional bioactivity	Chitin, antimicrobial peptides, immune-modulatory potential	Carotenoids, omega-3 fatty acids, antioxidant activity	Beta-glucans, immunostimulatory components	Fermentation can enhance prebiotic potential
Economic feasibility	Production cost decreasing but still higher than conventional plant meals	Often high production cost; improving with scale	Industrial scalability favorable; cost depends on substrate	Typically cost-effective; widely available
Regulatory maturity	Increasing approval in aquaculture; substrate restrictions apply in some regions	Generally accepted for feed use; species-specific regulations apply	Widely approved as feed additives; inclusion levels regulated	Long-standing approval in feed systems
Scalability	Expanding industrial capacity; still emerging globally	Established for select species; capital-intensive	Highly scalable fermentation infrastructure	Immediately scalable due to existing supply chains

heat treatment) can be seen as critical to making these alternative materials available for use in aqua-feeding.

3.1 Fermentation and microbial bioconversion

Microorganisms through microbial fermentation are able to effectively break down nondigestible antinutritional factors (ANF) such as phytic acids, polyphenols (tannins), and protease inhibitors; enhance protein digestibility; and increase the availability of amino acids from plant-based sources. A number of studies have demonstrated that both lactic acid bacteria and fungi used in fermentation processes can modify the nutritional quality of plant-based feed ingredients. These modifications occur through increased production of bioactive peptides and alteration of fiber structure (Ajila et al., 2012).

Improvements in fish performance (growth rate), gut anatomy, and reproduction were documented when fed plant-based feeds that had undergone fermentation relative to diets containing unfermented plant-based meals (Van Huis, 2013). Improved digestive health was observed due to the presence of a fermentation effect and the development of "probiotic" like gut flora populations which promote the development of the beneficial gut microbiome and support intestinal barrier integrity. In addition to providing a positive impact on animal health and productivity, fermentation of low value biomass provides a means to convert these materials into high value feed component products that will be more metabolically compatible with animals.

3.2 Enzymatic treatment and digestibility enhancement

The supplementation of enzymes (either via enzymatic treatment of ingredients during processing or through addition of enzymes to formulated rations) represents an additional method for increasing nutrient utilization. Commonly used enzymes include phytase, protease and carbohydrase, which are used to break down ANF's and the complex polysaccharides that are often associated with them. This action increases the bioavailability of minerals and enhances the assimilation of protein (Woodgate et al., 2022). Enzyme assisted treatments of oilseed cake and cereal by-product reduce fiber complexity and improve digestible coefficient values; especially in species that have poor endogenous enzyme production.

It is also important to consider the use of enzymatic technology when considering diets containing a high inclusion of plant based materials, because the presence of undigested fibers in these rations can reduce feed conversion ratios. In studies where enzyme added rations were compared to control rations, it was reported that there was increased retention of nutrients and decreased excretion of nitrogen; ultimately providing both economic advantages and environmental advantages (Woodgate et al., 2022). These results illustrate that while the substitution of ingredients is an effective method for optimizing feed quality, the optimization of biochemical pathways via the use of enzymatic supplements will enhance the nutritional quality of animal products.

3.3 Thermal processing and extrusion technologies

Pellet formation is just one of several ways that aqua feed manufacturers use to produce their products using commercially available methods such as extrusion cooking and controlled thermal processing. Extrusion cooking provides a method to improve the physical stability of the starch component in aqua feeds by improving gelatinization of starches during the extrusion process; it also may be able to inactivate some of the heat labile Anti-Nutritional Factors (ANF) present in many plant derived protein sources. Extrusion cooking also allows for optimization of the nutritional value and structure of pellets for long term utilization under aquatic conditions.

However, if there is too much thermal energy applied in the extrusion process then the amino acids that are released from the plant material during the extrusion process may be degraded into compounds with less nutritional value or the lipid components may undergo degradation resulting in loss of functionality; thus it is essential to have process controls in place to prevent overprocessing. Therefore, the combination of temperature, moisture content and mechanical shear are all factors that will determine how well nutrients are retained and utilized after consumption by fish. Thus, process engineering continues to be an important aspect of utilizing new aqua feed ingredient technologies at large scale in commercial aquaculture operations (Makkar et al., 2014).

3.4 Insect biomass processing optimization

Nutrient content of an insect meal depends on many factors such as: substrate used in rearing insects; stage at which the insects are harvested; and how they are processed after harvesting. Methods of defatting, drying, and grinding can also greatly impact: total amount of protein present; lipid type present; and the rate of nutrient digestion (Huyben et al., 2024; Albrektsen et al., 2022). Therefore, controlled drying processes are required to maintain the structural integrity of amino acids while preventing oxidation of lipids from occurring. Additionally, removing part or all of the chitin molecule from the insect's exoskeleton may be beneficial in increasing the digestive rates of proteins in those animals that have little to no chitinase enzyme activity. Consequently, it is important to establish standard insect processing procedures in order to provide a consistent nutrient quality product and to ensure regulatory compliance within feed production and distribution systems.

3.5 Microalgal cultivation and lipid extraction technologies

The feasibility of using microalgae instead of fish oil is influenced by both the efficiency of cultivating microalgae and extracting lipids from them. Heterotrophic fermentation of *Schizochytrium* sp. has increased the yield and economic value of DHA (Sprague et al., 2016; Venkata et al., 2020) through advances in the process. Additionally, the development of new processes for the solvent-free extraction of algal lipids (such as supercritical CO_2) are being studied to improve the chemical stability of extracted lipids while reducing the negative environmental impacts associated with traditional lipid extraction methods.

Whole algal biomass may provide a means of producing both protein and lipid fractions; however, this method of production remains problematic due to poor digestibility and difficulty in disrupting the cell walls. Optimizing the factors that affect the cultivation of microalgae (e.g., light intensity, nutrients available, etc.) directly affects the fatty acid composition of the biomass produced and the ability to scale-up the process.

3.6 Integrated circular production systems

The future of aquaculture feed production is a sustainable one where aqua-feed production will occur in the same ecosystem as aquaculture. For example, bioflocs are a type of microbe that forms in areas with high levels of nutrients in the water and produces protein rich biomass that can be converted to an aquaculture feed. Aquaponics has also shown great promise to recycle nutrients from fish to plants and vice versa in a closed loop system which would allow for the creation of biomass on site (Avnimelech, 2009).

The integration of aquaculture feed production into aquaculture itself will lead to reduced dependence on outside sources for inputs and improved efficiencies in the conversion of nutrients to biomass. These closed-loop approaches will embody many of the principles of a circular bio-economy and greatly reduce the environmental impacts associated with traditional aquaculture feed production practices.

To take advantage of this approach we need to continue to develop the technologies to convert non-traditional agricultural by-products into nutritionally complete and commercially viable aquaculture feeds. The use of processing technologies to enhance the nutritional quality of these alternative feed ingredients is necessary to ensure that they have sufficient digestibility, are free from antinutritional factors (i.e., phytochemicals), and have stable nutrient profiles that meet the needs of aquaculture species. Additionally, processing technologies can help to ensure that these alternative feed ingredients exhibit the functional properties needed to support healthy growth of aquatic animals. In order to advance this area of research further, it is essential to evaluate the total life cycle impact of different processing pathways to identify those that result in the least amount of waste and GHG emissions, optimize costs, and to develop and apply metrics to measure the sustainability of new technologies.

4. Functional impacts of non-conventional feed ingredients on fish growth, physiology, and product quality

The integration of non-conventional agricultural byproducts into aquafeed formulations ultimately requires validation through biological performance outcomes. Beyond ingredient substitution and technological processing, the critical determinant of feasibility lies in the physiological responses of cultured species. Growth performance, feed conversion efficiency, immune modulation, oxidative stability, gut microbiota composition, and fillet quality collectively define the functional adequacy of alternative feed systems.

4.1 Growth performance and feed utilization efficiency

Specific growth rate (SGR) and feed conversion ratio (FCR) continue to be the most common indicators for assessing performance from novel feed sources. The literature indicates that it is possible to maintain an SGR and FCR similar to those obtained using traditional fish meal based diets, if diets containing up to 50% of the fish meal content are replaced by alternative protein sources (e.g., insect meal, plant derived coproducts, microbial protein or low trophic resource), provided the diets are nutritionally optimal.

Examples include, the replacement of fish meal by black soldier fly larvae (BSFL) meal in Atlantic Salmon demonstrated that BSFL meal could be used at commercial scale when the formulation was appropriate for the level of inclusion (Huyben et al., 2024), and this has also been shown in tilapia and carp which have been fed insect based proteins and had an efficient use of nutrients and accumulated biomass (Albrektsen et al., 2022). Pumpkin seed cake and red vetchling are examples of plant derived residues that, through controlled dietary optimization, have produced improved or similar growth parameters compared to control diets (Mente et al., 2019; Hobbi et al., 2022). It is however important to note that completely replacing fish meal may result in species dependent thresholds for the replacement of fish meal, hence the need for adequate amino acid balance and lipid supplementation. As such digestibility coefficients, protein retention efficiency and nitrogen utilization should be evaluated in addition to growth indexes to ensure metabolic completeness.

4.2 Gut health and microbiota modulation

Intestinal health is now being seen as an essential component for sustainable aquaculture. Alternative feeds are known to have an effect on the gastrointestinal system of fish through the modification of the gut, its mucosa, and the microbes present in it. Modulation of the intestinal microbiota in comparison to conventional feeds has been observed in studies that evaluate alternative feed raw materials. Components from fermented plant material, insect derived feed and microalgae may provide prebiotic/immunomodulatory benefits in addition to enhance the growth performance of fish by providing the optimal environment for beneficial microbial communities and optimizing the uptake of nutrients by the fish (Van Huis, 2013; Woodgate et al., 2022; Ringø et al., 2010). These examples illustrate the complex relationship between feed, the host, and the microbiota as it relates to the ability of fish to be healthy over the long term and their ability to resist disease.

4.3 Immune response and oxidative stability

A great deal of research has shown that alternative protein sources also have bioactive compounds that can help stimulate an animal's immune system (e.g., its natural defense against pathogens). These studies show that the feeding of fish with insect-based feed has resulted in improved levels of expression of genes related to the immune response as well as improved markers of disease resistance in both carp and tilapia (Albrektsen et al., 2022). Additionally, feeding fish bioactive plant residue or algae-derived polysaccharides have increased the antioxidant activity of enzymes in the liver as well as maintained a healthy balance of oxidation in the liver compared to traditional feeds (Hobbi et al., 2022; Ringø et al., 2010).

While there are many potential mechanisms by which bioactive compounds from alternative protein sources may modulate an animal's immune function (i.e., influence immune cell development, cytokine production etc.) we believe that research into these mechanisms would be valuable. Nonetheless, current data indicate that if properly prepared, certain unconventional feed ingredients can provide stable immunity without negatively affecting physiological homeostasis.

4.4 Fillet quality, fatty acid profile, and consumer-relevant attributes

The use of substitute lipid sources instead of fish oil has raised several concerns about quality of the final product. Fish oil substitute microalgal oil containing high levels of DHA have shown to be as good at maintaining the fatty acid profile of tissues from salmonids and marine species as fish oil when using a large portion of fish oil replacement (Sprague et al., 2016; Venkata et al., 2020). This is especially important for the retention of the nutritional quality of farmed fish destined for human consumption. The inclusion of insect meal into feed can alter the fatty acid profile dependent upon the substrate and processing conditions of the insect meal. Plant-based lipids could potentially lower the level of long-chain omega-3's if they are not supplemented with an algal source. As such, it is becoming increasingly common for formulation strategies to include combinations of protein and lipid replacement to maintain the quality of both sensory and nutritional aspects of products.

Socio-cultural perceptions of consumers intersect with their perception of quality of products. Research studies evaluating consumer acceptance of fish produced through feeding insects have determined that consumers who are aware of the environmental benefits of insect production will accept this type of fish as a viable option as long as the quality of the fish remains unchanged (De Koning et al., 2020). Consequently, evaluations of product quality must assess the physical (chemical) and perceptual (sensory) aspects of the product.

4.5 Species-specific responses and nutritional precision

The responses of different taxonomic groups to alternative feed ingredients vary as a result of differences in their digestive physiology, metabolic capabilities, and trophic classifications. Therefore, carnivorous marine species generally are less tolerant of higher plant inclusion levels than are many of the omnivorous freshwater species. As a result, precision nutrition approaches (e.g., amino acid supplementation; enzyme inclusion; lipid balancing) are critical to maximize performance outcomes.

Therefore, future studies on this subject should focus on developing new and innovative integrative multi-omics approaches to further elucidate how fish adapt metabolically, how the innate immune system signals in response to novel diets, and how the gut microbiota interacts with novel feed systems. The findings from these types of mechanistic studies will be beneficial in creating more accurate predictive formulation models and in improving the translational scalability of those formulations.

In summary, while the results of the present study confirm that non-traditional agricultural by-products can provide nutritional support to grow-out fish without compromising either their growth or their physiological stability or product quality, there are still species-specific threshold levels of acceptable performance that must be established through a comprehensive biological evaluation process for each species being studied.

5. Environmental sustainability, life-cycle assessment, and circular bioeconomy implications

The transition toward non-conventional agricultural byproducts in aquafeed formulation is fundamentally driven by environmental sustainability considerations. As aquaculture production intensifies globally, feed production remains one of the most significant contributors to resource consumption, GHGemissions, eutrophication potential, and land-use pressure within aquaculture life-cycle assessments (LCAs) (Henriksson et al., 2012). Consequently, the environmental performance of alternative feed ingredients must be evaluated not only in terms of biological efficacy but also through comprehensive sustainability metrics.

5.1 Life-cycle environmental footprint of conventional versus alternative feeds

In fact, all life-cycle assessments have found that fish meal and fish oil production are among the most significant contributors to environmental impact in fish-based feed for carnivorous aquaculture systems (i.e., marine resource extraction, and energy consumption) (Henriksson et al., 2012; Pelletier et al., 2009). Although marine feed ingredients are nutritionally efficient, there is a constraint in obtaining them due to limited availability of wild fish and variability in ecosystems. Alternative sources of feed ingredients include plant-based byproducts, insect-based feed ingredients, and microbially-produced proteins. These alternative feed ingredients provide an opportunity for producing aquafeeds independently from dependence on direct marine resources.

The environmental performance of alternative feed ingredients will vary depending on how they are produced; including factors such as design of the production system, source of substrates used, and degree of processing of the final product. Insect production may reduce land demand when based on low-value side streams, but outcomes depend on energy inputs and processing; therefore, cradle-to-gate LCA is required (Huyben et al., 2024; De Koning et al., 2020). Another example includes the cultivation of microbially-produced proteins which could potentially alleviate land pressures associated with feed production, however, the energy required for fermentation conditions would need to be considered (Santos et al., 2025). Therefore, when assessing the sustainability of alternative feed ingredients, it is essential to conduct full life-cycle assessments of these products (from cradle-to-gate), as opposed to comparing individual ingredients.

5.2 Resource efficiency and nutrient recycling

One of the most significant benefits from utilizing the byproducts of the agroindustry is the increased efficiency of nutrient recovery through recycling. The process of valorizing oilseed cake, cereal bran, fruit residue, and other food-processing wastes reduces organic waste flow while also recovering the embedded protein, lipid, and micronutrient content of these materials for productive use (Yadav et al., 2024; Kusumowardani and Tjahjono, 2020). This approach is therefore directly aligned with the circular bioeconomic principle that seeks to reuse and repurpose biological resources to maximize their material lifecycle.

The integrated nature of systems such as biofloc technology and aquaponics provides an additional benefit by maximizing the capture and conversion of metabolic waste into microbial biomass or plant growth substrate (Avnimelech, 2009). These approaches provide two major environmental benefits, including reducing nitrogen and phosphorus discharge into surrounding water bodies and improving feed-use efficiency. Therefore, from a systems perspective, coupling innovations in animal feed production with waste management strategies increases the environmental sustainability and ability of agricultural production to be intensified sustainably.

5.3 Carbon footprint and land-use considerations

Plant-based feed ingredients' environmental benefits are often diminished by the large-scale monocultures required to produce them. These monocultures can lead to forest removals for land use, high water usage for irrigation, and GHG emissions from fertilizers used to support these monocultures. Therefore, whether or not plant-based protein is the more sustainable choice of protein will depend on how the plant-based protein is sourced and who governs the supply chain (Henriksson et al., 2012). Using by-products of agriculture rather than competing with humans for food, it creates a much more sustainable model.

Cultivating microalgae also has many land-use advantages compared to other forms of protein production because it can be produced in nonarable areas. Energy input into heterotrophic fermentation systems to produce algae oil for fish meal alternatives may significantly decrease land competition and preserve the omega-3 supply for aquaculture (Sprague et al., 2016; Venkata et al., 2020). However, there are still two major factors that determine the total carbon footprint of this form of production: energy input and scale of production.

5.4 Socioeconomic and market dimensions of sustainability

Fish farmers' choices for feed are influenced by their perceptions of economic risk and by how much they value time now compared to the future. When consumers feel informed about how the use of insects will make fish more sustainable, they may be more likely to accept the idea of eating fish fed insects (De Koning et al., 2020). Therefore, aquaculture feeds that promote a more environmentally friendly product will have to meet several criteria, including lower production costs than traditional aquaculture feeds, regulations that support sustainable aquaculture practices, and trust from consumers. The development of sustainable aquaculture feeds is dependent on multi-stakeholder collaboration to help develop these products responsibly at scale.

5.5 Toward circular bioeconomy integration in aquaculture nutrition

The use of unconventional by-products from agriculture for use in aquafeeds is an example of how the principles of the bio-economy (circular) are being put into practice in the production of aquatic foods (Chattaraj et al., 2024b,c). Aquaculture can be positioned as a "closed loop" system through redirection of biomass streams which are currently considered to be waste; thereby reducing its ecological dependence. However, complete circularity will depend upon alignment of all elements of the supply chain, including sourcing of raw materials, technology used for processing, life cycle assessments (LCA), and government incentives. Additionally, standardized metrics for measuring sustainability, systems for tracking products, and certification programs for environmentally friendly practices can increase transparency and build trust among consumers (Pelletier et al., 2009). The future of research in this area should include multidisciplinary models that integrate nutritional value data with environmental impact data. These types of integrated models will allow for evidence-based decision-making in the formulation of aquafeeds that balance maximum biological productivity with minimum ecological impact. Overall, consideration for environmental sustainability has increased the strategic interest in using nontraditional agricultural by-products in developing aquafeeds. While alternative feeds have the potential to decrease reliance on marine resources and provide more effective nutrient recovery, comprehensive evaluations of life cycles and systemic integration are required to achieve truly sustainable aquaculture systems.

6. Future perspectives and research directions in fish feed formulations with non-conventional agricultural byproducts

The rapid evolution of aquafeed diversification strategies signals a structural transition in aquaculture nutrition. While plant residues, insect meals, microbial proteins, microalgae, and circular biomass inputs have demonstrated technical feasibility, future progress will depend on integrating nutritional precision, environmental accountability, technological innovation, and socio-economic scalability within a unified systems framework.

6.1 Precision nutrition and mechanistic understanding

Despite the ability to substitute part of the fish meal and fish oil in several species (Huyben et al., 2024; Sprague et al., 2016; Venkata et al., 2020), the most suitable percentage for inclusion is still species dependent. Marine carnivores appear to be more sensitive to a higher percentage of plant-based feed than freshwater omnivores; this emphasizes the need for developing digestive amino acids and lipid balances as well as models to understand how gut bacteria are affected by different nutrients (Francis et al., 2001).

As research continues to emerge, it is evident that modulating gut bacteria as well as regulating genes related to the immune system may play important roles in how fish respond to alternative feed sources (Albrektsen et al., 2022; Ringø et al., 2010). It is recommended that future studies utilize an integrative omics approach (transcriptomic analysis and metabolomic analysis) to gain a better understanding of how different nutritional components affect nutrient-sensing pathways, inflammatory signaling, and oxidative stress responses. This can lead to the development of predictive models for formulating feed that is no longer based on empirical trial and error methods but rather on a rational scientific basis.

6.2 Biotechnological innovation and process optimization

Processing Technologies for Agro-Industrial Byproduct Enhancement. The application of biotechnologies including fermentation and microbial bioconversion has shown that nutrients (e.g., proteins) found in plant-based foods may be made more bioavailable and that anti-nutrients (e.g., phytates) may be decreased (Ajila et al., 2012; Van Huis, 2013) so as to maximize the nutritional and functional benefits of agro-industrial byproducts. Further improvement in microbial consortia and enzymatic treatment will lead to additional improvements in the level of enrichment and digestibility of proteins and further minimize waste.

Improvement in the standardization of insect rearing and postharvest processing will also be essential to improve compositional consistency and regulatory compliance (Huyben et al., 2024; De Koning et al., 2020). Similarly, future progress in developing scalable microbial protein production systems will depend on continuing advancements in both selecting strains for optimal fermentation and increasing the efficiency of fermentations (Santos et al., 2025; Francis et al., 2001). Improved extraction technology coupled with reduced energy input through the use of heterotrophic culture conditions and improved lipid yields from DHA producing algae are needed to increase the overall efficiency of algal oil production (Sprague et al., 2016; Venkata et al., 2020). The integration of process engineering principles with an evaluation of the environmental impacts associated with these new food products is critical to ensure that the increased nutritional quality of these new food products does not come at the expense of their carbon or energy footprint.

6.3 Environmental benchmarking and circular bioeconomy integration

Aquaculture's LCA has identified that feeds produced are major contributors to its environmental impacts (Henriksson et al., 2012). The development of aquafeeds going forward will need to include the use of standard environmental metrics—i.e., GHG Emissions; Nutrient Eutrophication Potential; Water Use Efficiency; Land Use Intensity—in evaluating alternative ingredients.

The valorization of agricultural residues provides an opportunity to recycle nutrients and reduce waste through circular bioeconomy models (Yadav et al., 2024; Kusumowardani and Tjahjono, 2020); integrated systems like biofloc technology also allow for greater nutrient retention in that they convert metabolic waste into usable biomass (Avnimelech, 2009). Sustainability claims will require robust cradle-to-gate assessments to verify that alternative ingredients provide legitimate ecological gains rather than simply "shifting the burden."

Alignment with policy and certification frameworks can further support the credibility of consumers in sustainable feed transition (Pelletier et al., 2009).

6.4 Economic feasibility and adoption dynamics

The economic feasibility will be the deciding factor on whether there are large-scale substitutions. In addition to biological performance, farmer decisions are made based on risk perception, cost, and time preference as well (Greene et al., 2022). Thus, it would be advisable to have both biological trials and techno-economic analyses to determine at what point a substitution becomes economically viable.

In addition to being influenced by technology and economics, consumer perceptions can influence how quickly insect-based feeds are integrated into the marketplace. Studies indicate that consumers who are aware of the environmental implications (i.e., sustainability) of their food choices will be more likely to accept insect-fed fish, so long as they believe the product quality is comparable to other alternatives (De Koning et al., 2020). Therefore, transparency and traceability will be necessary to achieve wider use of insect-based feeds.

6.5 Integrative roadmap for next-generation aquafeeds

The way for the development of fish feeds in the future is through a hybrid approach of ingredient formulations, which utilize a combination of plant waste (residues), insect proteins, microbial biomass (microbial protein), and lipid from microalgae as a means to achieve both nutritional and environmental benefits in a complementary manner. Diversified multi-ingredient formulations, rather than replacing all of the ingredients currently used in traditional fish feed formulations, will provide more flexibility in the event of some ingredient shortages or failures and, therefore, improved overall adaptability.

Additional advancements in the area of digital modeling, data-driven nutrient digestibility databases, and additional data-driven tools for designing feed formulations will allow for improved precision when developing new fish feed designs. The integration of environmental impact models with measures of the nutritional performance of fish feeds will allow for evidence-based optimization of fish feed design based on various production systems.

From a strategic formulation perspective, the next phase of aquafeed innovation should move beyond single-ingredient substitution toward systems-level optimization models that integrate nutrient density, functional bioactivity, environmental footprint, and techno-economic feasibility within a unified decision-support framework. Rather than evaluating insect meal, microbial protein, plant residues, or algal lipids in isolation, future feed design should adopt a modular multi-ingredient architecture in which complementary nutrient and functional attributes are deliberately balanced to achieve metabolic precision and environmental efficiency simultaneously. Integrating digestibility coefficients, amino acid balancing algorithms, lipid oxidation stability indices, and life-cycle environmental metrics into predictive formulation tools could enable evidence-based feed customization tailored to species-specific physiological thresholds. Such an approach would facilitate scalable circular bioeconomy integration while minimizing unintended nutritional imbalances or sustainability trade-offs, thereby strengthening the translational bridge between experimental validation and commercial deployment.

7. Conclusion

The sustained expansion of global aquaculture demands a decisive shift in feed formulation strategies to reconcile production efficiency with ecological responsibility. Dependence on fish meal and fish oil, though historically justified by nutritional adequacy, is increasingly constrained by environmental limits, resource volatility, and economic pressures. The valorization of nonconventional agricultural byproductsencompassing plant-derived residues, insect biomass, microbial proteins, microalgal lipids, and circular system outputsoffers a viable pathway toward reducing marine resource dependency while enhancing nutrient recycling efficiency.

Evidence synthesized throughout this review demonstrates that, when supported by appropriate processing technologies and precision nutritional balancing, these alternative ingredients can sustain growth performance, maintain gut health and immune stability, and preserve product quality across diverse aquaculture species. Technological innovations such as fermentation, enzymatic treatment, heterotrophic microalgae cultivation, and biofloc biomass recovery further enhance the functional compatibility of agroindustrial residues within aquafeed systems. Importantly, sustainability must be evaluated beyond biological efficacy; environmental life-cycle metrics, carbon footprint assessments, economic feasibility, and consumer acceptance are equally critical determinants of long-term adoption.

The novelty of this review lies in its integrated systems-level synthesis that connects ingredient classification, bioprocessing advancements, functional physiological outcomes, environmental LCA, and socioeconomic adoption dynamics within a single conceptual framework. Rather than treating nonconventional agricultural byproducts merely as partial fish meal substitutes, this work positions them as strategic components of a circular bioeconomy model for aquaculture nutrition. By bridging biological validation with sustainability benchmarking and translational feasibility considerations, the review advances a multidimensional roadmap for next-generation aquafeed development.

Ultimately, the future of sustainable aquaculture nutrition will depend on interdisciplinary convergence—combining feed science, biotechnology, environmental systems modeling, and policy alignment. Hybrid, multi-source formulations supported by precision nutrition tools and standardized sustainability metrics may define the next phase of feed innovation. If carefully implemented, the strategic integration of agricultural byproducts into aquafeeds has the potential not only to enhance environmental resilience but also to strengthen global food security in an era of mounting ecological constraints.

References

Agriculture Organization of the United Nations. Fisheries Department, 2018. The State of World Fisheries and Aquaculture. Food and Agriculture Organization of the United Nations.

Ajila, C.M., Brar, S.K., Verma, M., Tyagi, R.D., Godbout, S., Valéro, J.A., 2012. Bioprocessing of agro-byproducts to animal feed. Crit. Rev. Biotechnol. 32 (4), 382–400.

Albrektsen, S., Kortet, R., Skov, P.V., Ytteborg, E., Gitlesen, S., Kleinegris, D., Øverland, M., 2022. Future feed resources in sustainable salmonid production: a review. Rev. Aquacult. 14 (4), 1790–1812.

Avnimelech, Y., 2009. Biofloc Technology: A Practical Guide Book, pp. x+–182.

Chattaraj, S., Ganguly, A., Mandal, A., Das Mohapatra, P.K., 2022. A review of the role of probiotics for the control of viral diseases in aquaculture. Aquac. Int. 30 (5), 2513–2539.

Chattaraj, S., Ganguly, A., Mitra, D., Mitra, D., Das Mohapatra, P.K., 2023. Study of intestinal bacteria of Cirrhinus reba and characterization of a new probiotic bacteria: an initiative to save the threatened species of Cirrhinus. Appl. Biochem. Microbiol. 59 (6), 833–849.

Chattaraj, S., Das Mohapatra, P.K., 2024. Immunostimulatory and antagonistic potential of the methanolic extract of Oedogonium intermedium SCB in Cirrhinus reba challenged with Aeromonas hydrophila. Aquac. Int. 32 (3), 2523–2548.

Chattaraj, S., Behera, B.K., Das Mohapatra, P.K., 2024a. Effective valorization of Chlorella biomass and Brewers spent grain substituting fish meal and soybean meal in the diet of herbivorous fish Cirrhinus reba for higher growth, digestibility and sustainable cultivation. J. Appl. Phycol. 36 (1), 217–232.

Chattaraj, S., Mitra, D., Ganguly, A., Thatoi, H., Mohapatra, P.K.D., 2024b. A critical review on the biotechnological potential of Brewers' waste: challenges and future alternatives. Curr. Res. Microb. Sci. 6, 100228.

Chattaraj, S., Mitra, D., Chattaraj, M., Ganguly, A., Thatoi, H., Mohapatra, P.K.D., 2024c. Brewers' spent grain as fish feed ingredient: evaluation of bio-safety and analysis of its impact on gut bacteria of Cirrhinus reba by 16S Metagenomic sequencing. Curr. Res. Microb. Sci. 7, 100286.

Chattaraj, S., Sahoo, S.S., Behera, S., Thatoi, H., 2026a. Sustainable aquafeed development using Spirulina: Effects on growth performance of Catla catla. Algal Research 104447.

Chattaraj, S., Ganguly, A., Mitra, D., et al., 2026b. Sustainable valorization of algal biomass in fish nutrition and dietetics: a critical review and scientometric study. Waste Biomass Valor. https://doi.org/10.1007/s12649-026-03619-7.

Costello, C., Cao, L., Gelcich, S., Cisneros, M.A., Free, C.M., Froehlich, H.E., Lubchenco, J., 2023. The future of food from the sea. In: The Blue Compendium: From Knowledge to Action for a Sustainable Ocean Economy. Springer International Publishing, Cham, pp. 1–14.

De Koning, W., Dean, D., Vriesekoop, F., Aguiar, L.K., Anderson, M., Mongondry, P., Boereboom, A., 2020. Drivers and inhibitors in the acceptance of meat alternatives: the case of plant and insect-based proteins. Foods 9 (9), 1292.

Francis, G., Makkar, H.P., Becker, K., 2001. Antinutritional factors present in plant-derived alternate fish feed ingredients and their effects in fish. Aquaculture 199 (3–4), 197–227.

Ganguly, A., Chattaraj, S., Ganguly, M., Chattaraj, M., Banerjee, A., Mandal, A., Ali Khan, M., Sen, S.K., Das Mohapatra, P.K., 2025. Effect of three probiotic Bacillus strains supplemented feeds on growth, carcass composition and blood parameters of Clarias magur (Hamilton). J. Appl. Aquac. 37 (1), 19–44.

Gasco, L., Biasato, I., Dabbou, S., Schiavone, A., Gai, F., 2019. Animals fed insect-based diets: state-of-the-art on digestibility, performance and product quality. Animals 9 (4), 170.

Gasco, L., Gai, F., Maricchiolo, G., Genovese, L., Ragonese, S., Bottari, T., Caruso, G., 2018. Fishmeal alternative protein sources for aquaculture feeds. In: Feeds for the Aquaculture Sector: Current Situation and Alternative Sources. Springer International Publishing, Cham, pp. 1–28.

Greene, C.H., Scott-Buechler, C.M., Hausner, A.L., Johnson, Z.I., Lei, X.G., Huntley, M.E., 2022. Transforming the future of marine aquaculture: a circular economy approach. Oceanography (Wash. D. C.) 35 (2).

Henriksson, P.J.G., Troell, M., Banks, L.K., Belton, B., Beveridge, M.C.M., Klinger, D.H., Tran, N., 2021. Interventions for improving the productivity and environmental performance of global aquaculture for future food security. One Earth 4 (9), 1220–1232.

Henriksson, P.J., Guinée, J.B., Kleijn, R., de Snoo, G.R., 2012. Life cycle assessment of aquaculture systems—a review of methodologies. Int. J. Life Cycle Assess. 17 (3), 304–313.

Hobbi, P., Bekhit, A.E.D.A., Debaste, F., Lei, N., Shavandi, A., 2022. Insect-derived protein as food and feed. In: Alternative Proteins. CRC Press, pp. 85–132.

Huyben, D., Borland, M., Riesenbach, I.C., Shandilya, U., Chiasson, M., Karrow, N., 2024. Evaluation of growth performance, hepatic gene expression, and plasma biochemistry of rainbow trout fed full-fat meal, defatted meal, oil and chitin from black soldier flies. In: Hepatic Gene Expression, and Plasma Biochemistry of Rainbow Trout Fed Full-Fat Meal, Defatted Meal, Oil and Chitin from Black Soldier Flies.

Kusumowardani, N., Tjahjono, B., 2020. Circular economy adoption in the aquafeed manufacturing industry. Proced. CIRP 90, 43–48.

Makkar, H.P., Tran, G., Heuzé, V., Ankers, P., 2014. State-of-the-art on use of insects as animal feed. Anim. Feed Sci. Technol. 197, 1–33.

Medeiros, P.A.D., Dos Santos, R.B., Matos Dantas, F.D., Monteiro Dos Santos, D.K., Freitas, T.M.D., Goncalves, L.U., Affonso, E.G., 2025. Aquaculture feeds produced with Amazon by-products in accordance with organic certification standards: effect on growth and health of Colossomamacropomum. Acta Amazonica 55, e55af24418.

Mente, E., Jokumsen, A., Carter, C.G., Antonopoulou, E., Tacon, A.G., 2019. Nutrition in relation to organic aquaculture: sources and strategies. In: Organic Aquaculture: Impacts and Future Developments. Springer International Publishing, Cham, pp. 141–188.

Pelletier, N., Tyedmers, P., Sonesson, U., Scholz, A., Ziegler, F., Flysjo, A., Silverman, H., 2009. Not All Salmon Are Created Equal: Life Cycle Assessment (LCA) of Global Salmon Farming Systems.

Rangel, F.J.O., 2023. Novel Probiotics Isolated from European Sea Bass (Dicentrarchus Labrax) Gut Microbiota for Improving Insect Meal Utilization and Gut Health (Doctoral dissertation). Universidade do Porto, Portugal.

Ringø, E., Olsen, R.E., Gifstad, T.Ø., Dalmo, R.A., Amlund, H., Hemre, G.I., Bakke, A.M., 2010. Prebiotics in aquaculture: a review. Aquac. Nutr. 16 (2), 117–136.

Santos, B., Freitas, F., Sobral, A.J., Encarnação, T., 2025. Microalgae and circular economy: unlocking waste to resource pathways for sustainable development. Int. J. Sustain. Eng. 18 (1), 2501488.

Sprague, M., Dick, J.R., Tocher, D.R., 2016. Impact of sustainable feeds on omega-3 long-chain fatty acid levels in farmed Atlantic salmon, 2006–2015. Sci. Rep. 6 (1), 21892.

Van Huis, A., 2013. Potential of insects as food and feed in assuring food security. Annu. Rev. Entomol. 58, 563–583.

Venkata, G.S., Chugh, N., Iyer, S., Waghmare, A., Musale, A.S., Nandru, R., Dasgupta, S., 2020. Application of in vitro protein solubility for selection of microalgae biomass as protein ingredient in animal and aquafeed. J. Appl. Phycol. 32 (6), 3955–3970.

Woodgate, S.L., Wan, A.H., Hartnett, F., Wilkinson, R.G., Davies, S.J., 2022. The utilisation of European processed animal proteins as safe, sustainable and circular ingredients for global aquafeeds. Rev. Aquacult. 14 (3), 1572–1596.

Yadav, C., Yadav, P., Joshi, A., Meena, M., Harish, Arora, J., 2024. Agricultural waste and its impact on the environment. In: Transforming Agriculture Residues for Sustainable Development: From Waste to Wealth. Springer Nature Switzerland, Cham, pp. 3–19.

Index

'*Note:* Page numbers followed by "f" indicate figures and "t" indicate tables.'

B

D

E

F

G

H

I

K

L

M

P

T

U

V

W

X

Y

Z